Monografie Matematyczne

Instytut Matematyczny Polskiej Akademii Nauk (IMPAN)

Volume 69

(New Series)

Founded in 1932 by
S. Banach, B. Knaster, K. Kuratowski,
S. Mazurkiewicz, W. Sierpinski, H. Steinhaus

Volumes 31–62 of the series
Monografie Matematyczne were published by
PWN – Polish Scientific Publishers, Warsaw

T. V. Panchapagesan

The
Bartle-Dunford-Schwartz
Integral

Integration with Respect to a Sigma-Additive Vector Measure

Birkhäuser
Basel · Boston · Berlin

Author:

T.V. Panchapagesan
301, Shravanthi Palace
3rd Cross, C.R.Layout
J.P. Nagar, 1st Phase
Bangalore 560 078
India
e-mail: panchapa16@yahoo.com.mx

Formerly:
Departamento de Matemáticas
Facultad de Ciencias
Universidad de los Andes
Mérida 5101
Venezuela
e-mail: panchapa@ula.ve

2000 Mathematical Subject Classification: primary 28B05; secondary 46E40, 46G10

Library of Congress Control Number: 2007942613

Bibliographic information published by Die Deutsche Bibliothek. Die Deutsche Bibliothek lists this publication in the Deutsche Nationalbibliografie; detailed bibliographic data is available in the Internet at http://dnb.ddb.de

ISBN 978-3-7643-8601-6 Birkhäuser Verlag AG, Basel - Boston - Berlin

© 2008 Birkhäuser Verlag AG
Basel · Boston · Berlin
P.O. Box 133, CH-4010 Basel, Switzerland
Part of Springer Science+Business Media
Printed on acid-free paper produced from chlorine-free pulp. TCF ∞
Printed in Germany

ISBN 978-3-7643-8601-6

e-ISBN 978-3-7643-8602-3

9 8 7 6 5 4 3 2 1

www.birkhauser.ch

*The monograph is dedicated to the memory
of the author's spiritual guru*

Singaracharya Swamigal

*of Kondapalayam, Scholingur,
and to the memory of his eldest brother*

T.V. Gopalaiyer,

*a great Tamil scholar, of Ecole Française
D'extrême-Orient, Centre de Pondicherry.*

Contents

Preface

In 1953, Grothendieck [G] characterized locally convex Hausdorff spaces which have the Dunford-Pettis property and used this property to characterize weakly compact operators $u : C(K) \to F$, where K is a compact Hausdorff space and F is a locally convex Hausdorff space (briefly, lcHs) which is complete. Among other results, he also showed that there is a bijective correspondence between the family of all F-valued weakly compact operators u on $C(K)$ and that of all F-valued σ-additive Baire measures on K. But he did not develop any theory of integration to represent these operators.

Later, in 1955, Bartle, Dunford, and Schwartz [BDS] developed a theory of integration for scalar functions with respect to a σ-additive Banach-space-valued vector measure $\mathbf{m}$ defined on a σ-algebra of sets and used it to give an integral representation for weakly compact operators $u : C(S) \to X$, where S is a compact Hausdorff space and X is a Banach space. A modified form of this theory is given in Section 10 of Chapter IV of [DS1]. In honor of these authors, we call the integral introduced by them as well as its variants given in Section 2.2 of Chapter 2 and in Section 4.2 of Chapter 4, the Bartle-Dunford-Schwartz integral or briefly, the BDS-integral.

About fifteen years later, in [L1,L2] Lewis studied a Pettis type weak integral of scalar functions with respect to a σ-additive vector measure $\mathbf{m}$ having range in an lcHs X. This type of definition has also been considered by Kluvánek in [K2]. In honor of these mathematicians we call the integral introduced in [L1,L2] as well as its variants given in Section 2.1 of Chapter 2 and in Section 4.1 of Chapter 4, the Kluvánek-Lewis integral or briefly, the (KL)-integral. When the domain of the σ-additive vector measure $\mathbf{m}$ is a σ-algebra Σ and X is a Banach space, Theorem 2.4 of [L1] asserts that the (KL)-integral is the same as the (BDS)-integral. Though this result is true, its proof in [L1] lacks essential details. See Remark 2.2.6 of Chapter 2.

Let T be a locally compact Hausdorff space and $\mathcal{K}(T)$ (resp. $\mathcal{K}(T, \mathbb{R})$) be the vector space of all complex- (resp. real-) valued continuous functions on T with compact support, endowed with the inductive limit locally convex topology as in §1, Chapter III of [B]. In 1970, using the results of [G], Thomas [T] developed a theory of vectorial Radon integration with respect to a weakly compact bounded (resp. a prolongable) Radon operator u on $\mathcal{K}(T,\mathbb{R})$ with values in a real Banach space and more generally, with values in a real quasicomplete lcHs. (Thomas [T] calls them bounded weakly compact (resp. prolongable) Radon vector measures.) However, by making some modifications, it can be shown that his theory is equally valid for such operators u defined on $\mathcal{K}(T)$ with range contained in a complex Banach space or in a quasicomplete complex lcHs. See Section 7.1 of Chapter 7. Functions f integrable with respect to u are called u-integrable and the integral of f with respect to u is denoted by $\int_T f du$ in [T].

If S is a compact Hausdorff space, X a Banach space and $u : C(S) \to X$ is a weakly compact operator, then by the Bartle-Dunford-Schwartz representation theorem in [BDS] and in [DU] there exists a unique X-valued Borel regular σ-additive vector measure $\mathbf{m}_u$ on the σ-algebra $\mathcal{B}(S)$ of the Borel sets in S such that $uf = \int_S f d\mathbf{m}_u$ for $f \in C(S)$. If X is a real Banach space and if $u : C^r(S) \to X$ (where $C^r(S) = \{f \in C(S) : f \text{ real-valued}\}$) is a weakly compact operator, Thomas showed in [T] that a real function f on S is u-integrable if and only if it is $\mathbf{m}_u$-integrable in the sense of Section 10 of Chapter IV of [DS1] and in that case, $\int_S f du = (\text{BDS}) \int_S f d\mathbf{m}_u$. Moreover, such f is $\mathbf{m}_u$-measurable in the sense of Section 10 of Chapter IV of [DS1] though it is not necessarily $\mathcal{B}(S)$-measurable.

On the other hand, in [P3, P4] we have shown that there is a bijective correspondence Γ between the dual space $\mathcal{K}(T)^*$ and the family of all $\delta(\mathcal{C})$-regular complex measures on $\delta(\mathcal{C})$ such that $\Gamma\theta = \mu_\theta|_{\delta(\mathcal{C})}$ for $\theta \in \mathcal{K}(T)^*$, where μ_θ is the complex Radon measure determined by θ (in the sense of [P4]) and $\delta(\mathcal{C})$ is the δ-ring generated by the family $\mathcal{C}$ of all compact sets in T. As observed in [P13], the scalar-valued prolongable Radon operators on $\mathcal{K}(T)$ are precisely the continuous linear functionals on $\mathcal{K}(T)$ (i.e., the elements of $\mathcal{K}(T)^*$).

From the above results of Thomas [T] and of Panchapagesan [P3, P4, and P13] the following questions arise:

(Q1)   Similar to Section 10 of Chapter IV of [DS1], can a theory of integration of scalar functions be developed with respect to a σ-additive quasicomplete lcHs-valued vector measure $\mathbf{m}$ defined on a σ-algebra or on a σ-ring $\mathcal{S}$ of sets, permitting the integration of scalar functions (with respect to $\mathbf{m}$) which are not necessarily $\mathcal{S}$-measurable?

(Q2)   The same as in (Q1), excepting that the domain of $\mathbf{m}$ is a δ-ring $\mathcal{P}$ of sets and the $\mathcal{S}$-measurability of functions is replaced by $\sigma(\mathcal{P})$-measurability (where $\sigma(\mathcal{P})$ denotes the σ-ring generated by $\mathcal{P}$).

(Q3)   The Bartle-Dunford-Schwartz representation theorem has been generalized in [P9] for weakly compact operators on $C_0(T)$ and hence for weakly compact bounded Radon operators u on $\mathcal{K}(T)$ with values in a quasicomplete lcHs X asserting that u determines a unique $\mathcal{B}(T)$-regular X-valued σ-additive vector measure $\mathbf{m}_u$ on $\mathcal{B}(T)$, the σ-algebra of Borel sets in T. The question is: Is it possible to give a similar representation theorem for X-valued prolongable Radon operators u on $\mathcal{K}(T)$?

(Q4)   If (Q3) has an affirmative answer, does the prolongable operator u determine a σ-additive $\delta(\mathcal{C})$-regular (suitably defined) vector measure $\mathbf{m}_u$ on $\delta(\mathcal{C})$? (This question is suggested by the scalar analogue in [P13].)

(Q5)   Suppose (Q1) has an affirmative answer and suppose $u : \mathcal{K}(T) \to X$, X a quasicomplete lcHs, is a weakly compact bounded Radon operator determining the σ-additive vector measure $\mathbf{m}_u$ on $\mathcal{B}(T)$. Can it be shown that a scalar function f is u-integrable in T if and only if it is $\mathbf{m}_u$-integrable in T (in the sense of integration given for (Q1))? If so, is $\int_T f\,du = \int_T f\,d\mathbf{m}_u$?

(Q6)   If (Q2), (Q3) and (Q4) are answered in the affirmative and if $u : \mathcal{K}(T) \to X$, X a quasicomplete lcHs, is a prolongable Radon operator determining the σ-additive vector measure $\mathbf{m}_u$ on $\delta(\mathcal{C})$, then can it be shown that a scalar function f with compact support is u-integrable in T if and only if it is $\mathbf{m}_u$-integrable in T? If so, what is the relation between $\int_T f\,du$ and $\int_T f\,d\mathbf{m}_u$?

In the literature, integration of scalar functions with respect to a Banach-space-valued or a sequentially complete lcHs-valued σ-additive vector measure defined on a σ-algebra Σ of sets has been studied for Σ-measurable scalar functions in several papers such as [C4], [Del1, Del2, Del4], [FNR], [FMNP], [FN1,FN2], [JO], [K1, K2, K4], [KK], [L1], [N], [O], [OR1, OR2, OR3, OR4, OR5, OR6], [Ri1, Ri2, Ri3, Ri4, Ri5, Ri6] and [Shu1, Shu2]. A similar study has been done in [BD1], [L2], [Del3], and [MN2] for a σ-additive vector measure defined on a δ-ring $\mathcal{P}$. Recently, the vector measure integration on σ-algebras has been used to study the representation of real Banach lattices in [C1,C2, C3, C4]. Also see [CR1, CR2, CR3], [FMNSS1,FMNSS2], [MP], [OSV], [SP1,SP2,SP3,SP4] and [St]. Some other papers which can also be referred for various aspects of vector measures defined on rings, δ-rings and σ-rings are [Br], [BD1], [BD2] and [Del4]. None of the above papers considers the possibility of integrating non-Σ-measurable or non-$\sigma(\mathcal{P})$-measurable functions and hence the integration theory developed in the literature is not suitable for answering the above questions. Though the paper [BD2] treats the integration of non-$\sigma(\mathcal{P})$-measurable functions, its results do not answer the above questions.

However, adapting some of the concepts and techniques used by Dobrakov in [Do1, Do2] and by Dobrakov and Panchapagesan in [DP2] (in the study of integration of vector functions with respect to an operator-valued measure), the present **monograph answers all the questions raised above in the affirmative**. Moreover, a nice theory of L_p-spaces, $1 \leq p < \infty$, is also developed for a σ-additive Banach space-valued (resp. quasicomplete or sequentially complete lcHs-valued) vector measure $\mathbf{m}$ defined on a δ-ring $\mathcal{P}$ of sets and results similar to those known for the abstract Lebesgue and Bochner L_p-spaces are obtained. Compare with [FMNSS2] and [SP1].

The monograph consists of seven chapters. Chapter 1 is on Preliminaries and has two sections. Section 1.1 is devoted to fixing the notation and terminology and to give some definitions and results from the literature on Banach space-valued measures defined on a δ-ring $\mathcal{P}$ of subsets of a non-void set T and includes the theorem on interchange of limit and integral with proof. (See Proposition 1.1.21.) Moreover, motivated by [DP2], we introduce the concept of $\mathbf{m}$-measurability for functions $f : T \to \mathbb{K}$ or $[-\infty, \infty]$ where $\mathbb{K} = \mathbb{R}$ *or* $\mathbb{C}$ and this concept of $\mathbf{m}$-measurability is suitably generalized in Section 4.1 of Chapter 4 when $\mathbf{m}$ assumes values in an lcHs. This concept plays a crucial role in developing the theory of integration which permits integration of certain non-$\sigma(\mathcal{P})$-measurable functions too. Moreover, when T is a locally compact Hausdorff space and the vector measure $\mathbf{m}$ satisfies certain regularity conditions, it turns out that a scalar function f is $\mathbf{m}$-measurable if and only if f is Lusin $\mathbf{m}$-measurable with $N(f) = \{t \in T : f(t) \neq 0\}$ suitably restricted. (See Theorems 6.2.5 and 6.2.6 of Chapter 6.) Section 1.2 is devoted to giving some definitions and results on lcHs which are needed in the sequel.

In Section 2.1 of Chapter 2 we introduce the concept of (KL) $\mathbf{m}$-integrability for $\mathbf{m}$-measurable functions and study the properties of the integral. We give an $\mathbf{m}$-a.e. convergence version of the Lebesgue dominated (resp. bounded) convergence theorem for $\mathbf{m}$ and we briefly refer to it as LDCT (resp. LBCT). This theorem has been given in [L2] with an incorrect proof (see Remark 2.1.12 below). In Section 2.2 we define the (BDS) $\mathbf{m}$-integral and show that an $\mathbf{m}$-measurable function is (BDS) $\mathbf{m}$-integrable in T if and only if it is (KL) $\mathbf{m}$-integrable in T and in that case, both the integrals coincide. Hence we use the terminology of $\mathbf{m}$-integrability (resp. the symbol $\int_T f d\mathbf{m}$) to denote either integrability (resp. either of the integrals). When $\mathcal{P}$ is a σ-algebra, this result is given in Theorem 2.4 of [L1], but, as mentioned above, its proof lacks essential details (see Remark 2.2.6 below). Proposition 1.1.21 is generalized to the (BDS) $\mathbf{m}$-integral in Theorem 2.2.8.

Chapter 3 consists of Sections 3.1–3.5 and is devoted to the study of the spaces $\mathcal{L}_p(\mathbf{m})$, $1 \leq p \leq \infty$, for a Banach space-valued σ-additive measure $\mathbf{m}$ defined on a δ-ring of sets. Similar to [Do2], in Section 2.1 we introduce a seminormed space $\mathcal{L}_p\mathcal{M}(\mathbf{m})$ of $\mathbf{m}$-measurable scalar functions with its seminorm being denoted by $\mathbf{m}_p^\bullet(\cdot, T)$ for $1 \leq p < \infty$ and define the subspaces $\mathcal{L}_p\mathcal{I}(\mathbf{m})$, $\mathcal{L}_p\mathcal{I}_s(\mathbf{m})$ and $\mathcal{L}_p(\mathbf{m})$ of $\mathcal{L}_p\mathcal{M}(\mathbf{m})$ and show that all these subspaces are linear and coincide with

$$\mathcal{I}_p(\mathbf{m}) = \{f : T \to \mathbb{K}, f \ \mathbf{m}\text{-measurable and } |f|^p \text{ is (KL)}\mathbf{m}\text{-integrable in } T\}$$

for $1 \leq p < \infty$. If $\mathcal{L}_p(\sigma(\mathcal{P}), \mathbf{m}) = \{f \in \mathcal{L}_p(\mathbf{m}) : f \ \sigma(\mathcal{P})\text{-measurable}\}$, then Section 3.2 deals with the completeness of the spaces $\mathcal{L}_p(\mathbf{m})$ and $\mathcal{L}_p(\sigma(\mathcal{P}), \mathbf{m})$ for $1 \leq p \leq \infty$ where $\mathcal{L}_\infty(\mathbf{m})$ and $\mathcal{L}_\infty(\sigma(\mathcal{P}), \mathbf{m})$ are suitably defined (see Definition 3.2.10). Sections 3.3 and 3.4 study various versions of LDCT, LBCT and the Vitali convergence theorem for $\mathcal{L}_p(\mathbf{m})$, $1 \leq p < \infty$, and Section 3.5 obtains relations between the spaces $\mathcal{L}_p(\mathbf{m})$, $1 \leq p \leq \infty$, similar to those in the classical case. The study of $\mathcal{L}_p(\sigma(\mathcal{P}), \mathbf{m})$ for $1 \leq p < \infty$ for a σ-additive vector measure $\mathbf{m}$ defined on a σ-algebra with values in a real Banach space has recently been done in [SP1, SP2, SP4] and in [FMNSS2], for their study of the representation theory of real Banach lattices. Our results can also be compared with those in the literature for

$p = 1$ as noted in Remarks 2.1.6, 2.1.11, 2.1.12, 2.2.6, 3.1.5, 3.2.9, 3.3.5, 3.3.9 and 3.3.15. We would also like to emphasize the fact that the results of this chapter are much influenced by the techniques adopted by Dobrakov in [Do1, Do2].

Chapter 4 consists of Sections 4.1–4.6. Let X be an lcHs and let $\mathbf{m} : \mathcal{P} \to X$ be σ-additive, $\mathcal{P}$ being a δ-ring of subsets of a set T. Sections 4.1–4.6 generalize the results of Chapter 3 to such $\mathbf{m}$ when X is quasicomplete (resp. sequentially complete). For such $\mathbf{m}$, the concepts of (KL) $\mathbf{m}$-integrability and (KL) $\mathbf{m}$-integral are generalized in Section 4.1; Theorem 4.1.8 generalizes (i)–(iv) and (viii) of Theorem 2.1.5 while Theorems 4.1.9 and 4.1.11 (resp. Theorems 4.1.9′ and 4.1.11′ in Remark 4.1.15) generalize (v)–(vii) of Theorem 2.1.5 and Theorem 2.1.7 (LDCT) when X is quasicomplete (resp. sequentially complete with the functions considered being $\sigma(\mathcal{P})$-measurable). In Section 4.2 we generalize (BDS) $\mathbf{m}$-integrability and (BDS) $\mathbf{m}$-integral given in Section 2.2 to such X-valued $\mathbf{m}$ and by Theorems 4.2.2 and 4.2.3 (resp. by Theorem 4.2.2′ in Remark 4.2.12) an $\mathbf{m}$-measurable (resp. $\sigma(\mathcal{P})$-measurable) function f is (KL) $\mathbf{m}$-integrable in T if and only if it is (BDS) $\mathbf{m}$-integrable in T (with values in X) when X is quasicomplete (resp. sequentially complete) and in that case, both the integrals coincide. In the light of this result, we use the terminology of $\mathbf{m}$-integrability (resp. the symbol $\int_T f \, d\mathbf{m}$) to denote either integrability (resp. either of the integrals). For $1 \leq p < \infty$, we introduce in Section 4.3 a locally convex space $\mathcal{L}_p\mathcal{M}(\mathbf{m})$ (resp. $\mathcal{L}_p\mathcal{M}(\sigma(\mathcal{P}), \mathbf{m})$) of $\mathbf{m}$-measurable (resp. $\sigma(\mathcal{P})$-measurable) functions and introduce the subspaces $\mathcal{L}_p\mathcal{I}(\mathbf{m})$ and $\mathcal{L}_p(\mathbf{m})$ of $\mathcal{L}_p\mathcal{M}(\mathbf{m})$ (resp. $\mathcal{L}_p\mathcal{I}(\sigma(\mathcal{P}), \mathbf{m})$ and $\mathcal{L}_p(\sigma(\mathcal{P}), \mathbf{m})$ of $\mathcal{L}_p\mathcal{M}(\sigma(\mathcal{P}), \mathbf{m})$) and show that they are linear and coincide. When X is a Fréchet space, in Section 4.4 we show that these subspaces are pseudo-metrizable and complete. Section 4.5 is devoted to generalize the results in Sections 3.3, 3.4 and 3.5 when X is quasicomplete (resp. sequentially complete). We introduce a subspace $\mathcal{L}_p\mathcal{I}_s(\mathbf{m})$ (resp. $\mathcal{L}_p\mathcal{I}_s(\sigma(\mathcal{P}), \mathbf{m})$) of $\mathcal{L}_p\mathcal{M}(\mathbf{m})$ (resp. of $\mathcal{L}_p\mathcal{M}(\sigma\mathcal{P}), \mathbf{m})$ and show that it coincides with $\mathcal{L}_p\mathcal{I}(\mathbf{m})$ (resp. $\mathcal{L}_p\mathcal{I}(\sigma(\mathcal{P}), \mathbf{m})$). Section 4.6 gives some sufficient conditions for the separability of $\mathcal{L}_p(\mathbf{m})$ and $\mathcal{L}_p(\sigma(\mathcal{P}), \mathbf{m})$ for $1 \leq p < \infty$. To compare with some of the results in the literature for $p = 1$, see Remarks 4.1.16, 4.4.10, and 4.6.15. **The theory of the m-integral developed in Sections 4.1 and 4.2 for quasicomplete lcHs-valued σ-additive measures defined on $\mathcal{P}$ answers (Q1) and (Q2) in the affirmative.**

Chapters 5 and 6 give a vector measure treatment of the results of [T]. Chapter 5 consists of Sections 5.1, 5.2 and 5.3. Let T be a locally compact Hausdorff space. Section 5.1 gives some generalizations of the Vitali-Carathéodory integrability criterion for $\mathbf{m}$-measurable (resp. $\sigma(\mathcal{R})$-measurable) real functions where $\mathcal{R} = \mathcal{B}(T)$ or $\mathcal{R} = \delta(\mathcal{C})$, $\mathbf{m} : \mathcal{R} \to X$ is σ-additive and $\mathcal{R}$-regular and X is a quasicomplete (resp. sequentially complete) lcHs. These results play a key role in the study of the duals of $\mathcal{L}_1(\mathbf{m})$ and $\mathcal{L}_1(\mathbf{n})$ in Section 6.5 of Chapter 6, where $\mathbf{m} : \mathcal{B}(T) \to X$ is σ-additive and $\mathcal{B}(T)$-regular (resp. $\mathbf{n} : \delta(\mathcal{C}) \to X$ is σ-additive and $\delta(\mathcal{C})$-regular) and X is a Banach space. Let $\{\mu_n\}_1^\infty$ be a sequence of Borel-regular complex measures on T. By proving that, for each open set U in T, there exists an open Baire set $V \subset U$ such that $\mu_n(V) = \mu_n(U)$ for all n, it is shown in Section 5.2 that the boundedness hypothesis in Corollary 1 of [P8] is redundant. Using this result and adapting the proofs of [T] in the set-up of vector measures, the improved version of Corollary 1 of [P8] is generalized in Theorem 5.2.21 (resp. Theorem 5.2.23) to Banach space-valued (resp. sequentially complete lcHs-valued) σ-additive regular Borel measures. Section 5.3 deals with the weakly compact bounded and prolongable Radon operators on $\mathcal{K}(T)$ with values in a quasicomplete lcHs X. In Theorem 5.3.9 of Chapter 5, the Bartle-Dunford-Schwartz representation theorem is generalized to such an X-valued prolongable Radon operator u on $\mathcal{K}(T)$ and it is shown that u determines a unique X-valued $\delta(\mathcal{C})$-regular σ-additive measure $\mathbf{m}_u$ on $\delta(\mathcal{C})$. **Thus (Q3) and (Q4) are answered in the affirmative.** Using the results of [P9], 22 characterizations are given for an X-valued continuous linear map u on $\mathcal{K}(T)$ to be a prolongable Radon operator.

Chapter 6 consists of Sections 6.1–6.5. Let T be a locally compact Hausdorff space. Let $\mathbf{m} : \mathcal{B}(T) \to X$ (resp. $\mathbf{n} : \delta(\mathcal{C}) \to X$) be σ-additive and Borel regular (resp. and $\delta(\mathcal{C})$-regular). Section 6.1 deals with the generalized Lusin's theorem and its variants for $\mathbf{m}$ and for $\mathbf{n}$ with some applications. In Section 6.2 several characterizations of the $\mathbf{m}$-measurability (resp. $\mathbf{n}$-measurability) of a set A in T are given. The concepts of Lusin $\mathbf{m}$-measurability and Lusin $\mathbf{n}$-measurability for scalar functions are introduced and they are characterized in terms of $\mathbf{m}$-measurability and $\mathbf{n}$-measurability, respectively. The proofs of Lemmas 3.10 and 3.14, Propositions 2.17, 2.20 and 3.7 and Theorems 3.5, 3.13 and 3.20 of [T] are adapted here in the set-up of vector measures to improve Theorem 2.2.2 (resp. Theorem 4.2.2) when X is a Banach space (resp. a quasicomplete or complete lcHs) and when $\mathbf{m} : \delta(\mathcal{C}) \to X$ is σ-additive and $\delta(\mathcal{C})$-regular. See Theorems 6.3.4, 6.3.5 and 6.3.8. Section 6.4 deals with some additional convergence theorems. Section 6.5 is devoted to the study of the duals of $L_1(\mathbf{m})$ and $L_1(\mathbf{n})$ and it is shown that $L_1(\mathbf{m})$ and $L_1(\mathbf{n})$ are weakly sequentially complete Banach spaces when X is a Banach space with $c_0 \not\subset X$.

Chapter 7 consists of Sections 7.1–7.6. In Section 7.1 we briefly indicate how the results in Section 1 of [T] can be extended to complex functions in $\mathcal{K}(T)$. Section 7.2 is devoted to integration with respect to a weakly compact bounded Radon operator, improving the complex versions of Theorems 2.2, 2.7 and 2.7 bis of [T]. In Section 7.3 we improve most of the results such as the complex versions of Theorems 3.3, 3.4, 3.11, 3.13 and 3.20 of [T]. Section 7.4 studies the complex Baire versions of Proposition 4.8 and Theorem 4.9 of [T]. In Section 7.5 we introduce the concepts of weakly compact and prolongable Radon vector measures and generalize the results of [P3,P4] to such Radon vector measures. If u is a bounded Radon operator with values in a quasicomplete lcHs, we define $M_u = \{A \subset T : \chi_A \in \mathcal{L}_1(u)\}$ and $\boldsymbol{\mu}_u(A) = \int_A du$ for $A \in M_u$. The Radon vector measure induced by u is denoted by $\boldsymbol{\mu}_u$ and M_u is called the domain of $\boldsymbol{\mu}_u$. When u is a weakly compact bounded Radon operator, we show that M_u is a σ-algebra containing $\mathcal{B}(T)$ and $\boldsymbol{\mu}_u$ is the generalized Lebesgue completion of the representing measure $\mathbf{m}_u|_{\mathcal{B}(T)}$ of u (in the sense of 5.2.10). We give several characterizations of a weakly compact bounded (resp. a prolongable) Radon operator u and study the regularity properties of $\boldsymbol{\mu}_u$ in both the cases. Following [Si] we define the outer measure $\boldsymbol{\mu}_u^*$ induced by $\boldsymbol{\mu}_u$ and study its relation with $\boldsymbol{\mu}_u$ when u is a weakly compact bounded Radon operator. Introducing the concepts of Lebesgue-Radon completion and localized Lebesgue-Radon completion, we generalize Theorems 4.4 and 4.6 of [P4]. (See Theorems 7.5.24 and 7.5.27.) Thus Theorems 9.13, 9.14 and 9.17 of [P13] are proved here. In Section 7.6, we show that when u is a weakly compact bounded Radon operator on $\mathcal{K}(T)$ with values in a quasicomplete lcHs, $\mathcal{L}_p(u)$ is the same as $\mathcal{L}_p(\mathbf{m}_u)$ for $1 \leq p < \infty$; f is u-integrable if and only if f is $\mathbf{m}_u$-integrable in T and when f is u-integrable, $\int f du = \int_T f d\mathbf{m}_u$. (See Theorem 7.6.13.) When u is a prolongable Radon operator on $\mathcal{K}(T)$ with values in a quasicomplete lcHs and ω is a relatively compact open set in T, we show that $\mathcal{L}_p(u|_{\mathcal{K}(\omega)}) = \mathcal{L}_p(\mathbf{m}_u|_{\mathcal{B}(\omega)})$ for $1 \leq p < \infty$ and for $f \in \mathcal{L}_1(\mathbf{m}_u)$ with support $K \in \mathcal{C}$ and with $K \subset \omega$, $\int_T f d\mathbf{m}_u = \int f \chi_\omega du$. (See Theorem 7.6.19.). Thus **questions (Q5) and (Q6) are also answered in the affirmative.**

Chapter 1

Preliminaries

1.1 Banach space-valued measures

In this section we fix the notation and terminology with respect to Banach space-valued measures and state some definitions and results from the literature, often with proof.

Let T be a nonempty set. A collection $\mathcal{R}$ of subsets of T is called a ring of sets in T if $\emptyset \in \mathcal{R}$ and for A, B in $\mathcal{R}$, $A \cup B$ and $A \backslash B$ belong to $\mathcal{R}$. Moreover, if $\mathcal{R}$ is a ring of sets in T and if $T \in \mathcal{R}$, then $\mathcal{R}$ is called an algebra of sets in T. A ring of sets $\mathcal{R}$ in T is called a σ-ring if $\bigcup_1^\infty E_n \in \mathcal{R}$ whenever $(E_n)_1^\infty \subset \mathcal{R}$ and a σ-ring $\mathcal{R}$ of sets in T with $T \in \mathcal{R}$ is called a σ-algebra of sets in T. A ring of sets $\mathcal{P}$ in T is called a δ-ring if $\bigcap_1^\infty E_n \in \mathcal{P}$ whenever $(E_n)_1^\infty \subset \mathcal{P}$.

Let $\mathcal{C}$ be a nonvoid class of subsets of T. Then the smallest σ-ring of sets in T which contains $\mathcal{C}$ is called the σ-ring generated by $\mathcal{C}$ in T and is denoted by $\sigma(\mathcal{C})$. If $\mathcal{P}$ is a δ-ring of sets in T and if $E \in \sigma(\mathcal{P})$, then $A \cap E \in \mathcal{P}$ for each $A \in \mathcal{P}$. A nonvoid class $\mathcal{M}$ of subsets of T is called a monotone class in T if for an increasing sequence $(A_n)_1^\infty \subset \mathcal{M}$, $\bigcup_1^\infty A_n \in \mathcal{M}$ and for a decreasing sequence $(A_n)_1^\infty \subset \mathcal{M}$, $\bigcap_1^\infty A_n \in \mathcal{M}$. For a sequence $(A_n)_1^\infty$ of subsets of T, $\limsup_n A_n$ is the set $\bigcap_{n=1}^\infty \bigcup_{k \geq n} A_k$ and $\liminf_n A_n$ is the set $\bigcup_{n=1}^\infty \bigcap_{k \geq n} A_k$. For a sequence $(A_n)_1^\infty$ of subsets of T, we say that $(A_n)_1^\infty$ converges or $\lim_n A_n$ exists if $\liminf_n A_n = \limsup_n A_n$. If (A_n) is increasing, we denote it by $A_n \nearrow$, and if (A_n) is decreasing, we denote it by $A_n \searrow$. If $A_n \nearrow$ (resp. $A_n \searrow$), then $\lim_n A_n = \bigcup_{n=1}^\infty A_n$ (resp. $\lim_n A_n = \bigcap_{n=1}^\infty A_n$). For more information see §1 of [Din 1], where σ-ring is called a tribe and δ-ring is called a semitribe.

$\mathbb{K}$ denotes $\mathbb{C}$ or $\mathbb{R}$. X and Y are Banach spaces over the same scalar field $\mathbb{K}$ unless otherwise stated and $|\cdot|$ denotes their norm. $L(X,Y)$ is the Banach space of all continuous linear mappings $u : X \to Y$ with $|u| = \sup_{|x| \leq 1} |ux|$. $L(X, \mathbb{K}) = X^*$ is the dual of X. If $E \subset T$ and $f : T \to \mathbb{K}$ is bounded on E, then $\|f\|_E = \sup_{t \in E} |f(t)|$.

   A set function $\mathbf{m}$ on a δ-ring $\mathcal{P}$ with values in X is said to be an X-valued additive set function if $\mathbf{m}(E \cup F) = \mathbf{m}(E) + \mathbf{m}(F)$ and $E, F \in \mathcal{P}$ with $E \cap F = \emptyset$. Such $\mathbf{m}$ is called an X-valued vector measure on $\mathcal{P}$. Then its variation $v(\mathbf{m})$ is defined on $\sigma(\mathcal{P})$ by

$$v(\mathbf{m})(E) = \sup\left\{\sum_1^r |\mathbf{m}(E_i)| : (E_i)_1^r \subset \mathcal{P}, \text{ mutually disjoint with } \cup_1^r E_i \subset E\right\}$$

for $E \in \sigma(\mathcal{P})$. The semivariation $||\mathbf{m}||||\mathbf{m}||$ of $\mathbf{m}$ is defined on $\sigma(\mathcal{P})$ by

$$||\mathbf{m}||(E) = \sup\left\{\left|\sum_1^r \alpha_i \mathbf{m}(E_i)\right| : \alpha_i \in \mathbb{K}, |\alpha_i| \le 1, (E_i)_1^r \subset \mathcal{P}, E_i \cap E_j = \emptyset, i \ne j,\right.$$
$$\left. \text{with } \cup_1^r E_i \subset E\right\} \quad \text{for } E \in \sigma(\mathcal{P}).$$

Proposition 1.1.1. *Let $\mathcal{R}$ be a ring of sets in T and let $\mathbf{m} : \mathcal{R} \to X$ be additive. Then*

$$||\mathbf{m}||(E) = \sup_{|x^*| \le 1} v(x^* \mathbf{m})(E) \tag{1.1.1.1}$$

for $E \in \sigma(\mathcal{R})$.

Proof. Let π be a finite disjoint family $(E_i)_{i=1}^n \subset \mathcal{R}$ with $\cup_1^r E_i \subset E$ and let $(\alpha_i)_1^r \subset \mathbb{K}$ with $|\alpha_i| \le 1$ for $i = 1, 2, \ldots, r$. Then

$$\left|\sum_1^r \alpha_i \mathbf{m}(E_i)\right| = \sup\left\{\left|x^* \sum_1^r (\alpha_i \mathbf{m}(E_i))\right| : x^* \in X^*, |x^*| \le 1\right\}$$
$$\le \sup\left\{\left|\sum <nolimits_1^r \alpha_i x^*(\mathbf{m}(E_i))\right| : x^* \in X^*, |x^*| \le 1\right\}$$
$$\le \sup\left\{\sum_1^r |x^* \mathbf{m}(E_i)| : x^* \in X^*, |x^*| \le 1\right\}$$
$$\le \sup\{v(x^* \mathbf{m})(E) : x^* \in X^*, |x^*| \le 1\}$$

and hence

$$\sup_\pi \left\{\left|\sum_1^r \alpha_i \mathbf{m}(E_i)\right| : |\alpha_i| \le 1\right\} \le \sup\{v(x^* \mathbf{m})(E) : x^* \in X^*, |x^*| \le 1\}.$$
$$\tag{1.1.1.2}$$

   Conversely, let $x^* \in X^*$ with $|x^*| \le 1$ and let $\pi = \{E_i\}_1^r \subset \mathcal{R}$ with $E_i \cap E_j = \emptyset$ for $i \ne j$ and $\cup_1^r E_i \subset E$. Then

$$\sum_1^r |x^* \mathbf{m}(E_j)| = \sum_{j=1}^r e^{-i\theta_j} x^* \mathbf{m}(E_j) = \left|\sum_{j=1}^r e^{-i\theta_j} (x^* \mathbf{m})(E_j)\right|$$
$$= \left|x^* \left(\sum_{j=1}^r e^{-i\theta_j} \mathbf{m}(E_j)\right)\right| \le \left|\sum_{j=1}^r e^{-i\theta_j} \mathbf{m}(E_j)\right|$$

where $x^*\mathbf{m}(E_j) = |x^*\mathbf{m}(E_j)|e^{i\theta_j}$ and hence

$$\sup_{E_j\in\pi}\left\{\sum|x^*\mathbf{m}(E_j)| : x^* \in X^*, |x^*| \leq 1\right\} \leq \sup_{E_j\in\pi}\left\{\left|\sum\alpha_j\mathbf{m}(E_j)\right| : |\alpha_j| \leq 1\right\}.$$

Therefore,

$$\sup_{|x^*|\leq 1} v(x^*\mathbf{m})(E) \leq ||\mathbf{m}||(E) \tag{1.1.1.3}$$

for $E \in \sigma(\mathcal{P})$. Then by (1.1.1.2) and (1.1.1.3) we have

$$||\mathbf{m}||(E) = \sup_{|x^*|\leq 1} v(x^*\mathbf{m})(E)$$

for $E \in \sigma(\mathcal{P})$. □

Definition 1.1.2. Let $\mathcal{R}$ be a ring of sets and let $\mathbf{m} : \mathcal{R} \to X$ be σ-additive in the sense that $\mathbf{m}(\bigcup_1^\infty E_i) = \sum_1^\infty \mathbf{m}(E_i)$ for each disjoint sequence $(E_i)_1^\infty \subset \mathcal{R}$ with $\bigcup_1^\infty E_i \in \mathcal{R}$.

Then the results (i)–(v) in the following proposition are well known. See §2 of [Din2].

Proposition 1.1.3. *Let $\mathcal{R}$ be a ring of sets and let $\mathbf{m} : \mathcal{R} \to X$ be additive. Then:*

(i) $|\mathbf{m}(A)| \leq v(\mathbf{m})(A), A \in \mathcal{R}$.

(ii) *If $A \subset B$, $A, B \in \sigma(\mathcal{R})$, then $v(\mathbf{m})(A) \leq v(\mathbf{m})(B)$. Thus $v(\mathbf{m})$ is monotone on $\sigma(\mathcal{R})$.*

(iii) *$v(\mathbf{m})$ is additive on $\sigma(\mathcal{R})$. If $\mathbf{m}$ is σ-additive on $\mathcal{R}$, then $v(\mathbf{m})$ is σ-additive on $\sigma(\mathcal{R})$.*

(iv) *If $\mathcal{S}$ is a σ-ring and $\mathbf{m} : \mathcal{S} \to X$ is σ-additive, then $\mathbf{m}$ is bounded on $\mathcal{S}$ in the sense that $\sup_{E\in\mathcal{S}} |\mathbf{m}(E)| < \infty$.*

(v) *If $\mathbf{m} : \mathcal{R} \to X$ is additive and if $A \in \mathcal{R}$, then*

$$\sup\{|\mathbf{m}(B)| : B \in \mathcal{R}, B \subset A\} \leq v(\mathbf{m})(A) \leq 4\sup\{|\mathbf{m}(B)| : B \in \mathcal{R}, B \subset A\}.$$

Let $\mathcal{R}$ be a ring of sets. If $\mathbf{m} : \mathcal{R} \to X$ is additive, then $v(\mathbf{m})(N) = 0$ for $N \in \sigma(\mathcal{R})$ if and only if $||\mathbf{m}||(N) = 0$. If $\nu : \mathcal{R} \to \mathbb{K}$ is additive on $\mathcal{R}$, then $v(\nu)(E) = ||\nu||(E), E \in \sigma(\mathcal{R})$.

Definition 1.1.4. Let $\mathcal{R}$ be a ring of sets and let $\lambda : \mathcal{R} \to [0, \infty]$ be a set function. Then λ is called a submeasure if $\lambda(\emptyset) = 0$ and if λ is monotone and subadditive; λ is said to be continuous on $\mathcal{R}$ if $\lambda(E_n) \to 0$ whenever $E_n \searrow \emptyset$ in $\mathcal{R}$.

Proposition 1.1.5. *Let Σ be a σ-ring of sets. If $\gamma : \Sigma \to X$ is σ-additive and if (E_n) is a convergent sequence in Σ with its limit E, then $\lim_n ||\gamma||(E_n) = ||\gamma||(E)$.*

The proof of Theorem 1.3 of [L1] holds here to prove the above proposition.

Proposition 1.1.6. *Let $\mathcal{P}$ be a δ-ring of subsets of T and let $\mathbf{m} : \mathcal{P} \to X$ be σ-additive. Then $||\mathbf{m}||$ is a σ-subadditive submeasure on $\sigma(\mathcal{P})$ and $||\mathbf{m}||$ is continuous on $\mathcal{P}$. For $E \in \mathcal{P}$,*

$$\sup_{F \subset E, F \in \mathcal{P}} |\mathbf{m}(F)| \leq ||\mathbf{m}||(E) \leq 4 \sup_{F \subset E, F \in \mathcal{P}} |\mathbf{m}(F)|. \tag{1.1.6.1}$$

We define $||\mathbf{m}||(T) = \sup_{E \in \mathcal{P}} |\mathbf{m}(E)|$. Then $||\mathbf{m}||(E) < \infty$ for $E \in \mathcal{P}$ and $||\mathbf{m}||(T) < \infty$ whenever $\mathcal{P}$ is a σ-ring.

Proof. By Proposition 1.1.1, $||\mathbf{m}||$ is subadditive on $\sigma(\mathcal{P})$. Let $E \in \mathcal{P}$. Then $E \cap \mathcal{P} = \{E \cap F : F \in \mathcal{P}\}$ is a σ-algebra and hence for $E_n \searrow \emptyset$ in $\mathcal{P}$, $(E_n)_1^\infty \subset E_1 \cap \mathcal{P}$ and hence by Proposition 1.1.5, $\lim_n ||\mathbf{m}||(E_n) = 0$. Hence $||\mathbf{m}||$ is continuous on $\mathcal{P}$. For $E \in \mathcal{P}$, (1.1.6.1) holds by Theorem I.1(f) of [Ri5] or by Proposition 4, §1, capítulo 1 of [Bo] since $E \cap \mathcal{P}$ is an algebra of sets. Moreover, $||\mathbf{m}||(E) < \infty$ by Proposition 1.1.3(iv) since $E \cap \mathcal{P}$ is a σ-algebra in E and $\mathbf{m}$ is σ-additive on $E \cap \mathcal{P}$. Again by Proposition 1.1.3(iv), $||\mathbf{m}||(T) < \infty$ if $\mathcal{P}$ is a σ-ring. □

Proposition 1.1.7. *Let Σ be a σ-ring of sets and $\boldsymbol{\gamma}_i : \Sigma \to X$ be σ-additive for $i \in I$. $(\boldsymbol{\gamma}_i)_{i \in I}$ is said to be uniformly σ-additive on Σ if for any sequence $(E_n)_1^\infty$ of pairwise disjoint members of Σ, $\lim_n |\Sigma_{m=n}^\infty \boldsymbol{\gamma}_i(E_m)| = 0$ uniformly in $i \in I$. Then the following statements are equivalent:*

(i) *$\boldsymbol{\gamma}_i, i \in I$, are uniformly σ-additive.*

(ii) *Given $E_n \searrow \emptyset$ in Σ, $\lim_n ||\boldsymbol{\gamma}_i||(E_n) = 0$ uniformly in $i \in I$.*

(iii) *Given $E_n \searrow \emptyset$ in Σ, then $\lim_n \boldsymbol{\gamma}_i(E_n) = 0$ uniformly in $i \in I$.*

Proof. (i)⇒(ii) Let $E_n \searrow \emptyset$ in Σ and $F_n = E_n \backslash E_{n+1}$. Clearly, $(F_n)_1^\infty$ is a disjoint sequence in Σ and $E_n = \bigcup_{k=n}^\infty F_k$. Then for $x^* \in X^*$ with $|x^*| \leq 1$, $x^* \boldsymbol{\gamma}_i$ and hence $v(x^* \boldsymbol{\gamma}_i)$ is σ-additive on Σ for $i \in I$ by Proposition 1.1.3(iii). Then by (i), $v(x^* \boldsymbol{\gamma}_i)(E_n) = \sum_{k=n}^\infty v(x^* \boldsymbol{\gamma}_i)(F_k)$ converges to 0 uniformly in $i \in I$ and in $x^* \in B(X^*) = \{x^* \in X^* : |x^*| \leq 1\}$. Consequently, by Proposition 1.1.1. $\lim_n ||\boldsymbol{\gamma}_i||(E_n) = 0$ uniformly in $i \in I$ and hence (ii) holds.

Since $|\boldsymbol{\gamma}_i(E_n)| \leq ||\boldsymbol{\gamma}_i||(E_n)$, (ii)⇒(iii).

(iii)⇒(i) If (F_n) is a disjoint sequence in Σ, let $F = \bigcup_1^\infty F_n$ and let $E_k = F \backslash \bigcup_{n=1}^k F_n$. Then $(E_k)_1^\infty \subset \Sigma$ and $E_k \searrow \emptyset$. Hence by (iii), $\lim_n \boldsymbol{\gamma}_i(E_n) = \lim_n (\boldsymbol{\gamma}_i(F) - \sum_{k=1}^n \boldsymbol{\gamma}_i(F_k)) = 0$ uniformly in $i \in I$. Thus (i) holds. □

Proposition 1.1.8. (VHSN). *Let Σ be a σ-ring of sets. Let $\boldsymbol{\gamma}_n : \Sigma \to X, n \in \mathbb{N}$, be σ-additive and let $\lim_n \boldsymbol{\gamma}_n(E) = \boldsymbol{\gamma}(E)$ exist in X for each $E \in \Sigma$. Then $\boldsymbol{\gamma}_n, n \in \mathbb{N}$, are uniformly σ-additive on Σ and $\boldsymbol{\gamma} : \Sigma \to X$ is also σ-additive.*

Proof. The result holds for a σ-algebra $\mathcal{S}$ by Theorem I.4.8 of [DU] since X-valued σ-additive measures $\boldsymbol{\gamma}_n$ on a σ-algebra are strongly additive in the sense that $\lim_k \boldsymbol{\gamma}_n(E_k) = 0$ whenever $(E_k)_1^\infty$ is a disjoint sequence in $\mathcal{S}$. Suppose $(\boldsymbol{\gamma}_n)_1^\infty$ is not uniformly σ-additive on Σ. Then there exist an $\epsilon > 0$ and a sequence $A_k \searrow \emptyset$ in

Σ such that $|\gamma_{n_k}(A_k)| > \epsilon$ for a subsequence (n_k) of (n). Without loss of generality we can take $n_k = k, k \in \mathbb{N}$. Then

$$|\gamma_k(A_k)| > \epsilon \qquad (1.1.8.1)$$

for all $k \in \mathbb{N}$. Let $\mathcal{A} = A_1 \cap \Sigma$. Then $\mathcal{A}$ is a σ-algebra in A_1. Clearly, $\gamma_n : \mathcal{A} \to X$ is σ-additive for $n \in \mathbb{N}$ and $\lim_n \gamma_n(F) \in X$ for each $F \in \mathcal{A}$. Then by VHSN for σ-algebras, $(\gamma_n|_\mathcal{A})_1^\infty$ is uniformly σ-additive and hence by Proposition 1.1.7, for the sequence $(A_k)_1^\infty$, there exists k_0 such that $|\gamma_n(A_k)| < \epsilon$ for $k \geq k_0$ and for $n \in \mathbb{N}$. This contradicts (1.1.8.1) and hence VHSN holds for σ-rings too. $\qquad \square$

Remark 1.1.9. The above result is a particular case of Theorem 17, §3 of [Din2].

Definition 1.1.10. Let λ be a submeasure on a σ-ring Σ and let $\gamma : \Sigma \to X$ be σ-additive. Then γ is said to be absolutely continuous with respect to λ (in symbols, $\gamma \ll \lambda$) (resp. λ-continuous) if $\lambda(E) = 0$ implies $\gamma(E) = 0$ (resp. if $\lim_{\lambda(E) \to 0} \gamma(E) = 0$) for $E \in \Sigma$.

Proposition 1.1.11. (Pettis). *Let Σ be a σ-ring of subsets of T. Let $\lambda : \Sigma \to [0, \infty]$ be a σ-subadditive submeasure and let $\gamma : \Sigma \to X$ be σ-additive. Then $\gamma \ll \lambda$ if and only if γ is λ-continuous.*

Proof. Clearly the condition is sufficient. Suppose $\gamma \ll \lambda$ and γ is not λ-continuous. Then there exists an $\epsilon > 0$ such that, for each $n \in \mathbb{N}$, there exists a set $E_n \in \Sigma$ with $\lambda(E_n) < \frac{1}{2^n}$ for which $|\gamma(E_n)| \geq \epsilon$. Let $E = \limsup E_n$ and $A_n = \bigcup_{k=n}^\infty E_k$. Then $\lambda(E) = \lambda(\bigcap_{n=1}^\infty A_n) \leq \lambda(A_n) \leq \sum_{k=n}^\infty \lambda(E_k) < \frac{1}{2^{n-1}}$ for each n and hence $\lambda(E) = 0$. Then by hypothesis, $\gamma(E) = 0$. Clearly, $A_n \searrow E$ and hence by the continuity of $||\gamma||$, $\lim_n ||\gamma||(A_n \backslash E) = 0$. Thus there exists n_0 such that $||\gamma||(A_n \backslash E) < \epsilon$ for $n \geq n_0$. $\lambda(E) = 0$ implies $\lambda(F) = 0$ for all $F \subset E, F \in \Sigma$ and hence by hypothesis, $\gamma(F) = 0$ for all $F \subset E, F \in \Sigma$. Therefore, $||\gamma||(E) = 0$. Consequently, $||\gamma||(A_n) = ||\gamma||(A_n) - ||\gamma||(E) \leq ||\gamma||(A_n \backslash E) < \epsilon$ for $n \geq n_0$. This is impossible since $||\gamma||(A_n) \geq ||\gamma||(E_n) \geq |\gamma(E_n)| \geq \epsilon$ for $n \in \mathbb{N}$. Hence the condition is also necessary. $\qquad \square$

Proposition 1.1.12. *A continuous submeasure λ defined on a σ-ring S is σ-subadditive.*

Proof. Since λ is monotone, it suffices to show that $\lambda(\bigcup_1^\infty E_n) \leq \sum_{n=1}^\infty \lambda(E_n)$ for any disjoint sequence $(E_n)_1^\infty \subset S$. For such a sequence $(E_n)_1^\infty$, let $E = \bigcup_1^\infty E_n$ and let $F_n = \bigcup_{k=n}^\infty E_k$. Then $(F_n)_1^\infty \subset S$ and $F_n \searrow \emptyset$ so that by hypothesis

$$\lim_n \lambda(F_n) = 0. \qquad (1.1.12.1)$$

As λ is finitely subadditive, we have $\lambda(E) \leq \sum_{k=1}^{n-1} \lambda(E_k) + \lambda(F_n)$. Taking the limit as $n \to \infty$, we have

$$\lambda(E) \leq \sum_1^\infty \lambda(E_n) \quad \text{by (1.1.12.1).} \qquad \square$$

Proposition 1.1.13. *Let Σ be a σ-ring of sets and let $\mathcal{F} = \{\gamma_i : \Sigma \to X, i \in I\}$ be a family of uniformly σ-additive vector measures. Then there exists a σ-additive finite positive measure μ on Σ, called a control measure of the family $\mathcal{F}$, such that*

$$\mu(E) \leq \sup_{i \in I} ||\gamma_i||(E) \qquad and \qquad \lim_{\mu(E) \to 0} \sup_{i \in I} ||\gamma_i||(E) = 0, \ E \in \Sigma.$$

Proof. The above result is proved on pp. 13–14 of [DU] for a σ-algebra Σ when the family $\mathcal{F}$ is bounded and the proof clearly holds for all Banach spaces X over $\mathbb{C}$. Then by Proposition 1.1.12, Theorem I.2.1 of [DU] holds for σ-rings, since for any σ-additive X-valued measure γ on a σ-ring Σ, $\gamma \ll ||\gamma||$. If $\mathcal{F}$ is not bounded, as observed by Dobrakov on p. 690 of [Do2], one can replace

$$\frac{1}{n(m)} \sum_{j=1}^{n(m)} v(\mu_{\tau_j}^m)(E)$$

on p. 13 of [DU] by

$$\frac{A}{B}$$

where

$$A = \frac{1}{n(m)} \sum_{j=1}^{n(m)} v(\mu(\tau_j^m))(E) \qquad and \qquad B = 1 + \sup_{F \in \Sigma} v(\mu_{\tau_j}^m)(F)$$

Then the result follows.   □

In the proof of the classical Egoroff theorem with respect to a finite positive measure μ on a σ-ring of sets, only the continuity from above in the sense of [H] and the σ-subadditivity of μ are used. Thus, in the light of Proposition 1.1.12, we can adapt the proof of the classical Egoroff theorem to generalize it to the case of continuous submeasures. Thus we have

Proposition 1.1.14. (Egoroff). *Let $\lambda : \mathcal{S} \to [0, \infty]$ be a continuous submeasure on the σ-ring $\mathcal{S}$ and let $f, f_n : T \to X, n \in \mathbb{N}$, be $\mathcal{S}$-measurable. If $f_n \to f \lambda - a.e.$ in T, then given $\epsilon > 0$, there exists a set $E_\epsilon \in \mathcal{S}$ such that $\lambda(E_\epsilon) < \epsilon$ and $f_n \to f$ uniformly on $T \backslash E_\epsilon$.*

Let $\mathbf{m} : \mathcal{P} \to X$ be σ-additive. We denote by $\widetilde{\sigma(\mathcal{P})}$ the generalized Lebesgue-completion of $\sigma(\mathcal{P})$ with respect to the semivariation $||\mathbf{m}||$ in the sense of Definition 11 of [DP2]. That is, $\widetilde{\sigma(\mathcal{P})} = \{E = F \cup N : F \in \sigma(\mathcal{P}), N \subset M \in \sigma(\mathcal{P})$ with $||\mathbf{m}||(M) = 0\}$. As $||\mathbf{m}||$ is a σ-subadditive submeasure on $\sigma(\mathcal{P})$, it can easily be shown that $\widetilde{\sigma(\mathcal{P})}$ is a σ-ring and that $\widetilde{\sigma(\mathcal{P})}$ contains $\sigma(\mathcal{P})$. Let $E \in \widetilde{\sigma(\mathcal{P})}$ and define $\overline{||\mathbf{m}||}(E) = \overline{||\mathbf{m}||}(F \cup N) = ||\mathbf{m}||(F)$ where $E = F \cup N$ with $F \in \sigma(\mathcal{P})$ and $N \subset M \in \sigma(\mathcal{P})$ with $||\mathbf{m}||(M) = 0$. It can be shown that $\overline{||\mathbf{m}||}$ is well defined on $\widetilde{\sigma(\mathcal{P})}$, extends the semivariation $||\mathbf{m}||$ and is a σ-subadditive submeasure on

$\widetilde{\sigma(\mathcal{P})}$. **Hereafter we shall use the symbol** $||\mathbf{m}||$ **to denote** $\overline{||\mathbf{m}||}$ **also. Sets** $N \in \widetilde{\sigma(\mathcal{P})}$ with $||\mathbf{m}||(N) = 0$ are called $\mathbf{m}$-null.

If $\mathcal{P}$ is a σ-ring $\mathcal{S}$, then by Proposition 1.1.13 there exists a control measure μ on $\mathcal{S}$ such that $||\mathbf{m}||(E) = 0$ if and only if $\mu(E) = 0$ for $E \in \mathcal{S}$ and hence, the generalized Lebesgue completion of $\mathcal{S}$ with respect to $||\mathbf{m}||$ is the same as the Lebesgue completion of $\mathcal{S}$ with respect to μ.

Notation 1.1.15. For $f : T \to \mathbb{K}$ or $[-\infty, \infty]$, $N(f) = \{t \in T : f(t) \neq 0\}$. A property P is said to be true $\mathbf{m}$-a.e. in T if there exists $N \in \sigma(\mathcal{P})$ with $||\mathbf{m}||(N) = 0$ such that P holds for all $t \in T \backslash N$.

Definition 1.1.16. A function $f : T \to \mathbb{K}$ or $[-\infty, \infty]$ is said to be $\mathbf{m}$-measurable if f is $\widetilde{\sigma(\mathcal{P})}$-measurable; that is, if $f^{-1}(B) \cap N(f) \in \widetilde{\sigma(\mathcal{P})}$ for all Borel subsets B of $\mathbb{K}$ or of $[-\infty, \infty]$, respectively. The concepts of $\mathbf{m}$-essentially bounded functions and ess sup of an $\mathbf{m}$-measurable function are given with respect to $||\mathbf{m}||$ as in the case of positive measures. As $||\mathbf{m}||$ is a σ-subadditive submeasure on $\widetilde{\sigma(\mathcal{P})}$, it follows that there exists $N \in \sigma(\mathcal{P})$ with $||\mathbf{m}||(N) = 0$ such that ess sup $|f| = \sup_{t \in T \backslash N} |f(t)|$ whenever the $\mathbf{m}$-measurable function f is $\mathbf{m}$-essentially bounded on T. (See the proof of the claim in the proof of Theorem 4.1.9 of Chapter 4 for a more general situation.)

Notation 1.1.17. $\mathcal{I}_s$ (resp. $\mathcal{I}_{\tilde{s}}$) denotes the family of all $\mathcal{P}$-simple (resp. $\widetilde{\sigma(\mathcal{P})}$-simple) scalar functions on T.

Proposition 1.1.18. [1] *Let* $f : T \to \mathbb{K}$ *or* $[-\infty, \infty]$. *Then* f *is* $\mathbf{m}$*-measurable if and only if there exists a sequence* $(\tilde{s}_n) \subset \mathcal{I}_{\tilde{s}}$ *(resp.* $(s_n) \subset \mathcal{I}_s$*) such that* $\tilde{s}_n(t) \to f(t)$ *and* $|\tilde{s}_n(t)| \nearrow |f(t)|$ *for* $t \in T$ *(resp.* $s_n(t) \to f(t)$ $\mathbf{m}$*-a.e. in* T *and* $|s_n(t)| \nearrow |f(t)|$ $\mathbf{m}$*-a.e. in* T *so that there exists* $M \in \sigma(\mathcal{P})$ *with* $||\mathbf{m}||(M) = 0$ *such that* $f \chi_{T \backslash M}$ *is* $\sigma(\mathcal{P})$*-measurable).*

Proof. f is $\mathbf{m}$-measurable if and only if it is $\widetilde{\sigma(\mathcal{P})}$-measurable. Hence it is so if and only if there exists a sequence $(\tilde{s}_n) \subset \mathcal{I}_{\tilde{s}}$ such that $|\tilde{s}_n(t)| \nearrow |f(t)|$ and $\tilde{s}_n(t) \to f(t)$ for $t \in T$. Since each $E \in \widetilde{\sigma(\mathcal{P})}$ is of the form $E = F \cup N$, $F \in \sigma(\mathcal{P})$, $N \subset M \in \sigma(\mathcal{P})$ with $||\mathbf{m}||(M) = 0$, it is clear that, for each n, there exist $M_n \in \sigma(\mathcal{P})$ with $||\mathbf{m}||(M_n) = 0$ and a $\sigma(\mathcal{P})$-simple function ω_n such that $\tilde{s}_n(t) = \omega_n(t)$ for $t \in T \backslash M_n$. Let $M = \bigcup_1^\infty M_n$. Then $M \in \sigma(\mathcal{P})$, $||\mathbf{m}||(M) = 0$ and $|\omega_n(t)| \nearrow |f(t)|$ and $\omega_n(t) \to f(t)$ for $t \in T \backslash M$. Consequently, $f \chi_{T \backslash M}$ is $\sigma(\mathcal{P})$-measurable and hence $N(f) \cap (T \backslash M) \in \sigma(\mathcal{P})$. Then there exists $(E_n) \subset \mathcal{P}$ such that $E_n \nearrow N(f) \cap (T \backslash M)$ so that $s_n = \omega_n \chi_{E_n}$, $n \in \mathbb{N}$, belong to $\mathcal{I}_s$ and satisfy the desired properties. The proof of the converse is left to the reader. $\square$

---

[1] The relationship between $\mathbf{m}$-measurability and $\sigma(\mathcal{P})$-measurability is explained in the end of this section. See Example 1.1.23.

We refer to Chapter II of [Din1] for the theory of integration of scalar functions with respect to a (σ-additive) scalar measure defined on a δ-ring $\mathcal{P}$ of sets. The reader may also refer to [MN1].

Using Definition 1, §8 of [Din1] and modifying the argument in the proof of Theorem III.2.20(a) of [DS1], we obtain the following result. See also Theorem 2, §2 of [Din2].

Proposition 1.1.19. *Let $\mathcal{P}$ be a δ-ring of subsets of T and let ν be a (σ-additive) scalar measure on $\mathcal{P}$. If $f : T \to \mathbb{K}$ is ν-integrable, let $\gamma(E) = \int_E f d\nu$, $E \in \sigma(\mathcal{P})$. Then γ is σ-additive on $\sigma(\mathcal{P})$ (by Proposition 5, §7 of [Din1]), and*

$$v(\gamma)(E) = \int_E |f| dv(\nu), \ \ E \in \sigma(\mathcal{P}).$$

Proposition 1.1.20. (Egoroff-Lusin theorem). *Let $\mathcal{P}$ be a δ-ring of subsets of T and let $\lambda : \sigma(\mathcal{P}) \to [0, \infty]$ be a continuous submeasure. Let $f, f_n : T \to \mathbb{K}, n \in \mathbb{N}$, be $\sigma(\mathcal{P})$-measurable. Suppose $f_n(t) \to f(t)$ for $t \in T$. If $F = \bigcup_{n=1}^{\infty} N(f_n)$, then there exist $N \in \sigma(\mathcal{P}) \cap F$ with $\lambda(N) = 0$ and a sequence $(F_k)_1^{\infty} \subset \mathcal{P}$ with $F_k \nearrow F \backslash N$ such that $f_n \to f$ uniformly in every F_k.*

Proof. By hypothesis λ is a continuous submeasure and hence the Egoroff theorem holds by Proposition 1.1.14. Consequently, by applying the Egoroff theorem successively with $\epsilon = \frac{1}{n}$ in the nth step, we can construct a decreasing sequence $(G_n)_1^{\infty} \subset \sigma(\mathcal{P})$ such that $\lambda(G_n) < \frac{1}{n}$ and $f_k \to f$ uniformly on $G_{n-1} \backslash G_n$, $G_0 = F$. Let $N = \bigcap_1^{\infty} G_n$. Then $N \in \sigma(\mathcal{P})$ and $\lambda(N) = 0$. Moreover, $F \backslash N = \bigcup_1^{\infty} (F \backslash G_n)$ and $F \backslash G_n \nearrow$. Clearly, $f_l \to f$ uniformly on $F \backslash G_n = \bigcup_{k=1}^{n} (G_{k-1} \backslash G_k)$ for each n. As $F \backslash G_n \in \sigma(\mathcal{P})$, there exists an increasing sequence $(H_{n,m})_{m=1}^{\infty} \subset \mathcal{P}$ such that $\bigcup_{m=1}^{\infty} H_{n,m} = F \backslash G_n$. Let $F_n = \bigcup_{p,m=1}^{n} H_{p,m}$. Then $F_n \in \mathcal{P}$ for all n, $F_n \nearrow F \backslash N$ and $f_k \to f$ uniformly on each F_n. $\qquad\square$

The following result plays a key role in the proof of Theorem 2.2.2 and will be generalized to the case of σ-additive vector measures in Theorem 2.2.8. For the case of σ-algebras or σ-rings, it is an immediate consequence of the Vitali-Hahn-Saks theorem (see the proof of Lemma 2.3 of [L1]). However, in the case of δ-rings we have to give a different argument as $v(\nu)$ need not be finite-valued on $\sigma(\mathcal{P})$. We adapt the proof of the first part of Theorem 9 of [DP2] to prove the following

Proposition 1.1.21. (Theorem on the interchange of limit and integral). *Let $\mathcal{P}$ be a δ-ring of subsets of T and let ν be a (σ-additive) scalar measure on $\mathcal{P}$. Let $f : T \to \mathbb{K}$ and let $f_n, n \in \mathbb{N}$, be ν-integrable scalar functions on T. Let $\gamma_n(E) = \int_E f_n d\nu$, $E \in \sigma(\mathcal{P})$, $n \in \mathbb{N}$. If $f_n(t) \to f(t)$ ν-a.e. in T, then the following statements are equivalent:*

(i) $\lim_n \gamma_n(E) = \gamma(E)$ *exists in $\mathbb{K}$ for each $E \in \sigma(\mathcal{P})$.*

(ii) $\gamma_n, n \in \mathbb{N}$, *are uniformly σ-additive on $\sigma(\mathcal{P})$.*

(iii) $\lim_n \gamma_n(E) = \gamma(E)$ *exists in $\mathbb{K}$ uniformly with respect to $E \in \sigma(\mathcal{P})$.*

If any one of the above statements holds, then f is ν-integrable in T and

$$\int_E f d\nu = \lim_{n \to \infty} \int_E f_n d\nu, \quad E \in \sigma(\mathcal{P}),$$ (1.1.21.1)

the limit being uniform with respect to $E \in \sigma(\mathcal{P})$. Consequently, f is ν-integrable in T and (1.1.21) holds if and only if any one of (i), (ii) or (iii) holds.

Proof. (i)$\Rightarrow$(ii) by VHSN and (iii)$\Rightarrow$(i) obviously.

(ii)$\Rightarrow$(iii) By hypothesis there exists $M \in \sigma(\mathcal{P})$ with $v(\nu)(M) = 0$ such that $f_n \to f$ pointwise in $T \backslash M$ and $f_n \chi_{T \backslash M}$, $n \in \mathbb{N}$, are $\sigma(\mathcal{P})$-measurable (see the proof of Proposition 1.1.18). Then $f_n \chi_{T \backslash M} \to f \chi_{T \backslash M}$ pointwise in T and hence $f \chi_{T \backslash M}$ is $\sigma(\mathcal{P})$-measurable. Let $F = \bigcup_1^\infty N(f_n \chi_{T \backslash M}) = \bigcup_1^\infty N(f_n) \cap (T \backslash M)$. Then $F \in \sigma(\mathcal{P})$. Let $\epsilon > 0$. By (ii) and by Proposition 1.1.13 there exists a finite control measure μ on $\sigma(\mathcal{P})$ such that $\lim_{\mu(E) \to 0} \sup_n v(\gamma_n)(E) = 0$ for $E \in \sigma(\mathcal{P})$. Thus there exists $\delta > 0$ such that $v(\gamma_n)(E) < \frac{\epsilon}{3}$ for all n whenever $\mu(E) < \delta$.

By the Egoroff-Lusin theorem there exist $N \in F \cap \sigma(\mathcal{P})$ with $\mu(N) = 0$ and a sequence $(F_k)_1^\infty \subset \mathcal{P}$ with $F_k \nearrow F \backslash N$ such that $f_n \to f$ uniformly in each F_k. Since $\mu(N) = 0$, $v(\gamma_n)(N \cap E) = 0$ for all n and for all $E \in \sigma(\mathcal{P})$. As μ is continuous on $\sigma(\mathcal{P})$ and as $F \backslash N \backslash F_k \searrow \emptyset$, there exists k_0 such that $\mu(F \backslash N \backslash F_{k_0}) < \delta$. Then

$$v(\gamma_n)(F \backslash N \backslash F_{k_0}) < \frac{\epsilon}{3} \quad \text{for all } n \in \mathbb{N}.$$ (1.1.21.2)

Let $E \in \sigma(\mathcal{P})$. By Proposition 1.1.19,

$$\int_{E \cap N} |f_k| dv(\nu) = v(\gamma_k)(E \cap N) = 0$$

for all $k \in \mathbb{N}$ as $\mu(N) = 0$. Moreover, $\int_{E \cap M} |f_k| dv(\nu) = 0$ for all $k \in \mathbb{N}$ as $v(\nu)(M) = 0$. Consequently, by Proposition 4, §8 of [Din1], we have

$$\left| \int_E f_n d\nu - \int_E f_r d\nu \right| \le \int_E |f_n - f_r| dv(\nu)$$

$$\le \int_{E \cap F_{k_0}} |f_n - f_r| dv(\nu) + \int_{E \cap (F \backslash N \backslash F_{k_0})} |f_n| dv(\nu)$$

$$+ \int_{E \cap (F \backslash N \backslash F_{k_0})} |f_r| dv(\nu)$$ (1.1.21.3)

since $\int_{E \cap N} |f_k| dv(\nu) = 0$ and $\int_{E \cap M} |f_k| dv(\nu) = 0$ all all $k \in \mathbb{N}$.

Since $f_n \to f$ uniformly in F_{k_0} and since $v(\nu)(F_{k_0}) < \infty$ (as $F_{k_0} \in \mathcal{P}$), we can choose n_0 such that $\|f_n - f_r\|_{F_{k_0}} \cdot v(\nu)(F_{k_0}) < \frac{\epsilon}{3}$ for $n, r \ge n_0$. Then by Proposition 1.1.19 and by (1.1.21.2) we have

$$\int_{E \cap (F \backslash N \backslash F_{k_0})} |f_k| dv(\nu) = v(\gamma_k)(E \cap (F \backslash N \backslash F_{k_0})) < \frac{\epsilon}{3}$$ (1.1.21.4)

for all $k \in \mathbb{N}$.

Let us recall Corollary 3 of Proposition 8, §8 of [Din1] which essentially says that for $f \in \mathcal{L}_1(\nu)$ and for $A \in \sigma(\mathcal{P})$,

$$\int_A |f| dv(\nu) \leq ||f||_A v(\nu)(A). \qquad (*)$$

Then by (1.1.21.3) and (1.1.21.4) and by ($*$), we have

$$\left| \int_E f_n dv - \int_E f_r dv \right| \leq \int_E |f_n - f_r| dv(\nu)$$

$$\leq ||f_n - f_r||_{F_{k_0}} \cdot v(\nu)(F_{k_0}) + \frac{2\epsilon}{3} < \epsilon$$

for $n, r \geq n_0$ and for all $E \in \sigma(\mathcal{P})$. Hence (iii) holds.

If (i), (ii) or (iii) holds, then (ii) holds and hence by the above argument there exists n_0 such that

$$\int_E |f_n - f_r| dv(\nu) < \epsilon \qquad (1.1.21.5)$$

for all $n, r \geq n_0$ and for all $E \in \sigma(\mathcal{P})$. Hence $N_1(f_n - f_r) = \int_T |f_n - f_r| dv(\nu) < \epsilon$ for $n, r \geq n_0$, where $N_1(f) = \int |f| dv(\nu)$ as on p. 127 of [Din1] and hence by Proposition 17, §8 of [Din1], f is ν-integrable in T. Moreover, by (1.1.21.5) and by Fatou's lemma (Proposition 21, §8 of [Din1]), we have

$$\left| \int_E f dv - \int_E f_n dv \right| \leq \int_E |f - f_n| dv(\nu)$$

$$= \int_E \liminf_{r \to \infty} |f_r - f_n| dv(\nu)$$

$$\leq \liminf_{r \to \infty} \int_E |f_r - f_n| dv(\nu) < \epsilon$$

for $n \geq n_0$ and for all $E \in \sigma(\mathcal{P})$. Hence

$$\int_E f dv = \lim_{n \to \infty} \int_E f_n dv,$$

the limit being uniform with respect to $E \in \sigma(\mathcal{P})$. The last part is evident from the previous parts.   □

The reader may compare Proposition 1.1.13 with the result in the following remark.

Remark 1.1.22. Let $\mathcal{R}$ be a ring of sets and $\mathbf{m} : \mathcal{R} \to X$ be finitely additive. $\mathbf{m}$ is said to be strongly bounded if for each disjoint sequence $(E_n)_1^\infty \subset \mathcal{R}$, $\lim_n \mathbf{m}(E_n) = 0$. Theorem 1 of [Br] says that $\mathbf{m} : \mathcal{R} \to X$ is finitely additive and strongly bounded if and only if there exists a positive finitely additive bounded set function ν on $\mathcal{R}$ such that

$$\lim_{\nu(E)\to 0} \mathbf{m}(E) = 0$$

and

$$\nu(E) \leq \sup\{|\mathbf{m}(F)| : F \subset E, F \in \mathcal{R}\}, E \in \mathcal{R}.$$

The relationship between **m**-measurability (in Definition 1.1.16) and Proposition 1.1.18 plays a crucial role throughout the book. Since $\widetilde{\sigma(\mathcal{P})}$ contains $\sigma(\mathcal{P})$, $\sigma(\mathcal{P})$-measurability implies **m**-measurability. But, **m**-measurability does not imply $\sigma(\mathcal{P})$-measurability as is shown in the following example.

Example 1.1.23. Let $T = [0,1]$, and λ be the Lebesgue measure on $[0,1]$, $\mathcal{P}=$ $\mathcal{B}([0,1])$, the σ-algebra of the Borel sets in $[0,1]$ and $\mathbf{m}(\cdot) = \lambda(\cdot)x_0$, where x_0 is a nonzero vector in a Banach space X. Since $||\mathbf{m}||(E) = \sup_{|x^*|\leq 1} v(x^*\mathbf{m})(E)$, $||\mathbf{m}|| = \lambda|x_0|$. If C is the Cantor ternary set, then $||\mathbf{m}||(C) = 0$ and $C \in \sigma(\mathcal{P})$. Consequently, $||\mathbf{m}||(E) = 0$ for all $E \subset C$ and the cardinality of $\mathcal{P}(C)$ is $2^c > c$ =the cardinality of $\mathcal{B}([0,1])$=the cardinality of $\mathcal{B}(C)$. Hence there exist sets $E \subset C$ with $E \notin \sigma(\mathcal{P})$. Therefore, such E is not $\sigma(\mathcal{P})$-measurable, even though it is **m**-measurable.

1.2 lcHs-valued measures

As observed by Kluvánek in [K4], the study of spectral operators motivated the paper [BDS]. The σ-additivity of a spectral measure $E(\cdot)$ is considered in the strong operator topology which is a locally convex Hausdorff topology. See [DS2]. By the Orlicz-Pettis theorem (Theorem I.4 of [Ri5]) the study of σ-additive vector measures **m** on a σ-algebra Σ with values in a Banach space X is based on the σ-additivity of complex measures $x^*\mathbf{m}$ on Σ for each $x^* \in X^*$. See Proposition I.1 of [Ri5]. In fact, $\mathbf{m} : \Sigma \to X$ is σ-additive if and only if $x^*\mathbf{m}$ is a (σ-additive) complex measure on Σ for each $x^* \in X^*$. The proof of this result uses the Orlicz-Pettis theorem (see Theorem I.4 of [Ri5]) which says that **the series $\sum_1^\infty x_n$ of elements $(x_n)_1^\infty$ in the Banach space X is unconditionally norm convergent in X whenever it is weakly subseries convergent.** Thus the study of lcHs-valued σ-additive measures is indispensable in the study of Banach space-valued σ-additive vector measures. Hence the present section and Chapters 4–7 are devoted to the study of the integral in lcHs. See [KK],[K1], [K2],[L2], [OR3], [Ri1] and [Ri4]. [Ri5] has a 28 page Appendix which is just the references on vector- and operator-valued measures. [LC] treats an important case of σ-additive (for the strong operator topology) projection-valued measures (with unbounded range) on a δ-ring.

In the rest of this chapter X denotes an lcHs with topology τ and $\tilde{X}$ denotes its completion with the lcHs topology $\tilde{\tau}$ unless otherwise mentioned. Γ is the family of continuous seminorms on X so that Γ generates the topology τ on X. X^* denotes the topological dual of X.

For $q \in \Gamma$, let $\Pi_q : X \to X_q = X/q^{-1}(0)$ be the canonical quotient map. If we define $|x + q^{-1}(0)|_q = q(x)$, $x \in X$, then $|\cdot|_q$ is a well-defined norm on X_q and the Banach space completion of X_q with respect to $|\cdot|_q$ is denoted by $\widetilde{X}_q$ whose norm is also denoted by $|\cdot|_q$.

In the sequel, $\mathbf{m} : \mathcal{P} \to X$, $\mathcal{P}$ a δ-ring of subsets of T, is σ-additive. For $q \in \Gamma$, let $\mathbf{m}_q : \mathcal{P} \to X_q \subset \widetilde{X}_q$ be defined by $\mathbf{m}_q(A) = \Pi_q \circ \mathbf{m}(A)$, $A \in \mathcal{P}$. Then clearly $\mathbf{m}_q$ is σ-additive on $\mathcal{P}$ and the semivariation of $\mathbf{m}_q$ on $\sigma(\mathcal{P})$ is denoted by both $||\mathbf{m}_q||$ and $||\mathbf{m}||_q$; $v(\mathbf{m}_q)$ is the variation of $\mathbf{m}_q$ on $\sigma(\mathcal{P})$. Note that for defining $||\mathbf{m}||_q$ and $v(\mathbf{m}_q)$ it suffices that $\mathbf{m}$ be only additive on $\mathcal{P}$.

Definition 1.2.1. Let $q \in \Gamma$ be fixed. $\widetilde{\sigma(\mathcal{P})}_q$ is the generalized Lebesgue completion of $\sigma(\mathcal{P})$ with respect to $||\mathbf{m}||_q$ so that $\widetilde{\sigma(\mathcal{P})}_q = \{A = B_q \cup N_q : B_q \in \sigma(\mathcal{P}), N_q \subset M_q \in \sigma(\mathcal{P}) \text{ with } ||\mathbf{m}||_q(M_q) = 0\}$. Then $\sigma(\mathcal{P}) \subset \widetilde{\sigma(\mathcal{P})}_q$ and $\widetilde{\sigma(\mathcal{P})}_q$ is a σ-ring. We define $\overline{||\mathbf{m}||_q}(A) = ||\mathbf{m}||_q(B_q)$ if A, B_q are as in the above and clearly $\overline{||\mathbf{m}||_q}$ is well defined, extends $||\mathbf{m}||_q$ to $\widetilde{\sigma(\mathcal{P})}_q$ and is a σ-subadditive submeasure on $\widetilde{\sigma(\mathcal{P})}_q$. We shall use the symbol $||\mathbf{m}||_q$ to denote $\overline{||\mathbf{m}||_q}$ also. Sets $N \in \widetilde{\sigma(\mathcal{P})}_q$ with $||\mathbf{m}||_q(N) = 0$ are called $\mathbf{m}_q$-null . (See Section 1.1 above.)

By Proposition 1.1.6 we have the following

Proposition 1.2.2. *Let* $q \in \Gamma$. *Then* $||\mathbf{m}||_q$ *is a continuous submeasure on* $\mathcal{P}$. *If* $\mathcal{P}$ *is a* σ-*ring* $\mathcal{S}$, *then* $||\mathbf{m}||_q(T) < \infty$.

Definition 1.2.3. Let $\widetilde{\sigma(\mathcal{P})} = \{A = B \cup N : B \in \sigma(\mathcal{P}), N \subset M \in \sigma(\mathcal{P}) \text{ with } ||\mathbf{m}||_q(M) = 0 \text{ for all } q \in \Gamma\}$. Then $\widetilde{\sigma(\mathcal{P})}$ is called the generalized Lebesgue completion of $\sigma(\mathcal{P})$ with respect to $\mathbf{m}$. A set N in T is said to be $\mathbf{m}$-null if $N \in \bigcap_{q \in \Gamma} \widetilde{\sigma(\mathcal{P})}_q$ and if $||\mathbf{m}||_q(N) = 0$ for all $q \in \Gamma$.

Clearly, $\sigma(\mathcal{P}) \subset \widetilde{\sigma(\mathcal{P})} \subset \widetilde{\sigma(\mathcal{P})}_q$ for $q \in \Gamma$ and $\widetilde{\sigma(\mathcal{P})}$ is a σ-ring.

Definition 1.2.4. Let $q \in \Gamma$. A property P is said to be true $\mathbf{m}$-a.e. (resp.$\mathbf{m}_q$-a.e.) in T if there exists an $\mathbf{m}$-null (resp. $\mathbf{m}_q$-null) set $N \in \sigma(\mathcal{P})$ such that P holds for all $t \in T \backslash N$.

Remark 1.2.5. All the definitions and results in the sequel hold good if Γ is replaced by any subfamily of seminorms which generates the topology τ on X.

Definition 1.2.6. A function $f : T \to \mathbb{K}$ or $[-\infty, \infty]$ is said to be $\mathbf{m}$-measurable if, for each $q \in \Gamma$, f is $\mathbf{m}_q$-measurable (i.e., $\sigma(\mathcal{P})_q$-measurable). In that case, there exists $N_q \in \sigma(\mathcal{P})$ with $||\mathbf{m}||_q(N_q) = 0$ such that $f\chi_{T \backslash N_q}$ is $\sigma(\mathcal{P})$-measurable. (See Proposition 1.1.18 above.)

If X is a Banach space, $\sigma(X, X^*)$ is the weak topology on X and $(X, \sigma(X, X^*))$ is an lcHs. If $\mathbf{m} : \mathcal{P} \to X$ is σ-additive, then $\mathbf{m} : \mathcal{P} \to (X, \sigma(X, X^*))$ is also σ-additive. If $N \in \sigma(\mathcal{P})$ with $||\mathbf{m}||(N) = 0$, i.e., if N is $\mathbf{m}$-null, then

$$||\mathbf{m}||(N) = \sup_{|x^*| \leq 1} v(x^* \mathbf{m})(N) = 0 \qquad (1.2.6.1)$$

by Proposition 1.1.1 and hence $v(x^* \mathbf{m})(N) = 0$ for each $x^* \in X^*$. Thus N is weakly $\mathbf{m}$-null.

Conversely, if N is weakly $\mathbf{m}$-null (i.e., $v(x^* \mathbf{m})(N) = 0$ for each $x^* \in X^*$) then by Proposition 1.1.1., N is $\mathbf{m}$-null and **hence N is $\mathbf{m}$-null if and only if N is weakly $\mathbf{m}$-null**.

Example 1.2.7. Let X be a non-reflexive Banach space so that $X \subset X^{**}$ and $X \neq X^{**}$ where we identify X as a subspace of X^{**} (in the norm topology). Thus X is a proper closed subspace of X^{**}. If $\mathbf{m} : \mathcal{P} \to (X^*, \sigma(X^*, X^{**}))$ is σ-additive and if $q_x(x^*) = |x^*(x)|$ for $x \in X$ and $x^* \in X^*$, then let $||\mathbf{m}||_{q_x}(N) = 0$ for some $N \in \sigma(\mathcal{P})$ for each $x \in X$. Then by the Hahn-Banach theorem there exists $x_0^{**} \in X^{**}$ such that $x_0^{**}(x) = 0$ for $x \in X$ and $x_0^{**} \neq 0$. Then $||\mathbf{m}||_{q_{x_0^{**}}}(N) \neq 0$. Thus N is not $\mathbf{m}$-null for the topology $\sigma(X^*, X^{**})$ even though it is $\mathbf{m}$-null for the topology $\sigma(X^*, X)$. Consequently, if $f : T \to \mathbb{K}$ is $\mathbf{m}$-measurable for the topology $\sigma(X^*, X)$, it need not be $\mathbf{m}$-measurable for the topology $\sigma(X^*, X^{**})$. Note that $\mathbf{m} : \mathcal{P} \to (X^*, \sigma(X^*, X))$ is also σ-additive.

Thus the above example shows that the concept of a scalar function f being $\sigma(\mathcal{P})$-measurable is algebraic (nothing to do with the topology on X) while that of being m-measurable is topological.

Definition 1.2.8. Let X be an lcHs and $\mathbf{m} : \mathcal{P} \to X$ be σ-additive. An $\mathbf{m}$-measurable function $f : T \to \mathbb{K}$ or $[-\infty, \infty]$ is said to be $\mathbf{m}$-essentially bounded in T if there exists an $\mathbf{m}$-null set $N \in \sigma(\mathcal{P})$ such that f is bounded in $T \backslash N$. Then we define $\operatorname{ess\,sup}_{t \in T} |f(t)| = \inf\{\alpha > 0 : |f(t)| \leq \alpha \text{ for } t \in T \backslash N_\alpha, N_\alpha \in \sigma(\mathcal{P}), N_\alpha \text{ m-null}\}$.

Notation 1.2.9. $\mathcal{I}_{\widetilde{\sigma(\mathcal{P})_q}}$ denotes the family of all $\widetilde{\sigma(\mathcal{P})_q}$-simple functions on T for $q \in \Gamma$. However, as in Notation 1.1.17 above, $\mathcal{I}_s$ is the family of all $\mathcal{P}$-simple functions.

The proof of Proposition 1.1.18 above can be adapted to prove the following

Proposition 1.2.10. *Let $f : T \to \mathbb{K}$ or $[-\infty, \infty]$. For $q \in \Gamma$, f is $\mathbf{m}_q$-measurable if and only if there exists a sequence $(s_n^{(q)})_{n=1}^\infty \subset \mathcal{I}_{\widetilde{\sigma(\mathcal{P})_q}}$ (resp. $(s_n^{(q)})_1^\infty \subset \mathcal{I}_s$) such that $s_n^{(q)} \to f$ and $|s_n^{(q)}| \nearrow |f|$ pointwise in T (resp. $s_n^{(q)} \to f$ and $|s_n^{(q)}| \nearrow |f|$ pointwise in $T \backslash N_q$ where $N_q \in \sigma(\mathcal{P})$ with $||\mathbf{m}||_q(N_q) = 0$ – so that $f \chi_{T \backslash N_q}$ is $\sigma(\mathcal{P})$-measurable). The function f is $\sigma(\mathcal{P})$-measurable if and only if there exists a sequence $(s_n)_1^\infty \subset \mathcal{I}_s$ such that $s_n \to f$ and $|s_n| \nearrow |f|$ pointwise in T.*

Notation 1.2.11. $\mathcal{E}$ denotes the family of equicontinuous subsets of X^* and for each $E \in \mathcal{E}$, $q_E(x) = \sup_{x^* \in E} |x^*(x)|$, $x \in X$. $\Gamma_{\mathcal{E}} = \{q_E : E \in \mathcal{E}\}$.

By Proposition 7, §4, Chapter 3 of [Ho], we have the following

Proposition 1.2.12. $\Gamma_{\mathcal{E}} \subset \Gamma$ *and* $\Gamma_{\mathcal{E}}$ *generates the topology* τ *on* X.

Proposition 1.2.13. *Let* $\eta : \mathcal{P} \to X$ *be additive and let* $E \in \mathcal{E}$. *Then:*

(i) *For* $x^* \in E$, $\Psi_{x^*} : X_{q_E} \to \mathbb{K}$ *given by* $\Psi_{x^*}(x + q_E^{-1}(0)) = x^*(x)$, $x \in X$, *is well defined, linear and continuous. Thus* $\Psi_{x^*}(\Pi_{q_E}(x)) = x^*(x)$ *for* $x^* \in E$ *and for* $x \in X$.

(ii) $\{\Psi_{x^*} : x^* \in E\}$ *is a norm determining subset of the closed unit ball of* $(X_{q_E})^*$.

(iii) $\|\Pi_{q_E} \circ \eta\|(A) = \sup_{x^* \in E} v(x^*\eta)(A)$, $A \in \sigma(\mathcal{P})$.

(iv) *If* $E = \{x^*\}$, $x^* \in X^*$, *then* $\|\Pi_{q_E} \circ \eta\|(A) = v(x^*\eta)(A) = \|x^*\eta\|(A)$, $A \in \sigma(\mathcal{P})$.

Proof. (i) Clearly, Ψ_{x^*} is well defined and linear for $x^* \in E$. Moreover, for $x^* \in E$ we have $|\Psi_{x^*}(x + q_E^{-1}(0))| = |x^*(x)| \leq q_E(x) = |x + q_E^{-1}(0)|_{q_E}$, $x \in X$ and hence $\Psi_{x^*} \in (X_{q_E})^*$ with

$$|\Psi_{x^*}| \leq 1. \qquad (1.2.13.1)$$

(ii) $|x + q_E^{-1}(0)|_{q_E} = q_E(x) = \sup_{x^* \in E} |x^*(x)| = \sup_{x^* \in E} |\Psi_{x^*}(x + q_E^{-1}(0))|$ and hence by (1.2.13.1), (ii) holds.

(iii) Note that the proof of Proposition I.1.11 of [DU] holds for any additive set function γ on $\mathcal{P}$ with values in a normed space Y and for any norm determining subset of the closed unit ball of Y^*. Moreover, by replacing π on p. 5 of [DU] by $\pi = \{(A_i)_1^r \subset \mathcal{P}, A_i \cap A_j = \emptyset, \bigcup_1^r A_i \subset A\}$ for $A \in \sigma(\mathcal{P})$, continuing with the proof of I.1.11 of [DU] and using (i) and (ii) above we have

$$\|\Pi_{q_E} \circ \eta\|(A) = \sup_{x^* \in E} v(\Psi_{x^*}(\Pi_{q_E} \circ \eta))(A) = \sup_{x^* \in E} v(x^*\eta)(A), \ \ A \in \sigma(\mathcal{P}).$$

(iv) This is immediate from (iii) and from the fact that $\|\mu\|(A) = v(\mu)(A)$, $A \in \sigma(\mathcal{P})$ for a scalar-valued additive set function μ on $\mathcal{P}$. $\qquad\square$

Notation 1.2.14. For $q \in \Gamma$, U_q denotes the set $\{x \in X : q(x) \leq 1\}$ and U_q^o is the polar of U_q which is given by $U_q^0 = \{x^* \in X^* : |x^*(x)| \leq 1 \text{ for all } x \in U_q\}$.

Proposition 1.2.15. *For* $q \in \Gamma$ *the following hold:*

(i) $q(x) = \sup_{x^* \in U_q^0} |x^*(x)|$, $x \in X$.

(ii) *If* $\eta : \mathcal{P} \to X$ *is additive, then the following hold:*

(a) *For* $x^* \in U_q^o$, $\Psi_{x^*} : X_q \to \mathbb{K}$ *given by* $\Psi_{x^*}(x + q^{-1}(0)) = x^*(x)$, $x \in X$, *is well defined, linear and continuous. Thus* $(\Psi_{x^*} \circ \Pi_q)(x) = x^*(x)$, $x \in X$ *and* $x^* \in U_q^o$.

(b) $\{\Psi_{x^*} : x^* \in U_q^o\}$ *is a norm determining subset of the closed unit ball of* $(X_q)^*$.

(c) $\|\Pi_q \circ \eta\|(A) = \sup_{x^* \in U_q^o} v(x^*\eta)(A)$, $A \in \sigma(\mathcal{P})$.

Proof. (i) If $q(x) = 0$, then $q(nx) = 0$ for $n \in \mathbb{N}$ so that $nx \in U_q$. Then, for $x^* \in U_q^o$, we have $|x^*(nx)| = n|x^*(x)| \leq 1$ so that $|x^*(x)| \leq \frac{1}{n}$ for all n. Hence $\sup_{x^* \in U_q^o} |x^*(x)| = 0 = q(x)$. Thus (i) holds if $q(x) = 0$.

If $q(x) \neq 0$, then $\frac{x}{q(x)} \in U_q$ and hence $\sup_{x^* \in U_q^o} |x^*(x)| \leq q(x)$. We claim that $\sup_{x^* \in U_q^o} |x^*(x)| = q(x)$. Otherwise, $\alpha = \sup_{x^* \in U_q^o} |x^*(x)| < q(x)$. Then $q(\frac{x}{\alpha}) > 1$ and hence $\frac{x}{\alpha} \notin U_q$. As U_q is τ-closed, it is weakly closed. Moreover, U_q is absolutely convex. Hence by the bipolar theorem (see Theorem 1, §3, Chapter 3 of [Ho]), $U_q = U_q^{oo}$. Consequently, there exists $x^* \in U_q^o$ such that $|x^*(\frac{x}{\alpha})| > 1$ and hence $|x^*(x)| > \alpha$, a contradiction. Thus (i) holds if $q(x) \neq 0$. Therefore, (i) holds.

(ii) By Proposition 6, §4, Chapter 3 of [Ho], $E = U_q^0 \in \mathcal{E}$ and hence, by (i) and by Notation 1.2.11, $q = q_E$. Then (ii) holds by Proposition 1.2.12. $\qquad\square$

Notation and Convention 1.2.16. The lcHs completion $(\tilde{X}, \tilde{\tau})$ of the lcHs (X, τ) is unique up to a topological isomorphism, X is $\tilde{\tau}$ dense in $\tilde{X}$ and $\tilde{\tau}|_X = \tau$. Each $q \in \Gamma$ has a unique continuous extension $\tilde{q}$ to $\tilde{X}$ and $\tilde{q}$ is thus a $\tilde{\tau}$-continuous seminorm on $\tilde{X}$. Moreover, $\{\tilde{q} : q \in \Gamma\}$ generates the topology $\tilde{\tau}$ on $\tilde{X}$. By an abuse of notation, we denote its continuous extension to $\tilde{X}$ also by q itself and by an abuse of language we also say that Γ generates the topology of $\tilde{X}$. See Theorem 1, §9, Chapter 2 of [Ho] and pp. 134–135 of [Ho].

Notation 1.2.17. Let Γ and $\Gamma_{\mathcal{E}}$ be directed by the partial order $q_1 \leq q_2$ if $q_1(x) \leq q_2(x)$ for $x \in X$ and $q_{E_1} \leq q_{E_2}$ if $E_1 \subset E_2$, for $q_1, q_2 \in \Gamma$ and $E_1, E_2 \in \mathcal{E}$. For $q_1 \leq q_2$, $A_{q_1 q_2} : X_{q_2} \to X_{q_1}$ defined by $A_{q_1 q_2}(x + q_2^{-1}(0)) = x + q_1^{-1}(0)$ is a continuous onto linear mapping. Similarly, $A_{q_{E_1} q_{E_2}} : X_{q_{E_2}} \to X_{q_{E_1}}$ is defined if $E_1 \subset E_2$ (so that $q_{E_1} \leq q_{E_2}$) and is a continuous onto linear mapping. Then $\Pi_{q_1} = A_{q_1 q_2} \Pi_{q_2}$ (resp., $\Pi_{q_{E_1}} = A_{q_{E_1} q_{E_2}} \Pi_{q_{E_2}}$). Let $Y = \tilde{X}$, the completion of X. Let $\tilde{\Gamma} = \{\tilde{q} : q \in \Gamma\}$ be as in 1.2.16. For $q \in \Gamma$, $\Pi_{\tilde{q}} : Y \to Y_{\tilde{q}}$ and $\Pi_{\tilde{q}}|_X = \Pi_q$ so that $X_q = \Pi_{\tilde{q}}(X) \subset \widetilde{X}_q$. As X is dense in Y, it follows that $X_q \subset \Pi_{\tilde{q}}(Y) = Y_{\tilde{q}} \subset \widetilde{X}_q$ and hence the completion of $\Pi_{\tilde{q}}(Y)$ is equal to $Y_{\tilde{q}} = \widetilde{X}_q$ for $q \in \Gamma$. If $q_1, q_2 \in \Gamma$ with $q_1 \leq q_2$, then clearly, $\tilde{q}_1 \leq \tilde{q}_2$ and hence by 5.4, Chapter II of [Scha], Y is topologically isomorphic to the projective limit $\varprojlim A_{\tilde{q}_1 \tilde{q}_2} Y_{\tilde{q}_2}$. If $\widetilde{A_{\tilde{q}_1 \tilde{q}_2}}$ is the continuous extension of $A_{\tilde{q}_1 \tilde{q}_2}$ to $Y_{\tilde{q}_2} (= \widetilde{X}_{q_2})$ with values in $Y_{\tilde{q}_1}$, then again by 5.4, Chapter II of [Scha], Y is topologically isomorphic to $\varprojlim \widetilde{A_{\tilde{q}_1 \tilde{q}_2}} Y_{\tilde{q}_2} = \varprojlim \widetilde{A_{\tilde{q}_1 \tilde{q}_2}} \widetilde{X}_{q_2}$. Thus each $y \in Y$ is written as $y = \varprojlim x_{\tilde{q}}$ with $x_{\tilde{q}} = \Pi_{\tilde{q}}(y) \in Y_{\tilde{q}}$, $q \in \Gamma$ and is also written as $y = \varprojlim x_{\tilde{q}}$, $q \in \Gamma$ with $x_{\tilde{q}} \in \widetilde{X}_q$ (without mentioning the partial order in Γ and the transformations $A_{\tilde{q}_1 \tilde{q}_2}$ and $\widetilde{A_{\tilde{q}_1 \tilde{q}_2}}$). A similar description holds in terms of $(q_E : E \in \mathcal{E})$. For details, see pp. 53–54 of [Scha].

Chapter 2

Basic Properties of the Bartle-Dunford-Schwartz Integral

2.1 (KL) m-integrability

In [L1,L2] Lewis studied a Pettis type weak integral of scalar functions with respect to an lcHs-valued σ-additive vector measure. As noted on p. 27 of [KK], this type of definition has also been considered by Kluvánek in [K2]. In honor of these mathematicians, we call the integral introduced in [L1, L2] the Kluvánek-Lewis integral or briefly, the (KL)-integral. For a Banach space-valued σ-additive vector measure $\mathbf{m}$ defined on a δ-ring $\mathcal{P}$ of sets, we define the (KL) $\mathbf{m}$-integrability and the (KL) $\mathbf{m}$-integral of a scalar function f, study the basic properties of the (KL) $\mathbf{m}$-integral and show that the Lebesgue dominated convergence theorem (briefly, LDCT) is valid for the integral. When $\mathcal{P}$ is a σ-ring $\mathcal{S}$, the Lebesgue bounded convergence theorem (briefly, LBCT) is also valid. The reader can note that the version of LDCT given here is much stronger than the Banach space versions of Theorem 2.2(2) of [L1] and of Theorem 3.3 of [L2] (whose proof is incorrect – see Remark 2.2.12 below) and of Theorem II.5.2 of [KK]. The version of LDCT given in [Ri5] is the same as ours except that the domain of $\mathbf{m}$ in [Ri5] is a σ-algebra Σ and the functions considered there are Σ-measurable. (See Remark 2.1.12.) In Chapter 4 we define (KL) $\mathbf{m}$-integrability in T of $\mathbf{m}$-measurable functions (resp. $\sigma(\mathcal{P})$-measurable functions) when $\mathbf{m}$ is a quasicomplete (resp. sequentially complete) lcHs-valued σ-additive measure on $\mathcal{P}$ and the arguments and results of this section play a key role in generalizing them in Chapter 4.

Definition 2.1.1. Let $f : T \to \mathbb{K}$ or $[-\infty, \infty]$.

(i) Let f be $\sigma(\mathcal{P})$-measurable. Then f is said to be Kluvánek-Lewis integrable with respect to $\mathbf{m}$ in T or simply, (KL) $\mathbf{m}$-integrable in T if f is $x^*\mathbf{m}$-integrable in T for each $x^* \in X^*$ and if, for each $E \in \sigma(\mathcal{P})$, there exists a vector $x_E \in X$ such that $x^*(x_E) = \int_E f d(x^*\mathbf{m})$ for $x^* \in X^*$. In that case, we define (KL)$\int_E f d\mathbf{m} = x_E$ for $E \in \sigma(\mathcal{P})$ and (KL)$\int_T f d\mathbf{m} = $ (KL)$\int_{N(f)} f d\mathbf{m}$. (See Notation 1.1.15 for $N(f)$.) Note that by the Hahn-Banach theorem these integrals are well defined and unique.

(ii) [1] If f is $\mathbf{m}$-measurable, then there exists $M \in \sigma(\mathcal{P})$ with $||\mathbf{m}||(M) = 0$ such that $f\chi_{T\backslash M}$ is $\sigma(\mathcal{P})$-measurable (by Proposition 1.1.18). If $f\chi_{T\backslash M}$ is (KL) $\mathbf{m}$-integrable in T, then we say that f is (KL) $\mathbf{m}$-integrable in T, and in that case, we define

$$(\text{KL}) \int_E f d\mathbf{m} = (\text{KL}) \int_E f\chi_{T\backslash M} d\mathbf{m} \text{ for } E \in \sigma(\mathcal{P})$$

and

$$(\text{KL}) \int_T f d\mathbf{m} = (\text{KL}) \int_N f\chi_{T\backslash M} d\mathbf{m}.$$

Let f be $\mathbf{m}$-measurable and let it be (KL) $\mathbf{m}$-integrable in T with M as in (ii) of the above definition. If $M_1 \in \sigma(\mathcal{P})$ with $||\mathbf{m}||(M_1) = 0$ is such that $f\chi_{T\backslash M_1}$ is $\sigma(\mathcal{P})$-measurable, then it can easily be shown that $f\chi_{T\backslash M_1}$ is also (KL) $\mathbf{m}$-integrable in T and (KL)$\int_E f\chi_{T\backslash M} d\mathbf{m} = $ (KL)$\int_E f\chi_{T\backslash M_1} d\mathbf{m}$ for $E \in \sigma(\mathcal{P}) \cup \{T\}$. Consequently, (KL)$\int_E f d\mathbf{m}$ is well defined and does not depend on the particular choice of the $\mathbf{m}$-null set M in (ii) of the above definition.

Notation 2.1.2. $\mathcal{I}(\mathbf{m})$ denotes the class of all $\mathbf{m}$-measurable scalar functions on T which are (KL) $\mathbf{m}$-integrable in T.

Remark 2.1.3. Let $f : T \to [-\infty, \infty]$ be (KL) $\mathbf{m}$-integrable in T. Then $E = \{t \in T : |f(t)| = \infty\} \in \widetilde{\sigma(\mathcal{P})}$. We claim that : $||\mathbf{m}||(E) = 0$. In fact, let $E = F \cup N$, $F \in \sigma(\mathcal{P})$, $N \subset M \in \sigma(\mathcal{P})$ with $||\mathbf{m}||(M) = 0$. Then by Proposition 1.1.1, $||\mathbf{m}||(E) = ||\mathbf{m}||(F) = \sup_{|x^*| \leq 1} v(x^*\mathbf{m})(F)$. If $f = \hat{f}$ $\mathbf{m}$-a.e. in T with $\hat{f}$ $\sigma(\mathcal{P})$-measurable, then $\hat{f}$ is (KL) $\mathbf{m}$-integrable in T, and hence $\hat{f}$ is $x^*\mathbf{m}$-a.e. finite in T for each $x^* \in X^*$. Therefore, $v(x^*\mathbf{m})(F) = 0$ for each $x^* \in X^*$. Consequently, $||\mathbf{m}||(E) = 0$.

Notation 2.1.4. Functions integrable in the sense of [L1, L2] or [KK] (when $\mathbf{m}$ has range in a Banach space), are clearly (KL) $\mathbf{m}$-integrable in T with the same integrals. Hence we shall also refer to them as (KL) $\mathbf{m}$-integrable functions.

Theorem 2.1.5. *Let* $f : T \to \mathbb{K}$ *or* $[-\infty, \infty]$ *and let* s *be a* $\mathcal{P}$*-simple function. Then the following hold:*

(i) *If* $s = \sum_1^r \alpha_i \chi_{E_i}$ *is* $\mathcal{P}$*-simple with* $(\alpha_i)_1^r \subset \mathbb{K}$, $(E_i)_1^r \subset \mathcal{P}$ *and* $E_i \cap E_j = \emptyset$ *for* $i \neq j$, *then* s *is* (KL) $\mathbf{m}$*-integrable in* T *and* (KL) $\int_E s d\mathbf{m} = $

---

[1] Part (ii) is really needed and useful in general. See Remark 2.1.14.

$\sum_1^r \alpha_i \mathbf{m}(E_i \cap E)$ *for* $E \in \sigma(\mathcal{P})$. *We write* $\int_E s d\mathbf{m}$ *instead of* $(KL) \int_E s d\mathbf{m}$. *Consequently,* $||\mathbf{m}||(E) = \sup\{|\int_E s d\mathbf{m}| : s\, \mathcal{P}\text{-simple}, |s(t)| \le \chi_E(t), t \in T\}$.

(ii) *If* f *is* (KL) **m**-*integrable in* T, *then* $\boldsymbol{\gamma}(\cdot) : \sigma(\mathcal{P}) \to X$ *given by* $\boldsymbol{\gamma}(\cdot) = (KL) \int_{(\cdot)} f d\mathbf{m}$ *is* σ-*additive (in norm)*.

(iii) *If* $\boldsymbol{\gamma}(\cdot)$ *is as in* (ii), *then:*
   (a) $||\boldsymbol{\gamma}||(E) = \sup_{|x^*| \le 1} \int_E |f| dv(x^* \mathbf{m})$, $E \in \sigma(\mathcal{P})$.
   (b) $\lim_{||\mathbf{m}||(E) \to 0} \boldsymbol{\gamma}(E) = \lim_{||\mathbf{m}||(E) \to 0} ||\boldsymbol{\gamma}||(E) = 0$, $E \in \sigma(\mathcal{P})$.

(iv) $\mathcal{I}(\mathbf{m})$ *is a vector space over* $\mathbb{K}$ *with respect to pointwise addition and scalar multiplication. Moreover, for* $E \in \sigma(\mathcal{P})$ *fixed, the mapping* $f \to (KL) \int_E f d\mathbf{m}$ *is linear on* $\mathcal{I}(\mathbf{m})$. *Consequently,* $\int_E s d\mathbf{m} = \sum_{i=1}^r \alpha_i \mathbf{m}(E_i \cap E)$ *for* $E \in \sigma(\mathcal{P})$ *and for* $s = \sum_1^r \alpha_i \chi_{E_i}$ *with* $(E_i)_1^r \subset \mathcal{P}$ *not necessarily mutually disjoint.*

(v) *For a set* A *in* T, *let* $\mathcal{P} \cap A = \{B \cap A : B \in \mathcal{P}\}$. *If* f *is an* **m**-*measurable function on* T *and if* f *is* **m**-*essentially bounded in a set* $A \in \mathcal{P}$, *then* f *is* (KL) **m**-*integrable in each* $B \in \mathcal{P} \cap A$ *and*

$$|(KL) \int_B f d\mathbf{m}| \le (\text{ess} \sup_{t \in A} |f(t)|) \cdot ||\mathbf{m}||(B), \; B \in \sigma(\mathcal{P}) \cap A = \mathcal{P} \cap A.$$

Consequently, if $\mathcal{P}$ *is a* σ-*ring* $\mathcal{S}$ *and if* f *is* **m**-*essentially bounded in* T, *then* f *is* (KL) **m**-*integrable in* T *and*

$$\left|(KL) \int_A f d\mathbf{m}\right| \le (\text{ess} \sup_{t \in T} |f(t)|) \cdot ||\mathbf{m}||(A) \le (\text{ess} \sup_{t \in T} |f(t)|) \cdot ||\mathbf{m}||(T)$$

for each $A \in \mathcal{S}$. *(Note that* $||\mathbf{m}||(T) < \infty$ *by Proposition* 1.1.6.)

(vi) *If* φ *is an* **m**-*essentially bounded* **m**-*measurable scalar function on* T *and if* $f \in \mathcal{I}(\mathbf{m})$, *then* $\varphi \cdot f$ *is* (KL) **m**-*integrable in* T. *In particular, if* f *is* (KL) **m**-*integrable in* T *and if* $E \in \sigma(\mathcal{P})$, *then* $\chi_E f$ *is* (KL) **m**-*integrable in* T *and* $(KL) \int_E f d\mathbf{m} = (KL) \int_T \chi_E f d\mathbf{m}$.

(vii) **(domination principle).** *If* f *is an* **m**-*measurable scalar function on* T *and* $g \in \mathcal{I}(\mathbf{m})$ *with* $|f(t)| \le |g(t)|$ **m**-*a.e. in* T, *then* $f \in \mathcal{I}(\mathbf{m})$. *Consequently, an* **m**-*measurable function* $f : T \to \mathbb{K}$ *or* $[-\infty, \infty]$ *is* (KL) **m**-*integrable in* T *if and only if* $|f|$ *is* (KL) **m**-*integrable in* T. *Moreover, for an* **m**-*measurable function* $f : T \to \mathbb{K}$, *the following statements are equivalent:* (a) $f \in \mathcal{I}(\mathbf{m})$; (b) $|f| \in \mathcal{I}(\mathbf{m})$; (c) $\bar{f} \in \mathcal{I}(\mathbf{m})$; (d) $Ref \in I(\mathbf{m})$ *and* $Imf \in \mathcal{I}(\mathbf{m})$; (e) $(Ref)^+$, $(Imf)^+, (Ref)^-$ *and* $(Imf)^-$ *belong to* $\mathcal{I}(\mathbf{m})$. *Further, if* $f_1, f_2 : T \to \mathbb{R}$ *are* **m**-*measurable and* **m**-*integrable in* T, *then* $\max(f_1, f_2)$ *and* $\min(f_1, f_2)$ *are* **m**-*measurable and belong to* $\mathcal{I}(\mathbf{m})$.

(viii) *If* $u \in L(X, Y)$, *then* $u\mathbf{m} : \mathcal{P} \to Y$ *is* σ-*additive, and* f *is* (KL) $u\mathbf{m}$-*integrable in* T *whenever* f *is* (KL) **m**-*integrable in* T. *In that case,* $u((KL) \int_E f d\mathbf{m}) = (KL) \int_E f d(u\mathbf{m})$ *for* $E \in \sigma(\mathcal{P})$.

Proof. In the light of Definition 2.1.1, without loss of generality we shall assume that all the **m**-measurable functions considered here are further $\sigma(\mathcal{P})$-measurable.

(i) and (iv) are obvious. (ii) is due to the Orlicz-Pettis theorem. (See the first paragraph of Section 1.2.)

(iii)(a) By (ii) and by Propositions 1.1.1 and 1.1.19,

$$||\boldsymbol{\gamma}||(E) = \sup_{|x^*| \leq 1} v(x^* \boldsymbol{\gamma})(E) = \sup_{|x^*| \leq 1} \int_E |f| dv(x^* \mathbf{m}) \quad \text{for} \quad E \in \sigma(\mathcal{P}).$$

(iii)(b) If $||\mathbf{m}||(E) = 0, E \in \sigma(\mathcal{P})$, then $v(x^* \mathbf{m})(E) = 0$ for each $x^* \in X^*$ and hence by (iii)(a), $||\boldsymbol{\gamma}||(E) = 0$. As $||\mathbf{m}||$ and $||\boldsymbol{\gamma}||$ are nonnegative, monotone and σ-subadditive on the σ-ring $\sigma(\mathcal{P})$ and as $||\boldsymbol{\gamma}||$ is further continuous on $\sigma(\mathcal{P})$ (by (ii) and by Proposition 1.1.5), the second equality holds by Propositions 1.1.11 and 1.1.12. As $|\boldsymbol{\gamma}(E)| \leq ||\boldsymbol{\gamma}||(E), E \in \sigma(\mathcal{P})$, the first equality also holds.

(v) Let $A \in \mathcal{P}$ and let $\alpha = \text{ess sup}_{t \in A} |f(t)|$. Then there exists $M \in \sigma(\mathcal{P}) \cap A = \mathcal{P} \cap A$ with $||\mathbf{m}||(M) = 0$ such that $\sup_{t \in A \setminus M} |f(t)| = \alpha$. (See the proof of (4.1.9.1) for a more general situation.) Let $\Sigma_A = \mathcal{P} \cap A$. Then $\Sigma_A \subset \mathcal{P}$ and is a σ-algebra of subsets of A. Clearly, $f \chi_{A \setminus M}$ is Σ_A-measurable and bounded. Hence there exists a sequence $(s_n)_1^\infty$ of Σ_A-simple functions (so that $(s_n)_1^\infty \subset \mathcal{I}_s$) such that $s_n \to f \chi_{A \setminus M}$ uniformly in A with $|s_n| \nearrow |f|$ in A.

Let $B \in \Sigma_A$. Then, given $\epsilon > 0$, there exists n_0 such that $||s_n - s_r||_A \cdot ||\mathbf{m}||(A) < \epsilon$ for $n, r \geq n_0$. Then by the last part of (i), we have

$$\left| \int_B s_n d\mathbf{m} - \int_B s_r d\mathbf{m} \right| \leq ||s_n - s_r||_B ||\mathbf{m}||(B) \leq ||s_n - s_r||_A ||\mathbf{m}||(A) < \epsilon$$

for $n, r \geq n_0$. Hence, for each $B \in \Sigma_A$, there exists a vector $x_B \in X$ such that $\lim_n \int_B s_n d\mathbf{m} = x_B$.

On the other hand, by LDCT for scalar measures on the σ-algebra Σ_A, we have

$$x^*(x_B) = \lim_n \int_B s_n d(x^* \mathbf{m}) = \int_B f d(x^* \mathbf{m})$$

for each $x^* \in X^*$ and for each $B \in \Sigma_A$. Thus f is (KL) **m**-integrable in A and

$$(\text{KL}) \int_B f d\mathbf{m} = x_B$$

for each $B \in \Sigma_A$. Moreover, again by the last part of (i), we have

$$\left| (\text{KL}) \int_B f d\mathbf{m} \right| = |x_B| = \left| \lim_n \int_B s_n d\mathbf{m} \right| \leq \alpha \cdot ||\mathbf{m}||(B)$$

for $B \in \Sigma_A$.

The last part is immediate from the first part.

(vi) Let $\operatorname{ess\,sup}_{t\in T}|\varphi(t)| = \alpha$ and let $\boldsymbol{\nu}(\cdot) = (KL)\int_{(\cdot)} f d\mathbf{m}$ on $\sigma(\mathcal{P})$. By (ii), $\boldsymbol{\nu}$ is σ-additive on $\sigma(\mathcal{P})$ and by (iii)(b), the $\sigma(\mathcal{P})$-measurable function φ is also $\boldsymbol{\nu}$-essentially bounded. Then by the last part of (v), φ is (KL) $\boldsymbol{\nu}$-integrable in T. Let $(KL)\int_E \varphi d\boldsymbol{\nu} = x_E$, for $E \in \sigma(\mathcal{P})$. Then $x^*(x_E) = \int_E \varphi d(x^*\boldsymbol{\nu}) = \int_E \varphi f d(x^*\mathbf{m})$ for $E \in \sigma(\mathcal{P})$ and $x^* \in X^*$. (The last equality can be proved for $\mathcal{P}$-simple functions φ and then by applying Proposition 1.1.18 above and Theorem 3 (LDCT), §8 of [Din1], it can be extended to a general $\sigma(\mathcal{P})$-measurable $x^*\boldsymbol{\nu}$-integrable function φ. Also see Theorem 30, §2 of [Din2], which holds for $\mathbb{K}$ also.) Hence $\varphi \cdot f$ is (KL) **m**-integrable in T.

If $E \in \sigma(\mathcal{P})$ and if f is (KL) **m**-integrable in T, then by the first part $\chi_E f$ is (KL) **m**-integrable in T. If $(KL)\int_E f d\mathbf{m} = x_E \in X$, then $x^*(x_E) = \int_E f d(x^*\mathbf{m}) = \int_T \chi_E f d(x^*\mathbf{m})$ for $x^* \in X^*$ and hence the last part also holds.

(vii) Let $h(t) = \frac{f(t)}{g(t)}$, $t \in N(g)$ and $h(t) = 0$ otherwise. Then h is $\sigma(\mathcal{P})$-measurable and $|h(t)| \le 1$ **m**-a.e. in T. Moreover, $f = gh$ **m**-a.e. in T and consequently, by (vi), $f \in \mathcal{I}(\mathbf{m})$. If $f : T \to \mathbb{K}$ or $[-\infty,\infty]$ is (KL) **m**-integrable in T, then by Remark 2.1.3 there exists an **m**-null set $N \in \sigma(\mathcal{P})$ such that f is finite in $T\backslash N$ and $g = f\chi_{T\backslash N} \in \mathcal{I}(\mathbf{m})$ and $|f| \le |g|$ **m**-a.e. in T. Hence by the above case, $|f|$ is (KL) **m**-integrable in T. Conversely, if $|f|$ is (KL) **m**-integrable in T, then by Definition 2.1.1 and Remark 2.1.3, there exists an **m**-null set $N \in \sigma(\mathcal{P})$ such that $g = |f|\chi_{T\backslash N}$ is a scalar-valued $\sigma(\mathcal{P})$-measurable (KL) **m**-integrable function. Since $|f| \le |g|$ **m**-a.e. in T, by the above part, f is (KL) **m**-integrable in T. The other parts follow from the first, from (iv) and from the fact that $\max(f_1,f_2) = \frac{1}{2}(f_1 + f_2 + |f_1 - f_2|)$ and $\min(f_1,f_2) = \frac{1}{2}(f_1 + f_2 - |f_1 - f_2|)$.

(viii) Clearly, $u\mathbf{m} : \mathcal{P} \to Y$ is σ-additive. For $y^* \in Y^*$, $y^*u \in X^*$ and hence the result holds. $\qquad\square$

Remark 2.1.6. The proofs of (vi) and (vii) of Theorem 2.1.5 are respectively similar to those of Lemma II.3.1 and Theorem II.3.1 of [KK] given for a real lcHs-valued σ-additive measure on a σ-algebra of sets. A lengthy argument using the Banach space version of Theorem 3.3 (LDCT) of [L2] is given on pp. 145–147 of [MN2] to prove Theorem 2.1.5(vii) for $\sigma(\mathcal{P})$-measurable functions f. But, the result is not really established there since the proofs of the results of [L2] used in their argument are incorrect. See Remark 2.1.12 below.

Theorem 2.1.7. (Lebesgue dominated convergence theorem-(LDCT- m-a.e. convergence version)). Let f_n, $n \in \mathbb{N}$, be **m**-*measurable with values in* $\mathbb{K}$ *or in* $[-\infty,\infty]$ *and let* $g : T \to \mathbb{K}$ *be (KL)* **m**-*integrable in* T. *Suppose* $|f_n(t)| \le |g(t)|$ **m**-*a.e. in* T *for all* n. *If* $f : T \to \mathbb{K}$ *and if* $f_n(t) \to f(t)$ **m**-*a.e. in* T, *then* f, f_n, $n \in \mathbb{N}$, *are (KL)* **m**-*integrable in* T *and*

$$(KL)\int_E f d\mathbf{m} = \lim_n (KL)\int_E f_n d\mathbf{m}, \ E \in \sigma(\mathcal{P})$$

where the limit is uniform with respect to $E \in \sigma(\mathcal{P})$. Moreover,

$$\lim_n \sup_{|x^*| \leq 1} \int_T |f_n - f| dv(x^*\mathbf{m}) = 0.$$

Proof. In the light of Definition 2.1.1, without loss of generality we shall assume that all the $\mathbf{m}$-measurable functions considered here are further $\sigma(\mathcal{P})$-measurable. Let $\boldsymbol{\nu}(E) = (KL) \int_E g d\mathbf{m}$, $E \in \sigma(\mathcal{P})$. Then by Theorem 2.1.5(ii), $\boldsymbol{\nu}$ is σ-additive on $\sigma(\mathcal{P})$. By hypothesis and by Theorem 2.1.5(vii), f_n, $n \in \mathbb{N}$, are (KL) $\mathbf{m}$-integrable in T. As $f_n \to f$ $\mathbf{m}$-a.e. in T, clearly f is $\widetilde{\sigma(\mathcal{P})}$-measurable and hence is $\mathbf{m}$-measurable. Then in view of Definition 2.1.1 we shall assume that f is also $\sigma(\mathcal{P})$-measurable. Moreover, $|f| \leq |g|$ $\mathbf{m}$-a.e. in T and hence by Theorem 2.1.5(vii), f is also (KL) $\mathbf{m}$-integrable in T. By hypothesis and by Remark 2.1.3 there exists $M \in \sigma(\mathcal{P})$ with $||\mathbf{m}||(M) = 0$ such that g is finite on $T\backslash M$ and $f_n(t) \to f(t)$ for $t \in T\backslash M$. Let $F = \bigcup_{n=1}^{\infty} N(f_n) \cap (T\backslash M) = \bigcup_{n=1}^{\infty} N(f_n \chi_{T\backslash M})$. Then $F \in \sigma(\mathcal{P})$ and $f_n \chi_{T\backslash M} \to f \chi_{T\backslash M}$ pointwise in T. By Proposition 1.1.13, $\boldsymbol{\nu}$ has a control measure $\mu : \sigma(\mathcal{P}) \to [0, \infty)$. Thus, given $\epsilon > 0$, there exists $\delta > 0$ such that $\mu(E) < \delta$ implies $||\boldsymbol{\nu}||(E) < \frac{\epsilon}{3}$. Now by the Egoroff-Lusin theorem there exist a set $N \in \sigma(\mathcal{P}) \cap F$ with $\mu(N) = 0$ and a sequence $(F_k)_1^{\infty} \subset \mathcal{P}$ with $F_k \nearrow F\backslash N$ such that $f_n \to f$ uniformly in each F_k. As $\mu(F\backslash N\backslash F_k) \searrow 0$, there exists k_0 such that $\mu(F\backslash N\backslash F_k) < \delta$ for $k \geq k_0$. Since $\mu(N) = 0$, $||\boldsymbol{\nu}||(N) = 0$ and since $||\mathbf{m}||(M) = 0$, by Theorem 2.1.5(iii) we have $||\boldsymbol{\nu}||(M) = 0$. As $f_n \to f$ uniformly in F_{k_0}, there exists n_0 such that $||f_n - f||_{F_{k_0}} \cdot ||\mathbf{m}||(F_{k_0}) < \frac{\epsilon}{3}$ for $n \geq n_0$. Then, for $E \in \sigma(\mathcal{P})$ and for $|x^*| \leq 1$, by Proposition 4, §8 of [Din1], we have

$$\left| x^*(KL) \int_E f_n d\mathbf{m} - x^*(KL) \int_E f d\mathbf{m} \right| = \left| \int_E (f_n - f) d(x^*\mathbf{m}) \right|$$

$$\leq \int_E |f_n - f| dv(x^*\mathbf{m})$$

$$\leq \int_{E \cap F_{k_0}} |f_n - f| dv(x^*\mathbf{m}) + \int_{E \cap (T\backslash F_{k_0})} |f_n - f| dv(x^*\mathbf{m}).$$

Now, by Proposition 8, §8 of [Din1],

$$\int_{E \cap F_{k_0}} |f_n - f| dv(x^*\mathbf{m}) \leq ||f_n - f||_{F_{k_0}} \cdot v(x^*\mathbf{m})(F_{k_0}) < \frac{\epsilon}{3}$$

for $n \geq n_0$ since $v(x^*\mathbf{m})(F_{k_0}) \leq ||\mathbf{m}||(F_{k_0})$ for $|x^*| \leq 1$.

By (iii)(a) of Theorem 2.1.5 we have

$$\int_{E \cap (T\backslash F_{k_0})} |f_n - f| dv(x^*\mathbf{m})$$

$$\leq 2 \int_{E \cap (F\backslash N\backslash F_{k_0})} |g| dv(x^*\mathbf{m}) + 2 \left(\int_{E \cap N} + \int_{E \cap M} |g| dv(x^*\mathbf{m}) \right)$$

$$\leq 2||\boldsymbol{\nu}||(F\backslash N\backslash F_{k_0}) + 2||\boldsymbol{\nu}||(N) + 2||\boldsymbol{\nu}||(M) < \frac{2\epsilon}{3}$$

since $\mu(F \backslash N \backslash F_{k_0}) < \delta$ and $||\boldsymbol{\nu}||(N) = ||\boldsymbol{\nu}||(M) = 0$. Consequently, we have

$$\left| (KL) \int_E f_n d\mathbf{m} - (KL) \int_E f d\mathbf{m} \right| \leq \sup_{|x^*| \leq 1} \int_E |f_n - f| dv(x^* \mathbf{m}) \leq \epsilon$$

for $n \geq n_0$ and for all $E \in \sigma(\mathcal{P})$. □

Corollary 2.1.8. (Lebesgue bounded convergence theorem)-(LBCT-m a.e. conver-gence version)). *Let $\mathcal{P}$ be a σ-ring $\mathcal{S}$. Let f_n, $n \in \mathbb{N}$, be **m**-measurable with values in $\mathbb{K}$ or in $[-\infty, \infty]$ and let K be a positive constant such that $|f_n(t)| \leq K$ **m**-a.e. in T for all n. If $f : T \to \mathbb{K}$ and if $f_n(t) \to f(t)$ **m**-a.e. in T, then f, f_n, $n \in \mathbb{N}$, are (KL) **m**-integrable in T and*

$$(KL) \int_E f d\mathbf{m} = \lim_n (KL) \int_E f_n d\mathbf{m}$$

for $E \in \mathcal{S}$, the limit being uniform with respect to $E \in \mathcal{S}$.

Proof. Since $\mathcal{P} = \mathcal{S}$ is a σ-ring, by the last part of Theorem 2.1.5(v) constant functions are (KL) **m**-integrable in T and hence the result follows from Theorem 2.1.7. □

Corollary 2.1.9. *Let f be (KL) **m**-integrable in T. Then there exists a sequence $(s_n) \subset \mathcal{I}_s$ such that $s_n(t) \to f(t)$ **m**-a.e. in T and $|s_n(t)| \nearrow |f(t)|$ **m**-a.e. in T. Then for any such (s_n), $(KL) \int_E f d\mathbf{m} = \lim_n \int_E s_n d\mathbf{m}$ for $E \in \sigma(\mathcal{P})$, and the limit is uniform with respect to $E \in \sigma(\mathcal{P})$. Consequently, for $E \in \sigma(\mathcal{P})$,*

$$||\mathbf{m}||(E) = \sup \left\{ \left| (KL) \int_E h d\mathbf{m} \right| : h \in \mathcal{I}(\mathbf{m}), |h(t)| \leq 1 \text{ } \mathbf{m}\text{-a.e. } in E \right\}$$

$$= \sup \left\{ \left| (KL) \int_E h d\mathbf{m} \right| : h \in \mathcal{I}(\mathbf{m}), |h(t)| \leq 1 \text{ } in E \right\}.$$

Proof. The existence of such (s_n) is guaranteed by Proposition 1.1.18. The second result is due to Theorem 2.1.7(LDCT). By the last part of Theorem 2.1.5(i) and by the fact that $\mathcal{I}_s \subset \mathcal{I}(\mathbf{m})$, we have

$$||\mathbf{m}||(E) = \sup \left\{ \left| \int_E s d\mathbf{m} \right| : s \in \mathcal{I}_s, |s| \leq \chi_E \right\}$$

$$\leq \sup \left\{ \left| (KL) \int_E f d\mathbf{m} \right| : f \in \mathcal{I}(\mathbf{m}), |f| \leq \chi_E \right\}$$

$$\leq \sup \left\{ \left| (KL) \int_E f d\mathbf{m} \right| : f \in \mathcal{I}(\mathbf{m}), |f| \leq \chi_E \text{ } \mathbf{m}\text{-a.e. } in T \right\}.$$

But, by the last part of Theorem 2.1.5(i) and by the second part of the present corollary, $|(KL) \int_E f d\mathbf{m}| \leq ||\mathbf{m}||(E)$ for $f \in \mathcal{I}(\mathbf{m})$ with $|f| \leq \chi_E$ **m**-a.e. in T. Hence the last part of the corollary also holds. □

Corollary 2.1.10. (Simple function approximation). *Given $f \in \mathcal{I}(\mathbf{m})$, there exists a sequence $(s_n)_1^\infty \subset \mathcal{I}_s$ such that $s_n \to f$ $\mathbf{m}$-a.e. in T and such that*

$$\lim_n \sup_{|x^*| \leq 1} \int_T |f - s_n| dv(x^*\mathbf{m}) = 0.$$

(In other words, in terms of Definition 3.1.1 and Theorem 3.1.3 of Chapter 3, $\lim_n \mathbf{m}_1^\bullet(f - s_n, T) = 0$.)

Proof. This follows from the first part of Corollary 2.1.9 and the last part of Theorem 2.1.7. $\qquad\square$

Remark 2.1.11. The hypothesis of (KL) $\mathbf{m}$-integrability of the dominated functions in Theorem 2.2(2) of [L1] is redundant due to Theorem II.3.1 of [KK] which holds for complex lcHs too. If $\mathcal{P}$ is a σ-ring, the proof of the said theorem of [L1] can be adapted here. Theorem 2.1.5(vii) and the proof of the said theorem in [L1] have been used in [Ri5] to obtain LDCT in the case of σ-algebras. If $\mathcal{P}$ is a δ-ring, $\|\mathbf{m}\|(E)$ can be infinite for some $E \in \sigma(\mathcal{P})$ and hence the argument in the proof of Theorem 2.2(2) of [L1] cannot be used here.

When $\mathbf{m}$ is an operator-valued measure on $\mathcal{P}$ as in [Do1], for an $\mathbf{m}$-integrable vector function $\mathbf{f}$ we can find a sequence of X-valued $\mathcal{P}$-simple functions $(\mathbf{s}_n)$ such that $\mathbf{s}_n(t) \to \mathbf{f}(t)$ $\mathbf{m}$-a.e. in T and $|\mathbf{s}_n(t)| \nearrow |\mathbf{f}(t)|$ $\mathbf{m}$-a.e. in T and such that $\lim_n \int_E \mathbf{s}_n d\mathbf{m} = \int_E \mathbf{f} d\mathbf{m}$, $E \in \sigma(\mathcal{P})$, the limit being uniform with respect to $E \in \sigma(\mathcal{P})$. But, this result cannot be asserted for all sequences $(\boldsymbol{\omega}_n)$ of X-valued $\mathcal{P}$-simple functions with $\boldsymbol{\omega}_n \to \mathbf{f}$ $\mathbf{m}$-a.e. in T and with $|\boldsymbol{\omega}_n| \nearrow |\mathbf{f}|$ $\mathbf{m}$-a.e. in T. See Remark 9 of [DP2] and [P5]. However, the analogue of Corollary 2.1.9 holds for $\mathbf{m}$-integrable vector functions $\mathbf{f}$ if and only if $\mathbf{f} \in \mathcal{L}_1(\mathbf{m})$. See [P5].

Remark 2.1.12. Theorem 3.3 of [L2] gives LDCT (with the hypothesis of (KL) $\mathbf{m}$-integrability for the dominated $\sigma(\mathcal{P})$-measurable functions) for σ-additive vector measures μ defined on a δ-ring τ of subsets of a set S with values in a sequentially complete lcHs E. Its proof is based on Lemma 3.4 of [L2] which states as follows. *Lemma* 3.4 (*of* [L2]). If g is μ-integrable, U is a zero neighborhood in E and $\epsilon > 0$, there is an $A \in \tau$ such that

$$\sup_{x' \in U^0} \int_{S \backslash A} |f(t)| |V(x'\mu, dt) < \epsilon.$$

Unfortunately, the proof of the said lemma is incorrect as it is impossible to have $A_n \subset A_{n+1}$, $V(x'_n \lambda, A_n) > \epsilon$ and $V(x'_n \lambda, A_{n+1}) < \frac{\epsilon}{2}$ (in the notation of [L2]). Therefore, LDCT for the case of δ-rings remains unestablished in [L2].

However, we can provide a correct proof of Lemma 3.4 of [L2] as follows. Following the notation of [L2], let τ be a δ-ring of subsets of a set S, $C(\tau)$ the σ-algebra of sets locally in τ, E a sequentially complete lcHs, $\mu : \tau \to E$ a σ-additive vector measure and g a (KL) μ-integrable function (see Notation 2.1.4).

From Definition 3.1 of [L2], it follows that all (KL) **m**-integrable functions are $\sigma(\tau)$-measurable and hence $N(g) \in \sigma(\tau)$. Let E' be the topological dual of E.

For a neighborhood U of 0 in E, let p_{U° be the seminorm on E given by $p_{U^\circ}(x) = \sup_{x' \in U^0} |x'(x)|$, where $U^0 = \{x' \in E' : |x'(x)| \leq 1 \text{ for } x \in U\}$. Let

$$\lambda(\cdot) = \int_{(\cdot)} g d\mu$$

on $C(\tau)$. Then λ is σ-additive on $C(\tau)$ with values in E and the semivariation of λ with respect to the seminorm p_{U°, denoted by $||\lambda||_{p_{U^\circ}}$, is given by

$$||\lambda||_{p_{U^\circ}}(\cdot) = \sup_{x' \in U^\circ} \int_{(\cdot)} |g(t)| V(x'\mu, dt).$$

Then $||\lambda||_{p_{U^\circ}}$ is continuous on $C(\tau)$ and vanishes on $S \backslash N(g)$. As $N(g) \in \sigma(\tau)$, there is an increasing sequence $(A_n)_1^\infty \subset \tau$ such that $N(g) = \bigcup_1^\infty A_n$. Then, given $\epsilon > 0$, there exists n_0 such that $||\lambda||_{p_U^0}(N(g) \backslash A_n) < \epsilon$ for $n \geq n_0$. Let $A = A_{n_0}$. Then

$$\sup_{x' \in U^\circ} \int_{S \backslash A} |g(t)| V(x'\mu, dt) = \sup_{x' \in U^\circ} \int_{N(g) \backslash A} |g(t)| V(x'\mu, dt)$$
$$= ||\lambda||_{p_{U^\circ}}(N(g) \backslash A) < \epsilon.$$

Hence Lemma 3.4 of [L2] (where $|f(t)|$ should be replaced by $|g(t)|$) holds and consequently, Theorem 3.3 of [L2] also holds.

The claim in line 3 from the top on p. 298 of [L2] implies that the (KL) μ-integrable function g is bounded on A and this is not true in general, even though $A \in \tau$. Hence the proof of Theorem 3.5 of [L2] is also incorrect.

Now we shall prove Theorem 3.5 of [L2]. Let $\epsilon > 0$. By Lemma 3.4 of [L2] (which has been proved above) there exists $A \in \tau$ such that

$$\sup_{x' \in U^\circ} \int_{S \backslash A} |g(t)| V(x'\mu, dt) = \sup_{x' \in U^\circ} \int_{N(g) \backslash A} |g(t)| V(x'\mu, dt) < \frac{\epsilon}{3}.$$

As g is $\sigma(\tau)$-measurable, there exists a sequence (s_n) of τ-simple functions such that $s_n \to g$ and $|s_n| \nearrow |g|$ pointwise in S. As μ is σ-additive on the σ-algebra $\sigma(\tau) \cap A$, by an argument similar to that on p. 161 of [L1] we can choose n_0 such that

$$\sup_{x' \in U^\circ} \int_A |g(t) - s_{n_0}(t)| V(x'\mu, dt) < \frac{\epsilon}{3}.$$

Let $f = s_{n_0}$. Then

$$\sup_{x' \in U^\circ} \int_S |f(t) - g(t)| V(x'\mu, dt) = \sup_{x' \in U^\circ} \int_{N(g)} |f(t) - g(t)| V(x'\mu, dt)$$
$$\leq 2 \sup_{x' \in U^0} \int_{N(g) \backslash A} |g(t)| V(x'\mu, dt) + \sup_{x' \in U^\circ} \int_A |f(t) - g(t)| V(x'\mu, dt) < \epsilon$$

and hence Theorem 3.5 of [L2] holds. i.e., *If g is μ-integrable, U is a zero neighborhood in E and $\epsilon > 0$, then there is a τ-simple function f satisfying*

$$\sup_{x' \in U^0} \int |f(t) - g(t)| V(x'\mu, dt) < \epsilon.$$

Masani and Niemi use the Banach space versions of Theorems 3.3 and 3.5 and Lemma 3.4 of [L2] in Section 4 of [MN2]. Since we have proved them above, their results in the said section are restored.

Remark 2.1.13. In Chapter 3 we give not only several versions of LDCT (mean^p) and LBCT(mean^p) but also those of the Vitali convergence theorems in (mean^p) for $1 \le p < \infty$. See Theorems 3.3.1, 3.3.2, 3.3.6, 3.4.5, 3.4.6, 3.4.7 and 3.4.10 and Corollary 3.4.11 of Chapter 3. Further generalizations to quasicomplete (resp. sequentially complete) lcHs-valued **m** are given in Section 4.5 of Chapter 4.

Remark 2.1.14. By Part (ii) of Definition 2.1.1, the **m**-a.e. limit of a sequence of **m**-measurable functions is also **m**-measurable.

In fact, let $f_n : T \to \mathbb{K}$ be **m**-measurable for $n \in \mathbb{N}$ and let $f : T \to \mathbb{K}$ be such that $f_n \to f$ **m**-a.e. in T. Then by hypothesis, there exist $M_n \in \sigma(\mathcal{P})$ with $||\mathbf{m}||(M_n) = 0$ such that $f_n \chi_{T \setminus M_n}$ is $\sigma(\mathcal{P})$-measurable for $n \in \mathbb{N}$ and a set $M_0 \in \sigma(\mathcal{P})$ with $||\mathbf{m}||(M_0) = 0$ such that $f_n \chi_{T \setminus M_0} \to f \chi_{T \setminus M_0}$ pointwise in T. Let $M = \bigcup_{n=0}^{\infty} M_n$. Then $M \in \sigma(\mathcal{P})$, $||\mathbf{m}||(M) = 0$ and $f_n \chi_{T \setminus M} \to f \chi_{T \setminus M}$ pointwise in T. Consequently, $f \chi_{T \setminus M}$ is $\sigma(\mathcal{P})$-measurable and hence f is **m**-measurable.

2.2   (BDS) m-integrability

In this section we give the definition of **m**-integrability (in the sense of Bartle-Dunford-Schwartz [BDS] even in the case of **m** defined on a δ-ring of sets unlike the original work in [BDS]) and show that the concepts of (KL) **m**-integrability and of (BDS) **m**-integrability and the respective integrals coincide (see Theorem 2.2.2). (This result was first mentioned in Theorem 2.4 of [L1] for the case of σ-algebras, but its proof lacks essential details. See Remark 2.2.6 below.) Hence Theorems 2.1.5 and 2.1.7 and Corollaries 2.1.8, 2.1.9 and 2.1.10 remain valid for **m**-integrable functions. Following Dobrakov [Do1], we characterize **m**-integrability in Theorem 2.2.4 and then give necessary and sufficient conditions in Theorem 2.2.8 for the validity of the interchange of limit and integral. The proof of the latter result is based on Theorem 2.2.2 and Proposition 1.1.13 and is much simpler than that of the corresponding result known for the Dobrakov integral of vector functions as given in [Do1] and [DP2]. Finally we deduce Theorem IV.10.9 of [DS1] as a simple corollary of Theorem 2.2.8. In Chapter 4 we study the (BDS)-integral of scalar functions with respect to a σ-additive quasicomplete (resp. sequentially complete) lcHs-valued measure and generalize the results of this section and those of Sections 3.2, 3.3, 3.4 and 3.5 of Chapter 3 in Chapter 4.

Before concluding the chapter we give some examples of σ-additive Banach space-valued measures defined on a δ-ring of sets. See Examples 2.2.13–2.2.16 below.

Definition 2.2.1. Let $\mathbf{m} : \mathcal{P} \to X$ be σ-additive and let $f : T \to \mathbb{K}$ or $[-\infty, \infty]$ be **m**-measurable. Then there exists $N \in \sigma(\mathcal{P})$ with $||\mathbf{m}||(N) = 0$ such that $f\chi_{T\setminus N}$ is $\sigma(\mathcal{P})$-measurable. If there exists a sequence $(s_n) \subset \mathcal{I}_s$ such that $s_n\chi_{T\setminus N} \to f\chi_{T\setminus N}$ pointwise in T and such that $\lim_n \int_E s_n d\mathbf{m} = \lim_n \int_{E\setminus N} s_n$ exists in X for each $E \in \sigma(\mathcal{P})$, then f is said to be integrable in the sense of Bartle-Dunford-Schwartz with respect to **m** in T or (BDS) **m**-integrable in T or simply, **m**-integrable in T, where $\int_E s d\mathbf{m}$, $s \in \mathcal{I}_s$, is the same as that in Theorem 2.1.5(i). In that case, we define

$$(BDS)\int_E f d\mathbf{m} = \lim_n \int_{E\setminus N} s_n d\mathbf{m} = \lim_n \int_E s_n d\mathbf{m}, \quad E \in \sigma(\mathcal{P})$$

and

$$(BDS)\int_T f d\mathbf{m} = (BDS)\int_{N(f)\setminus N} f d\mathbf{m} = \lim_n \int_{N(f)\setminus N} s_n d\mathbf{m}.$$

(Note that $N(f)\setminus N = N(f\chi_{T\setminus N}) \in \sigma(\mathcal{P})$ and $\int_{N(f)\setminus N} s_n d\mathbf{m}$ is well defined for all n.)

Theorem 2.2.2. *Let f be an **m**-measurable function on T with values in $\mathbb{K}$ or in $[-\infty, \infty]$. Then f is (KL) **m**-integrable in T if and only if it is **m**-integrable in T and in that case, $(BDS)\int_E f d\mathbf{m} = (KL)\int_E f d\mathbf{m}$ for $E \in \sigma(\mathcal{P})$. Thus if f is **m**-integrable in T, then $(BDS)\int_E f d\mathbf{m}$ is well defined for $E \in \sigma(\mathcal{P})$ and* **hereafter, we shall denote either of the integrals of f over $E \in \sigma(\mathcal{P})$ by $\int_E f d\mathbf{m}$.** *Moreover, if $\mathcal{P}$ is a σ-algebra, then the concepts of **m**-measurability and **m**-integrability and the value of the (BDS) **m**-integral coincide with the corresponding ones given in Section IV.10 of* [DS1].

Proof. By the first part of Corollary 2.1.9, (KL) **m**-integrability of f implies that f is **m**-integrable in T and that $(KL)\int_E f d\mathbf{m} = (BDS)\int_E f d\mathbf{m}$ for $E \in \sigma(\mathcal{P})$. Conversely, let f be **m**-integrable in T. Then by the hypothesis of **m**-measurability and of **m**-integrability of f, there exist an **m**-null set $M \in \sigma(\mathcal{P})$ such that $f\chi_{T\setminus M}$ is $\sigma(\mathcal{P})$-measurable and a sequence $(s_n) \subset \mathcal{I}_s$ such that $s_n\chi_{T\setminus M} \to f\chi_{T\setminus M}$ pointwise in T and such that $\lim_n \int_E s_n d\mathbf{m} = x_E$ (say) exists in X for each $E \in \sigma(\mathcal{P})$. Then $(BDS)\int_E f d\mathbf{m} = x_E$ for $E \in \sigma(\mathcal{P})$. On the other hand, $x^*(x_E) = \lim_n \int_E s_n d(x^*\mathbf{m})$ for each $E \in \sigma(\mathcal{P})$ and for each $x^* \in X^*$. Consequently, by Proposition 1.1.21 we have

$$x^*(x_E) = \lim_n \int_E s_n d(x^*\mathbf{m}) = \int_E f d(x^*\mathbf{m}) = \lim_n \int_E f\chi_{T\setminus M} d(x^*\mathbf{m})$$

for $x^* \in X^*$ and for $E \in \sigma(\mathcal{P})$. Thus by the second part of Definition 2.1.1, f is (KL) **m**-integrable in T and $(KL)\int_E f d\mathbf{m} = x_E$. Hence $(BDS)\int_E f d\mathbf{m} = x_E =$

(KL) $\int_E f d\mathbf{m}$, $E \in \sigma(\mathcal{P})$. Since (KL) $\int_E f d\mathbf{m}$ is well defined, (BDS)$\int_E f d\mathbf{m}$ is also well defined and does not depend on the particular choice of the sequence (s_n) in Definition 2.2.1.

The last part follows from the first and from the paragraph preceding Notation 1.1.15. $\qquad\square$

Remark 2.2.3. In the light of Theorem 2.2.2, **Theorems 2.1.5 and 2.1.7 and Corollaries 2.1.8, 2.1.9 and 2.1.10 hold for m-integrable functions and moreover, $\mathcal{I}(\mathbf{m})$ is the same as the family of all m-measurable scalar functions on T which are m-integrable in T. Consequently, Theorem 2.1.7 gives a stronger version of Theorem IV.10.10 of [DS1] since in Theorem 2.1.7 not only the dominated functions are not assumed to be m-integrable but also the domain of m is only a δ-ring of sets.** Moreover, the reader can note that the original proof of Theorem IV.10.8(a) in [DS1] to show that the (BDS) **m**-integral is well defined is very lengthy, while ours is immediate through its equivalence with (KL) **m**-integral.

The following theorem plays a key role in the proof of Theorem 3.2.7 in Chapter 3, and the latter is used to prove many important results in Chapter 3. For example, one can mention an analogue of the Riesz-Fischer theorem (Theorem 3.2.8) and the theorem of characterizations of $\mathcal{L}_p \mathcal{I}(\mathbf{m})$, $1 \le p < \infty$, (Theorem 3.3.6) in Chapter 3, which are needed to obtain several versions of the Vitali convergence theorem and LDCT for $\mathcal{L}_p \mathcal{I}(\mathbf{m})$ for $1 \le p < \infty$. See also Remark 2.2.9.

Theorem 2.2.4. *Let f be an **m**-measurable function on T with values in $\mathbb{K}$. Let $(s_n)_1^\infty \subset \mathcal{I}_s$ be such that $s_n \to f$ **m**-a.e. in T and let $\boldsymbol{\gamma}_n : \sigma(\mathcal{P}) \to X$ be given by $\boldsymbol{\gamma}_n(\cdot) = \int_{(\cdot)} s_n d\mathbf{m}$, $n \in \mathbb{N}$. Then the following statements are equivalent:*

(i) $\lim_n \boldsymbol{\gamma}_n(E) = \boldsymbol{\gamma}(E)$ *exists in X for each $E \in \sigma(\mathcal{P})$.*

(ii) $\boldsymbol{\gamma}_n(\cdot)$, $n \in \mathbb{N}$, *are uniformly σ-additive on $\sigma(\mathcal{P})$.*

(iii) $\lim_n \boldsymbol{\gamma}_n(E)$ *exists in X uniformly with respect to $E \in \sigma(\mathcal{P})$.*

If any one of (i), (ii) *or* (iii) *holds, then f is **m**-integrable in T and $\int_E f d\mathbf{m} = \boldsymbol{\gamma}(E)$ for $E \in \sigma(\mathcal{P})$.*

Proof. Since $\boldsymbol{\gamma}_n$ is σ-additive on $\sigma(\mathcal{P})$ for each $n \in \mathbb{N}$, (i)$\Rightarrow$(ii) by VHSN and (iii)$\Rightarrow$(i) obviously.

(ii)$\Rightarrow$(iii) By the hypothesis (ii) and by Proposition 1.1.13 there exists a control measure $\mu : \sigma(\mathcal{P}) \to [0, \infty)$ for $(\boldsymbol{\gamma}_n)_1^\infty$. Thus, given $\epsilon > 0$, there exists $\delta > 0$ such that $\sup_n \|\boldsymbol{\gamma}_n\|(E) < \frac{\epsilon}{3}$ whenever $\mu(E) < \delta$. As $s_n(t) \to f(t)$ **m**-a.e. in T, there exists $M \in \sigma(\mathcal{P})$ with $\|\mathbf{m}\|(M) = 0$ such that $s_n(t) \to f(t)$ for $t \in T \backslash M$. Clearly, by the definition of semivariation, $\|\boldsymbol{\gamma}_n\|(M) = 0$ for all n. Let $F = \bigcup_{n=1}^\infty N(s_n) \cap (T \backslash M)$. Then $F \in \sigma(\mathcal{P})$ and $s_n \chi_{T \backslash M} \to f \chi_{T \backslash M}$ pointwise in T. Therefore, by the Egoroff-Lusin theorem, there exist $N \in \sigma(\mathcal{P}) \cap F$ with $\mu(N) = 0$ and a sequence $(F_k) \subset \mathcal{P}$ with $F_k \nearrow F \backslash N$ such that $f_n \to f$ uniformly in each F_k. As $F \backslash N \backslash F_k \searrow \emptyset$, there exists k_0 such that $\mu(F \backslash N \backslash F_k) < \delta$ for $k \ge k_0$.

Choose n_0 such that $||s_n - s_r||_{F_{k_0}} \cdot ||\mathbf{m}||(F_{k_0}) < \frac{\epsilon}{3}$ for $n, r \geq n_0$. Then by the last part of Theorem 2.1.5(i), for all $E \in \sigma(\mathcal{P})$, we have

$$\left| \int_{E \cap F_{k_0}} s_n d\mathbf{m} - \int_{E \cap F_{k_0}} s_r d\mathbf{m} \right| \leq ||s_n - s_r||_{F_{k_0}} \cdot ||\mathbf{m}||(E \cap F_{k_0}) < \frac{\epsilon}{3}$$

for $n, r \geq n_0$. Moreover, $||\gamma_n||(F \backslash N \backslash F_{k_0}) < \frac{\epsilon}{3}$ for all $n \in \mathbb{N}$ as $\mu(F \backslash N \backslash F_{k_0}) < \delta$. As $\mu(N) = 0$, $||\gamma_n||(N \cap E) = 0$ for all n and for all $E \in \sigma(\mathcal{P})$. As observed above, $||\gamma_n||(M \cap E) = 0$ for all n and for all $E \in \sigma(\mathcal{P})$. Consequently,

$$\left| \int_E s_n d\mathbf{m} - \int_E s_r d\mathbf{m} \right| \leq \left| \int_{E \cap F_{k_0}} (s_n - s_r) d\mathbf{m} \right|$$
$$+ ||\gamma_n||(E \cap (F \backslash N \backslash F_{k_0})) + ||\gamma_r||(E \cap (F \backslash N \backslash F_{k_0})) < \epsilon$$

for $n, r \geq n_0$ and for all $E \in \sigma(\mathcal{P})$. Hence (iii) holds.

If any one of (i), (ii) or (iii) holds, then (i) holds and hence by Definition 2.2.1, f is **m**-integrable in T and $\int_E f d\mathbf{m} = \gamma(E) = \lim_n \int_E s_n d\mathbf{m}$ for $E \in \sigma(\mathcal{P})$. □

Theorem 2.2.5. *Let $\mathcal{P}$ be a σ-ring $\mathcal{S}$. Then an **m**-measurable function f is **m**-integrable in T if and only if there exists a sequence (f_n) of bounded **m**-measurable functions on T such that $f_n(t) \to f(t)$ **m**-a.e. in T and such that $\lim_n \int_E f_n d\mathbf{m}$ exists in X for each $E \in \mathcal{S}$. (Note that by Theorem 2.2.2 and by the last part of Theorem 2.1.5(v), such f_n are **m**-integrable in T.) In that case, $\int_E f d\mathbf{m} = \lim_n \int_E f_n d\mathbf{m}$ for $E \in \mathcal{S}$, the limit being uniform with respect to $E \in \mathcal{S}$.*

Proof. The condition is necessary by Definition 2.2.1. As $\mathcal{S}$ is a σ-ring, by Theorem 2.2.2 and by the last part of Theorem 2.1.5(v), bounded **m**-measurable functions are **m**-integrable in T. In the light of Definition 2.1.1, without loss of generality we shall assume that all the functions f_n and f are further $\mathcal{S}$-measurable. Let $E \in \mathcal{S}$ and let $\lim_n \int_E f_n d\mathbf{m} = x_E \in X$. For $x^* \in X^*$,

$$x^*(x_E) = x^* \left(\lim_n \int_E f_n d\mathbf{m} \right) = \lim_n \int_E f_n d(x^* \mathbf{m}) = \int_E f d(x^* \mathbf{m})$$

by Theorems 2.2.2 and 2.1.5(viii) and by Proposition 1.1.21. Therefore, f is (KL) **m**-integrable in T and hence **m**-integrable in T. Moreover, $\int_E f d\mathbf{m} = x_E = \lim_n \int_E f_n d\mathbf{m}$ for $E \in \sigma(\mathcal{P})$. Let $\gamma(E) = \int_E f d\mathbf{m}$. By Theorem 2.1.5(ii), $\gamma_n(\cdot) = \int_{(\cdot)} f_n d\mathbf{m}$, $n \in \mathbb{N}$, are σ-additive on $\sigma(\mathcal{P})$. Moreover, as observed above, $\lim_n \gamma_n(E) = \gamma(E)$ for $E \in \sigma(\mathcal{P})$. Hence by VHSN $\gamma_n, n \in \mathbb{N}$, are uniformly σ-additive on $\sigma(\mathcal{P})$. Consequently, the proof of (ii)⇒(iii) of Theorem 2.2.4 holds here verbatim, if we replace s_n by f_n. Therefore, the last part also holds. □

Remark 2.2.6. When the domain of **m** is a σ-algebra, the equivalence of the (KL)-integral and the (BDS)-integral was first obtained in Theorem 2.4 of [L1] and has been cited by various authors like [C1],[JO],[KK], [Ri5], etc. Theorem 2.4 of [L1] is as follows:

Suppose X is a sequentially complete lcHs, $\mu : \mathcal{S} \to X$ is σ-additive where $\mathcal{S}$ is a σ-algebra of subsets of a set S and f is a complex-valued function on S ($\mathcal{S}$-measurable). Then the following are equivalent:

(1) *f is μ-integrable.*

(2) *There is a sequence (f_n) of bounded measurable functions which converge pointwise to f and for which $(\int_E f_n(t)\mu(dt))$ is Cauchy uniformly with respect to $E \in \Sigma$.*

(3) *There is a sequence (f_n) of simple functions which converge pointwise to f and for which $(\int_E f_n(t)\mu(dt))$ is Cauchy for each $E \in \Sigma$.*

In the proof of the above theorem the author claims that (2) clearly implies (3)in [L1]. But we do not see how it is so, since bounded measurable functions need not be simple functions. However, by using Theorem 2.2(2) (LDCT) of [L1], we can prove the said theorem as follows. It has been proved on p. 162 of [L1] that (1) implies (2) and (3) implies (1). Using an argument similar to that in the proof of (3) implies (1) in [L1], one can show that (2) implies (1). By Theorem 2(2) (LDCT) of [L1], by Proposition 1.1.18 above and by the fact that simple functions are (KL) **m**-integrable (even when **m** has values in an lcHs) (1) implies (3) and hence Theorem 2.4 of [L1] holds.

Remark 2.2.7. The proof of Theorem 2.2.4 is similar to that of Theorem 7 of [DP2], but here we use the control measure of $(\gamma_n)_1^\infty$ and invoke the Egoroff-Lusin theorem with respect to it.

The next theorem generalizes Theorem 2.2.4 to **m**-integrable functions and is also known as the closure theorem. Compare with Remark 4.2.10 of Chapter 4.

Theorem 2.2.8. (Theorem on the interchange of limit and integral). *Let $\mathbf{m} : \mathcal{P} \to X$ be σ-additive. Let $f_n : T \to \mathbb{K}$ be **m**-measurable and **m**-integrable in T for $n \in \mathbb{N}$, and let $f : T \to \mathbb{K}$. If $f_n \to f$ **m**-a.e. in T, then f is **m**-measurable. Let $\gamma_n(\cdot) = \int_{(\cdot)} f_n d\mathbf{m}$, $n \in \mathbb{N}$. Then the following statements are equivalent:*

(i) *$\lim_n \gamma_n(E) = \gamma(E)$ exists in X for each $E \in \sigma(\mathcal{P})$.*

(ii) *γ_n, $n \in \mathbb{N}$, are uniformly σ-additive on $\sigma(\mathcal{P})$.*

(iii) *$\lim_n \gamma_n(E) = \gamma(E) \in X$ exists uniformly with respect to $E \in \sigma(\mathcal{P})$.*

*If any one of (i), (ii) or (iii) holds, then f is **m**-integrable in T and*

$$\int_E f d\mathbf{m} = \int_E \lim_n f_n d\mathbf{m} = \lim_n \int_E f_n d\mathbf{m}, \ E \in \sigma(\mathcal{P}),$$

the limit being uniform with respect to $E \in \sigma(\mathcal{P})$. Conversely, if $f \in \mathcal{I}(\mathbf{m})$ and if

$$\int_E f d\mathbf{m} = \lim_n \int_E f_n d\mathbf{m}, \ \text{for } E \in \sigma(\mathcal{P}),$$

then (i), (ii) and (iii) hold.

Proof. Clearly, f is **m**-measurable. In the light of Definition 2.1.1, without loss of generality we shall assume that all the functions considered here are further $\sigma(\mathcal{P})$-measurable. (i)$\Rightarrow$(ii) by Theorem 2.2.2, by Theorem 2.1.5(ii) and by VHSN and (iii)$\Rightarrow$(i) obviously. The proof of (ii)$\Rightarrow$(iii) is almost the same as that of (ii)$\Rightarrow$(iii) of Theorem 2.2.4, with small modifications as below. Let us use the same notation as in the proof of Theorem 2.2.4, changing s_n to f_n. By Theorem 2.2.2 and by the last part of Corollary 2.1.9, we have the inequality

$$\left| \int_{E\cap F_{k_0}} f_n d\mathbf{m} - \int_{E\cap F_{k_0}} f_r d\mathbf{m} \right| \leq ||f_n - f_r||_{F_{k_0}} \cdot ||\mathbf{m}||(F_{k_0}).$$

Moreover, by Theorem 2.2.2 and by Theorem 2.1.5(iii)(a) we have $||\gamma_n||(M) = \sup_{|x^*|\leq 1} \int_M |f_n| dv(x^*\mathbf{m}) = 0$ for all $n \in \mathbb{N}$ since $v(x^*\mathbf{m})(M) \leq ||\mathbf{m}||(M) = 0$ for $|x^*| \leq 1$. Then arguing as in the proof of Theorem 2.2.4 we have (ii)$\Rightarrow$(iii).

If anyone of (i),(ii) or (iii) holds, then by the first part (i) holds and hence $\lim_n \int_E f_n d\mathbf{m} = \gamma(E) \in X$ for each $E \in \sigma(\mathcal{P})$. Thus, for $x^* \in X^*$ and for $E \in \sigma(\mathcal{P})$, by Proposition 1.1.21, by Theorem 2.2.2 and by Theorem 2.1.5(viii) we have

$$x^*\gamma(E) = x^*\left(\lim_n \int_E f_n d\mathbf{m} \right) = \lim_n \int_E f_n d(x^*\mathbf{m}) = \int_E f d(x^*\mathbf{m}).$$

This shows that f is (KL) **m**-integrable in T and hence **m**-integrable in T, $\gamma(E) = \int_E f d\mathbf{m}$, and

$$\lim_n \int_E f_n d\mathbf{m} = \gamma(E) = \int_E f d\mathbf{m}, \ E \in \sigma(\mathcal{P}).$$

Then by (iii) the limit is uniform with respect to $E \in \sigma(\mathcal{P})$. Conversely, if $f \in \mathcal{I}(\mathbf{m})$ and if $\lim_n \int_E f_n d\mathbf{m} = \int_E f d\mathbf{m}$, $E \in \sigma(\mathcal{P})$, then clearly (i) holds and hence by the first part, (ii) and (iii) also hold. $\square$

Remark 2.2.9. The analogue of Theorem 2.2.8 is not valid for the Bochner integral. See [P5]. In Section 4.3 of Chapter 4, Theorems 2.2.2, 2.2.4, 2.2.5 and 2.2.8 are generalized to (BDS)-integrals with respect to a quasicomplete lcHs-valued σ-additive **m** on $\mathcal{P}$.

Remark 2.2.10. In the proof of the second part of the above theorem, we used the fact that a function g is (KL) **m**-integrable in T if and only if it is **m**-integrable in T and this has reduced the arguments considerably in comparison with the proof of the last part of Theorem 15 of [Do1] or of Theorem 9 of [DP2] for the corresponding result on vector functions.

As an immediate corollary of Theorem 2.2.8, we deduce the following result which is an improved version of Theorem IV.10.9 of [DS1] and the latter is used to prove LDCT (Theorem IV.10.10) in [DS1]. When $\mathcal{P}$ is a σ-ring, one can deduce Proposition 1.1.21 (which is used in the proof of Theorem 2.2.8) as an easy consequence of the Vitali-Hahn-Sacks theorem (see the proof of Lemma 2.2 of [L1]) and consequently, our proof of Theorem IV.10.9 of [DS1] given below is undoubtedly much shorter and more elegant than that of [DS1].

Corollary 2.2.11. *Let Σ be a σ-ring of sets and let $\boldsymbol{\mu} : \Sigma \to X$ be σ-additive. Let $f_n : T \to \mathbb{K}$, $n \in \mathbb{N}$, be $\boldsymbol{\mu}$-measurable and $\boldsymbol{\mu}$-integrable in T and let $f : T \to \mathbb{K}$. If $f_n \to f$ $\boldsymbol{\mu}$-a.e. in T and if*

$$\lim_{||\boldsymbol{\mu}||(E) \to 0} \int_E f_n d\boldsymbol{\mu} = 0, \;\; E \in \Sigma$$

uniformly for $n \in \mathbb{N}$, then f is $\boldsymbol{\mu}$-integrable in T and

$$\int_E f d\boldsymbol{\mu} = \lim_n \int_E f_n d\boldsymbol{\mu}, \;\; E \in \Sigma,$$

the limit being uniform with respect to $E \in \Sigma$.

Proof. Let $\boldsymbol{\gamma}_n(\cdot) = \int_{(\cdot)} f_n d\boldsymbol{\mu}$, $n \in \mathbb{N}$. Let $\epsilon > 0$. Then by hypothesis there exists $\delta > 0$ such that $\sup_n |\boldsymbol{\gamma}_n(E)| < \epsilon$ whenever $||\boldsymbol{\mu}||(E) < \delta$, $E \in \Sigma$. Let $F_k \searrow \emptyset$ in Σ. As $\boldsymbol{\mu}$ is σ-additive on Σ and as Σ is a σ-ring, $||\boldsymbol{\mu}||$ is continuous on Σ by Proposition 1.1.5 and hence there exists k_0 such that $||\boldsymbol{\mu}||(F_k) < \delta$ for $k \geq k_0$ and consequently, $\sup_n |\boldsymbol{\gamma}_n(F_k)| < \epsilon$ for $k \geq k_0$. Then, by Proposition 1.1.7, $\boldsymbol{\gamma}_n$, $n \in \mathbb{N}$, are uniformly σ-additive on Σ. Now the result holds by Theorem 2.2.8. $\qquad\square$

Now we give some examples of Banach space-valued σ-additive measures on a δ-ring or on a σ-ring.

Definition 2.2.12. A collection $(x_\alpha) \subset X$ is said to be *summable* if the net $(\sum_{\alpha \in A} x_\alpha)$ formed by sums over finite sets of indices and directed by inclusion, is convergent in X. The collection (x_α) is said to be *absolutely summable* if $\sum_\alpha |x_\alpha| < \infty$. (As X is a Banach space, it follows that (x_α) is summable in X whenever it is absolutely summable.) Finally, a sequence $(x_n)_1^\infty \subset X$ is said to be *subseries convergent* if $\sum_{k \leq n} \chi_A(k) x_k$ exists in X for each $A \subset \mathbb{N}$. See [D].

Example 2.2.13. For a nonvoid set S, let F_S be the δ-ring of all finite subsets of S. A collection $(x_\alpha)_{\alpha \in S} \subset X$ defines an X-valued σ-additive measure $\mathbf{m}$ on F_S if we define $\mathbf{m}(\emptyset) = 0$ and $\mathbf{m}(A) = \sum_{\alpha \in A} x_\alpha$ if $A \neq \emptyset$. We refer to the measure $\mathbf{m}$ defined in this manner as the measure defined by $(x_\alpha)_{\alpha \in S}$.

Example 2.2.14. Let $(x_n)_1^\infty$ be a summable sequence in X. Let $\mathbf{m} : \mathcal{P}(\mathbb{N}) \to X$ be defined by $\mathbf{m}(E) = \sum_{n \in E} x_n$. Then $\mathbf{m}$ is an X-valued σ-additive measure on $\mathcal{P}(\mathbb{N})$.

By Theorem 2.2 of [L2] we have:

Example 2.2.15. Any X-valued σ-additive measure $\mathbf{m}$ defined on a δ-ring $\mathcal{P}$ of sets is of finite variation if and only if $(\mathbf{m}(A_n))_1^\infty$ is absolutely summable to $\mathbf{m}(\bigcup_1^\infty A_n)$ for each pairwise disjoint sequence $(A_n)_1^\infty \subset \mathcal{P}$ with $\bigcup_1^\infty A_n \in \mathcal{P}$.

By Corollary 4.3 of [L2] we have:

Example 2.2.16. If $\mathbf{m}$ is an X-valued σ-additive measure defined on a δ-ring $\mathcal{P}$ of subsets of T and if every summable sequence (x_n) in X is absolutely summable, then every $\mathbf{m}$-integrable function in T is $v(\mathbf{m})$-integrable in T.

Chapter 3

$\mathcal{L}_p$-spaces, $1 \leq p \leq \infty$

We develop the theory of $\mathcal{L}_p$-spaces ($1 \leq p < \infty$) for the Bartle-Dunford-Schwartz integral with respect to a Banach space-valued σ-additive vector measure $\mathbf{m}$ defined on a δ-ring of sets and obtain results analogous to those known for such spaces in the theory of the abstract Lebesgue and Bochner integrals. For this we adapt some of the techniques employed by Dobrakov in the study of integration with respect to operator-valued measures. Though a few of these results are already there in the literature for $p = 1$ (sometimes with incorrect proofs as observed in Chapter 2)), they are treated here differently with simpler proofs.

3.1   The seminorms $\mathbf{m}_p^{\bullet}(\cdot, T)$ on $\mathcal{L}_p\mathcal{M}(\mathbf{m})$, $1 \leq p < \infty$

Dobrakov introduced a seminorm $\hat{\mathbf{m}}(\cdot, T)$ in [Do2] to define and study exhaustively the $\mathcal{L}_1$-spaces associated with an operator-valued measure $\mathbf{m}$ which is σ-additive in the strong operator topology on a δ-ring of sets and studied very briefly the corresponding $\mathcal{L}_p$-seminorms, $1 < p < \infty$, indicating the difficulties in developing analogous theory for the $\mathcal{L}_p$-spaces. Similar seminorms, for $p = 1$, were introduced in [KK], [MN2], [Ri5], etc., for studying the space of (KL) $\mathbf{m}$-integrable functions when $\mathbf{m}$ is a σ-additive Banach space-valued vector measure. In the present section, we introduce the seminorms $\mathbf{m}_p^{\bullet}(\cdot, T)$, $1 \leq p < \infty$, similar to those in [Do2], define the spaces $\mathcal{L}_p\mathcal{M}(\mathbf{m})$, $\mathcal{I}_p(\mathbf{m})$ and $\mathcal{L}_p\mathcal{I}(\mathbf{m})$ and study their basic properties. As in [Do1, Do2] we introduce various notions of convergence similar to those used in the theory of the abstract Lebesgue integral and study their interrelations. We also study relations between the spaces $\mathcal{L}_p(\mathbf{m})$ for $1 \leq p \leq \infty$, where $\mathcal{L}_\infty(\mathbf{m})$ is suitably defined.

Hereafter, most often, without mentioning Theorem 2.2.2 of Chapter 2, we shall directly apply Theorems 2.1.5 and 2.1.7 and Corollaries 2.1.8, 2.1.9, and 2.1.10 of Chapter 2 to functions m-integrable in T. (See Remark 2.2.3 of Chapter 2.)

Definition 3.1.1. Let $g : T \to \mathbb{K}$ or $[-\infty, \infty]$ be an **m**-measurable function. Let $1 \leq p < \infty$ and let $E \in \sigma(\mathcal{P})$. Then

$$\mathbf{m}_p^\bullet(g, E) = \sup \left\{ \left| \int_E s d\mathbf{m} \right|^{\frac{1}{p}} : s \in \mathcal{I}_s, |s(t)| \leq |g(t)|^p \text{ } \mathbf{m}\text{-a.e. in } E \right\}$$

is called the L_p-gauge of g on E. We define

$$\mathbf{m}_p^\bullet(g, T) = \sup_{E \in \sigma(\mathcal{P})} \mathbf{m}_p^\bullet(g, E).$$

By Proposition 1.1.18, $\mathbf{m}_p^\bullet(g, E)$ and $\mathbf{m}_p^\bullet(g, T)$ are well defined and belong to $[0, \infty]$.

Lemma 3.1.2. *Let $g : T \to \mathbb{K}$ or $[-\infty, \infty]$ be $\mathbf{m}$-measurable and let $1 \leq p < \infty$. Let $x^* \in X^*$ and let $E \in \sigma(\mathcal{P})$. Then:*

(i) $(x^*\mathbf{m})_p^\bullet(g, E) = \sup \left\{ | \int_E s d(x^*\mathbf{m}) |^{\frac{1}{p}} : s \in \mathcal{I}_s, |s(t)| \leq |g(t)|^p \text{ } \mathbf{m}\text{-a.e. in } E \right\}.$

(ii) $(x^*\mathbf{m})_p^\bullet(g, E) = \left(\int_E |g|^p dv(x^*\mathbf{m}) \right)^{\frac{1}{p}}.$

Proof. (i) By Definition 3.1.1,

$$(x^*\mathbf{m})_p^\bullet(g, E) = \sup \left\{ \left| \int_E s d(x^*\mathbf{m}) \right|^{\frac{1}{p}} : s \in \mathcal{I}_s, |s(t)| \leq |g(t)|^p \text{ } (x^*\mathbf{m})\text{-a.e. in } E \right\}.$$

Let $\alpha = \sup \left\{ | \int_E s d(x^*\mathbf{m}) |^{\frac{1}{p}} : s \in \mathcal{I}_s, |s(t)| \leq |g(t)|^p \text{ } \mathbf{m}\text{-a.e. in } E \right\}$. Clearly, $\alpha \leq (x^*\mathbf{m})_p^\bullet(g, E)$. To prove the reverse inequality, let $0 \leq c < (x^*\mathbf{m})_p^\bullet(g, E)$. Then there exist $s_0 \in \mathcal{I}_s$ and $M \in \sigma(\mathcal{P})$ with $v(x^*\mathbf{m})(M) = 0$ such that $|s_0(t)| \leq |g(t)|^p$ for $t \in T \backslash M$ and $c < | \int_E s_0 d(x^*\mathbf{m}) |^{\frac{1}{p}}$. If $s = s_0 \chi_{E \backslash M}$, then $s \in \mathcal{I}_s$ and $|s(t)| \leq |g(t)|^p$ for $t \in E$ and hence $|s(t)| \leq |g(t)|^p$ $\mathbf{m}$-a.e. in E. Moreover, $c < | \int_E s d(x^*\mathbf{m}) |^{\frac{1}{p}}$. Hence (i) holds.

(ii) Since $| \int_E s d(x^*\mathbf{m}) | \leq \int_E |g|^p dv(x^*\mathbf{m})$ for $s \in \mathcal{I}_s$ with $|s(t)| \leq |g(t)|^p$ $(x^*\mathbf{m})$-a.e. in E, we have $(x^*\mathbf{m})_p^\bullet(g, E) \leq \left(\int_E |g|^p dv(x^*\mathbf{m}) \right)^{\frac{1}{p}}$. To prove the reverse inequality, let $0 \leq c < \left(\int_E |g|^p dv(x^*\mathbf{m}) \right)^{\frac{1}{p}}$. Then by Proposition 1.1.18 of Chapter 1, there exists $s = \sum_1^r a_i \chi_{E_i}$, $(E_i)_1^r \subset \mathcal{P}$, $E_i \cap E_j = \emptyset$ for $i \neq j$, with $|s(t)| \leq |g(t)|^p$ $(x^*\mathbf{m})$-a.e. in E such that $c < (\sum_1^r |a_k| v(x^*\mathbf{m})(E \cap E_k))^{\frac{1}{p}}$. Hence there exist $(E_{kj})_{j=1}^{\ell_k} \subset \mathcal{P}$, pairwise disjoint, such that $\bigcup_{j=1}^{\ell_k} E_{kj} \subset E \cap E_k$, $k = 1, 2, \ldots, r$ and such that

$$c^p < \sum_{k=1}^r |a_k| \sum_{j=1}^{\ell_k} |(x^*\mathbf{m})(E_{kj})| = \int_E h d(x^*\mathbf{m})$$

where $h = \sum_{k=1}^r \sum_{j=1}^{\ell_k} |a_k| \mathrm{sgn}\left((x^*\mathbf{m})(E_{kj})\right) \chi_{E_{kj}}$. Then $h \in \mathcal{I}_s$ and $|h(t)| \leq |g(t)|^p$ $(x^*\mathbf{m})$-a.e. in E. Hence (ii) holds. $\square$

Theorem 3.1.3. *Let $g : T \to \mathbb{K}$ or $[-\infty, \infty]$ and let $1 \le p < \infty$. If g is $\mathbf{m}$-measurable and $E \in \sigma(\mathcal{P})$, then*

$$\mathbf{m}_p^\bullet(g, E) = \sup_{|x^*| \le 1} \left(\int_E |g|^p dv(x^*\mathbf{m}) \right)^{\frac{1}{p}} = \sup_{|x^*| \le 1} (x^*\mathbf{m})_p^\bullet(g, E)$$

$$= \sup \left\{ \left| \int_E f d\mathbf{m} \right|^{\frac{1}{p}} : f \in \mathcal{I}(\mathbf{m}), |f| \le |g|^p \ \mathbf{m}\text{-a.e. in } E \right\}.$$

Consequently, $|\int_E |f|^p d\mathbf{m}|^{\frac{1}{p}} \le \mathbf{m}_p^\bullet(f, E)$ for $|f|^p \in \mathcal{I}(\mathbf{m})$ and for $E \in \sigma(\mathcal{P})$. Moreover,

$$\left| \int_E f d\mathbf{m} \right| \le \mathbf{m}_1^\bullet(f, E) \tag{3.1.3.1}$$

for $f \in \mathcal{I}(\mathbf{m})$ and for $E \in \sigma(\mathcal{P})$.

Proof. By Definition 3.1.1 and Lemma 3.1.2(i) we have

$$\mathbf{m}_p^\bullet(g, E) = \sup \left\{ \left| \int_E s d\mathbf{m} \right|^{\frac{1}{p}} : s \in \mathcal{I}_s, |s(t)| \le |g(t)|^p \ \mathbf{m}\text{-a.e. in } E \right\}$$

$$= \sup \left\{ \left(\sup_{|x^*| \le 1} \left| \int_E s d(x^*\mathbf{m}) \right|^{\frac{1}{p}} \right) : s \in \mathcal{I}_s, |s(t)| \le |g(t)|^p \ \mathbf{m}\text{-a.e. in } E \right\}$$

$$= \sup_{|x^*| \le 1} (x^*\mathbf{m})_p^\bullet(g, E).$$

Then by Lemma 3.1.2(ii),

$$\mathbf{m}_p^\bullet(g, E) = \sup_{|x^*| \le 1} \left(\int_E |g|^p dv(x^*\mathbf{m}) \right)^{\frac{1}{p}}. \tag{3.1.3.2}$$

Since $\mathcal{I}_s \subset \mathcal{I}(\mathbf{m})$, by (3.1.3.2) we have

$$\mathbf{m}_p^\bullet(g, E) \le \sup \left\{ \left| \int_E f d\mathbf{m} \right|^{\frac{1}{p}}, f \in \mathcal{I}(\mathbf{m}), |f| \le |g|^p \ \mathbf{m}\text{-a.e. in } E \right\}$$

$$= \sup_{|x^*| \le 1} \sup \left\{ \left| \int_E f d(x^*\mathbf{m}) \right|^{\frac{1}{p}}, f \in \mathcal{I}(\mathbf{m}), |f| \le |g|^p \ \mathbf{m}\text{-a.e. in } E \right\}$$

$$\le \sup_{|x^*| \le 1} \int_E |g|^p dv(x^*\mathbf{m}) = \mathbf{m}_p^\bullet(g, E)$$

and hence the first part holds.

The other parts are immediate from the first. $\qquad\square$

Definition 3.1.4. Let $g : T \to \mathbb{K}$ or $[-\infty, \infty]$ be $\mathbf{m}$-measurable. Let $E \in \widetilde{\sigma(\mathcal{P})}$ (see paragraphs preceding Notation 1.1.15 in Section 1.1 of Chapter 1), and let $1 \le p < \infty$. Then E is of the form $E = F \cup N$, $F \in \sigma(\mathcal{P})$, $N \subset M \in \sigma(\mathcal{P})$ with $\|\mathbf{m}\|(M) = 0$. We define $\mathbf{m}_p^\bullet(g, E) = \mathbf{m}_p^\bullet(g, F)$. We also define $\int_E |g|^p dv(x^*\mathbf{m}) = \int_F |g|^p dv(x^*\mathbf{m})$. Then

$$\mathbf{m}_p^\bullet(g, E) = \sup_{|x^*| \le 1} \left(\int_E |g|^p dv(x^*\mathbf{m}) \right)^{\frac{1}{p}}. \tag{3.1.4.1}$$

To verify that $\mathbf{m}_p^\bullet(g, E)$ and $\int_E |g|^p dv(x^*\mathbf{m})$ are well defined, let $E = F_1 \cup N_1 = F_2 \cup N_2$ with $F_i \in \sigma(\mathcal{P})$, $N_i \subset M_i \in \sigma(\mathcal{P})$ and $\|\mathbf{m}\|(M_i) = 0$ for $i = 1, 2$. Let $M = M_1 \cup M_2$. Then $\|\mathbf{m}\|(M) = 0$ and $F_1 \cup M = F_2 \cup M$. Hence $\int_{F_1} s d\mathbf{m} = \int_{F_1 \cup M} s d\mathbf{m} = \int_{F_2 \cup M} s d\mathbf{m} = \int_{F_2} s d\mathbf{m}$ for $s \in \mathcal{I}_s$ and therefore, $\mathbf{m}_p^\bullet(g, E)$ is well defined. Moreover, $\int_{F_1} |g|^p dv(x^*\mathbf{m}) = \int_{F_1 \cup M} |g|^p dv(x^*\mathbf{m}) = \int_{F_2 \cup M} |g|^p dv(x^*\mathbf{m}) = \int_{F_2} |g|^p dv(x^*\mathbf{m})$ and hence $\int_E |g|^p dv(x^*\mathbf{m})$ is also well defined. Then (3.1.4.1) holds by Theorem 3.1.3.

Remark 3.1.5. When $\mathcal{P}$ is a σ-algebra $\mathcal{S}$, the gauge in Definition 3.1.1 above is given in [KK] for $p = 1$ and that too for (KL) $\mathbf{m}$-integrable $\mathcal{S}$-measurable functions only. Of course, there $\mathbf{m}$ has values in a real lcHs. The analogue of Theorem 3.1.3 for $p = 1$ is given in Lemma II.2.2 of [KK]. But our proof is more general and elementary than that of the said lemma in [KK] and is adaptable to the case of lcHs-valued vector measures defined on $\mathcal{P}$ (see Definition 4.3.1 and Theorem 4.3.2 of Chapter 4). A similar result for operator-valued measures is given in Theorem $4'$ of [Do2] without proof. Also see [SP1].

Theorem 3.1.6. *Let* $1 \le p < \infty$. *Let* $f : T \to \mathbb{K}$ *or* $[-\infty, \infty]$ *be* $\mathbf{m}$-*measurable and let* $|f|^p$ *be* $\mathbf{m}$-*integrable in* T. *Let* $\boldsymbol{\gamma}(\cdot) = \int_{(\cdot)} |f|^p d\mathbf{m}$. *Then* $\mathbf{m}_p^\bullet(f, E) = (\|\boldsymbol{\gamma}\|(E))^{\frac{1}{p}}$, $E \in \sigma(\mathcal{P})$ *and consequently,* $\mathbf{m}_p^\bullet(f, T) = (\|\boldsymbol{\gamma}\|(T))^{\frac{1}{p}} < \infty$. *Moreover,* $\mathbf{m}_p^\bullet(f, \cdot)$ *is continuous on* $\sigma(\mathcal{P})$ *(in the sense of Definition* 1.1.4 *of Chapter* 1).

Proof. For $x^* \in X^*$ and for $E \in \sigma(\mathcal{P})$, $v(x^*\boldsymbol{\gamma})(E) = \int_E |f|^p dv(x^*\mathbf{m})$ by Theorem 2.1.5(ii) of Chapter 2 and Proposition 1.1.19 of Chapter 1. Therefore, by Theorem 3.1.3 and by (iii)(a) of Theorem 2.1.5 we have

$$\left(\mathbf{m}_p^\bullet(f, E) \right)^p = \sup_{|x^*| \le 1} v(x^*\boldsymbol{\gamma})(E) = \|\boldsymbol{\gamma}\|(E).$$

Consequently,

$$\left(\mathbf{m}_p^\bullet(f, T) \right)^p = \sup_{E \in \sigma(\mathcal{P})} \left(\mathbf{m}_p^\bullet(f, E) \right)^p = \sup_{E \in \sigma(\mathcal{P})} \|\boldsymbol{\gamma}\|(E) = \|\boldsymbol{\gamma}\|(T) < \infty$$

as $\boldsymbol{\gamma}$ is an X-valued σ-additive vector measure on $\sigma(\mathcal{P})$ by Theorem 2.1.5(ii) of Chapter 2. The continuity of $\mathbf{m}_p^\bullet(f, \cdot) = (\|\boldsymbol{\gamma}\|(\cdot))^{\frac{1}{p}}$ is due to Proposition 1.1.5 of Chapter 1. $\qquad\square$

The converse of Theorem 3.1.6 is not true in general; i.e., if $f : T \to \mathbb{K}$ or $[-\infty, \infty]$ is $\mathbf{m}$-measurable with $\mathbf{m}_p^\bullet(f, T) < \infty$ for some p, $1 \le p < \infty$, then $|f|^p$ need not be $\mathbf{m}$-integrable in T. In fact, we have the following counter-example (see also p. 31 of [KK]).

Counter-example 3.1.7. Let $T = \mathbb{N}$ and $\mathcal{S} = \sigma$-algebra of all subsets of T. Let $X = c_0$. Let $1 \le p < \infty$ and let $f(t) = t^{\frac{1}{p}}$, $t \in T$. For $E \in \mathcal{S}$, let $\mathbf{m}(E) = (a_n) \in c_0$, where $a_n = \frac{1}{n}$ if $n \in E$ and $a_n = 0$ otherwise. Clearly $\mathbf{m}$ is σ-additive on $\mathcal{S}$. If $x^* \in c_0^* = l_1$, let $x^* = (x_n)$. Then $|x^*| = \sum_1^\infty |x_n| < \infty$. Now $\left(\mathbf{m}_p^\bullet(f, T)\right)^p = \sup_{|x^*| \le 1} \int_T |f|^p dv(x^*\mathbf{m}) = \sup_{|x^*| \le 1} \sum_1^\infty nv(x^*\mathbf{m})(\{n\})) = \sup_{|x^*| \le 1} \sum_1^\infty n\frac{1}{n}|x_n| \le 1 < \infty$. But $|f|^p$ is not $\mathbf{m}$-integrable in T. In fact, on the contrary we would have $\int_{\{n\}} |f|^p d\mathbf{m} = e_n$ for $n \in T$, where $e_n = (\delta_{nj})_{j=1}^\infty$ with $\delta_{nj} = 1$ if $j = n$ and $= 0$ otherwise; and if $\gamma(\cdot) = \int_{(\cdot)} |f|^p d\mathbf{m}$, then by Theorem 2.1.5(ii) of Chapter 2, γ would be σ-additive on $\mathcal{S}$ with $\gamma(T) \in c_0$ so that $\gamma(T) = \sum_1^\infty \gamma(\{n\}) = \sum_1^\infty e_n = (1, 1, 1, \dots) \notin c_0$. This contradiction shows that $f^p = |f|^p \notin \mathcal{I}(\mathbf{m})$.

However, when $c_0 \not\subset X$, we have the following characterization of $\mathbf{m}$-integrability of $|f|^p$.

Theorem 3.1.8. *Let $c_0 \not\subset X$ and let $1 \le p < \infty$. Then for an $\mathbf{m}$-measurable function f on T with values in $\mathbb{K}$ or $[-\infty, \infty]$, $|f|^p$ is $\mathbf{m}$-integrable in T if and only if $\mathbf{m}_p^\bullet(f, T) < \infty$. Moreover, if $|f|^p$ is $(x^*\mathbf{m})$-integrable for each $x^* \in X^*$, then $|f|^p$ is $\mathbf{m}$-integrable in T and $\mathbf{m}_p^\bullet(f, T) < \infty$.*

Proof. In the light of Theorem 3.1.6, it is enough to show that the condition is sufficient. Let $\mathbf{m}_p^\bullet(f, T) < \infty$. Then $\sup_{|x^*| \le 1} \int_T |f|^p dv(x^*\mathbf{m}) = (\mathbf{m}_p^\bullet(f, T))^p < \infty$ by Theorem 3.1.3. By Proposition 1.1.18 of Chapter 1 there exists a sequence $(s_n) \subset \mathcal{I}_s$ such that $0 \le s_n \nearrow |f|^p$ $\mathbf{m}$-a.e. in T. If $u_n = s_n - s_{n-1}$, for $n \ge 1$, where $s_0 = 0$, then $\sum_1^\infty u_n = |f|^p$ $\mathbf{m}$-a.e. in T. Let $E \in \sigma(\mathcal{P})$. Then, for each $x^* \in X^*$, by the Beppo-Levi theorem for positive measures we have $\sum_1^\infty \int_E |u_n| dv(x^*\mathbf{m}) = \sum_1^\infty \int_E u_n dv(x^*\mathbf{m}) = \int_E |f|^p dv(x^*\mathbf{m}) < \infty$ and consequently, by Proposition 4, §8 of [Din1], $\sum_1^\infty |x^*(\int_E u_n d\mathbf{m})| < \infty$. As $c_0 \not\subset X$, by the Bessaga-Pelczyński theorem there exists a vector $x_E \in X$ such that $x_E = \sum_1^\infty \int_E u_n d\mathbf{m} = \lim_n \int_E s_n d\mathbf{m}$. Since this holds for each $E \in \sigma(\mathcal{P})$, by Definition 2.1.1 and by Theorem 2.2.2 of Chapter 2 $|f|^p$ is $\mathbf{m}$-integrable in T. $\qquad \square$

In the above proof we observe that $(x^*\mathbf{m})$-integrability of $|f|^p$ for each $x^* \in X^*$ implies that $|f|^p$ is $\mathbf{m}$-integrable in T and hence the last part holds by Theorem 3.1.6.

Similar to [Do2] we give the following

Definition 3.1.9. Let $1 \le p < \infty$. Then we define

$$\mathcal{L}_p\mathcal{M}(\mathbf{m}) = \{f : T \to \mathbb{K}, \ f \ \mathbf{m}\text{-measurable with } \mathbf{m}_p^\bullet(f, T) < \infty\};$$
$$\mathcal{I}_p(\mathbf{m}) = \{f : T \to \mathbb{K}, f \ \mathbf{m}\text{-measurable and } |f|^p \in \mathcal{I}(\mathbf{m})\}$$

(so that $\mathcal{I}_1(\mathbf{m}) = \mathcal{I}(\mathbf{m})$ by Theorem 2.1.5(vii) of Chapter 2 – see Notation 2.1.2 of Chapter 2); and

$$\mathcal{L}_p\mathcal{I}(\mathbf{m}) = \{f \in \mathcal{I}_p(\mathbf{m}) : \mathbf{m}_p^\bullet(f, T) < \infty\}.$$

The following result is immediate from Theorems 3.1.6 and 3.1.8.

Theorem 3.1.10. *Let* $1 \leq p < \infty$. *Then* $\mathcal{L}_p\mathcal{I}(\mathbf{m}) = \mathcal{I}_p(\mathbf{m}) \subset \mathcal{L}_p\mathcal{M}(\mathbf{m})$. *If* $c_0 \not\subset X$, *then* $\mathcal{L}_p\mathcal{M}(\mathbf{m}) = \mathcal{L}_p\mathcal{I}(\mathbf{m}) = \mathcal{I}_p(\mathbf{m})$.

As an immediate consequence of Definition 3.1.4 and Theorem 3.1.3 we have the following theorem, whose easy proof is omitted.

Theorem 3.1.11. *Let* $g : T \to \mathbb{K}$ *or* $[-\infty, \infty]$ *be* $\mathbf{m}$-*measurable and let* $E \in \widetilde{\sigma(\mathcal{P})}$. *Let* $1 \leq p < \infty$. *Then the following assertions hold:*

(i) $\mathbf{m}_p^\bullet(g, \cdot) : \widetilde{\sigma(\mathcal{P})} \to [0, \infty]$ *is monotone and subadditive and vanishes on* $\emptyset$.

(ii) $\mathbf{m}_p^\bullet(ag, E) = |a|\mathbf{m}_p^\bullet(g, E)$ *for* $a \in \mathbb{K}$, *where* $0.\infty = 0$.

(iii) $(\inf_{t \in E} |g(t)|) \cdot (\|\mathbf{m}\|(E))^{\frac{1}{p}} \leq \mathbf{m}_p^\bullet(g, E) \leq (\sup_{t \in E} |g(t)|) \cdot (\|\mathbf{m}\|(E))^{\frac{1}{p}}$.

(iv) *If* h *is* $\mathbf{m}$-*measurable and if* $|h| \leq |g|$ $\mathbf{m}$-*a.e. in* E, *then* $\mathbf{m}_p^\bullet(h, E) \leq \mathbf{m}_p^\bullet(g, E)$.

(v) $\mathbf{m}_p^\bullet(g, E) = \mathbf{m}_p^\bullet(g, E \cap N(g))$. *(See Notation 1.1.15 of Chapter 1.)*

(vi) $\mathbf{m}_p^\bullet(g, E) = 0$ *if and only if* $\|\mathbf{m}\|(E \cap N(g)) = 0$.

Theorem 3.1.12. *Let* $g : T \to \mathbb{K}$ *or* $[-\infty, \infty]$ *be* $\mathbf{m}$-*measurable,* η *a positive real number and* $E \in \widetilde{\sigma(\mathcal{P})}$. *Let* $1 \leq p < \infty$. *Then:*

(i) **(The Tschebyscheff inequality).**

$$\|\mathbf{m}\|(\{t \in E : |g(t)| \geq \eta\}) \leq \frac{1}{\eta^p} \left(\mathbf{m}_p^\bullet(g, E)\right)^p.$$

(ii) *If* $\mathbf{m}_p^\bullet(g, E) < \infty$, *then* g *is finite* $\mathbf{m}$-*a.e. in* E. *Consequently, if* $\mathbf{m}_p^\bullet(g, T) < \infty$, *then* g *is finite* $\mathbf{m}$-*a.e. in* T.

Proof. (i) Let $F = \{t \in E : |g(t)| \geq \eta\}$. Then $F \in \widetilde{\sigma(\mathcal{P})}$ and by (i) and (iii) of Theorem 3.1.11 we have

$$\eta \left(\|\mathbf{m}\|(F)\right)^{\frac{1}{p}} \leq \left(\inf_{t \in F} |g(t)|\right) \cdot (\|\mathbf{m}\|(F))^{\frac{1}{p}} \leq \mathbf{m}_p^\bullet(g, F) \leq \mathbf{m}_p^\bullet(g, E)$$

and hence (i) holds.

(ii) Let $F = \{t \in E : |g(t)| = \infty\}$. Then $F \in \widetilde{\sigma(\mathcal{P})}$ and $F \subset F_n = \{t \in E : |g(t)| \geq n\} \in \widetilde{\sigma(\mathcal{P})}$, $n \in \mathbb{N}$. Now by (i), by Theorem 3.1.11(i) and by hypothesis, we have $\|\mathbf{m}\|(F) \leq \|\mathbf{m}\|(F_n) \leq (\frac{1}{n})^p \cdot \left(\mathbf{m}_p^\bullet(g, E)\right)^p \to 0$ as $n \to \infty$ and hence g is finite $\mathbf{m}$-a.e. in E . Since $\mathbf{m}_p^\bullet(g, T) = \mathbf{m}_p^\bullet(g, N(g))$, the last part of (ii) also holds. $\qquad\square$

Theorem 3.1.13. *Let f, g be $\mathbf{m}$-measurable on T with values in $\mathbb{K}$ and let $E \in \widetilde{\sigma(\mathcal{P})}$. Let $1 \leq p < \infty$ and if $1 < p < \infty$, let $\frac{1}{p} + \frac{1}{q} = 1$. Then:*

(i) $\mathbf{m}_p^\bullet(f + g, E) \leq \mathbf{m}_p^\bullet(f, E) + \mathbf{m}_p^\bullet(g, E)$.

(ii) $\mathbf{m}_p^\bullet(f + g, T) \leq \mathbf{m}_p^\bullet(f, T) + \mathbf{m}_p^\bullet(g, T)$.

(iii) *Let $1 < p < \infty$. Then $\mathbf{m}_1^\bullet(fg, T) \leq \mathbf{m}_p^\bullet(f, T) \cdot \mathbf{m}_q^\bullet(g, T)$. Consequently, if $f \in \mathcal{L}_p\mathcal{M}(\mathbf{m})$ and $g \in \mathcal{L}_q\mathcal{M}(\mathbf{m})$, then $fg \in \mathcal{L}_1\mathcal{M}(\mathbf{m})$. (The last part is improved in Theorem 3.5.1.)*

Proof. In the light of Definition 3.1.4, without loss of generality we shall assume that $E \in \sigma(\mathcal{P})$. Then by Theorem 3.1.3 we have

$$\mathbf{m}_p^\bullet(f + g, E) = \sup_{|x^*| \leq 1} \left(\int_E |f + g|^p dv(x^*\mathbf{m}) \right)^{\frac{1}{p}}$$

$$\leq \sup_{|x^*| \leq 1} \left(\int_E |f|^p dv(x^*\mathbf{m}) \right)^{\frac{1}{p}} + \sup_{|x^*| \leq 1} \left(\int_E |g|^p dv(x^*\mathbf{m}) \right)^{\frac{1}{p}}$$

$$= \mathbf{m}_p^\bullet(f, E) + \mathbf{m}_p^\bullet(g, E).$$

Consequently,

$$\mathbf{m}_p^\bullet(f + g, T) = \sup_{E \in \sigma(\mathcal{P})} \mathbf{m}_p^\bullet(f + g, E)$$

$$\leq \sup_{E \in \sigma(\mathcal{P})} \mathbf{m}_p^\bullet(f, E) + \sup_{E \in \sigma(\mathcal{P})} \mathbf{m}_p^\bullet(g, E)$$

$$= \mathbf{m}_p^\bullet(f, T) + \mathbf{m}_p^\bullet(g, T).$$

Hence (i) and (ii) hold.

(iii) By Theorem 3.1.3 and by Hölder's inequality we have

$$\mathbf{m}_1^\bullet(fg, T) = \sup_{|x^*| \leq 1, E \in \sigma(\mathcal{P})} \int_E |fg| dv(x^*\mathbf{m})$$

$$\leq \sup_{|x^*| \leq 1, E \in \sigma(\mathcal{P})} \left\{ \left(\int_E |f|^p dv(x^*\mathbf{m}) \right)^{\frac{1}{p}} \cdot \left(\int_E |g|^q dv(x^*\mathbf{m}) \right)^{\frac{1}{q}} \right\}$$

$$\leq \mathbf{m}_p^\bullet(f, T) \cdot \mathbf{m}_q^\bullet(g, T).$$

Consequently, if $f \in \mathcal{L}_p\mathcal{M}(\mathbf{m})$ and $g \in \mathcal{L}_q\mathcal{M}(\mathbf{m})$, then $fg \in \mathcal{L}_1\mathcal{M}(\mathbf{m})$. $\qquad\square$

Theorem 3.1.14. *If $1 \leq p < \infty$, then $(\mathcal{L}_p\mathcal{M}(\mathbf{m}), \mathbf{m}_p^\bullet(\cdot, T))$ and $(\mathcal{L}_p\mathcal{I}(\mathbf{m}), \mathbf{m}_p^\bullet(\cdot, T))$ are seminormed spaces.*

Proof. By Theorems 3.1.11(ii) and 3.1.13(ii), $(\mathcal{L}_p\mathcal{M}(\mathbf{m}), \mathbf{m}_p^\bullet(\cdot, T))$ is a seminormed space. Let $f, g \in \mathcal{L}_p\mathcal{I}(\mathbf{m})$ and let α be a scalar. Then $|f|^p \in \mathcal{I}(\mathbf{m})$ and $|g|^p \in \mathcal{I}(\mathbf{m})$. Since $|f + g|^p \leq 2^p \max(|f|^p, |g|^p) \leq 2^p(|f|^p + |g|^p)$, by (iv) and (vii) of Theorem

2.1.5 of Chapter 2 $|f + g|^p \in \mathcal{I}(\mathbf{m})$ and hence $f + g \in \mathcal{L}_p\mathcal{I}(\mathbf{m})$ by Theorem 3.1.10. Clearly, $|\alpha f|^p \in \mathcal{I}(\mathbf{m})$ and hence $\mathcal{L}_p\mathcal{I}(\mathbf{m})$ is a vector space. Then by Theorem 3.1.10 and by the fact that $(\mathcal{L}_p\mathcal{M}(\mathbf{m}), \mathbf{m}_p(\cdot, T))$ is a seminormed space, it follows that $(\mathcal{L}_p\mathcal{I}(\mathbf{m}), \mathbf{m}_p^\bullet(\cdot, T))$ is also a seminormed space.   □

Notation 3.1.15. By $\mathcal{L}_p\mathcal{M}(\mathbf{m})$ and $\mathcal{L}_p\mathcal{I}(\mathbf{m})$, $1 \leq p < \infty$, we mean the seminormed spaces $(\mathcal{L}_p\mathcal{M}(\mathbf{m}), \mathbf{m}_p^\bullet(\cdot, T))$ and $(\mathcal{L}_p\mathcal{I}(\mathbf{m}), \mathbf{m}_p^\bullet(\cdot, T))$, respectively.

In Chapter 4 we generalize Theorems 3.1.10 and 3.1.14 to lcHs-valued σ-additive $\mathbf{m}$ on $\mathcal{P}$. See Definition 4.3.7 and Theorem 4.3.8 of Chapter 4.

Similar to [Do1, Do2] we give the following concepts of convergence for $\mathbf{m}$-measurable functions.

Definition 3.1.16. Let $f, (f_n)_{n \in \mathbb{N}}, (f_\alpha)_{\alpha \in (D, \geq)}$ be $\mathbf{m}$-measurable scalar functions on T where $(D, \geq)$ is a directed set. Then:

(i) The sequence $(f_n)_1^\infty$ is said to converge in measure to f in T (resp. to be Cauchy in measure in T) with respect to $\mathbf{m}$ (when there is no ambiguity of the measure in question we drop $\mathbf{m}$) if, for each $\eta > 0$,

$$\lim_n \|\mathbf{m}\| (\{t \in T : |f_n(t) - f(t)| \geq \eta\}) = 0$$

(resp.

$$\lim_{n,r \to \infty} \|\mathbf{m}\| (\{t \in T : |f_n - f_r| \geq \eta\}) = 0).$$

Similarly we define convergence in measure to f in E and Cauchy in measure in E for $E \in \sigma(\mathcal{P})$. Similar definitions are given for the net $(f_\alpha)_{\alpha \in (D, \geq)}$.

(ii) The sequence $(f_n)_1^\infty$ is said to converge to f almost uniformly (resp. to be Cauchy for almost uniform convergence) in T (with respect to $\mathbf{m}$) if, given $\epsilon > 0$, there exists a set $E_\epsilon \in \sigma(\mathcal{P})$ with $\|\mathbf{m}\|(E_\epsilon) < \epsilon$ such that the sequence (f_n) converges to f uniformly in $T \backslash E_\epsilon$ (resp. is Cauchy for uniform convergence in $T \backslash E_\epsilon$). Similarly we define almost uniform convergence and Cauchy for almost uniform convergence in $E \in \sigma(\mathcal{P})$.

(iii) Let $1 \leq p < \infty$. If we further assume that $f, f_n, n \in \mathbb{N}$ (resp. $f_\alpha, \alpha \in (D, \geq)$), are in $\mathcal{L}_p\mathcal{M}(\mathbf{m})$, then the sequence (resp. the net) is said to converge to f in (mean^p) (with respect to $\mathbf{m}$) if $\lim_{n \to \infty} \mathbf{m}_p^\bullet(f_n - f, T) = 0$ (resp. $\lim_\alpha \mathbf{m}_p^\bullet(f_\alpha - f, T) = 0$). It is said to be Cauchy in (mean^p) if $\lim_{n,r \to \infty} \mathbf{m}_p^\bullet(f_n - f_r, T) = 0$ (resp. if $\lim_{\alpha, \beta \to \infty} \mathbf{m}_p^\bullet(f_\alpha - f_\beta, T) = 0$).

Definition 3.1.17. Two functions $f, g \in \mathcal{L}_p\mathcal{M}(\mathbf{m})$ are said to be $\mathbf{m}$-equivalent or simply, equivalent if $f = g$ $\mathbf{m}$-a.e. in T. In that case, we write $f \sim g$.

Theorem 3.1.18. *Let* $1 \leq p < \infty$. *Let* $f, g, (f_n)_{n \in \mathbb{N}}$ *and the net* $(f_\alpha)_{\alpha \in (D, \geq)}$ *be* $\mathbf{m}$-*measurable scalar functions on* T. *Then:*

(i) *'$\sim$' is an equivalence relation on* $\mathcal{L}_p\mathcal{M}(\mathbf{m})$ *and we denote* $\mathcal{L}_p\mathcal{M}(\mathbf{m})/\sim$ *by* $L_p\mathcal{M}(\mathbf{m})$.

(ii) *For $f, g \in \mathcal{L}_p\mathcal{M}(\mathbf{m})$, $f \sim g$ if and only if $\mathbf{m}_p^\bullet(f - g, T) = 0$.*

(iii) *For $f \in \mathcal{L}_p\mathcal{M}(\mathbf{m})$, $\mathbf{m}_p^\bullet(f, T) = 0$ if and only if $||\mathbf{m}||(N(f)) = 0$ and hence if and only if $f = 0$ $\mathbf{m}$-a.e. in T.*

(iv) *If (f_n) (resp. $(f_\alpha)_{\alpha \in (D, \geq)}$) converges in measure to f and g in T, then $f = g$ $\mathbf{m}$-a.e. in T. If (f_n) (resp. (f_α)) converges in measure to f, then (f_n) (resp. (f_α)) is Cauchy in measure.*

(v) *Let $(f_n)_1^\infty \subset \mathcal{L}_p\mathcal{M}(\mathbf{m})$ (resp. $(f_\alpha)_{\alpha \in (D, \geq)} \subset \mathcal{L}_p\mathcal{M}(\mathbf{m})$). If (f_n) (resp. $(f_\alpha)_{\alpha \in (D, \geq)}$) converges in $(mean^p)$ to f and g, then $f = g$ $\mathbf{m}$-a.e. in T. If (f_n) (resp. (f_α)) converges to f in $(mean^p)$, then it is Cauchy in $(mean^p)$.*

(vi) *Let $(f_n)_1^\infty \subset \mathcal{L}_p\mathcal{M}(\mathbf{m})$ (resp. $(f_\alpha)_{\alpha \in (D, \geq)} \subset \mathcal{L}_p\mathcal{M}(\mathbf{m})$). If (f_n) (resp. $(f_\alpha)_{\alpha \in (D, \geq)}$) converges to f in $(mean^p)$, then it converges to f in measure in T.*

(vii) *If $f_n \to f$ in measure in T (resp. if $(f_n)_1^\infty \subset \mathcal{L}_p\mathcal{M}(\mathbf{m})$ and if f_n converges to f in $(mean^p)$), then every subsequence of (f_n) also converges to f in measure in T (resp. in $(mean^p)$).*

(viii) **(Generalized Egoroff's theorem).** *Let $\mathcal{P}$ be a σ-ring $\mathcal{S}$ and let $(h_n)_1^\infty$ be $\mathbf{m}$-measurable scalar functions on T. Let $h : T \to \mathbb{K}$. If $h_n \to h$ $\mathbf{m}$-a.e. in T, then h is $\mathbf{m}$-measurable, $h_n \to h$ in measure in T and also almost uniformly in T.*

Proof. The easy verification of (i)–(iii) (using Theorem 3.1.11) is left to the reader. Since $||\mathbf{m}||$ is nonnegative, monotone and σ-subadditive, the classical proof holds here to prove (iv). For example, see the proof of Theorem 22C of [H].

(v) Since $\mathbf{m}_p^\bullet(\cdot, T)$ is subadditive by Theorem 3.1.13(ii), $\mathbf{m}_p^\bullet(f - g, T) \leq \mathbf{m}_p^\bullet(f - f_n, T) + \mathbf{m}_p^\bullet(f_n - g, T) \to 0$ as $n \to \infty$ (resp. $\leq \mathbf{m}_p^\bullet(f - f_\alpha, T) + \mathbf{m}_p^\bullet(f_\alpha - g, T) \to 0$ as $\alpha \to \infty$). Hence the first part holds by (ii). Similarly the second part is proved.

(vi) is immediate from the hypothesis and the Tschebyscheff inequality (see Theorem 3.1.12(i)) while (vii) follows from the respective definitions of convergence.

(viii) Clearly, h is $\mathbf{m}$-measurable. By hypothesis and by Proposition 1.1.18 of Chapter 1, there exists $N \in \sigma(\mathcal{P}) = \mathcal{S}$ with $||\mathbf{m}||(N) = 0$ such that $h_n(t) \to h(t)$ for $t \in T \backslash N$ and such that $h_n \chi_{T \backslash N}$, $n \in \mathbb{N}$, and $h \chi_{T \backslash N}$ are $\sigma(\mathcal{P})$-measurable. Then, given $\eta > 0$, let $E_n = \{t \in T \backslash N : |h_n(t) - h(t)| \geq \eta\}$. Then $(E_n)_1^\infty \subset \mathcal{S}$. As $h_n \to h$ pointwise in $T \backslash N$, $\lim_n E_n = \emptyset$. Then by the hypothesis that $\mathcal{S}$ is a σ-ring and by Proposition 1.1.5 of Chapter 1, $\lim_n ||\mathbf{m}||(E_n) = 0$. If $E_n(\eta) = \{t \in T : |h_n(t) - h(t)| \geq \eta\}$, then $E_n(\eta) \subset E_n \cup N$ and hence $\lim_n ||\mathbf{m}||(E_n(\eta)) = 0$. Therefore, $h_n \to h$ in measure in T. Since $||\mathbf{m}||$ is continuous and σ-subadditive on $\mathcal{S}$, the proof of Theorem 1 in §21 of [Be] can be adapted here to show that $h_n \to h$ almost uniformly in T. Hence (viii) holds. $\square$

As $||\mathbf{m}||$ is nonnegative, monotone and σ-subadditive on $\sigma(\mathcal{P})$, the following theorem can be proved by an argument similar to the numerical case (for example, see [H, pp. 92–94]). The proof is left to the reader.

Theorem 3.1.19. *Let $(f_n)_1^\infty$ be a sequence of $\mathbf{m}$-measurable scalar functions on T. If it is Cauchy in measure in T, then there exists a subsequence $(f_{n_k})_{k=1}^\infty$ of $(f_n)_1^\infty$ such that $(f_{n_k})_{k=1}^\infty$ is almost uniformly Cauchy in T, thus there exists an $\mathbf{m}$-measurable scalar function f on T such that $\lim_k f_{n_k}(t) = f(t)$ almost uniformly in T, and hence $f_n \to f$ in measure in T and $f_{n_k} \to f$ $\mathbf{m}$-a.e. in T.*

3.2 Completeness of $\mathcal{L}_p\mathcal{M}(\mathbf{m})$ and $\mathcal{L}_p\mathcal{I}(\mathbf{m}), 1 \leq p < \infty$, and of $\mathcal{L}_\infty(\mathbf{m})$

We first give a generalized Fatou's lemma in Theorem 3.2.1 and then use it to show that the seminormed space $\mathcal{L}_p\mathcal{M}(\mathbf{m})$ is complete for $1 \leq p < \infty$ (Theorem 3.2.3). Then, similar to the spaces given in Dobrakov [Do2] (for $p = 1$), we introduce the spaces $\mathcal{L}_p\mathcal{I}_s(\mathbf{m})$ and $\mathcal{L}_p(\mathbf{m})$ for $1 \leq p < \infty$ and show that $\mathcal{L}_p(\mathbf{m})$ is complete. By proving that $\mathcal{L}_p\mathcal{I}(\mathbf{m}) = \mathcal{L}_p(\mathbf{m})$, we deduce that $\mathcal{L}_p\mathcal{I}(\mathbf{m})$ is complete for $1 \leq p < \infty$. When $p = 1$ we also give an alternative proof for the completeness of $\mathcal{L}_1(\mathbf{m})$; this proof is based on inequality (3.1.3.1) and is similar to the proof of the corresponding part in Theorem 7 of [Do2]. The proof of Theorem 3.2.3 is analogous to that of the Riesz-Fischer theorem given in [Ru1] and can be adapted to give an alternative simpler proof of Theorem 7 of [Do2]. Finally we define $(\mathcal{L}_\infty(\mathbf{m}), ||\cdot||_\infty)$ and show that it is also a complete seminormed space.

Theorem 3.2.1. *Let f, f_n, $n \in \mathbb{N}$, be $\mathbf{m}$-measurable on T with values in $\mathbb{K}$ or in $[-\infty, \infty]$. Let $1 \leq p < \infty$. Then:*

(i) **(The Fatou property of $\mathbf{m}_p^\bullet(\cdot, E)$).** *If $|f_n| \nearrow |f|$ $\mathbf{m}$-a.e. in $E \in \widetilde{\sigma(\mathcal{P})}$, then $\mathbf{m}_p^\bullet(f, E) = \sup_n \mathbf{m}_p^\bullet(f_n, E) = \lim_n \mathbf{m}_p^\bullet(f_n, E)$.*

(ii) **(Generalized Fatou's lemma).** *$\mathbf{m}_p^\bullet(\liminf_n |f_n|, E) \leq \liminf_n \mathbf{m}_p^\bullet(f_n, E)$ for $E \in \widetilde{\sigma(\mathcal{P})}$.*

Proof. In the light of Definition 3.1.4, without loss of generality we shall assume that $E \in \sigma(\mathcal{P})$.

(i) By hypothesis and by Proposition 1.1.18 of Chapter 1, there exists $M \in \sigma(\mathcal{P})$ with $||\mathbf{m}||(M) = 0$ such that $|f_n(t)|\chi_{E\setminus M}(t) \nearrow |f(t)|\chi_{E\setminus M}(t)$ for $t \in T$ and $f_n\chi_{E\setminus M}$, $n \in \mathbb{N}$, and $f\chi_{E\setminus M}$ are $\sigma(\mathcal{P})$-measurable. Now, by Theorem 3.1.3 and by MCT for positive measures we have

$$\mathbf{m}_p^\bullet(f, E) = \mathbf{m}_p^\bullet(f, E\setminus M) = \sup_{|x^*| \leq 1} \left(\int_{E\setminus M} |f|^p dv(x^*\mathbf{m}) \right)^{\frac{1}{p}}$$

$$= \sup_{|x^*| \leq 1} \sup_n \left(\int_{E \backslash M} |f_n|^p dv(x^* \mathbf{m}) \right)^{\frac{1}{p}} = \sup_n \sup_{|x^*| \leq 1} \left(\int_{E \backslash M} |f_n|^p dv(x^* \mathbf{m}) \right)^{\frac{1}{p}}$$

$$= \sup_n \mathbf{m}_p^\bullet(f_n, E \backslash M) = \sup_n \mathbf{m}_p^\bullet(f_n, E) = \lim_n \mathbf{m}_p^\bullet(f_n, E)$$

as $\mathbf{m}_p^\bullet(f_n, E) \nearrow$.

(ii) Let $g_n = \inf_{k \geq n} |f_k|$. Then $g_n \nearrow$ and $\liminf_n |f_n| = \lim_n g_n = \sup_n g_n$. Therefore, by (i) we have

$$\mathbf{m}_p^\bullet(\liminf_n |f_n|, E) = \mathbf{m}_p^\bullet(\lim_n g_n, E) = \lim_n \mathbf{m}_p^\bullet(g_n, E) \leq \liminf_n \mathbf{m}_p^\bullet(f_n, E)$$

as $g_n = |g_n| \leq |f_n|$, $n \in \mathbb{N}$. $\square$

Remark 3.2.2. When $p = 1$, the generalized Fatou's lemma for operator-valued measures is stated without proof in Lemma 1 on p. 614 of [Do4].

Theorem 3.2.3. *Let $1 \leq p < \infty$. Then:*

(i) $\mathcal{L}_p\mathcal{M}(\mathbf{m})$ *is a complete seminormed space.*

(ii) *If $c_0 \not\subset X$, then $\mathcal{L}_p\mathcal{I}(\mathbf{m})$ is a complete seminormed space. (This result holds even if X is arbitrary. See Theorem 3.2.8.)*

Proof. (i) Let $(f_n)_1^\infty$ be Cauchy in (mean^p) in $\mathcal{L}_p\mathcal{M}(\mathbf{m})$. Then we can choose a subsequence (f_{n_k}) of (f_n) such that $\mathbf{m}_p^\bullet(f_{n_{k+1}} - f_{n_k}, T) < \frac{1}{2^k}$, $k \in \mathbb{N}$. Let $h_k = f_{n_k}$, $k \in \mathbb{N}$. Let

$$g_k = \sum_{i=1}^k |h_{i+1} - h_i| \text{ and } g = \sum_{i=1}^\infty |h_{i+1} - h_i|.$$

Then by Theorem 3.1.13(ii), $\mathbf{m}_p^\bullet(g_k, T) \leq \sum_{i=1}^\infty \mathbf{m}_p^\bullet(h_{i+1} - h_i, T) < 1$ for $k \in \mathbb{N}$ and $g_k \nearrow g$ in T. Hence by the Fatou property we have $\mathbf{m}_p^\bullet(g, T) = \sup_k \mathbf{m}_p^\bullet(g_k, T) \leq 1$. Then by Theorem 3.1.12(ii), g is finite $\mathbf{m}$-a.e. in T. By this fact and by Definition 2.1.1 of Chapter 2, we can choose $N \in \sigma(\mathcal{P})$ with $\|\mathbf{m}\|(N) = 0$ such that g is finite in $T \backslash N$ and such that $h_i \chi_{T \backslash N}$, $i \in \mathbb{N}$, are $\sigma(\mathcal{P})$-measurable. Then the series $\sum_{i=1}^\infty (h_{i+1} - h_i)$ is absolutely convergent in $T \backslash N$. Let $h_0 = 0$. Define

$$f(t) = \begin{cases} \sum_{k=0}^\infty (h_{k+1}(t) - h_k(t)) = \lim_k h_k(t), & \text{for } t \in T \backslash N \\ 0, & \text{for } t \in N. \end{cases}$$

Then f is finite in T, is $\sigma(\mathcal{P})$-measurable and is $\mathbf{m}$-a.e. pointwise limit of $(h_k)_1^\infty$ in T. Let $\epsilon > 0$. By hypothesis there exists n_0 such that $\mathbf{m}_p^\bullet(f_n - f_r, T) < \epsilon$ for $n, r \geq n_0$ and hence $\mathbf{m}_p^\bullet(h_k - h_\ell, T) < \epsilon$ for $n_k, n_\ell \geq n_0$. Let $F = (\bigcup_{k=1}^\infty N(h_k)) \cap (T \backslash N)$.

Then $F \in \sigma(\mathcal{P})$ and $N(f) \subset F$. Now by the generalized Fatou's lemma we have

$$\mathbf{m}_p^\bullet(f - h_k, T) = \mathbf{m}_p^\bullet(|f - h_k|, T) = \mathbf{m}_p^\bullet(|f - h_k|, T\backslash N)$$

$$= \mathbf{m}_p^\bullet\left(\lim_\ell |h_\ell - h_k|, T\backslash N\right) = \mathbf{m}_p^\bullet\left(\liminf_{\ell\to\infty}|h_\ell - h_k|, T\right)$$

$$\le \liminf_{\ell\to\infty}\mathbf{m}_p^\bullet(h_\ell - h_k, T) < \epsilon$$

for $n_k \ge n_0$. Then by the triangle inequality, $f \in \mathcal{L}_p\mathcal{M}(\mathbf{m})$ and moreover, $\lim_k \mathbf{m}_p^\bullet(f - h_k, T) = 0$. Consequently, $\mathbf{m}_p^\bullet(f - f_n, T) \le \mathbf{m}_p^\bullet(f - f_{n_k}, T) + \mathbf{m}_p^\bullet(f_{n_k} - f_n, T) \to 0$ as $n_k, n \to \infty$ and hence $\lim_n \mathbf{m}_p^\bullet(f_n - f, T) = 0$. Thus (i) holds.

(ii) As $c_0 \not\subset X$, by Theorem 3.1.10 we have $\mathcal{L}_p\mathcal{I}(\mathbf{m}) = \mathcal{L}_p\mathcal{M}(\mathbf{m})$ and hence the result holds by (i). $\square$

The following corollary is immediate from the last part of the proof of Theorem 3.2.3(i).

Corollary 3.2.4. *Let $1 \le p < \infty$. If $(f_n)_1^\infty \subset \mathcal{L}_p\mathcal{M}(\mathbf{m})$ is Cauchy in $(mean^p)$ and if there exist a scalar function f on T and a subsequence (f_{n_k}) of (f_n) such that $f_{n_k}(t) \to f(t)$ $\mathbf{m}$-a.e. in T, then f is $\mathbf{m}$-measurable, $f \in \mathcal{L}_p\mathcal{M}(\mathbf{m})$ and $\mathbf{m}_p^\bullet(f_n - f, T) \to 0$ as $n \to \infty$.*

Similar to the spaces in Dobrakov [Do2], we define two more spaces as below.

Definition 3.2.5. *Let $1 \le p < \infty$. Then we define $\mathcal{L}_p\mathcal{I}_s(\mathbf{m}) =$ the closure of $\mathcal{I}_s$ in $\mathcal{L}_p\mathcal{M}(\mathbf{m})$ with respect to $\mathbf{m}_p^\bullet(\cdot, T)$; and*

$$\mathcal{L}_p(\mathbf{m}) = \{f \in \mathcal{L}_p\mathcal{M}(\mathbf{m}) : \mathbf{m}_p^\bullet(f, \cdot) \text{ is continuous on } \sigma(\mathcal{P})\}.$$

If we consider equivalence classes of functions, then we shall write L_p instead of $\mathcal{L}_p$.

Clearly, $\mathcal{L}_p\mathcal{I}_s(\mathbf{m})$ is a linear subspace of $\mathcal{L}_p\mathcal{M}(\mathbf{m})$. If $f, g \in \mathcal{L}_p(\mathbf{m})$, then by Theorem 3.1.13(i), $\mathbf{m}_p^\bullet(f + g, \cdot) \le \mathbf{m}_p^\bullet(f, \cdot) + \mathbf{m}_p^\bullet(g, \cdot)$ and hence $f + g \in \mathcal{L}_p(\mathbf{m})$. Obviously, $\alpha f \in \mathcal{L}_p(\mathbf{m})$ for $f \in \mathcal{L}_p(\mathbf{m})$ and for any scalar α. Hence $\mathcal{L}_p(\mathbf{m})$ is also a linear subspace of $\mathcal{L}_p\mathcal{M}(\mathbf{m})$.

Notation 3.2.6. By $\mathcal{L}_p\mathcal{I}_s(\mathbf{m})$ and $\mathcal{L}_p(\mathbf{m})$ we mean the seminormed spaces

$$\left(\mathcal{L}_p\mathcal{I}_s(\mathbf{m}), \mathbf{m}_p^\bullet(\cdot, T)\right) \qquad \text{and} \qquad \left(\mathcal{L}_p(\mathbf{m}), \mathbf{m}_p^\bullet(\cdot, T)\right),$$

respectively.

Theorem 3.2.7. *Let $1 \le p < \infty$. If $f : T \to \mathbb{K}$ is $\mathbf{m}$-measurable and $\mathbf{m}_p^\bullet(f, \cdot)$ is continuous on $\sigma(\mathcal{P})$, then $f \in \mathcal{L}_p\mathcal{M}(\mathbf{m})$ and hence $f \in \mathcal{L}_p(\mathbf{m})$. Moreover, $\mathcal{L}_p\mathcal{I}(\mathbf{m}) = \mathcal{L}_p(\mathbf{m})$.*

Proof. Let $\mathbf{m}_p^\bullet(f, \cdot)$ be continuous on $\sigma(\mathcal{P})$. By Proposition 1.1.18 of Chapter 1 there exists a sequence $(s_n) \subset \mathcal{I}_s$ such that $0 \le s_n(t) \nearrow |f(t)|^p$ $\mathbf{m}$-a.e. in T.

Let $\gamma_n(\cdot) = \int_{(\cdot)} s_n d\mathbf{m}, n \in \mathbb{N}$. Then $||\gamma_n||(\cdot) = \sup_{|x^*|\leq 1} \left(\int_{(\cdot)} s_n dv(x^*\mathbf{m})\right) \leq$ $\sup_{|x^*|\leq 1} \left(\int_{(\cdot)} |f|^p dv(x^*\mathbf{m})\right) = (\mathbf{m}_p^\bullet(f, \cdot))^p$ by Theorem 2.1.5(iii)(a) of Chapter 2 and by Theorem 3.1.3 above and hence $||\gamma_n||, n \in \mathbb{N}$, are uniformly continuous on $\sigma(\mathcal{P})$. Then, by Proposition 1.1.7 of Chapter 1, $\gamma_n, n \in \mathbb{N}$, are uniformly σ-additive on $\sigma(\mathcal{P})$. Consequently, by Theorem 2.2.4 of Chapter 2, $|f|^p$ is $\mathbf{m}$-integrable in T. Then by Theorem 3.1.6, $\mathbf{m}_p^\bullet(f, T) < \infty$ and hence $f \in \mathcal{L}_p\mathcal{M}(\mathbf{m})$. Then by hypothesis and by Definition 3.2.5, $f \in \mathcal{L}_p(\mathbf{m})$. Moreover, as $|f|^p \in \mathcal{I}(\mathbf{m})$, by Theorem 3.1.10 $f \in \mathcal{L}_p\mathcal{I}(\mathbf{m})$. This also proves that $\mathcal{L}_p(\mathbf{m}) \subset \mathcal{L}_p\mathcal{I}(\mathbf{m})$.

Conversely, let $f \in \mathcal{L}_p\mathcal{I}(\mathbf{m})(= \mathcal{I}_p(\mathbf{m}))$. Then $|f|^p$ is $\mathbf{m}$-integrable in T and hence by Theorem 3.1.6, not only $f \in \mathcal{L}_p\mathcal{M}(\mathbf{m})$ but also $\mathbf{m}_p^\bullet(f, \cdot)$ is continuous on $\sigma(\mathcal{P})$. Hence $f \in \mathcal{L}_p(\mathbf{m})$. Thus $\mathcal{L}_p\mathcal{I}(\mathbf{m}) = \mathcal{L}_p(\mathbf{m})$. □

Theorem 3.2.8. (Analogue of the Riesz-Fischer theorem) *Let* $1 \leq p < \infty$. *Then* $\mathcal{L}_p(\mathbf{m})$ *is closed in* $\mathcal{L}_p\mathcal{M}(\mathbf{m})$ *and hence* $\mathcal{L}_p\mathcal{I}(\mathbf{m})$ $(= \mathcal{L}_p(\mathbf{m}))$ *is a complete seminormed space. If* $\mathcal{L}_p(\sigma(\mathcal{P}), \mathbf{m}) = \{f \in \mathcal{L}_p(\mathbf{m}) : f \sigma(\mathcal{P})$-*measurable*$\}$, *then* $\mathcal{L}_p(\sigma(\mathcal{P}), \mathbf{m})$ *is closed in* $\mathcal{L}_p(\mathbf{m})$ *and hence is a complete seminormed space. Consequently,* $L_p(\mathbf{m})$ *and* $L_p(\sigma(\mathcal{P}), \mathbf{m})$ *are Banach spaces and are treated as function spaces in which two functions which are equal* $\mathbf{m}$-*a.e. in* T *are identified.*

Proof. Let $(f_n)_1^\infty \subset \mathcal{L}_p(\mathbf{m})$ and let $\lim_n \mathbf{m}_p^\bullet(f_n - f, T) = 0$ for some $f \in \mathcal{L}_p\mathcal{M}(\mathbf{m})$. Let $\epsilon > 0$. Then there exists n_0 such that $\mathbf{m}_p^\bullet(f_n - f, T) < \frac{\epsilon}{2}$ for $n \geq n_0$. Let $(E_k)_1^\infty \searrow \emptyset$ in $\sigma(\mathcal{P})$. As $f_{n_0} \in \mathcal{L}_p(\mathbf{m})$, by Definition 3.2.5 $\mathbf{m}_p^\bullet(f_{n_0}, \cdot)$ is continuous on $\sigma(\mathcal{P})$ and hence there exists k_0 such that $\mathbf{m}_p^\bullet(f_{n_0}, E_k) < \frac{\epsilon}{2}$ for $k \geq k_0$. Consequently,

$$\mathbf{m}_p^\bullet(f, E_k) \leq \mathbf{m}_p^\bullet(f - f_{n_0}, E_k) + \mathbf{m}_p^\bullet(f_{n_0}, E_k) \leq \mathbf{m}_p^\bullet(f - f_{n_0}, T) + \mathbf{m}_p^\bullet(f_{n_0}, E_k) < \epsilon$$

for $k \geq k_0$. Hence $f \in \mathcal{L}_p(\mathbf{m})$ and thus $\mathcal{L}_p(\mathbf{m})$ is closed in $\mathcal{L}_p\mathcal{M}(\mathbf{m})$. Now by Theorems 3.2.3(i) and 3.2.7, $\mathcal{L}_p\mathcal{I}(\mathbf{m})$ is a complete seminormed space.

Clearly, $\mathcal{L}_p(\sigma(\mathcal{P}), \mathbf{m})$ is a seminormed space. Now, let $(f_n)_1^\infty \subset \mathcal{L}_p(\sigma(\mathcal{P}), \mathbf{m})$ be Cauchy in (mean^p). As $\mathcal{L}_p(\sigma(\mathcal{P}), \mathbf{m}) \subset \mathcal{L}_p(\mathbf{m})$ and as $\mathcal{L}_p(\mathbf{m})$ is complete, there exists $f \in \mathcal{L}_p(\mathbf{m})$ such that $\mathbf{m}_p^\bullet(f_n - f, T) \to 0$ as $n \to \infty$. As f is $\mathbf{m}$-measurable, there exists $N \in \sigma(\mathcal{P})$ with $||\mathbf{m}||(N) = 0$ such that $f\chi_{T\backslash N}$ is $\sigma(\mathcal{P})$-measurable and clearly $f\chi_{T\backslash N} \in \mathcal{L}_p(\sigma(\mathcal{P}), \mathbf{m})$ and $\lim_n (\mathbf{m})_p^\bullet(f_n - f\chi_{T\backslash N}, T) = 0$. Hence $\mathcal{L}_p(\sigma(\mathcal{P}), \mathbf{m})$ is closed in $\mathcal{L}_p(\mathbf{m})$ and thus is a complete seminormed space. Then the last part is immediate from the previous parts. □

For $p = 1$, we give below an alternative proof similar to that in the proof of Theorem 7 of [Do2].

By Theorem 3.1.10, $\mathcal{L}_1\mathcal{I}(\mathbf{m}) = \mathcal{I}(\mathbf{m}) \subset \mathcal{L}_1\mathcal{M}(\mathbf{m})$. Let $(f_n)_1^\infty \subset \mathcal{I}(\mathbf{m})$ and let $f \in \mathcal{L}_1\mathcal{M}(\mathbf{m})$ be such that $\lim_n \mathbf{m}_1^\bullet(f_n - f, T) = 0$. We shall show that $f \in \mathcal{I}(\mathbf{m})$. By the proof of Theorem 3.2.3(i) there exist a subsequence $(f_{n_k})_{k=1}^\infty$ of $(f_n)_1^\infty$ and a function $g \in \mathcal{L}_1\mathcal{M}(\mathbf{m})$ such that $f_{n_k} \to g$ $\mathbf{m}$-a.e. in T as $k \to \infty$ and

$\lim_k \mathbf{m}_1^\bullet(f_{n_k} - g, T) = 0$. Then, by Theorem 3.1.18(v), $f = g$ **m**-a.e. in T. Moreover, by inequality (3.1.3.1),

$$\left| \int_E f_{n_k} d\mathbf{m} - \int_E f_{n_\ell} d\mathbf{m} \right| \leq \mathbf{m}_1^\bullet(f_{n_k} - f_{n_\ell}, T) \to 0 \text{ as } k, \ell \to \infty$$

for $E \in \sigma(\mathcal{P})$. Thus $(\int_E f_{n_k} d\mathbf{m})$ is Cauchy and hence converges to a vector x_E (say) in X for each $E \in \sigma(\mathcal{P})$. Consequently, by Theorem 2.2.8 of Chapter 2, f is **m**-integrable in T and hence $f \in \mathcal{I}(\mathbf{m})$. Thus $\mathcal{L}_1\mathcal{I}(\mathbf{m})$ is closed in $\mathcal{L}_1\mathcal{M}(\mathbf{m})$ and hence $\mathcal{L}_1\mathcal{I}(\mathbf{m})$ is complete by Theorems 3.2.3(i).

Remark 3.2.9. When $\mathcal{P}$ is a σ-algebra Σ and X is a real Banach space, the completeness of $\mathcal{L}_1(\sigma(\mathcal{P}), \mathbf{m})$ follows from Theorems IV.7.1 and IV.4.1 of [KK], where the proof is indirect and complicated. A somewhat direct proof of the completeness of $\mathcal{L}_1(\sigma(\mathcal{P}), \mathbf{m})$ when $\mathcal{P}$ is a σ-algebra Σ and when $\mathbf{m}$ has values in a complex Fréchet space is given in [FNR]. Under the σ-algebra hypothesis for a Banach space-valued σ-additive $\mathbf{m}$, [Ri5] gives a nice proof of the completeness of $\mathcal{L}_1(\sigma(\mathcal{P}), \mathbf{m})$ where the Rybakov theorem plays a vital role. In our case we cannot use his argument as the Rybakov theorem is not available for σ-additive vector measures defined on δ-rings. The completeness of $\mathcal{L}_1(\sigma(\mathcal{P}), \mathbf{m})$ is treated in Theorem 4.7 of [MN2] for the δ-ring case, and his proof is based on Lemma 3.4 and Theorem 3.5 of [L2]. As noted in Remark 2.1.12 of Chapter 2, the proofs of these results as given in [L2] are incorrect. However, we have provided a correct proof of these results in the said remark and hence Theorem 4.7 of [MN2] is restored. The proof of the first part of Theorem 3.2.3 for $p = 1$ is similar to that of Lemma 3.13(a) of [MN2]. Now let $\mathbf{m} : \mathcal{P} \to X$ be σ-additive where X is an lcHs. When X is a metrizable lcHs, Theorem 3.2.3(i) is generalized in Theorem 4.4.2(i) of Chapter 4 ; the first part of Definition 3.2.5 is generalized in Definition 4.5.4 of Chapter 4 for a quasicomplete (resp. sequentially complete) lcHs-valued $\mathbf{m}$ on $\mathcal{P}$ and for such $\mathbf{m}$, Theorem 3.3.6 is generalized in Theorem 4.5.5 (resp. in Theorem 4.5.6) of Chapter 4. When $\mathbf{m}$ has values in a Fréchet space, Theorem 3.2.8 is generalized in Theorem 4.4.8 of Chapter 4. (Also see Remark 4.4.10 of Chapter 4.) Also see [SP1].

Definition 3.2.10. Let $\mathbf{m} : \mathcal{P} \to X$ be σ-additive. We define $\mathcal{L}_\infty(\mathbf{m}) = \{f : T \to \mathbb{K}, f \text{ **m**-essentially bounded in } T\}$ (see Definition 1.1.16 of Chapter 1) and $||f||_\infty = \text{ess sup}_{t \in T} |f(t)|$ for $f \in \mathcal{L}_\infty(\mathbf{m})$. We define $\mathcal{L}_\infty(\sigma(\mathcal{P}), \mathbf{m}) = \{f \in \mathcal{L}_\infty(\mathbf{m}) : f \ \sigma(\mathcal{P})\text{-measurable}\}$ with the seminorm $|| \cdot ||_\infty$.

Theorem 3.2.11. *Let* $\mathbf{m} : \mathcal{P} \to X$ *be σ-additive. Then* $(\mathcal{L}_\infty(\mathbf{m}), || \cdot ||_\infty)$ *or simply,* $\mathcal{L}_\infty(\mathbf{m})$ *is a complete seminormed space. Consequently,* $L_\infty(\mathbf{m}) = \mathcal{L}_\infty(\mathbf{m})/ \sim$ *is a Banach space. (Also see Theorem 4.5.11.)*

Proof. Since $||\mathbf{m}||$ is a σ-subadditive submeasure on $\sigma(\mathcal{P})$, the last part of the proof of Theorem 3.11 of [Ru1] holds here and hence the theorem. □

Remark 3.2.12. Relations between the spaces $\mathcal{L}_p(\mathbf{m})$, $1 \leq p \leq \infty$, are studied in Section 3.5 below.

3.3 Characterizations of $\mathcal{L}_p\mathcal{I}(\mathbf{m})$, $1 \le p < \infty$

Let $1 \le p < \infty$. If $(f_n)_1^\infty \subset \mathcal{L}_p(\mathbf{m})$, $f_0 : T \to \mathbb{K}$ and if $f_n \to f_0$ $\mathbf{m}$-a.e. in T, then Theorem 3.3.1 gives a characterization for (f_n) to converge to f_0 in (mean^p). Theorem 3.3.2 gives the analogue of an improved version of the Vitali convergence theorem in Halmos ([H], Theorem 26C) for $\mathcal{L}_p(\mathbf{m})$. From Theorems 3.3.1 and 3.3.2 are deduced a.e. convergence (mean^p)-versions as well as convergence in measure (mean^p)-versions of LDCT and LBCT for $\mathcal{L}_p(\mathbf{m})$. Using the present a.e. convergence (mean^p)-version of LDCT it is shown that an $\mathbf{m}$-measurable scalar function f on T belongs to $\mathcal{L}_p\mathcal{I}(\mathbf{m})$ if and only if f belongs to $\mathcal{L}_p\mathcal{I}_s(\mathbf{m})$ (see Definition 3.2.5). In other words, by this result and by Theorems 3.1.10 and 3.2.7 it follows that $\mathcal{I}_p(\mathbf{m}) = \mathcal{L}_p\mathcal{I}(\mathbf{m}) = \mathcal{L}_p\mathcal{I}_s(\mathbf{m}) = \mathcal{L}_p(\mathbf{m})$ for $p \in [1, \infty)$ and this is the main result of this section. As in the cases of the abstract Lebesgue and Bochner integrals (see [DS1],[H]), it is also shown that the class of $\mathbf{m}$-integrable functions f can be defined in terms of convergence in measure to f of a sequence $(s_n) \subset \mathcal{I}_s$ which is Cauchy in (mean). This result does not hold for the general Dobrakov integral. See Remark 3.3.15.

To obtain the analogue of an improved version of the Vitali convergence theorem in Halmos ([H], Theorem 26C) for $\mathcal{L}_p(\mathbf{m})$ we prove the following

Theorem 3.3.1. *Let $1 \le p < \infty$. Let $(f_n)_1^\infty \subset \mathcal{L}_p(\mathbf{m})$ and let $f_0 : T \to \mathbb{K}$. Suppose $(f_n)_1^\infty$ converges to f_0 $\mathbf{m}$-a.e. in T. Then $\mathbf{m}_p^\bullet(f_n - f_0, T) \to 0$ as $n \to \infty$ if and only if $\mathbf{m}_p^\bullet(f_n, \cdot)$, $n \in \mathbb{N}$, are uniformly continuous on $\sigma(\mathcal{P})$. In that case, $f_0 \in \mathcal{L}_p(\mathbf{m})$. Moreover, in such case, for $p = 1$,*

$$\int_E f_0 d\mathbf{m} = \lim_n \int_E f_n d\mathbf{m}, \quad E \in \sigma(\mathcal{P})$$

the limit being uniform with respect to $E \in \sigma(\mathcal{P})$.

Proof. Clearly f_0 is $\mathbf{m}$-measurable. If $\mathbf{m}_p^\bullet(f_n - f_0, T) \to 0$, then, given $\epsilon > 0$, there exists n_0 such that $\mathbf{m}_p^\bullet(f_n - f_\ell, T) < \frac{\epsilon}{2}$ for $n, \ell \ge n_0$. Let $E_k \searrow \emptyset$ in $\sigma(\mathcal{P})$. Then by Definition 3.2.5, there exists k_0 such that $\mathbf{m}_p^\bullet(f_n, E_k) < \frac{\epsilon}{2}$ for $n = 1, 2, \ldots, n_0$ and for $k \ge k_0$. Consequently, $\mathbf{m}_p^\bullet(f_n, E_k) \le \mathbf{m}_p^\bullet(f_n - f_{n_0}, E_k) + \mathbf{m}_p^\bullet(f_{n_0}, E_k) \le \mathbf{m}_p^\bullet(f_n - f_{n_0}, T) + \mathbf{m}_p^\bullet(f_{n_0}, E_k) < \epsilon$ for $n \ge n_0$ and for $k \ge k_0$. Hence $\mathbf{m}_p^\bullet(f_n, \cdot)$, $n \in \mathbb{N}$, are uniformly continuous on $\sigma(\mathcal{P})$.

Conversely, let $\mathbf{m}_p^\bullet(f_n, \cdot)$, $n \in \mathbb{N}$ be uniformly continuous on $\sigma(\mathcal{P})$. Let $E_k \searrow \emptyset$ in $\sigma(\mathcal{P})$ and let $\epsilon > 0$. By hypothesis there exists k_0 such that $\mathbf{m}_p^\bullet(f_n, E_k) < \epsilon$ for all $n \in \mathbb{N}$ and for $k \ge k_0$. Consequently, by the generalized Fatou's lemma (Theorem 3.2.1(ii)), $\mathbf{m}_p^\bullet(f_0, E_k) = \mathbf{m}_p^\bullet(\liminf_n |f_n|, E_k) \le \liminf_{n \to \infty} \mathbf{m}_p^\bullet(f_n, E_k) < \epsilon$ for $k \ge k_0$. Hence $\mathbf{m}_p^\bullet(f_0, \cdot)$ is continuous on $\sigma(\mathcal{P})$ and consequently, by Theorem 3.2.7, $f_0 \in \mathcal{L}_p(\mathbf{m})$.

By Theorems 3.1.10 and 3.2.7, $|f_n|^p$, $n \in \mathbb{N}$, and $|f_0|^p$ are $\mathbf{m}$-integrable in T. Now, let $\boldsymbol{\gamma}_n(\cdot) = \int_{(\cdot)} |f_n|^p d\mathbf{m}$ for $n \in \mathbb{N} \cup \{0\}$. By Theorem 3.1.6, $(\|\boldsymbol{\gamma}_n\|(\cdot))^{\frac{1}{p}} =$

$\mathbf{m}_p^\bullet(f_n, \cdot)$, $n \in \mathbb{N} \cup \{0\}$. Then by hypothesis and by the fact that $f_0 \in \mathcal{L}_p(\mathbf{m})$, $||\boldsymbol{\gamma}_n||$, $n \in \mathbb{N} \cup \{0\}$, are uniformly continuous on $\sigma(\mathcal{P})$. Therefore, by Proposition 1.1.7 of Chapter 1, $\boldsymbol{\gamma}_n$, $n \in \mathbb{N} \cup \{0\}$, are uniformly σ-additive on $\sigma(\mathcal{P})$. Then by Proposition 1.1.13 of Chapter 1 there exists a σ-additive control measure μ : $\sigma(\mathcal{P}) \to [0, \infty)$ such that $||\boldsymbol{\gamma}_n||, n \in \mathbb{N} \cup \{0\}$, are uniformly μ-continuous. Thus, given $\epsilon > 0$, there exists $\delta > 0$ such that $||\boldsymbol{\gamma_n}||(E) < (\frac{\epsilon}{3})^p$, for $n \in \mathbb{N} \cup \{0\}$, whenever $\mu(E) < \delta$. By hypothesis and by Proposition 1.1.18 of Chapter I, there exists $M \in \sigma(\mathcal{P})$ with $||\mathbf{m}||(M) = 0$ such that $f_n(t) \to f_0(t)$ for $t \in T \backslash M$ and such that $f_n \chi_{T \backslash M}$, $n \in \mathbb{N}$, and $f_0 \chi_{T \backslash M}$ are $\sigma(\mathcal{P})$-measurable. Let $F = (\bigcup_{n=1}^\infty N(f_n)) \cap (T \backslash M)$. Clearly, $F \in \sigma(\mathcal{P})$. Then by the Egoroff-Lusin theorem there exist a set $N \in \sigma(\mathcal{P}) \cap F$ with $\mu(N) = 0$ and a sequence $(F_k) \subset \mathcal{P}$ with $F_k \nearrow F \backslash N$ such that $f_n \to f_0$ uniformly in each F_k. As $F \backslash N \backslash F_k \searrow \emptyset$, there exists k_0 such that $\mu(F \backslash N \backslash F_{k_0}) < \delta$. Then $||\boldsymbol{\gamma}_n||(F \backslash N \backslash F_{k_0}) < (\frac{\epsilon}{3})^p$ for $n \in \mathbb{N} \cup \{0\}$. Choose n_0 such that $||f_n - f||_{F_{k_0}} \cdot (||\mathbf{m}||(F_{k_0}))^{\frac{1}{p}} < \frac{\epsilon}{3}$ for $n \geq n_0$. Note that $||\boldsymbol{\gamma}_n||(M \cup N) = 0$ for $n \in \mathbb{N} \cup \{0\}$. Then by (i) and (iii) of Theorem 3.1.11, by (i) of Theorem 3.1.13 and by Theorem 3.1.6 we have $\mathbf{m}_p^\bullet(f_n - f_0, T) \leq \mathbf{m}_p^\bullet(f_n - f_0, F_{k_0}) + \mathbf{m}_p^\bullet(f_n - f_0, F \backslash N \backslash F_{k_0}) + \mathbf{m}_p^\bullet(f_n, N) + \mathbf{m}_p^\bullet(f_0, N) + \mathbf{m}_p^\bullet(f_0, M) + \mathbf{m}_p^\bullet(f_n, M) \leq ||f_n - f_0||_{F_{k_0}} \cdot (||\mathbf{m}||(F_{k_0}))^{\frac{1}{p}} + (||\boldsymbol{\gamma}_n||(F \backslash N \backslash F_{k_0}))^{\frac{1}{p}} + (||\boldsymbol{\gamma}_0||(F \backslash N \backslash F_{k_0}))^{\frac{1}{p}} + (||\boldsymbol{\gamma}_n||(N))^{\frac{1}{p}} + (||\boldsymbol{\gamma}_0||(N))^{\frac{1}{p}} + (||\boldsymbol{\gamma}_n||(M))^{\frac{1}{p}} + (||\boldsymbol{\gamma}_0||(M))^{\frac{1}{p}} < \epsilon$ for $n \geq n_0$. Hence $\lim_n \mathbf{m}_p^\bullet(f_n - f_0, T) = 0$. The last part is immediate from inequality (3.1.3.1) since $\mathcal{L}_1(\mathbf{m}) = \mathcal{L}_1 \mathcal{I}(\mathbf{m}) = \mathcal{I}_1(\mathbf{m}) = \mathcal{I}(\mathbf{m})$ by Theorem 2.1.5(vii) of Chapter 2 and by Theorems 3.2.7 and 3.1.10.   □

Theorem 3.3.2. (Analogue of an improved version of the Vitali convergence theorem in Halmos ([H], Theorem 26C) for $\mathcal{L}_p(\mathbf{m})$) *Let $1 \leq p < \infty$ and let $f_0 : T \to \mathbb{K}$ be $\mathbf{m}$-measurable. A sequence $(f_n)_1^\infty \subset \mathcal{L}_p(\mathbf{m})$ converges to f_0 in $(mean^p)$ if and only if (f_n) converges to f_0 in measure in T, $f_0 \in \mathcal{L}_p \mathcal{M}(\mathbf{m})$ and $\mathbf{m}_p^\bullet(f_n, \cdot), n \in \mathbb{N}$, are uniformly continuous on $\sigma(\mathcal{P})$. In that case, $f_0 \in \mathcal{L}_p(\mathbf{m})$. Moreover, in such case, for $p = 1$, the result in the last part of Theorem 3.3.1 holds here verbatim.*

Proof. Let $\lim_n \mathbf{m}_p^\bullet(f_n - f_0, T) = 0$. Then by the triangle inequality $f_0 \in \mathcal{L}_p \mathcal{M}(\mathbf{m})$ and then by Theorem 3.1.18(vi), $f_n \to f_0$ in measure in T. Moreover, by the necessity part of Theorem 3.3.1, $\mathbf{m}_p^\bullet(f_n, \cdot), n \in \mathbb{N}$, are uniformly continuous on $\sigma(\mathcal{P})$.

Conversely, let $f_n \to f_0$ in measure in T, $f_0 \in \mathcal{L}_p \mathcal{M}(\mathbf{m})$ and $\mathbf{m}_p^\bullet(f_n, \cdot), n \in \mathbb{N}$, be uniformly continuous on $\sigma(\mathcal{P})$. If possible, let $\mathbf{m}_p^\bullet(f_n - f_0, T) \not\to 0$ as $n \to \infty$. Then there would exist an $\epsilon > 0$ and a subsequence (g_n) of (f_n) such that $\mathbf{m}_p^\bullet(g_n - f_0, T) > \epsilon$ for $n \in \mathbb{N}$. On the other hand, by Theorem 3.1.18(iv), by the first part of Theorem 3.1.18(vii) and by Theorem 3.1.19, there would exist a subsequence (g_{n_k}) of (g_n) such that $g_{n_k} \to f_0$ $\mathbf{m}$-a.e. in T. Then, as $\mathbf{m}_p^\bullet(g_{n_k}, \cdot), k \in \mathbb{N}$, are uniformly continuous on $\sigma(\mathcal{P})$, by Theorem 3.3.1 there would exist k_0 such that $\mathbf{m}_p^\bullet(g_{n_k} - f_0, T) < \epsilon$ for all $k \geq k_0$. This is a contradiction and hence $\lim_n \mathbf{m}_p^\bullet(f_n - f_0, T) = 0$.   □

Remark 3.3.3. In the light of Theorem 3.3.2, Theorem 26C of [H] can be improved as below.

Let Σ be a σ-ring of subsets of T and let $\mu : \Sigma \to [0, \infty]$ be a positive measure. Let $\mathcal{P} = \{E \in \Sigma : \mu(E) < \infty\}$. Then $\mathcal{P}$ is a δ-ring. If f is Σ-measurable and integrable with respect to μ, then $\int_T f d\mu = \int_T f \chi_{N(f)} d\mu$, $N(f)$ is of σ-finite measure and $f\chi_{N(f)}$ is $\sigma(\mathcal{P})$-measurable and integrable with respect to $\mu|_{\mathcal{P}}$. Consequently, by Theorem 3.3.2, **a sequence (f_n) of Σ-measurable scalar functions on T which are integrable with respect to μ converges in the mean to the integrable function f if and only if (f_n) converges in measure to f in T and $(\nu_n(\cdot))_1^\infty$ are equicontinuous from above at 0 (in the sense of p. 108 of Halmos [H])**, where $\nu_n(\cdot) = \int_{(\cdot)} |f_n| d\mu$. (Note that in our terminology, $\nu_n(\cdot)$ is the same as $\mu_1^\bullet(f_n, \cdot)$ and equicontinuity of $(\nu_n)_1^\infty$ from above at 0 is the same as uniform continuity of $(\mu_1^\bullet(f_n, \cdot))_1^\infty$ on $\sigma(\mathcal{P})$.) Clearly, this is an improved version of Theorem 26C of Halmos [H].

The following versions of LDCT and LBCT are immediate from Theorems 3.3.1 and 3.3.2.

Theorem 3.3.4. (a.e. convergence and convergence in measure versions of LDCT and LBCT for $\mathcal{L}_p(\mathbf{m})$) *Let $1 \le p < \infty$. Let f_n, $n \in \mathbb{N}$, be $\mathbf{m}$-measurable scalar functions on T and let $g \in \mathcal{L}_p(\mathbf{m})$ such that $|f_n(t)| \le |g(t)|$ $\mathbf{m}$-a.e. in T (resp. let $\mathcal{P}$ be a σ-ring $\mathcal{S}$ and let M be a finite constant such that $|f_n(t)| \le M$ $\mathbf{m}$-a.e. in T) for all n. If $f_n(t) \to f(t)$ $\mathbf{m}$-a.e. in T where f is a scalar function on T or if f is an $\mathbf{m}$-measurable scalar function on T with $f \in \mathcal{L}_p\mathcal{M}(\mathbf{m})$ and if $f_n \to f$ in measure in T, then f, f_n, $n \in \mathbb{N}$, belong to $\mathcal{L}_p(\mathbf{m})$ and $\lim_n \mathbf{m}_p^\bullet(f_n - f, T) = 0$. When $p = 1$,*

$$\lim_n \int_E f_n d\mathbf{m} = \int_E f d\mathbf{m}, \quad E \in \sigma(\mathcal{P}) \ (resp.\, E \in \mathcal{S})$$

where the limit is uniform with respect to $E \in \sigma(\mathcal{P})$ (resp. $E \in \mathcal{S}$).

Proof. As $g \in \mathcal{L}_p(\mathbf{m})$, $\mathbf{m}_p^\bullet(g, \cdot)$ is continuous on $\sigma(\mathcal{P})$. By hypothesis and by Theorem 3.1.11(i), $\mathbf{m}_p^\bullet(f_n, \cdot) \le \mathbf{m}_p^\bullet(g, \cdot)$ for all n and hence $\mathbf{m}_p^\bullet(f_n, \cdot), n \in \mathbb{N}$, are uniformly continuous on $\sigma(\mathcal{P})$. If $f_n \to f$ $\mathbf{m}$-a.e. in T (resp. if $f_n \to f$ in measure in T with $f \in \mathcal{L}_p\mathcal{M}(\mathbf{m})$), then LDCT holds by Theorem 3.3.1 (resp. by Theorem 3.3.2).

Now let $\mathcal{P}$ be a σ-ring $\mathcal{S}$ and let $|f_n(t)| \le M$ $\mathbf{m}$-a.e. in T for $n \in \mathbb{N}$. As $\mathcal{L}_p(\mathbf{m}) = \mathcal{I}_p(\mathbf{m})$ by Theorems 3.1.10 and 3.2.7, and as $\mathcal{S}$ is a σ-ring, by the last part of Theorem 2.1.5(v) of Chapter 2 the constant function $M \in \mathcal{L}_p(\mathbf{m})$ and hence the present versions of LBCT follow from the corresponding versions of LDCT obtained above. □

Remark 3.3.5. For $\mathbb{K} = \mathbb{R}$ and $\mathcal{P}$ a σ-algebra, the case of $\mathbf{m}$-a.e. convergence of Theorem 3.3.4 is Proposition 2.1 of [FMNSS2].

Using Theorem 3.3.4 we prove the following main theorem of the present section.

Theorem 3.3.6. (Characterizations of $\mathcal{L}_p\mathcal{I}(\mathbf{m})$) *Let $1 \leq p < \infty$ and let $f : T \to \mathbb{K}$ be $\mathbf{m}$-measurable. Then the following statements are equivalent:*

(i) $f \in \mathcal{I}_p(\mathbf{m})$.

(ii) $\mathbf{m}_p^\bullet(f, \cdot)$ *is continuous on* $\sigma(\mathcal{P})$ *(so that* $f \in \mathcal{L}_p(\mathbf{m})$ *by Theorem 3.2.7).*

(iii) **(Simple Function Approximation).** *There exists a sequence* $(s_n) \subset \mathcal{I}_s$ *such that* $s_n \to f$ $\mathbf{m}$-*a.e. in* T *and* $\lim_n \mathbf{m}_p^\bullet(s_n - f, T) = 0$.

Consequently,

$$\mathcal{L}_p\mathcal{I}(\mathbf{m}) = \mathcal{I}_p(\mathbf{m}) = \mathcal{L}_p\mathcal{I}_s(\mathbf{m}) = \mathcal{L}_p(\mathbf{m}).$$

If $c_0 \not\subset X$, then

$$\mathcal{L}_p\mathcal{M}(\mathbf{m}) = \mathcal{L}_p\mathcal{I}(\mathbf{m}) = \mathcal{I}_p(\mathbf{m}) = \mathcal{L}_p\mathcal{I}_s(\mathbf{m}) = \mathcal{L}_p(\mathbf{m}).$$

Proof. (i)$\Leftrightarrow$(ii) by Theorems 3.1.10 and 3.2.7.

(ii)$\Rightarrow$(iii) By Proposition 1.1.18 of Chapter 1 there exists $(s_n) \subset \mathcal{I}_s$ such that $s_n \to f$ $\mathbf{m}$-a.e. in T and $|s_n| \nearrow |f|$ $\mathbf{m}$-a.e. in T. As $f \in \mathcal{L}_p(\mathbf{m})$ by hypothesis and by Theorem 3.2.7, Theorem 3.3.4 implies $\lim_n \mathbf{m}_p^\bullet(f - s_n, T) = 0$. (For $p = 1$, one can use (i) and Theorem 2.1.7 of Chapter 2.)

(iii)$\Rightarrow$(ii) Let $\epsilon > 0$ and let $E_k \searrow \emptyset$ in $\sigma(\mathcal{P})$. By hypothesis, there exists n_0 such that $\mathbf{m}_p^\bullet(s_n - f, T) < \frac{\epsilon}{2}$ for $n \geq n_0$. By Theorem 3.1.6, $\mathbf{m}_p^\bullet(s_{n_0}, \cdot)$ is continuous on $\sigma(\mathcal{P})$ and hence there exists k_0 such that $\mathbf{m}_p^\bullet(s_{n_0}, E_k) < \frac{\epsilon}{2}$ for $k \geq k_0$. Consequently, $\mathbf{m}_p^\bullet(f, E_k) \leq \mathbf{m}_p^\bullet(f - s_{n_0}, E_k) + \mathbf{m}_p^\bullet(s_{n_0}, E_k) \leq \mathbf{m}_p^\bullet(f - s_{n_0}, T) + \mathbf{m}_p^\bullet(s_{n_0}, E_k) < \epsilon$ for $k \geq k_0$. Hence $\mathbf{m}_p^\bullet(f, \cdot)$ is continuous on $\sigma(\mathcal{P})$.

Thus (i)$\Leftrightarrow$(ii)$\Leftrightarrow$(iii). $\qquad\square$

Since $\mathcal{I}_s \subset \mathcal{I}_p(\mathbf{m}) = \mathcal{L}_p\mathcal{I}(\mathbf{m})$ by Theorem 3.1.10, $\mathcal{L}_p\mathcal{I}_s(\mathbf{m}) \subset$ closure of $\mathcal{L}_p\mathcal{I}(\mathbf{m})$ in $\mathcal{L}_p\mathcal{M}(\mathbf{m}) =$ closure of $\mathcal{L}_p(\mathbf{m})$ in $\mathcal{L}_p\mathcal{M}(\mathbf{m})$ (by Theorem 3.2.7) . But by Theorem 3.2.8, $\mathcal{L}_p(\mathbf{m})$ is closed in $\mathcal{L}_p\mathcal{M}(\mathbf{m})$ and hence $\mathcal{L}_p\mathcal{I}_s(\mathbf{m}) \subset \mathcal{L}_p(\mathbf{m})$. On the other hand, $\mathcal{I}_s(\mathbf{m})$ is dense in $\mathcal{L}_p(\mathbf{m})$ by (iii) and hence $\mathcal{L}_p\mathcal{I}_s(\mathbf{m}) \supset \mathcal{L}_p(\mathbf{m})$. Therefore, $\mathcal{L}_p\mathcal{I}_s(\mathbf{m}) = \mathcal{L}_p(\mathbf{m})$. Consequently, by Theorems 3.1.10 and 3.2.7 we have

$$\mathcal{L}_p\mathcal{I}(\mathbf{m}) = \mathcal{I}_p(\mathbf{m}) = \mathcal{L}_p\mathcal{I}_s(\mathbf{m}) = \mathcal{L}_p(\mathbf{m}).$$

If $c_0 \not\subset X$, use the second part of Theorem 3.1.10 along with the previous part.

Notation 3.3.7. In the light of Theorem 3.3.6, **we shall hereafter use the symbol $\mathcal{L}_p(\mathbf{m})$ not only to denote the space given in the second part of Definition 3.2.5 but also any one of the spaces $\mathcal{L}_p\mathcal{I}_s(\mathbf{m})$, $\mathcal{L}_p\mathcal{I}(\mathbf{m})$ or $\mathcal{I}_p(\mathbf{m})$.** The quotient $\mathcal{L}_p(\mathbf{m})/\sim$ is denoted by $L_p(\mathbf{m})$, and is treated as a function space in which two functions which are equal $\mathbf{m}$-a.e. in T are identified.

The following theorem is immediate from Theorem 3.3.6.

Theorem 3.3.8. *$\mathcal{I}_s$ is dense in $\mathcal{L}_p(\mathbf{m})$ for $1 \leq p < \infty$.*

Remark 3.3.9. For $\mathbb{K} = \mathbb{R}$ and $\mathcal{P}$ a σ-algebra, Theorems 3.2.8 and 3.3.8 occur as Proposition 4 of [SP1].

Remark 3.3.10. For $p = 1$, Theorem 3.3.8 subsumes Theorem 4.5 of [MN2], which is assumed valid by Theorem 3.5 in [L2]. But the proof of the said theorem in [L2] is incorrect as observed in Remark 2.1.12 of Chapter 2 , where a correct proof for the said theorem of [L2] (with the vector measure having values in a sequentially complete lcHs) is also given.

Remark 3.3.11. Let X be an lcHs and let $\mathbf{m} : \mathcal{P} \to X$ be σ-additive. Definition 3.2.5 is generalized to such $\mathbf{m}$ in Definition 4.5.4 of Chapter 4 where Theorems 3.3.1, 3.3.2, 3.3.4 and 3.3.6 are generalized in Theorems 4.5.1, 4.5.2, 4.5.3 and 4.5.5, respectively, when X is quasicomplete (resp. sequentially complete and the functions considered are $\sigma(\mathcal{P})$-measurable).

Convention 3.3.12. Let $\mathbf{m} : \mathcal{P} \to X$ be σ-additive. If $f \in \mathcal{L}_\infty(\mathbf{m})$, then there exists $M \in \sigma(\mathcal{P})$ such that $||\mathbf{m}||(M) = 0$ and such that $||f||_\infty = \sup_{t \in T\backslash M} |f(t)|$. If we define $g(t) = f(t)$ for $t \in T\backslash M$ and $g(t) = 0$ for $t \in M$, then $f = g$ $\mathbf{m}$-a.e. in T and $||f||_\infty = \sup_{t \in T} |g(t)|$. Thus, for the equivalence class $\tilde{f}$ determined by $f \in \mathcal{L}_\infty(\mathbf{m})$, there exists a bounded $\mathbf{m}$-measurable function $g_{\tilde{f}} \in \tilde{f}$ such that $\sup_{t \in T} |g_{\tilde{f}}(t)| = ||f||_\infty$ and hence we make the convention to define $L_\infty(\mathbf{m}) = \{g_{\tilde{f}} :$ only one representative from $\tilde{f}$ for $f \in \mathcal{L}_\infty(\mathbf{m})\}$. Thus, for $f \in L_\infty(\mathbf{m}), ||f||_\infty = \sup_{t \in T} |f(t)|$. Similarly, the spaces $L_p(\mathbf{m})$, $1 \leq p < \infty$, are treated as function spaces (see the last part of Notation 3.3.7).

Theorem 3.3.13. *Let $\mathbf{m} : \mathcal{P} \to X$ be σ-additive. Let $\mathcal{L}_p^r(\mathbf{m}) = \{f \in \mathcal{L}_p(\mathbf{m}), f$ real-valued$\}$ for $1 \leq p \leq \infty$. Then $L_p^r(\mathbf{m})$ and $L_p^r(\sigma(\mathcal{P}), \mathbf{m})$, $1 \leq p \leq \infty$, are Banach lattices under the partial order $f \leq g$ defined by $f(t) \leq g(t)$ $\mathbf{m}$-a.e. in T.*

Proof. Under the given partial order, clearly $L_\infty^r(\mathbf{m})$ is a vector lattice. If $|f| \leq |g|$, $f, g \in L_\infty^r(\mathbf{m})$, then $||f||_\infty = \sup_{t \in T} |f(t)| \leq \sup_{t \in T} |g(t)| = ||g||_\infty$ (see Convention 3.3.12). Then by Theorem 3.2.11, $L_\infty^r(\mathbf{m})$ is a Banach lattice.

Let $1 \leq p < \infty$. Then $L_p^r(\mathbf{m})$ is a lattice by the facts that $L_p^r(\mathbf{m})$ is a vector space and that $\max(f, g) = \frac{1}{2}(f + g + |f - g|)$ and $\min(f, g) = \frac{1}{2}(f + g - |f - g|)$ belong to $L_p^r(\mathbf{m})$ whenever $f, g \in L_p^r(\mathbf{m})$ (see Theorem 2.1.5(vii) of Chapter 2). Clearly, $f \leq g$ implies $f + h \leq g + h$ and $\alpha f \leq \alpha g$ for $f, g, h \in L_p^r(\mathbf{m})$ and $\alpha \geq 0$. Hence $L_p^r(\mathbf{m})$ is a vector lattice. Consequently, by Theorems 3.1.11(iv) and 3.2.8, $L_p^r(\mathbf{m})$ is a Banach lattice.

Similarly, the results for $L_p^r(\sigma(\mathcal{P}), \mathbf{m})$ are proved for $1 \leq p \leq \infty$. $\qquad \square$

Using the convergence in measure of a sequence of simple scalar (resp. vector) functions which are Cauchy in mean, the abstract Lebesgue integral (resp. the Bochner integral) is defined in [H] (resp. in [DS1]). The following theorem asserts that an analogous result holds for $\mathbf{m}$-integrable functions.

Theorem 3.3.14. *Let $f : T \to \mathbb{K}$ be $\mathbf{m}$-measurable and let $1 \leq p < \infty$. Then a scalar function $f \in \mathcal{L}_p(\mathbf{m})$ if and only if there exists a sequence $(s_n) \subset \mathcal{I}_s$ (resp.*

$(f_n) \subset \mathcal{L}_p(\mathbf{m}))$ such that $s_n \to f$ (resp. $f_n \to f$) in measure in T and (s_n) (resp. (f_n)) is Cauchy in (mean^p). In that case, $\lim_n \mathbf{m}_p^\bullet(s_n - f, T) = 0$ (resp. $\lim_n \mathbf{m}_p^\bullet(f_n - f, T) = 0$). When $p = 1$, $f \in \mathcal{L}_1(\mathbf{m})$, and

$$\int_E f d\mathbf{m} = \lim_n \int_E s_n d\mathbf{m} \ (resp. \ \lim_n \int_E f_n d\mathbf{m}), \ E \in \sigma(\mathcal{P})$$

the limit being uniform with respect to $E \in \sigma(\mathcal{P})$.

Proof. Let $f \in \mathcal{L}_p(\mathbf{m})$. Then by Theorem 3.3.6 there exists a sequence $(s_n) \subset \mathcal{I}_s \subset \mathcal{L}_p(\mathbf{m})$ such that $s_n \to f$ $\mathbf{m}$-a.e. in T and $\mathbf{m}_p^\bullet(s_n - f, T) \to 0$ as $n \to \infty$. Then by Theorem 3.1.18(vi), (s_n) converges to f in measure in T. Clearly, (s_n) is Cauchy in (mean^p).

Conversely, let $(s_n)_1^\infty \subset \mathcal{I}_s$ (resp. $(f_n)_1^\infty \subset \mathcal{L}_p(\mathbf{m})$) satisfy the hypothesis. Let $u_n = s_n$ for all n or $u_n = f_n$ for all n, as the case may be. By hypothesis, $u_n \to f$ in measure in T and (u_n) is Cauchy in (mean^p). Then by Theorem 3.1.19 there exist a subsequence $(u_{n_k})_{k=1}^\infty$ of $(u_n)_1^\infty$ and an $\mathbf{m}$-measurable scalar function g on T such that $u_{n_k} \to g$ $\mathbf{m}$-a.e. in T and $u_{n_k} \to g$ in measure in T. Then by (vii) and (iv) of Theorem 3.1.18, $f = g$ $\mathbf{m}$-a.e. in T. Consequently, by Corollary 3.2.4, $f \in \mathcal{L}_p\mathcal{M}(\mathbf{m})$ and $\mathbf{m}_p^\bullet(u_n - f, T) \to 0$ as $n \to \infty$. As $\mathcal{L}_p(\mathbf{m})$ is closed in $\mathcal{L}_p\mathcal{M}(\mathbf{m})$ by Theorem 3.2.8, $f \in \mathcal{L}_p(\mathbf{m})$.

Now let $p = 1$. Given $\epsilon > 0$, there exists n_0 such that $\mathbf{m}_1^\bullet(u_n - f, T) < \epsilon$ for $n \ge n_0$. Then, by inequality (3.1.3.1) we have

$$\left| \int_E u_n d\mathbf{m} - \int_E f d\mathbf{m} \right| \le \mathbf{m}_1^\bullet(u_n - f, T) < \epsilon$$

for $n \ge n_0$ and for all $E \in \sigma(\mathcal{P})$. Hence the last part also holds. $\square$

Remark 3.3.15. The above theorem (for $p = 1$) fails for the Dobrakov integral of vector functions. See p. 530 of [Do1].

Following [Del3] we give some examples of σ-additive Banach space-valued vector measures defined on a δ-ring.

Example 3.3.16. (See Example 2.1 of [Del3]). Let $\mathcal{P}$ be the δ-ring of Borel subsets of $\mathbb{R}$ having finite Lebesgue measure which is denoted by λ. Let $1 \le p < \infty$ and let $X = L_p(\mathbb{R})$ with respect to λ. Let $\mathbf{m} : \mathcal{P} \to X$ be given by $\mathbf{m}(A) = \chi_A \in X$ for $A \in \mathcal{P}$. Clearly, $\mathbf{m}$ is a σ-additive vector measure. $\mathbf{m}$ is not bounded since

$$|\mathbf{m}(A)|_p = \left(\int (\chi_A)^p d\lambda \right)^{\frac{1}{p}} = (\lambda(A))^{\frac{1}{p}}$$

for $A \in \mathcal{P}$. Let $s = \sum_1^r \alpha_i \chi_{E_i} \in \mathcal{I}_s$.

By Definition 3.1.1,

$$\mathbf{m}_1^\bullet(s, \mathbb{R}) = \sup \left\{ \left\| \int_{\mathbb{R}} s' d\mathbf{m} \right\|_p : s' \in \mathcal{I}_s, |s'| \leq |s|\lambda - a.e. \right\}$$

$$= \left(\int_{\mathbb{R}} \left| \sum_{i=1}^r \alpha_i \mathbf{m}(E_i) \right|^p d\lambda \right)^{\frac{1}{p}} = \left(\int_{\mathbb{R}} \left| \sum_{i=1}^r \alpha_i \chi_{E_i} \right|^p d\lambda \right)^{\frac{1}{p}} = ||s||_p.$$

Since $\mathcal{I}_s$ is dense in $L_1(\mathbf{m})$ by Theorem 3.3.8 and since $\mathcal{I}_s$ is also dense in $L_1(\mathbb{R})$, we conclude that $L_1(\mathbf{m}) = L_p(\mathbb{R})$. Since $X = L_p(\mathbb{R})$ does not contain a copy of c_0, by Theorem 3.1.10 and Notation 3.3.7 we have $L_1(\mathbf{m}) = L_1\mathcal{M}(\mathbf{m})$ and hence $L_1\mathcal{M}(\mathbf{m}) = L_1(\mathbf{m}) = L_p(\mathbb{R})$.

Example 3.3.17. (see Example 2.2 of [Del3]). Let Γ be a non-void abstract set and let $\mathcal{P}$ be the δ-ring F_Γ of all finite subsets of Γ (see Example 2.2.13 of Chapter 2). Given $1 \leq p < \infty$, let $X_p = \ell_p(\Gamma)$ and let $\mathbf{m} : \mathcal{P} \to X_p$ be defined by $\mathbf{m}(A) = \sum_{\gamma \in A} e_\gamma$, where e_γ is the characteristic function of the point $\gamma \in \Gamma$. Clearly, $\mathbf{m}$ is σ-additive on $\mathcal{P}$. The only $\mathbf{m}$-null set is the empty set.

Claim. $L_1(\mathbf{m}) = X_p = \ell_p(\Gamma)$.

In fact, for $s \in \mathcal{I}_s$,

$$\int_\Gamma s d\mathbf{m} = \sum s(\gamma) e_\gamma = s \in X_p. \tag{3.3.17.1}$$

Each $x^* \in X_p^*$ is identified with some $(x_\gamma)_{\gamma \in \Gamma} \in \ell_q(\Gamma)$, where $\frac{1}{p} + \frac{1}{q} = 1$. So $x^*\mathbf{m}(A) = \sum_{\gamma \in A} x_\gamma$ and $|x^*\mathbf{m}|(A) = \sum_{\gamma \in A} |x_\gamma|$ for $A \in \mathcal{P}$. As X_p does not contain a copy of c_0, by Theorem 3.1.10 and by Notation 3.3.7 we have $L_1\mathcal{M}(\mathbf{m}) = L_1(\mathbf{m})$. Given $s \in \mathcal{I}_s$, for each $x^* = (x_\gamma)_{\gamma \in \Gamma} \in X_p^* = X_q$, we have

$$\int_\Gamma |s| d|x^*\mathbf{m}| = \sum_{\gamma \in \Gamma} |s(\gamma)||x_\gamma| \leq ||s||_p ||x^*||_q$$

and so

$$\mathbf{m}_1^\bullet(s, \Gamma) \leq ||s||_p. \tag{3.3.17.2}$$

On the other hand, by (3.1.3.1) and (3.3.17.1) we have

$$\left\| \int_\Gamma s d\mathbf{m} \right\| = ||s||_p \leq \mathbf{m}_1^\bullet(s, \Gamma). \tag{3.3.17.3}$$

Hence by (3.3.17.2) and (3.3.17.3) we have

$$||s||_p = \mathbf{m}_1^\bullet(s, \Gamma).$$

Since $\mathcal{I}_s$ is dense in X_p and since $\mathcal{I}_s$ is dense in $L_1(\mathbf{m})$ by Theorem 3.3.8, we conclude that $L_1(\mathbf{m}) = X_p = \ell_p(\Gamma)$.

Example 3.3.18. (see Example 2.2 of [Del3]). In Example 3.3.17, let $p = \infty$ and let $X_\infty = c_0(\Gamma)$. Let $\mathcal{P} = F_\Gamma$ and let $\mathbf{m}(A) = \sum_{\gamma \in A} e_\gamma$. Then $\mathbf{m}$ is σ-additive on $\mathcal{P}$ and the only $\mathbf{m}$-null set is the empty set. Let $x^* = (x_\gamma)_{\gamma \in \gamma} \in c_0(\Gamma)^* = \ell_1(\gamma)$ so that x^* has countable support. If f is a $\sigma(\mathcal{P})$-measurable scalar function, then

$$\int_\Gamma |f| d|x^* \mathbf{m}| = \sum |f(\gamma)| \|x_\gamma\| \leq \|f\|_\infty \|x^*\|_1$$

and hence

$$\sup_{|x^*| \leq 1} \left\{ \int_\gamma |f| d|x^* \mathbf{m}| \right\} = \mathbf{m}_1^\bullet(f, \Gamma) \leq \|f\|_\infty.$$

On the other hand,

$$|f(\gamma)| = \int_\gamma |f| d|e_\gamma \mathbf{m}| \leq \mathbf{m}_1^\bullet(f, \Gamma)$$

for each $\gamma \in \Gamma$ and hence

$$\|f\|_\infty = \sup_{\gamma \in \Gamma} |f(\gamma)| \leq \mathbf{m}_1^\bullet(f, \Gamma).$$

Therefore, $\|f\|_\infty = \mathbf{m}_1^\bullet(f, \Gamma)$. Hence $\ell_\infty(\Gamma) = L_1 \mathcal{M}(\mathbf{m})$.

Since $\mathcal{I}_s$ is dense in $c_0(\Gamma)$ and since $\mathcal{I}_s$ is dense in $L_1(\mathbf{m})$ by Theorem 3.3.8, we conclude that $c_0(\Gamma) = L_1(\mathbf{m})$. Hence $L_1(\mathbf{m})$ is a proper closed subspace of $L_1 \mathcal{M}(\mathbf{m})$.

Remark 3.3.19. For further examples of σ-additive Banach space-valued vector measures defined on δ-rings which arise from the Hilbert transform on the real line and the Volterra convolution operators, see Section 4 of [Del3].

Remark 3.3.20. See Chapter V of [KK] for examples of Banach space-valued σ-additive vector measures defined on σ-algebras.

3.4   Other convergence theorems for $\mathcal{L}_p(\mathbf{m})$, $1 \leq p < \infty$

In this section we give two versions of the Vitali convergence theorem analogous to Theorems III.3.6 and III.6.15 of [DS1]. We also give LDCT for nets analogous to Theorem III.3.7 of [DS1]. Finally we include some results of Dobrakov [Do3,Do4] specialized to vector measures.

Notation 3.4.1. For $f \in \mathcal{L}_p \mathcal{M}(\mathbf{m})$, $1 \leq p < \infty$, $\mathbf{m}_p^\bullet(f, T \backslash A) = \mathbf{m}_p^\bullet(f, N(f) \backslash A)$ for $A \in \sigma(\mathcal{P})$. (See Notation 1.1.15.)

For $p = 1$, the following lemma is essentially the Banach space version of Lemma 3.4 of [L2] whose proof is corrected in Remark 2.1.12 of Chapter 2.

Lemma 3.4.2. *Let* $1 \leq p < \infty$. *If* $f \in \mathcal{L}_p(\mathbf{m})$, *then, for each* $\epsilon > 0$, *there exists* $A_\epsilon \in \mathcal{P}$ *such that* $\sup_{|x^*| \leq 1} \int_{T \backslash A_\epsilon} |f|^p dv(x^* \mathbf{m}) < \epsilon$ *or equivalently,* $\mathbf{m}_p^\bullet(f, T \backslash A_\epsilon) < \epsilon^{\frac{1}{p}}$.

Proof. Let $\hat{f}$ be a $\sigma(\mathcal{P})$-measurable function such that $\hat{f} = f$ $\mathbf{m}$-a.e. in T. Then $\hat{f} \in \mathcal{L}_p(\mathbf{m})$ and $\int_E |\hat{f}|^p d\mathbf{m} = \int_E |f|^p d\mathbf{m}$ for $E \in \sigma(\mathcal{P})$. Let $\boldsymbol{\gamma} : \sigma(\mathcal{P}) \to X$ be given by $\boldsymbol{\gamma}(\cdot) = \int_{(\cdot)} |\hat{f}|^p d\mathbf{m}$. As $N(\hat{f}) \in \sigma(\mathcal{P})$, there exists an increasing sequence $(E_n) \subset \mathcal{P}$ such that $N(\hat{f}) = \bigcup_1^\infty E_n$. By Theorem 2.1.5(ii) of Chapter 2, $\boldsymbol{\gamma}$ is an X-valued σ-additive vector measure on $\sigma(\mathcal{P})$, and hence by Proposition 1.1.5 of Chapter 1, $||\boldsymbol{\gamma}||(N(\hat{f})\backslash E_n) \searrow 0$ as $n \to \infty$. Thus, there exists n_0 such that $||\boldsymbol{\gamma}||(N(\hat{f})\backslash E_n) < \epsilon$ for $n \geq n_0$. Let $A_\epsilon = E_{n_0}$. Then by Definition 3.1.4, Notation 3.4.1 and Theorem 3.1.3, we have

$$\mathbf{m}_p^\bullet(f, T\backslash A_\epsilon) = \mathbf{m}_p^\bullet(f, N(f)\backslash A_\epsilon)$$

$$= \sup_{|x^*|\leq 1} \left(\int_{N(f)\backslash A_\epsilon} |f|^p dv(x^*\mathbf{m}) \right)^{\frac{1}{p}} = \sup_{|x^*|\leq 1} \left(\int_{T\backslash A_\epsilon} |f|^p dv(x^*\mathbf{m}) \right)^{\frac{1}{p}}$$

$$= \sup_{|x^*|\leq 1} \left(\int_{N(\hat{f})\backslash A_\epsilon} |\hat{f}|^p dv(x^*\mathbf{m}) \right)^{\frac{1}{p}} = \left(||\boldsymbol{\gamma}||(N(\hat{f})\backslash A_\epsilon) \right)^{\frac{1}{p}} < \epsilon^{\frac{1}{p}}$$

and hence

$$\sup_{|x^*|\leq 1} \int_{T\backslash A_\epsilon} |f|^p dv(x^*\mathbf{m}) < \epsilon. \qquad \square$$

Definition 3.4.3. A set function $\lambda : \sigma(\mathcal{P}) \to [0, \infty]$ is said to be $\mathbf{m}$-continuous on $\sigma(\mathcal{P})$ if, given $\epsilon > 0$, there exists $\delta > 0$ such that $\lambda(E) < \epsilon$ whenever $||\mathbf{m}||(E) < \delta$, $E \in \sigma(\mathcal{P})$.

Lemma 3.4.4. Let $1 \leq p < \infty$. For $f \in \mathcal{L}_p(\mathbf{m})$, $\mathbf{m}_p^\bullet(f, \cdot)$ is $\mathbf{m}$-continuous on $\sigma(\mathcal{P})$.

Proof. By Theorem 3.3.6 there exists a sequence $(s_n) \subset \mathcal{I}_s$ such that $s_n \to f$ $\mathbf{m}$-a.e. in T and such that $\mathbf{m}_p^\bullet(f - s_n, T) \to 0$ as $n \to \infty$. Thus, given $\epsilon > 0$, there exists n_0 such that $\mathbf{m}_p^\bullet(f - s_{n_0}, T) < \frac{\epsilon}{2}$. Let $s = s_{n_0} = \sum_1^r a_i \chi_{E_i}$ with $(E_i)_1^r \subset \mathcal{P}$, and let $M = ||s||_T$. Let $A_\epsilon = \bigcup_1^r E_i$. Let $E \in \sigma(\mathcal{P})$ with $||\mathbf{m}||(E) < (\frac{\epsilon}{2(M+1)})^p$. Then by Theorem 3.1.11(iii) we have $\mathbf{m}_p^\bullet(s, E) \leq ||s||_E \cdot (||\mathbf{m}||(E))^{\frac{1}{p}} \leq \frac{M\epsilon}{2(M+1)} < \frac{\epsilon}{2}$. Consequently, by Theorems 3.1.11(i) and 3.1.13(ii) we have

$$\mathbf{m}_p^\bullet(f, E) \leq \mathbf{m}_p^\bullet(f - s, E) + \mathbf{m}_p^\bullet(s, E) \leq \mathbf{m}_p^\bullet(f - s_{n_0}, T) + \mathbf{m}_p^\bullet(s, E) < \epsilon.$$

Hence $\mathbf{m}_p^\bullet(f, \cdot)$ is $\mathbf{m}$-continuous on $\sigma(\mathcal{P})$. $\qquad \square$

The following theorem is an analogue of Theorem III.3.6 of [DS1] for $\mathcal{L}_p(\mathbf{m})$.

Theorem 3.4.5. (Analogue of the convergence in measure version of the Vitali convergence theorem of [DS1] for $\mathcal{L}_p(\mathbf{m})$) Let $1 \leq p < \infty$. Let $(f_n)_1^\infty \subset \mathcal{L}_p(\mathbf{m})$ and let $f : T \to \mathbb{K}$ be $\mathbf{m}$-measurable. Then $f \in \mathcal{L}_p(\mathbf{m})$ and $\lim_n \mathbf{m}_p^\bullet(f_n - f, T) = 0$ if and only if the following conditions hold:

(i) $f_n \to f$ in measure in each $E \in \mathcal{P}$.

(ii) $\mathbf{m}_p^\bullet(f_n, \cdot)$, $n \in \mathbb{N}$, are uniformly $\mathbf{m}$-continuous on $\sigma(\mathcal{P})$, in the sense that, given $\epsilon > 0$, there exists $\delta > 0$ such that $\mathbf{m}_p^\bullet(f_n, E) < \epsilon$ for all $n \in \mathbb{N}$ whenever $E \in \sigma(\mathcal{P})$ with $||\mathbf{m}||(E) < \delta$.

(iii) For each $\epsilon > 0$, there exists $A_\epsilon \in \mathcal{P}$ such that $\mathbf{m}_p^\bullet(f_n, T\backslash A_\epsilon) < \epsilon$ for all $n \in \mathbb{N}$. (See Notation 3.4.1.)

In such case, when $p = 1$, $\int_E f d\mathbf{m} = \lim_n \int_E f_n d\mathbf{m}$ for $E \in \sigma(\mathcal{P})$ and the limit is uniform with respect to $E \in \sigma(\mathcal{P})$.

Proof. Let $f \in \mathcal{L}_p(\mathbf{m})$ and let $\lim_n \mathbf{m}_p^\bullet(f_n - f, T) = 0$. Then by Theorem 3.1.18(vi), $f_n \to f$ in measure in T and hence (i) holds. Let $\epsilon > 0$ and let $f_0 = f$. By hypothesis there exists n_0 such that $\mathbf{m}_p^\bullet(f_n - f_0, T) < \frac{\epsilon}{2}$ for $n \ge n_0$. By Lemma 3.4.4 there exists $\delta > 0$ such that $\mathbf{m}_p^\bullet(f_i, E) < \frac{\epsilon}{2}$ for $i = 0, 1, 2, \ldots, n_0$, whenever $E \in \sigma(\mathcal{P})$ with $||\mathbf{m}||(E) < \delta$. Consequently, for such E we also have

$$\mathbf{m}_p^\bullet(f_n, E) \le \mathbf{m}_p^\bullet(f_n - f_0, E) + \mathbf{m}_p^\bullet(f_0, E) \le \mathbf{m}_p^\bullet(f_n - f_0, T) + \mathbf{m}_p^\bullet(f_0, E) < \epsilon$$

for $n \ge n_0$. Hence (ii) holds. By Lemma 3.4.2 there exists $A_\epsilon \in \mathcal{P}$ such that $\mathbf{m}_p^\bullet(f_n, T \backslash A_\epsilon) < \frac{\epsilon}{2}$ for $n = 0, 1, 2, \ldots, n_0$. For $n \ge n_0$, $\mathbf{m}_p^\bullet(f_n, T \backslash A_\epsilon) \le \mathbf{m}_p^\bullet(f_n - f_0, T \backslash A_\epsilon) + \mathbf{m}_p^\bullet(f_0, T \backslash A_\epsilon) \le \mathbf{m}_p^\bullet(f_n - f_0, T) + \mathbf{m}_p^\bullet(f_0, T \backslash A_\epsilon) < \epsilon$, by Definition 3.1.4 and Theorems 3.1.13(i) and 3.1.11(i). Hence (iii) holds.

Conversely, let (i), (ii) and (iii) hold and let $\epsilon > 0$. By (iii) there exists $A_\epsilon \in \mathcal{P}$ such that $\mathbf{m}_p^\bullet(f_n, T \backslash A_\epsilon) < \frac{\epsilon}{6}$ for $n \in \mathbb{N}$. By (ii) there exists $\delta > 0$ such that $\mathbf{m}_p^\bullet(f_n, E) < \frac{\epsilon}{6}$ for $n \in \mathbb{N}$ whenever $E \in \sigma(\mathcal{P})$ with $||\mathbf{m}||(E) < \delta$. Let $\delta_0 = \dfrac{\epsilon}{3(||\mathbf{m}||(A_\epsilon)+1)^{\frac{1}{p}}}$. Let $E_{n,r} = \{t \in A_\epsilon : |f_n(t) - f_r(t)| > \delta_0\}$. By (i), $f_n \to f$ in measure in A_ϵ and hence by Theorem 3.1.18(iv), (f_n) is Cauchy in measure in A_ϵ. Then there exists n_0 such that $||\mathbf{m}||(E_{n,r}) < \delta$ for $n, r \ge n_0$. Therefore, $\mathbf{m}_p^\bullet(f_k, E_{n,r}) < \frac{\epsilon}{6}$ for $k \in \mathbb{N}$ and for $n, r \ge n_0$. Since $||f_n - f_r||_{A_\epsilon \backslash E_{n,r}} \le \delta_0$, by Theorem 3.1.11(iii) we have

$$\mathbf{m}_p^\bullet(f_n - f_r, A_\epsilon \backslash E_{n,r}) \le \delta_0 (||\mathbf{m}||(A_\epsilon \backslash E_{n,r}))^{\frac{1}{p}} < \epsilon/3$$

for all $n, r \in \mathbb{N}$. Then by Definition 3.1.4 and by Theorem 3.1.13(i) we have

$$\mathbf{m}_p^\bullet(f_n - f_r, T) \le \mathbf{m}_p^\bullet(f_n - f_r, T \backslash A_\epsilon) + \mathbf{m}_p^\bullet(f_n - f_r, A_\epsilon \backslash E_{n,r}) + \mathbf{m}_p^\bullet(f_n - f_r, E_{n,r}) < \epsilon$$

for $n, r \ge n_0$ and hence $\lim_{n,r} \mathbf{m}_p^\bullet(f_n - f_r, T) = 0$. Therefore, (f_n) is Cauchy in (mean^p) and consequently, by Theorem 3.2.8 there exists $g \in \mathcal{L}_p(\mathbf{m})$ such that $\lim_n \mathbf{m}_p^\bullet(f_n - g, T) = 0$. Then by (vi) and (iv) of Theorem 3.1.18, $f = g$ $\mathbf{m}$-a.e. in each $E \in \mathcal{P}$. As $N(f) \in \sigma(\mathcal{P})$, there exists $(E_n)_1^\infty \subset \mathcal{P}$ such that $N(f) = \bigcup_1^\infty E_n$. For each n, $\{t \in E_n : f(t) \ne g(t)\} \in \sigma(\mathcal{P})$ and is $\mathbf{m}$-null. Hence $N(f - g) = \bigcup_1^\infty \{t \in E_n : f(t) - g(t) \ne 0\}$ is $\mathbf{m}$-null and $f = g$ $\mathbf{m}$-a.e. in T. Hence $f \in \mathcal{L}_p(\mathbf{m})$ and $\lim_n \mathbf{m}_p^\bullet(f_n - f, T) = 0$.

The last part is due to the first, Theorem 3.1.11(i) and inequality (3.1.3.1). $\square$

The following theorem is an analogue of Theorem III.6.15 of [DS1] for $\mathcal{L}_p(\mathbf{m})$.

Theorem 3.4.6. (Analogue of the a.e. convergence version of the Vitali convergence theorem of [DS1] for $\mathcal{L}_p(\mathbf{m})$) *Let $1 \le p < \infty$. Let $(f_n) \subset \mathcal{L}_p(\mathbf{m})$ and let $f : T \to \mathbb{K}$. Suppose $f_n \to f$ $\mathbf{m}$-a.e. in T. Then $f \in \mathcal{L}_p(\mathbf{m})$ and $\lim_n \mathbf{m}_p^\bullet(f_n - f, T) = 0$ if and only if the following conditions are satisfied:*

(a) *$\mathbf{m}_p^\bullet(f_n, \cdot)$, $n \in \mathbb{N}$, are uniformly $\mathbf{m}$-continuous on $\sigma(\mathcal{P})$.*

(b) *For each $\epsilon > 0$, there exists $A_\epsilon \in \mathcal{P}$ such that $\mathbf{m}_p^\bullet(f_n, T \backslash A_\epsilon) < \epsilon$ for all n.*

In such case, for $p = 1$, $\int_E f d\mathbf{m} = \lim_n \int_E f_n d\mathbf{m}$ for $E \in \sigma(\mathcal{P})$ and the limit is uniform with respect to $E \in \sigma(\mathcal{P})$.

Proof. Let $f \in \mathcal{L}_p(\mathbf{m})$ and let $\lim_n \mathbf{m}_p^\bullet(f_n - f, T) = 0$. Then (a) and (b) hold by Theorem 3.4.5.

Conversely, let (a) and (b) hold. By hypothesis, f is $\mathbf{m}$-measurable. Let $\epsilon > 0$. By (b) there exists $A \in \mathcal{P}$ such that

$$\mathbf{m}_p^\bullet(f_n, T\backslash A) < \frac{\epsilon}{3} \qquad (3.4.6.1)$$

for all n. Let $\Sigma = \sigma(\mathcal{P}) \cap A$. Then Σ is a σ-algebra of subsets of A and by hypothesis $f_n \to f$ $\mathbf{m}$-a.e. in A. Therefore, by Theorem 3.1.18(viii), $f_n \to f$ in measure in A. Moreover, (a) implies that $\mathbf{m}_p^\bullet(f_n, \cdot)$, $n \in \mathbb{N}$, are uniformly $\mathbf{m}|_\Sigma$-continuous on Σ. Hence conditions (i) and (ii) of Theorem 3.4.5 are satisfied with $\mathcal{P}$ and $\sigma(\mathcal{P})$ being replaced by Σ. Further, as $\|\mathbf{m}\|(A) < \infty$, condition (iii) of the said theorem also holds with $A_\epsilon = A$. Hence by Theorem 3.4.5, there exists n_0 such that $\mathbf{m}_p^\bullet(f_n - f, A) < \frac{\epsilon}{3}$ for $n \ge n_0$. By (3.4.6.1) and by the generalized Fatou's lemma (Theorem 3.2.1(ii)), $\mathbf{m}_p^\bullet(f, T\backslash A) = \mathbf{m}_p^\bullet(\liminf_n f_n, T\backslash A) \le \liminf_n \mathbf{m}_p^\bullet(f_n, T\backslash A) < \frac{\epsilon}{3}$. Consequently, by (3.4.6.1) we have $\mathbf{m}_p^\bullet(f_n - f, T) \le \mathbf{m}_p^\bullet(f_n - f, A) + \mathbf{m}_p^\bullet(f_n, T\backslash A) + \mathbf{m}_p^\bullet(f, T\backslash A) < \epsilon$ for $n \ge n_0$ and hence $\lim_n \mathbf{m}_p^\bullet(f_n - f, T) = 0$. Then by the triangle inequality, $f \in \mathcal{L}_p\mathcal{M}(\mathbf{m})$. As $\mathcal{L}_p(\mathbf{m})$ is closed in $\mathcal{L}_p\mathcal{M}(\mathbf{m})$ by Theorem 3.2.8, it follows that $f \in \mathcal{L}_p(\mathbf{m})$. The last part is due to the first part, Theorem 3.1.11(i) and inequality (3.1.3.1). $\qquad \square$

Theorem 3.4.7. *LDCT and LBCT as given in Theorem 3.3.4 are deducible from Theorems 3.4.5 and 3.4.6.*

Proof. For the dominating function g in LDCT, Lemmas 3.4.2 and 3.4.4 hold and hence (ii) and (iii) of Theorem 3.4.5 (resp. (a) and (b) of Theorem 3.4.6) hold. Thus, if $f_n \to f$ in measure in T (resp. $\mathbf{m}$-a.e. in T), then LDCT holds by Theorem 3.4.5 (resp. by Theorem 3.4.6). If $\mathcal{P}$ is a σ-ring, then constant functions belong to $\mathcal{L}_p(\mathbf{m})$ by the last part of Theorem 2.1.5(v) of Chapter 2 and hence both the versions of LBCT follow from the corresponding versions of LDCT. $\qquad \square$

Now we define a translation invariant pseudo metric ρ (similar to that in Section 2, Chapter III of [DS1]) in the set of all $\mathbf{m}$-measurable scalar functions. Then the following lemma says that $f_n \to f$ in measure in T if and only if $\rho(f_n, f) \to 0$. Using this lemma, we obtain an analogue of Theorem III.3.7 of [DS1] in Theorem 3.4.10.

Definition 3.4.8. Let $\mathcal{M}(\widetilde{\sigma(\mathcal{P})})$ be the set of all $\widetilde{\sigma(\mathcal{P})}$-measurable (i.e., $\mathbf{m}$-measurable) scalar functions on T. For $f \in \mathcal{M}(\widetilde{\sigma(\mathcal{P})})$ and $c > 0$, c real, let $\{t \in T : |f(t)| > c\} = T(|f| > c)$. Let φ be a continuous strictly increasing real function on $[0, \infty)$ such that $\varphi(0) = 0$, $\varphi(x + y) \le \varphi(x) + \varphi(y)$ if $0 \le x \le y$ and $\varphi(\infty) = \lim_{x \to \infty} \varphi(x)$ exists as a real number. (For example, φ given by

$\varphi(x) = \frac{x}{1+x}$ satisfies these conditions.) For $f, g \in \mathcal{M}(\widetilde{\sigma(\mathcal{P})})$, we define $\rho(f, g) = \inf_{c>0}\{c + \varphi\left(||\mathbf{m}||\left(T(|f - g| > c)\right)\right)\}$.

It is easy to verify that ρ is a translation invariant pseudo metric on $\mathcal{M}(\widetilde{\sigma(\mathcal{P})})$. (See p. 102 of [DS1].)

Lemma 3.4.9. *Let* $(f_\alpha)_{\alpha \in (D, \geq)}$ *be a net of* **m**-*measurable scalar functions on* T *and let* $f : T \to \mathbb{K}$ *be* **m**-*measurable. Then* $f_\alpha \to f$ *in measure in* T *if and only if* $\rho(f_\alpha, f) \to 0$ *as* $\alpha \to \infty$.

Proof. Let $f_\alpha \to f$ in measure in T. Let $\epsilon > 0$. By the continuity of φ in $t = 0$, there exists $\delta > 0$ such that $0 \leq \varphi(t) < \epsilon$ if $0 \leq t < \delta$. As $f_\alpha \to f$ in measure in T (see Definition 3.1.16(i)), there exists α_0 such that $||\mathbf{m}||(T(|f_\alpha - f| > \epsilon)) < \delta$ for $\alpha \geq \alpha_0$ so that $\varphi\left(||\mathbf{m}||(T|f_\alpha - f| > \epsilon)\right) < \epsilon$ for $\alpha \geq \alpha_0$. Thus $\rho(f_\alpha, f) \leq \epsilon + \varphi\left(||\mathbf{m}||\left(T(|f_\alpha - f| > \epsilon)\right)\right) < 2\epsilon$ for $\alpha \geq \alpha_0$. This shows that $\lim_\alpha \rho(f_\alpha, f) = 0$.

Conversely, let $\lim_\alpha \rho(f_\alpha, f) = 0$. Let $\epsilon > 0$ and let $\varphi\left(||\mathbf{m}||\left(T(|f_\alpha - f| > \epsilon)\right)\right) > \delta > 0$. Then, for $0 < \eta \leq \epsilon$, $T(|f_\alpha - f| > \eta) \supset T(|f_\alpha - f| > \epsilon)$ and hence $\varphi\left(||\mathbf{m}||\left(T(|f_\alpha - f| > \eta)\right)\right) > \delta$ and thus $\eta + \varphi(||\mathbf{m}||(T(|f_\alpha - f| > \eta))) > \delta$ for $0 < \eta \leq \epsilon$. If $\eta > \epsilon$, then $\eta + \varphi(||\mathbf{m}||(T(|f_\alpha - f| > \eta))) \geq \eta > \epsilon$. Thus $\rho(f_\alpha, f) \geq \min\{\epsilon, \delta\}$. If (f_α) does not converge to f in measure in T, then there would exist $\epsilon > 0$, $\delta > 0$ such that for each $\alpha \in (D, \geq)$ there would exist $\beta_\alpha \in (D, \geq)$ with $\beta_\alpha \geq \alpha$ such that $\varphi(||\mathbf{m}||(T(|f_{\beta_\alpha} - f| > \epsilon))) > \delta$. Then by the above argument we would have $\rho(f_{\beta_\alpha}, f) \geq \min\{\epsilon, \delta\}$ for each β_α. This is a contradiction since by hypothesis there exists an $\alpha_0 \in (D, \geq)$ such that $\rho(f_\alpha, f) < \min\{\epsilon, \delta\}$ for all $\alpha \geq \alpha_0$. Hence the lemma holds. $\qquad\square$

Theorem 3.4.10. (LDCT for net-convergence in measure version for $\mathcal{L}_p(\mathbf{m})$) *Let* $1 \leq p < \infty$. *Let* $(f_\alpha)_{\alpha \in (D, \geq)}$ *be a net of* **m**-*measurable scalar functions on* T *and let* $f : T \to \mathbb{K}$ *be* **m**-*measurable. Let* $g \in \mathcal{L}_p(\mathbf{m})$. *If* $|f_\alpha(t)| \leq |g(t)|$ **m**-*a.e. in* T *for each* α, *then* $f_\alpha \to f$ *in measure in* T *if and only if* $f \in \mathcal{L}_p(\mathbf{m})$ *and* $\lim_{\alpha \in (D, \geq)} \mathbf{m}_p^\bullet(f_\alpha - f, T) = 0$. *Under the above hypothesis, if* $p = 1$ *and if* $f_\alpha \to f$ *in measure in* T, *then* $f \in \mathcal{L}_1(\mathbf{m})$ *and*

$$\int_E f d\mathbf{m} = \lim_\alpha \int_E f_\alpha d\mathbf{m}, \quad E \in \sigma(\mathcal{P})$$

the limit being uniform with respect to $E \in \sigma(\mathcal{P})$.

Proof. First let us consider the case of a sequence (f_n) satisfying $|f_n(t)| \leq |g(t)|$ **m**-a.e. in T, with $g \in \mathcal{L}_p(\mathbf{m})$. If $f_n \to f$ in measure in T, then by Theorem 3.3.4 or by Theorem 3.4.7, $f \in \mathcal{L}_p(\mathbf{m})$ and $\lim_n \mathbf{m}_p^\bullet(f_n - f, T) = 0$. Conversely, if $\lim_n \mathbf{m}_p^\bullet(f_n - f, T) = 0$, then by Theorem 3.1.18(vi), $f_n \to f$ in measure in T. Hence the first part of the theorem holds for sequences.

Now let us pass on to the case of nets. Let $(f_\alpha)_{\alpha \in (D, \geq)}$ satisfy the hypothesis of domination and let $f_\alpha \to f$ in measure in T. Then $\mathbf{m}_p^\bullet(f_\alpha, T) \leq \mathbf{m}_p^\bullet(g, T)$ for all α and hence $(f_\alpha)_{\alpha \in (D, \geq)} \subset \mathcal{L}_p(\mathbf{m})$ by Theorem 2.1.5(vii) of Chapter 2. If

$\mathbf{m}_p^{\bullet}(f_\alpha - f, T) \nrightarrow 0$, then there would exist an $\epsilon > 0$ such that for each $\alpha \in (D, \geq)$ there would exist $\beta_\alpha \geq \alpha$ in $(D, \geq)$ such that $\mathbf{m}_p^{\bullet}(f_{\beta_\alpha} - f, T) > \epsilon$. As $f_\alpha \to f$ in measure in T, (f_{β_α}) also converges to f in measure in T. Then by Lemma 3.4.9, for each $n \in \mathbb{N}$, there exists $f_{\beta_{\alpha_n}}$ with $\rho(f_{\beta_{\alpha_n}}, f) < \frac{1}{n}$ and hence $\rho(f_{\beta_{\alpha_n}}, f) \to 0$ as $n \to \infty$. Therefore, again by Lemma 3.4.9, $f_{\beta_{\alpha_n}} \to f$ in measure in T and consequently, by the previous case of sequences $\mathbf{m}_p^{\bullet}(f_{\beta_{\alpha_n}} - f, T) \to 0$, which is a contradiction. Hence

$$\lim_\alpha \mathbf{m}_p^{\bullet}(f_\alpha - f, T) = 0. \qquad (3.4.10.1)$$

As $(f_\alpha) \subset \mathcal{L}_p(\mathbf{m})$, by (3.4.10.1) and by the triangle inequality we have $f \in \mathcal{L}_p\mathcal{M}(\mathbf{m})$. As $\mathcal{L}_p(\mathbf{m})$ is closed in $\mathcal{L}_p\mathcal{M}(\mathbf{m})$ by Theorem 3.2.8, it follows from (3.4.10.1) that $f \in \mathcal{L}_p(\mathbf{m})$. Thus the necessity part of the theorem holds.

Conversely, if $f \in \mathcal{L}_p(\mathbf{m})$ and if $\mathbf{m}_p^{\bullet}(f_\alpha - f, T) \to 0$ as $\alpha \to \infty$, then by Theorem 3.1.18(vi) $f_\alpha \to f$ in measure in T as $\alpha \to \infty$.

Let $p = 1$. Then $f \in \mathcal{L}_1(\mathbf{m})$ and given $\epsilon > 0$, there exists α_0 such that $\mathbf{m}_1^{\bullet}(f_\alpha - f, T) < \epsilon$ for $\alpha \geq \alpha_0$. Then, by inequality (3.1.3.1) and by Theorem 3.1.11(i), we have $|\int_E f d\mathbf{m} - \int_E f_\alpha d\mathbf{m}| \leq \mathbf{m}_1^{\bullet}(f - f_\alpha, E) \leq \mathbf{m}_1^{\bullet}(f - f_\alpha, T) < \epsilon$ for all $\alpha \geq \alpha_0$ and for all $E \in \sigma(\mathcal{P})$. Hence the last part also holds. $\qquad \square$

Corollary 3.4.11. (LBCT for net-convergence in measure version for $\mathcal{L}_p(\mathbf{m})$) *Let $1 \leq p < \infty$. Let $\mathcal{P}$ be a σ-ring $\mathcal{S}$ and let $0 < M < \infty$. Let $(f_\alpha)_{\alpha \in (D, \geq)}$ be a net of $\mathbf{m}$-measurable functions on T with values in $\mathbb{K}$ and let $f : T \to \mathbb{K}$ be $\mathbf{m}$-measurable. If $|f_\alpha(t)| \leq M$ $\mathbf{m}$-a.e. in T for each α, then $f_\alpha \to f$ in measure in T if and only if $f \in \mathcal{L}_p(\mathbf{m})$ and $\lim_\alpha \mathbf{m}_p^{\bullet}(f_\alpha - f, T) = 0$. Under the above hypothesis, if $p = 1$ and if $f_\alpha \to f$ in measure in T, then $f \in \mathcal{L}_1(\mathbf{m})$ and*

$$\int_E f d\mathbf{m} = \lim_\alpha \int_E f_\alpha d\mathbf{m}, \quad E \in \mathcal{S},$$

the limit being uniform with respect to $E \in \mathcal{S}$.

Proof. As $\mathcal{P}$ is a σ-ring $\mathcal{S}$, by the last part of Theorem 2.1.5(v) of Chapter 2, constant functions are in $\mathcal{I}_p(\mathbf{m}) = \mathcal{L}_p(\mathbf{m})$ (by Theorems 3.1.10 and 3.3.6) and hence the result is immediate from Theorem 3.4.10. $\qquad \square$

Remark 3.4.12. Theorems III.3.6 and III.3.7 of [DS1] hold for any complex-valued or extended real-valued finitely additive set function μ defined on a σ-algebra of sets Σ, even though the spaces $\mathcal{L}_p(\mu)$, $1 \leq p < \infty$, are not complete. But, our proofs of the analogues of these theorems for $\mathbf{m}$ are based on the facts that $\mathcal{L}_p(\mathbf{m})$ is complete and that $\mathbf{m}$ is σ-additive. When X is an lcHs and $\mathbf{m} : \mathcal{P} \to X$ is σ-additive, Theorems 3.4.5, 3.4.6, 3.4.7 and 3.4.10 and Corollary 3.4.11 are generalized in Theorem 4.6.12 of Chapter 4 for quasicomplete X (resp. sequential complete X for $\sigma(\mathcal{P})$-measurable scalar functions).

If $c_0 \not\subset X$, then the hypothesis that $\mathcal{P}$ is a σ-ring in LBCT can be weakened as follows.

Theorem 3.4.13. (LBCT) *Let* $\mathbf{m} : \mathcal{P} \to X$ *be* σ-*additive and let* $||\mathbf{m}||(T) < \infty$. *If* $c_0 \not\subset X$, *then the versions of LBCT in Theorem 3.3.4 and Corollary 3.4.11 hold.*

Proof. By hypothesis and by Theorem 2.2.2 of Chapter 2 and by Theorems 3.1.11(iii) and 3.1.8, constant functions are $\mathbf{m}$-integrable in T and hence the results hold. (See the proofs of Theorem 3.3.4 and Corollary 3.4.11). $\qquad\square$

The following result is an analogue of Theorem 1 of [Do4] for $\mathcal{L}_p(\mathbf{m})$, $1 \leq p < \infty$.

Theorem 3.4.14. (Extended Vitali convergence theorem) *Let* $1 \leq p < \infty$ *and let* $(f_n)_1^\infty \subset \mathcal{L}_p(\mathbf{m})$. *Suppose* $f_n \to f$ $\mathbf{m}$-*a.e. in* T *where* f *is a scalar function on* T, *or suppose there exists an* $\mathbf{m}$-*measurable scalar function* f *on* T *such that* $f_n \to f$ *in measure in* T. *Then the following statements are equivalent:*

(i) $\mathbf{m}_p^\bullet(f_n - f, T) \to 0$ *as* $n \to \infty$.

(ii) $f \in \mathcal{L}_p(\mathbf{m})$ *and* $\mathbf{m}_p^\bullet(f_n, E) \to \mathbf{m}_p^\bullet(f, E)$ *as* $n \to \infty$, *for each* $E \in \sigma(\mathcal{P})$.

(iii) $\mathbf{m}_p^\bullet(f_n, \cdot)$, $n \in \mathbb{N}$, *are uniformly continuous on* $\sigma(\mathcal{P})$.

If any one of the above statements holds for $p = 1$, *then*

$$\lim_n \int_E f_n d\mathbf{m} = \int_E f d\mathbf{m}, \quad E \in \sigma(\mathcal{P}),$$

the limit being uniform with respect to $E \in \sigma(\mathcal{P})$.

Proof. (i)$\Rightarrow$(ii) Let $\lim_n \mathbf{m}_p^\bullet(f_n - f, T) = 0$. Then by Theorem 3.1.13(ii), $f \in \mathcal{L}_p\mathcal{M}(\mathbf{m})$ and consequently, by Theorem 3.2.8, $f \in \mathcal{L}_p(\mathbf{m})$. Moreover, by Theorems 3.1.13(i) and 3.1.11(i) we have

$$|\mathbf{m}_p^\bullet(f_n, E) - \mathbf{m}_p^\bullet(f, E)| \leq \mathbf{m}_p^\bullet(f_n - f, E) \leq \mathbf{m}_p^\bullet(f_n - f, T) \to 0 \text{ as } n \to \infty$$

for each $E \in \sigma(\mathcal{P})$.

(ii)$\Rightarrow$(iii) Let $E_k \searrow \emptyset$ in $\sigma(\mathcal{P})$ and let $\epsilon > 0$. As $f \in \mathcal{L}_p(\mathbf{m})$, by Definition 3.2.5 there exists k_0 such that $\mathbf{m}_p^\bullet(f, E_k) < \frac{\epsilon}{2}$ for $k \geq k_0$. By hypothesis, there exists n_0 such that $|\mathbf{m}_p^\bullet(f_n, E_{k_0}) - \mathbf{m}_p^\bullet(f, E_{k_0})| < \frac{\epsilon}{2}$ for $n \geq n_0$. Then $\mathbf{m}_p^\bullet(f_n, E_{k_0}) \leq |\mathbf{m}_p^\bullet(f_n, E_{k_0}) - \mathbf{m}_p^\bullet(f, E_{k_0})| + \mathbf{m}_p^\bullet(f, E_{k_0}) < \epsilon$ for $n \geq n_0$. As $f_n \in \mathcal{L}_p(\mathbf{m})$ for $n = 1, 2, \ldots, n_0$, there exists $k_1 > k_0$ such that $\mathbf{m}_p^\bullet(f_n, E_k) < \epsilon$ for $n = 1, 2, \ldots, n_0$ and for $k \geq k_1$. Then by Theorem 3.1.11(i), (iii) holds.

(iii)$\Rightarrow$(i) by Theorem 3.3.1 if $f_n \to f$ $\mathbf{m}$-a.e in T and by Theorem 3.3.2 if $f_n \to f$ in measure in T.

The last part is immediate from (i), inequality (3.1.3.1) and Theorem 3.1.11(i). $\qquad\square$

The following result is an analogue of Theorem 2 of [Do4] for $\mathcal{L}_p(\mathbf{m})$, $1 \leq p < \infty$.

Theorem 3.4.15. (Monotone convergence theorem for $\mathcal{L}_p(\mathbf{m})$) *Let $c_0 \not\subset X$ and let $1 \leq p < \infty$. Let f_n, $n \in \mathbb{N}$, be $\mathbf{m}$-measurable scalar functions on T and let $f : T \to \mathbb{K}$. Suppose $f_n \to f$ $\mathbf{m}$-a.e. in T and $|f_n| \nearrow |f|$ $\mathbf{m}$-a.e. in T. Then the following statements are equivalent:*

(i) $\sup_n \mathbf{m}_p^\bullet(f_n, T) < \infty$.

(ii) $f \in \mathcal{L}_p(\mathbf{m})$.

If (i) or (ii) holds, then f, f_n, $n \in \mathbb{N}$, belong to $\mathcal{L}_p(\mathbf{m})$ and $\lim_n \mathbf{m}_p^\bullet(f_n - f, T) = 0$. In such case, for $p = 1$,

$$\lim_n \int_E f_n d\mathbf{m} = \int_E f d\mathbf{m}, \quad E \in \sigma(\mathcal{P}),$$

the limit being uniform with respect to $E \in \sigma(\mathcal{P})$.

Proof. By hypothesis and by Theorem 3.2.1(i),

$$\sup_n \mathbf{m}_p^\bullet(f_n, T) = \mathbf{m}_p^\bullet(f, T). \tag{3.4.15.1}$$

Then (i)$\Rightarrow$(ii) by (3.4.15.1) and by Theorem 3.1.8; (ii)$\Rightarrow$(i) by Theorems 3.1.8 and by (3.4.15.1). If (i) or (ii) holds, then (ii) holds and hence the last part holds by Theorem 3.3.4. $\quad\square$

The hypothesis that $c_0 \not\subset X$ in the above theorem cannot be omitted as shown in the following counter-example.

Counter-example 3.4.16. Let T, S, X, $\mathbf{m}$ and f be as in Counter-example 3.1.7. Let $f_n = f\chi_{\{t \leq n\}}$ for $n \in \mathbb{N}$. Then $f_n(t) = |f_n(t)| \nearrow |f(t)| = f(t)$ for $t \in T$. As shown in the discussion of the said counter-example, $\mathbf{m}_p^\bullet(f, T) < \infty$ and by Theorem 3.2.1(i), $\sup_n \mathbf{m}_p^\bullet(f_n, T) = \mathbf{m}_p^\bullet(f, T)$. But, by Counter-example 3.1.7, $f \notin \mathcal{L}_p(\mathbf{m}) = \mathcal{I}_p(\mathbf{m})$ for $1 \leq p < \infty$.

The following result is an analogue of Theorem 6 of [Do3] for $\mathcal{L}_p(\mathbf{m})$, $1 \leq p < \infty$.

Theorem 3.4.17. *Let $1 \leq p < \infty$. Suppose $\mathbf{m}_n : \mathcal{P} \to X$ is σ-additive for $n \in \mathbb{N}$. Let $(f_k)_{k=1}^\infty$ be $\mathbf{m}_n$-measurable scalar functions for $n \in \mathbb{N}$, let f_0 be a scalar function on T and let $\lim_k f_k(t) = f_0(t)$ for $t \in T\backslash M$, where $M \in \sigma(\mathcal{P})$ with $\sup_n \|\mathbf{m}_n\|(M) = 0$. Suppose $\sup_n \|\mathbf{m}_n\|(E) < \infty$ for each $E \in \mathcal{P}$. Then:*

(i) *If $(\mathbf{m}_n)_p^\bullet(f_k, \cdot)$, $k, n \in \mathbb{N}$, are uniformly continuous on $\sigma(\mathcal{P})$, then*

$$\lim_{k \to \infty} \sup_n (\mathbf{m}_n)_p^\bullet(f_k - f_0, T) = 0. \tag{3.4.17.1}$$

Moreover, if $p = 1$, then

$$\lim_k \int_E f_k d\mathbf{m}_n = \int_E f_0 d\mathbf{m}_n, \quad E \in \sigma(\mathcal{P}),$$

the limit being uniform with respect to $n \in \mathbb{N}$ and $E \in \sigma(\mathcal{P})$.

(ii) **(Extended LDCT for $\mathcal{L}_p(\mathbf{m})$).** *Let $g \in \bigcap_{n=1}^{\infty} \mathcal{L}_p(\mathbf{m}_n)$ be such that $\mathbf{m}_n^{\bullet}(g, \cdot)$, $n \in \mathbb{N}$, are uniformly continuous on $\sigma(\mathcal{P})$. If, for each $n \in \mathbb{N}$, $|f_k(t)| \le |g(t)|$ $\mathbf{m}_n$-a.e. in T for all $k \in \mathbb{N}$, then the conclusions of (i) hold.*

Proof. (i) In the light of Definition 2.2.1 of Chapter 2, without loss of generality we shall assume that the functions $(f_k)_1^{\infty}$ are $\sigma(\mathcal{P})$-measurable. Then by hypothesis and by Theorems 3.2.7 and 3.3.1, $(f_k)_{k=0}^{\infty} \subset \mathcal{L}_p(\mathbf{m}_n)$ and $\lim_k (\mathbf{m}_n)_p^{\bullet}(f_k - f_0, T) = 0$ for each $n \in \mathbb{N}$. Let $\gamma_{n,k}(\cdot) = \int_{(\cdot)} |f_k|^p d\mathbf{m}_n$ for $k \in \mathbb{N} \cup \{0\}$. Then by Theorem 3.1.6 and by hypothesis,

$$||\gamma_{n,k}||(\cdot) = ((\mathbf{m}_n)_p^{\bullet}(f_k, \cdot))^p, \ k \in \mathbb{N} \cup \{0\} \tag{3.4.17.2}$$

are uniformly continuous on $\sigma(\mathcal{P})$ for each $n \in \mathbb{N}$ and hence, in particular, by Proposition 1.1.7 of Chapter 1, $(\gamma_{n,k})_{k=0}^{\infty}$ are uniformly σ-additive on $\sigma(\mathcal{P})$ for each $n \in \mathbb{N}$. Then, for each $n \in \mathbb{N}$, by Proposition 1.1.13 of Chapter 1 there exists a control measure $\mu_n : \sigma(\mathcal{P}) \to [0, \infty)$ for $(\gamma_{n,k})_{k=0}^{\infty}$. Let $K_n = \sup_{E \in \sigma(\mathcal{P})} \mu_n(E)$. Let

$$\mu(E) = \sum_{n=1}^{\infty} \frac{1}{2^n} \left(\frac{\mu_n(E)}{1 + K_n} \right) \text{ for } E \in \sigma(\mathcal{P}).$$

Then $\mu : \sigma(\mathcal{P}) \to [0, \infty)$ is σ-additive, and $\mu(N_0) = 0$ implies

$$\sup_{n \in \mathbb{N}, \, k \in \mathbb{N} \cup \{0\}} ||\gamma_{n,k}||(N_0) = 0$$

so that, by (3.4.17.2),

$$(\mathbf{m}_n)_p^{\bullet}(f_k, N_0) = 0 \tag{3.4.17.3}$$

for all $n \in \mathbb{N}$ and for all $k \in \mathbb{N} \cup \{0\}$.

Let $F = \bigcup_{k=1}^{\infty} N(f_k) \cap (T \backslash M)$. Then $F \in \sigma(\mathcal{P})$. By hypothesis, $f_k \chi_{T \backslash M} \to f_0 \chi_{T \backslash M}$ pointwise in T. Then by the Egoroff-Lusin theorem (see Proposition 1.1.20 of Chapter 1) there exist $N \in \sigma(\mathcal{P}) \cap F$ with $\mu(N) = 0$ and a sequence $(F_\ell)_1^{\infty} \subset \mathcal{P}$ with $F_l \nearrow F \backslash N$ such that $f_k \to f_0$ uniformly in each F_ℓ.

Let $\epsilon > 0$. Since $F \backslash N \backslash F_\ell \searrow \emptyset$ and since by hypothesis $(\mathbf{m}_n)_p^{\bullet}(f_k, \cdot)$, $k \in \mathbb{N} \cup \{0\}$, $n \in \mathbb{N}$, are uniformly continuous on $\sigma(\mathcal{P})$, there exists ℓ_0 such that $\sup_{n \in \mathbb{N}, \, k \in \mathbb{N} \cup \{0\}} (\mathbf{m}_n)_p^{\bullet}(f_k, F \backslash N \backslash F_\ell) < \frac{\epsilon}{3}$ for $\ell \ge \ell_0$. By hypothesis, there exists a finite constant K such that $\sup_n ||\mathbf{m}_n||(F_{\ell_0}) \le K$. As $f_k \to f_0$ uniformly in F_{ℓ_0}, there exists k_0 such that $||f_k - f_0||_{F_{\ell_0}} \cdot K^{\frac{1}{p}} < \frac{\epsilon}{3}$ for all $k \ge k_0$. Then, by Theorems 3.1.11 and 3.1.13 and by (3.4.17.3), we have

$$(\mathbf{m}_n)_p^{\bullet}(f_k - f_0, T) = (\mathbf{m}_n)_p^{\bullet}(f_k - f_0, F) = (\mathbf{m}_n)_p^{\bullet}(f_k - f_0, F \backslash N)$$
$$\le (\mathbf{m}_n)_p^{\bullet}(f_k - f_0, F_{\ell_0}) + (\mathbf{m}_n)_p^{\bullet}(f_k - f_0, F \backslash N \backslash F_{\ell_0})$$
$$< ||f_k - f_0||_{F_{\ell_0}} \cdot (||\mathbf{m}_n||(F_{\ell_0}))^{\frac{1}{p}} + (\mathbf{m}_n)_p^{\bullet}(f_k, F \backslash N \backslash F_{\ell_0}) + (\mathbf{m}_n)_p^{\bullet}(f_0, F \backslash N \backslash F_{\ell_0}) < \epsilon$$

for $k \ge k_0$ and for all $n \in \mathbb{N}$. Hence (3.4.17.1) holds. The last part of (i) is due to (3.4.17.1) and inequality (3.1.3.1).

(ii) is immediate from (i). $\qquad \square$

3.5   Relations between the spaces $\mathcal{L}_p(\mathbf{m})$

In this section we obtain results analogous to those in Section 5, §12 of [Din1] for the spaces $\mathcal{L}_p(\mathbf{m})$. The following theorem plays a key role in this section.

Theorem 3.5.1. (Hölder's inequality) *Let* $\mathbf{m} \to X$ *be* σ-*additive. Let* $1 < p < \infty$ *and let* $\frac{1}{p} + \frac{1}{q} = 1$. *If* $f \in \mathcal{L}_p(\mathbf{m})$ *and* $g \in \mathcal{L}_q(\mathbf{m})$, *then* $fg \in \mathcal{L}_1(\mathbf{m})$ *and*

$$\mathbf{m}_1^{\bullet}(fg, T) \leq \mathbf{m}_p^{\bullet}(f, T) \cdot \mathbf{m}_q^{\bullet}(g, T). \tag{3.5.1.1}$$

Proof. By Proposition 1.1.18 of Chapter 1 there exist $(s_n)_1^{\infty}$, $(\omega_n)_1^{\infty} \subset \mathcal{I}_s$ such that $s_n \to f$ and $|s_n| \nearrow |f|$ $\mathbf{m}$-a.e. in T and $\omega_n \to g$ and $|\omega_n| \nearrow |g|$ $\mathbf{m}$-a.e. in T. Then by hypothesis and by LDCT (Theorem 3.3.4), $\lim_n \mathbf{m}_p^{\bullet}(f - s_n, T) = 0$ and $\lim_n \mathbf{m}_q^{\bullet}(g - \omega_n, T) = 0$. Hence, given $\epsilon > 0$, there exists n_0 such that $\mathbf{m}_p^{\bullet}(s_n - s_r, T) \cdot \mathbf{m}_q^{\bullet}(g, T) < \frac{\epsilon}{2}$ and $\mathbf{m}_q^{\bullet}(\omega_n - \omega_r, T) \cdot \mathbf{m}_p(f, T) < \frac{\epsilon}{2}$ for $n, r \geq n_0$. Let $E \in \sigma(\mathcal{P})$. Then by Theorems 3.1.3 and 3.1.13(iii) we have

$$\left| \int_E s_n \omega_n d\mathbf{m} - \int_E s_r \omega_r d\mathbf{m} \right|$$

$$\leq \sup_{|x^*| \leq 1} \int_E |s_n(\omega_n - \omega_r)| dv(x^* \mathbf{m}) + \sup_{|x^*| \leq 1} \int_E |\omega_r(s_n - s_r)| dv(x^* \mathbf{m})$$

$$\leq \mathbf{m}_1^{\bullet}(s_n(\omega_n - \omega_r), T) + \mathbf{m}_1^{\bullet}(\omega_r(s_n - s_r), T)$$

$$\leq \mathbf{m}_p^{\bullet}(s_n, T) \cdot \mathbf{m}_q^{\bullet}(\omega_n - \omega_r, T) + \mathbf{m}_q^{\bullet}(\omega_r, T) \cdot \mathbf{m}_p^{\bullet}(s_n - s_r, T)$$

$$\leq \mathbf{m}_p^{\bullet}(f, T) \cdot \mathbf{m}_q^{\bullet}(\omega_n - \omega_r, T) + \mathbf{m}_q^{\bullet}(g, T) \cdot \mathbf{m}_p^{\bullet}(s_n - s_r, T)$$

$$< \epsilon$$

for $n, r \geq n_0$ and for all $E \in \sigma(\mathcal{P})$. Thus there exists $x_E \in X$ such that $\lim_n \int_E s_n \omega_n d\mathbf{m} = x_E$ for $E \in \sigma(\mathcal{P})$. Since $(s_n \omega_n)_1^{\infty} \subset \mathcal{I}_s$ and since $s_n \omega_n \to fg$ $\mathbf{m}$-a.e. in T such that $\lim_n \int_E s_n \omega_n d\mathbf{m} = x_E \in X$ for $E \in \sigma(\mathcal{P})$, by Definition 2.1.1 of Chapter 2, $fg \in \mathcal{L}_1(\mathbf{m})$. Inequality (3.5.1.1) holds by Theorem 3.1.13(iii). $\square$

Theorem 3.5.2. *Let* $\mathbf{m} : \mathcal{P} \to X$ *be* σ-*additive. Then:*

(i) *If* $1 \leq r < p < s \leq \infty$, *then* $\mathcal{L}_r(\mathbf{m}) \bigcap \mathcal{L}_s(\mathbf{m}) \subset \mathcal{L}_p(\mathbf{m})$.

(ii) *If* $f : T \to \mathbb{K}$ *is* $\mathbf{m}$-*measurable, then the set* $\mathcal{I}_f = \{p : 1 \leq p \leq \infty, f \in \mathcal{L}_p(\mathbf{m})\}$ *is either void or an interval, where singletons are considered as intervals.*

(iii) *For* f *in* (ii) *with* $\mathcal{I}_f \neq \emptyset$, *the function* $p \to \log \mathbf{m}_p^{\bullet}(f, T)$ *is convex on* $\mathcal{I}_f$ *and the function* $p \to \mathbf{m}_p^{\bullet}(f, T)$ *is continuous on* $Int \, \mathcal{I}_f$.

Proof. The proof of Proposition 21, §12 of [Din1] holds here in virtue of Theorems 2.1.5(vi) and 2.2.2 of Chapter 2 and of Theorems 3.1.3 and 3.5.1. Details are left to the reader. $\square$

Theorem 3.5.3. *Let* $\mathbf{m} : \mathcal{P} \to X$ *be* σ-*additive and let* $A \in \widetilde{\sigma(\mathcal{P})}$ *such that* χ_A *is* $\mathbf{m}$-*integrable in* T *(so that by Theorem 3.1.6,* $\mathbf{m}_1^{\bullet}(A, T) = ||\mathbf{m}||(A) < \infty$). *Then the set*

$\mathcal{I}_f(A) = \{p : 1 \leq p \leq \infty, \, f\chi_A \in \mathcal{L}_p(\mathbf{m})\}$ *is either void or an interval containing* 1 *($\mathcal{I}_f(A) = \{1\}$ is permitted) and the function* $p \to \mathbf{m}_p^{\bullet}(f\chi_A, T) \cdot (||\mathbf{m}||(A))^{-\frac{1}{p}}$ *is increasing in* $\mathcal{I}_f(A)$, *where* $\mathbf{m}_{\infty}^{\bullet}(f\chi_A, T) = ||f\chi_A||_{\infty}$.

Proof. By (vi) and (vii) (Domination principle) of Theorem 2.1.5 and by Theorem 2.2.2 of Chapter 2 and by Theorems 3.1.3 and 3.5.1, the proof of Proposition 22, §12 of [Din1] holds here verbatim and the details are left to the reader.   □

Corollary 3.5.4. *Let* $\mathbf{m} : \mathcal{S} \to X$ *be* σ-additive, *where* $\mathcal{S}$ *is a* σ-ring of subsets of T. *Then:*

(i) *If* $1 \leq r < s \leq \infty$, *then* $\mathcal{L}_s(\mathbf{m}) \subset \mathcal{L}_r(\mathbf{m})$ *and the topology of* $\mathcal{L}_s(\mathbf{m})$ *is finer than that of* $\mathcal{L}_r(\mathbf{m})$.

(ii) *If* $f : T \to \mathbb{K}$ *is* $\mathbf{m}$-measurable, then the set $\mathcal{I}_f = \{p : 1 \leq p \leq \infty, \, f \in \mathcal{L}_p(\mathbf{m})\}$ *is either void or an interval containing* 1 *($\mathcal{I}_f = \{1\}$ is permitted).*

(iii) *If* $\mathcal{I}_f \neq \emptyset$, *then the function* $p \to \mathbf{m}_p^{\bullet}(f, T) \cdot (||\mathbf{m}||(N(f))^{-\frac{1}{p}}$ *is an increasing function on* $\mathcal{I}_f$.

Proof. By the hypothesis on $\mathcal{S}$ and by the last part of Theorem 2.1.5(v) of Chapter 2, $\mathcal{L}_{\infty}(\mathbf{m}) \subset \mathcal{L}_p(\mathbf{m})$ for $1 \leq p \leq \infty$. Let $s > r$ and let $f \in \mathcal{L}_s(\mathbf{m})$. If $E = \{t \in T : |f(t)| > 1\}$ and $F = \{t \in N(f) : |f(t)| \leq 1\}$, then E and F belong to $\widetilde{\sigma(\mathcal{P})}$, $|f\chi_E|^r \leq |f\chi_E|^s \leq |f|^s \in \mathcal{L}_1(\mathbf{m})$ and $|f\chi_F|^r \leq \chi_F \in \mathcal{L}_1(\mathbf{m})$. Hence by Theorems 2.2.2 and 2.1.5(vii) (Domination principle) of Chapter 2, $f\chi_E$ and $f\chi_F$ belong to $\mathcal{L}_r(\mathbf{m})$ and hence $f \in \mathcal{L}_r(\mathbf{m})$. Therefore, the first part of (i) holds. The second part of (i) holds by Theorem 3.5.3, as $||\mathbf{m}||(N(f)) \leq ||\mathbf{m}||(T) < \infty$.

(ii) is due to Theorem 3.5.2(ii) and (i) of the present corollary.

(iii) is immediate if we take $A = N(f)$ in Theorem 3.5.3 since by the hypothesis on $\mathcal{S}$, $||\mathbf{m}||(T) < \infty$.   □

Remark 3.5.5. Theorems 3.5.2 and 3.5.3 and Corollary 3.5.4 are generalized to an lcHs-valued $\mathbf{m}$ in Theorem 4.5.13 of Chapter 4.

Chapter 4

Integration With Respect to lcHs-valued Measures

4.1 (KL) m-integrability (m lcHs-valued)

Let X be an lcHs and let $\mathbf{m} : \mathcal{P} \to X$ be σ-additive. The concepts of $\mathbf{m}$-measurable functions and (KL) $\mathbf{m}$-integrable functions given in Chapters 1 and 2 are suitably generalized here to lcHs-valued σ-additive measures. Theorem 4.1.4 below plays a key role in the subsequent theory of (KL) $\mathbf{m}$-integrability. While (i)–(iv) and (viii) of Theorem 2.1.5 of Chapter 2 are generalized in Theorem 4.1.8 to an arbitrary lcHs-valued σ-additive measure $\mathbf{m}$ on $\mathcal{P}$, the remaining parts of Theorem 2.1.5, Theorem 2.1.7 and Corollaries 2.1.8, 2.1.9 and 2.1.10 of Chapter 2 are generalized in Theorems 4.1.9 and 4.1.11 and in Corollaries 4.1.12, 4.1.13 and 4.1.14, respectively, when X is quasicomplete. Finally, the above mentioned results are also generalized to $\sigma(\mathcal{P})$-measurable functions in Remark 4.1.15 when X is sequentially complete.

Let X be an lcHs and let $\mathbf{m} : \mathcal{P} \to X$ be σ-additive. If $x^* \in X^*$, then q_{x^*} given by $q_{x^*}(x) = |x^*(x)|$, $x \in X$, belongs to Γ and by Proposition 1.2.13(iv), $\|\mathbf{m}_{q_{x^*}}\|(A) = v(x^*\mathbf{m})(A)$ for $A \in \sigma(\mathcal{P})$. Using this observation, we give the following

Definition 4.1.1. Let X be an lcHs and $\mathbf{m} : \mathcal{P} \to X$ be σ-additive. Let $f : T \to \mathbb{K}$ or $[-\infty, \infty]$ be $\mathbf{m}$-measurable. Then, for each $x^* \in X^*$, by Definition 1.2.6 there exists $N_{x^*} \in \sigma(\mathcal{P})$ with $v(x^*\mathbf{m})(N_{x^*}) = 0$ such that $f\chi_{T \setminus N_{x^*}}$ is $\sigma(\mathcal{P})$-measurable. We say that f is $x^*\mathbf{m}$-integrable if $f\chi_{T \setminus N_{x^*}}$ is $x^*\mathbf{m}$-integrable and in that case, we define

$$\int_A f \, d(x^*\mathbf{m}) = \int_A f\chi_{T \setminus N_{x^*}} \, d(x^*\mathbf{m})$$

for $A \in \sigma(\mathcal{P})$ and

$$\int_T f d(x^* \mathbf{m}) = \int_{N(f) \setminus N_{x^*}} f d(x^* \mathbf{m}).$$

(**Note that** $N(f) \setminus N_{x^*} = N(f\chi_{T \setminus N_{x^*}}) \in \sigma(\mathcal{P}).$)

It is easy to check that the above integrals are well defined and do not depend on the set N_{x^*}.

The following definition generalizes the second part of Definition 2.1.1 of Chapter 2 to lcHs-valued σ-additive $\mathbf{m}$ on $\mathcal{P}$.

Definition 4.1.2. Let X be an lcHs and $\mathbf{m} : \mathcal{P} \to X$ be σ-additive. Let $f : T \to \mathbb{K}$ or $[-\infty, \infty]$ be $\mathbf{m}$-measurable. Then f is said to be (KL) $\mathbf{m}$-integrable in T if it is $x^*\mathbf{m}$-integrable for each $x^* \in X^*$ and if, for each $A \in \sigma(\mathcal{P}) \cup \{T\}$, there exists a vector $x_A \in X$ such that

$$x^*(x_A) = \int_A f d(x^* \mathbf{m})$$

for each $x^* \in X^*$. In that case, we define

$$\text{(KL)} \int_A f d\mathbf{m} = x_A, \ A \in \sigma(\mathcal{P}) \cup \{T\}. \tag{4.1.2.1}$$

By the Hahn-Banach theorem the integral in (4.1.2.1) is well defined for each $A \in \sigma(\mathcal{P}) \cup \{T\}$. Also note that the above definition includes the definition of (KL) $\mathbf{m}$-integrability given in [L2] when f is $\sigma(\mathcal{P})$-measurable.

In the sequel, X is an lcHs and $\mathbf{m} : \mathcal{P} \to X$ is σ-additive.

Proposition 4.1.3. *If* $f : T \to [-\infty, \infty]$ *is (KL)* $\mathbf{m}$-*integrable in* T, *then* f *is finite* $\mathbf{m}$-*a.e. in* T.

Proof. Let $A = \{t \in T : |f(t)| = \infty\}$. As f is $\mathbf{m}$-measurable, for each $q \in \Gamma$, there exist B_q, N_q, M_q such that $A = B_q \cup N_q, N_q \subset M_q, B_q, M_q \in \sigma(\mathcal{P})$ and $||\mathbf{m}||_q(M_q) = 0$. Then by Proposition 1.2.15(ii)(c),

$$||\mathbf{m}||_q(A) = ||\mathbf{m}||_q(B_q) = \sup_{x^* \in U_q^o} v(x^* \mathbf{m})(B_q). \tag{4.1.3.1}$$

On the other hand, for each $x^* \in X^*$, by Definition 4.1.1 there exists $N_{x^*} \in \sigma(\mathcal{P})$ with $v(x^* \mathbf{m})(N_{x^*}) = 0$ such that $f\chi_{T \setminus N_{x^*}}$ is $\sigma(\mathcal{P})$-measurable and $x^*\mathbf{m}$-integrable. Then $f\chi_{T \setminus N_{x^*}}$ and hence f are $(x^* \mathbf{m})$-a.e. finite in T. Consequently, by (4.1.3.1) we have $||\mathbf{m}||_q(A) = 0$ for all $q \in \Gamma$ and hence A is $\mathbf{m}$-null. $\qquad \square$

The second part of the following theorem and Remark 4.1.5 below play a key role in the sequel.

Theorem 4.1.4. *Let* $f : T \to \mathbb{K}$ *or* $[-\infty, \infty]$ *be* $\mathbf{m}$-*measurable and let* X *be an lcHs. Let* $\mathbf{m} : \mathcal{P} \to X$ *be* σ-*additive. Then:*

(i) *If f is (KL) **m**-integrable in T, then f is (KL) $\mathbf{m}_q$-integrable in T with values in X_q (in the sense that the integrals of f assume values in X_q) for each $q \in \Gamma$ (resp. (KL) $\mathbf{m}_{q_E}$-integrable in T with values in X_{q_E} for each $E \in \mathcal{E}$).*

(ii) *If f is (KL) $\mathbf{m}_q$-integrable in T with values in $\widetilde{X_q}$ (i.e., the integrals of f assume values in $\widetilde{X_q}$) for each $q \in \Gamma$ (resp. if f is (KL) $\mathbf{m}_{q_E}$-integrable in T with values in $\widetilde{X_{q_E}}$ for each $E \in \mathcal{E}$), then f is (KL) **m**-integrable in T with values in $\tilde{X}$ and*

$$(KL)\int_A f d\mathbf{m} = \varprojlim (KL)\int_A f d\mathbf{m}_q (resp. = \varprojlim (KL)\int_A f d\mathbf{m}_{q_E})$$

for $A \in \sigma(\mathcal{P}) \cup \{T\}$.

Proof. (i) Suppose f is (KL) **m**-integrable in T. Let $A \in \sigma(\mathcal{P}) \cup \{T\}$. Then, for each $x^* \in X^*$, f is x^***m**-integrable and there exists $x_A \in X$ such that $x^*(x_A) = \int_A f d(x^* \mathbf{m})$. Let $q \in \Gamma$ and let $y^* \in (X_q)^*$. Then $y^* \Pi_q \in X^*$ and hence f is $y^* \Pi_q \mathbf{m}$-integrable and $(y^* \Pi_q)(x_A) = \int_A f d(y^* \Pi_q \mathbf{m}) = \int_A f d(y^* \mathbf{m}_q)$. Hence f is (KL) $\mathbf{m}_q$-integrable in T with values in X_q. Since $q_E \in \Gamma$ for $E \in \mathcal{E}$, f is (KL) $\mathbf{m}_{q_E}$-integrable in T with values in X_{q_E} for $E \in \mathcal{E}$.

(ii) Suppose

$$f \text{ is (KL) } \mathbf{m}_q\text{-integrable in } T \text{ with values in } \widetilde{X_q} \text{ for each } q \in \Gamma. \qquad (4.1.4.1)$$

Let $A \in \sigma(\mathcal{P}) \cup \{T\}$ and let $q \in \Gamma$. Then there exists $y_A^{(q)} \in \widetilde{X_q}$ such that

$$y^*(y_A^{(q)}) = \int_A f d(y^* \mathbf{m}_q) \qquad (4.1.4.2)$$

for $y^* \in (X_q)^*$ so that

$$y_A^{(q)} = (KL)\int_A f d\mathbf{m}_q. \qquad (4.1.4.3)$$

By the proof of Theorem 5.4, Chapter II of [Scha], there exists $x_A \in \tilde{X}$ such that

$$x_A = \varprojlim y_A^{(q)}. \qquad (4.1.4.4)$$

(See Notation 1 2.17.)

Let $x^* \in X^*$ be arbitrary. Then q_{x^*} given by $q_{x^*}(x) = |x^*(x)|$, $x \in X$, belongs to Γ and clearly, $\widetilde{q_{x^*}}(x) = |x^*(x)|$ for $x \in \tilde{X}$. (See Notation 1.2.16.) In the representation of the dual of $\tilde{X}$ given in §22, 6.(6) of [KÖ], take $u = (u_q)_{q \in \Gamma} \in \tilde{X}^*$ where $u_{q_{x^*}} = \Psi_{x^*} \in (X_{\widetilde{q_{x^*}}})^* = \widetilde{(X_{q_{x^*}})^*}$ as in Proposition 1.2.13(i) and $u_q = 0$ for $q \in \Gamma \backslash \{q_{x^*}\}$. Then, taking $E = \{u\} \in \mathcal{E}$, by (4.1.4.2) and (4.1.4.4) and by Proposition 1.2.13(i) we have $u(x_A) = u(\varprojlim y_A^{(q)}) = \Psi_{x^*}(\Pi_{q_{x^*}}(x_A)) =$

$\Psi_{x^*}(y_A^{(q_{x^*})}) = \int_A f d(\Psi_{x^*}\Pi_{q_{x^*}} \circ \mathbf{m}) = \int_A f d(u\mathbf{m})$. Hence f is (KL) $\mathbf{m}$-integrable in T with (KL)$\int_A f d\mathbf{m} = x_A \in \tilde{X}$ so that by (4.1.4.3) and (4.1.4.4) we have

(KL)$\int_A f d\mathbf{m} = x_A = \varprojlim y_A^{(q)} = \varprojlim (\mathrm{KL}) \int_A f d\mathbf{m}_q$ for $A \in \sigma(\mathcal{P}) \cup \{T\}$.

If condition (4.1.4.1) holds for $\{q_E : E \in \mathcal{E}\}$, then it holds all $q \in \Gamma$ by Proposition 1.2.15(i) since $q = q_{U_q^o}$ and $U_q^o \in \mathcal{E}$ by Proposition 6, §4, Chapter 3 of [Ho]. Hence f is (KL) $\mathbf{m}$-integrable in T with values in $\tilde{X}$ and other results also hold.   □

Remark 4.1.5. By Theorems 4.2.2 and 4.2.3 below, (ii) of the above theorem can be strengthened as follows:

If X is a quasicomplete lcHs and if f is (KL) $\mathbf{m}_q$-integrable (resp. (KL) $\mathbf{m}_{q_E}$-integrable) in T with values in $\widetilde{X_q}$ (resp. $\widetilde{X_{q_E}}$) for each $q \in \Gamma$ (resp. $E \in \mathcal{E}$), then f is (KL) $\mathbf{m}$-integrable in T with values in X and $(\mathbf{KL})\int_A f d\mathbf{m} = \varprojlim (\mathbf{KL}) \int_A f d\mathbf{m}_q$ (resp. $= \varprojlim (\mathbf{KL}) \int_A f d\mathbf{m}_{q_E}$) belongs to X for $A \in \sigma(\mathcal{P}) \cup \{T\}$.

Notation 4.1.6. Let X be an lcHs and let $\mathbf{m} : \mathcal{P} \to X$ be σ-additive. Then $\mathcal{I}(\mathbf{m})$ denotes the class of all $\mathbf{m}$-measurable scalar functions f on T which are (KL) $\mathbf{m}$-integrable in T with (KL) $\int_A f d\mathbf{m} \in X$ for all $A \in \sigma(\mathcal{P}) \cup \{T\}$.

The following lemma is needed to generalize Theorem 2.1.5(viii) of Chapter 2 to lcHs.

Lemma 4.1.7. *Let X, Y be lcHs over the same scalar field $\mathbb{K}$ and let $L(X, Y)$ be the vector space of all continuous linear mappings from X into Y. If $\mathbf{m} : \mathcal{P} \to X$ is σ-additive and $u \in L(X, Y)$, then $u\mathbf{m} : \mathcal{P} \to Y$ is σ-additive. If $f : T \to \mathbb{K}$ or $[-\infty, \infty]$ is $\mathbf{m}$-measurable, then f is also $u\mathbf{m}$-measurable.*

Proof. As u is linear and continuous, $u\mathbf{m}$ is σ-additive. Let $\mathcal{F}$ be the family of equicontinuous subsets of Y^*. Let $F \in \mathcal{F}$. Then, given $\epsilon > 0$, there exists a neighborhood W of 0 in Y such that $\sup_{y^* \in F} |y^*(y)| < \epsilon$ for $y \in W$. As u is continuous and linear, there exists a neighborhood U of 0 in X such that $u(U) \subset W$. Then $\sup_{y^* \in F} |y^*(ux)| < \epsilon$ for $x \in U$. Let u^* be the adjoint of u. Then $\sup_{y^* \in F} |(u^*y^*)(x)| = \sup_{y^* \in F} |(y^*u)(x)| < \epsilon$ for $x \in U$ and hence $u^*F \in \mathcal{E}$. As f is $\mathbf{m}$-measurable, there exists $N_{u^*F} \in \sigma(\mathcal{P})$ with $||\mathbf{m}||_{q_{u^*F}}(N_{u^*F}) = 0$ such that $f\chi_{T \setminus N_{u^*F}}$ is $\sigma(\mathcal{P})$-measurable. Then by Proposition 1.2.13(iii) we have

$$||u\mathbf{m}||_{q_F}(N_{u^*F}) = ||\Pi_{q_F} \circ u\mathbf{m}||(N_{u^*F}) = \sup_{y^* \in F} v(y^* u\mathbf{m})(N_{u^*F})$$

$$= \sup_{y^* \in F} v(u^*y^*\mathbf{m})(N_{u^*F}) = \sup_{x^* \in u^*F} v(x^*\mathbf{m})(N_{u^*F}) = ||\mathbf{m}||_{q_{u^*F}}(N_{u^*F}) = 0$$

and hence f is $(u\mathbf{m})_{q_F}$-measurable. Then, in the light of Remark 1.2.5 of Chapter 1, f is $u\mathbf{m}$-measurable.   □

We generalize below (i)–(iv) and (viii) of Theorem 2.1.5 of Chapter 2 to a quasicomplete lcHs-valued σ-additive **m**.

Theorem 4.1.8. *Let X be a quasicomplete lcHs and* $\mathbf{m} : \mathcal{P} \to X$ *be σ-additive. Then:*

(i) *A $\mathcal{P}$-simple function $s = \sum_1^r \alpha_i \chi_{A_i}$ with $(\alpha_i) \subset \mathbb{K}$, $(A_i)_1^r \subset \mathcal{P}$ and $A_i \cap A_j = \emptyset$ for $i \neq j$, is (KL) **m**-integrable in T and $(KL)\int_A s d\mathbf{m} = \sum_1^r \alpha_i \mathbf{m}(A_i \cap A)$ for $A \in \sigma(\mathcal{P})$. We write $\int_A s d\mathbf{m}$ instead of $(KL)\int_A s d\mathbf{m}$. Consequently, $||\mathbf{m}||_q(A) = \sup\{q(\int_A s d\mathbf{m}) = |\int_A s d\mathbf{m}_q|_q : s \,\mathcal{P}\text{-simple}, |s(t)| \leq \chi_A(t), t \in T\}$ for $q \in \Gamma$.*

(ii) *If $f : T \to \mathbb{K}$ or $[-\infty,\infty]$ is (KL) **m**-integrable in T, then $\gamma(\cdot) = (KL)\int_{(\cdot)} f d\mathbf{m}$ is σ-additive (in τ) on $\sigma(\mathcal{P})$.*

(iii) *If γ is as in (ii), $q \in \Gamma$ and $E \in \mathcal{E}$, then:*
  (a) $||\gamma||_q(A) = \sup_{x^* \in U_q^o} \int_A |f| dv(x^*\mathbf{m})$, $A \in \sigma(\mathcal{P})$.
  (b) $||\gamma||_{q_E}(A) = \sup_{x^* \in E} \int_A |f| dv(x^*\mathbf{m})$, $A \in \sigma(\mathcal{P})$.
  (c) $\lim_{||\mathbf{m}||_q(A) \to 0} \gamma_q(A) = \lim_{||\mathbf{m}||_q(A) \to 0} ||\gamma||_q(A) = 0$, $A \in \sigma(\mathcal{P})$.

(iv) *$\mathcal{I}(\mathbf{m})$ is a vector space over $\mathbb{K}$ with respect to pointwise addition and scalar multiplication. For $A \in \sigma(\mathcal{P})$ fixed, the mapping $f \to (KL)\int_A f d\mathbf{m}$ is linear on $\mathcal{I}(\mathbf{m})$ with values in X by Remark 4.1.5. Consequently, if s in (i) is with (A_i) not necessarily mutually disjoint, then also $\int_A s d\mathbf{m} = \sum_1^r \alpha_i \mathbf{m}(A \cap A_i)$, $A \in \sigma(\mathcal{P})$.*

(v) *Let Y, $L(X,Y)$, u and f be as in Lemma 4.1.7. If f is **m**-measurable and (KL) **m**-integrable in T, then f is $u\mathbf{m}$-measurable and (KL) $u\mathbf{m}$-integrable in T and $u((KL)\int_A f d\mathbf{m}) = (KL)\int_A f du\mathbf{m}$, $A \in \sigma(\mathcal{P}) \cup \{T\}$.*

(vi) *(i), (ii), (iii) and (v) hold for an arbitrary lcHs X.*

Proof. (i) and (iv) are obvious and (ii) is due to the Orlicz-Pettis theorem (for example, see [McA]).

(iii)(a) (resp.(b)) By (ii) γ is σ-additive on $\sigma(\mathcal{P})$ and hence by Proposition 1.1.19 and by Theorem 2.1.5(iii) of Chapter 2, by Proposition 1.2.15(ii)(c) (resp. by Proposition 1.2.13(iii)) and by the fact that $\gamma_q = \Pi_q \circ \gamma$ (resp. $\gamma_{q_E} = \Pi_{q_E} \circ \gamma$) the result holds.

(iii)(c) If $||\mathbf{m}||_q(A) = 0$, $A \in \sigma(\mathcal{P})$, then by Proposition 1.2.15(ii)(c), $v(x^*\mathbf{m})(A) = 0$ for $x^* \in U_q^o$ and hence by (a), $||\gamma||_q(A) = 0$. As $||\mathbf{m}||_q$ is a σ-subadditive submeasure on $\sigma(\mathcal{P})$ and as $\gamma_q = \Pi_q \circ \gamma : \sigma(\mathcal{P}) \to \widetilde{X_q}$ is σ-additive, the second equality holds by Proposition 1.1.11. As $|\gamma(E)|_q \leq ||\gamma||_q(E)$, $E \in \sigma(\mathcal{P})$, the first equality also holds.

(v) By Lemma 4.1.7, $u\mathbf{m}$ is σ-additive on $\sigma(\mathcal{P})$ and f is $u\mathbf{m}$-measurable. Let $A \in \sigma(\mathcal{P}) \cup \{T\}$ and let $(KL)\int_A f d\mathbf{m} = x_A \in X$. For $y^* \in Y^*$, $u^*y^* \in X^*$ and hence f is $u^*y^*\mathbf{m}$-integrable and $y^*(ux_A) = u^*y^*(x_A) = \int_A f d(u^*y^*\mathbf{m}) = \int_A f d(y^*u\mathbf{m})$. Hence the result holds. $\square$

The following theorem generalizes (v)–(vii) of Theorem 2.1.5 of Chapter 2 to quasicomplete lcHs-valued σ-additive vector measures.

Theorem 4.1.9. *Let X be a quasicomplete lcHs and $\mathbf{m} : \mathcal{P} \to X$ be σ-additive. Then:*

(i) *Let $f : T \to \mathbb{K}$ or $[-\infty, \infty]$ be $\mathbf{m}$-measurable.*

    (a) *If f is $\mathbf{m}$-essentially bounded in $A \in \mathcal{P}$, then f is $(KL)\,\mathbf{m}$-integrable in A with values in $\tilde{X}$ (in fact, with values in X by Remark 4.1.5) and, for each $q \in \Gamma$,*

$$q\left((KL)\int_B f\,d\mathbf{m}\right) = |(KL)\int_B f\,d\mathbf{m}_q|_q \le (\text{ess}\sup_{t \in A}|f(t)|) \cdot ||\mathbf{m}||_q(B)$$

    for $B \in A \cap \mathcal{P}\,(= \sigma(\mathcal{P}) \cap A)$.

    (b) *If f is $\mathbf{m}$-essentially bounded in T and if $\mathcal{P}$ is a σ-ring $\mathcal{S}$, then f is $(KL)\,\mathbf{m}$-integrable in T with values in $\tilde{X}$ (in fact, in X by Remark 4.1.5) and, for each $q \in \Gamma$,*

$$q\left((KL)\int_A f\,d\mathbf{m}\right) = |(KL)\int_A f\,d\mathbf{m}_q|_q \le (\text{ess}\sup_{t \in T}|f(t)|) \cdot ||\mathbf{m}||_q(A)$$

$$\le (\text{ess}\sup_{t \in T}|f(t)|) \cdot ||\mathbf{m}||_q(T)$$

    for $A \in \mathcal{S} \cup \{T\}$.

(ii) *If φ is an $\mathbf{m}$-essentially bounded $\mathbf{m}$-measurable scalar function on T and if $f \in \mathcal{I}(\mathbf{m})$, then $\varphi f \in \mathcal{I}(\mathbf{m})$. Consequently, for $f \in \mathcal{I}(\mathbf{m})$ and for $A \in \sigma(\mathcal{P}) \cup \{T\}$, $f\chi_A \in \mathcal{I}(\mathbf{m})$ and $(KL)\int_A f\,d\mathbf{m} = (KL)\int_T f\chi_A d\mathbf{m}$.*

(iii) **(Domination principle).** *If f is an $\mathbf{m}$-measurable scalar function on T and if $g \in \mathcal{I}(\mathbf{m})$ with $|f| \le |g|$ $\mathbf{m}$-a.e. in T, then $f \in \mathcal{I}(\mathbf{m})$. Consequently, an $\mathbf{m}$-measurable function $f : T \to \mathbb{K}$ or $[-\infty, \infty]$ is $(KL)\,\mathbf{m}$-integrable in T (with values in X) if and only if $|f|$ is so. Moreover, for an $\mathbf{m}$-measurable scalar function f on T the following statements are equivalent:*

    (a) $f \in \mathcal{I}(\mathbf{m})$;   (b) $|f| \in \mathcal{I}(\mathbf{m})$;   (c) $\bar{f} \in \mathcal{I}(\mathbf{m})$;

    (d) $Re\,f \in \mathcal{I}(\mathbf{m})$ and $Im\,f \in \mathcal{I}(\mathbf{m})$;

    (e) $(Re\,f)^+$, $(Im\,f)^+$, $(Re\,f)^-$ and $(Im\,f)^-$ belong to $\mathcal{I}(\mathbf{m})$.

    Moreover, if $f_1, f_2 : T \to \mathbb{R}$ belong to $\mathcal{I}(\mathbf{m})$, then $\max(f_1, f_2)$ and $\min(f_1, f_2)$ belong to $\mathcal{I}(\mathbf{m})$.

Proof. (i)(a) Let $A \in \mathcal{P}$ and let $\alpha = \text{ess}\sup_{t \in A}|f(t)|$.

Claim. There is an $\mathbf{m}$-null set $M \in \sigma(\mathcal{P})$ such that

$$\alpha = \sup_{t \in A \setminus M}|f(t)|. \tag{4.1.9.1}$$

In fact, for each $n \in \mathbb{N}$, there exists $M_n \in \sigma(\mathcal{P})$ with $|f(t)| \leq \alpha + \frac{1}{n}$ for $t \in A \backslash M_n$ and $||\mathbf{m}_i||(M_n) = 0$ $i \in \Gamma$. Then by Propositions 1.1.5 and 1.1.12, $||\mathbf{m}_i||(\bigcup_1^\infty M_n) = 0$ for for each $i \in \Gamma$. Moreover, $|f(t)| \leq \alpha$ for $t \in A \backslash \bigcup_1^\infty M_n$). Then it follows that $\alpha = \sup_{t \in A \backslash \bigcup_1^\infty A_n} |f(t)|$. Hence (4.1.9.1) holds.

Thus by (4.1.9.1) there exists an **m**-null set $M \in \sigma(\mathcal{P})$ such that $\alpha = \sup_{t \in A \backslash M} |f(t)|$. Then $f\chi_{A \backslash M}$ is a bounded **m**-measurable function on T. Let $q \in \Gamma$. Then there exists $N_q \in \sigma(\mathcal{P})$ with $||\mathbf{m}||_q(N_q) = 0$ such that $f\chi_{A \backslash M \backslash N_q}$ is $\sigma(\mathcal{P})$-measurable. Let $\Sigma_A = A \cap \mathcal{P}$. Then Σ_A is a σ-algebra of subsets of A, $\Sigma_A \subset \mathcal{P}$ and $f\chi_{A \backslash M \backslash N_q}$ is a bounded Σ_A-measurable function. Therefore, there exists a sequence $(s_n^{(q)})_1^\infty$ of Σ_A-simple functions such that $s_n^{(q)} \to f\chi_{A \backslash M \backslash N_q}$ uniformly in A with $|s_n^{(q)}| \nearrow |f\chi_{A \backslash M \backslash N_q}|$. Moreover, $\mathbf{m}_q$ is σ-additive on Σ_A with values in $X_q \subset \widetilde{X_q}$, and hence by Proposition 1.1.6 of Chapter 1 , $||\mathbf{m}||_q(A) < \infty$. Then by an argument similar to that in the proof of Theorem 2.1.5(v) of Chapter 2, for each $B \in \Sigma_A$, f is (KL) $\mathbf{m}_q$-integrable in B and there exists $x_B^{(q)} \in \widetilde{X_q}$ such that $(KL)\int_B f d\mathbf{m}_q = x_B^{(q)} = \lim_n \int_B s_n^{(q)} d\mathbf{m}_q$ and

$$\left| (KL) \int_B f d\mathbf{m}_q \right|_q = \left| \lim_n \int_B s_n^{(q)} d\mathbf{m}_q \right|_q \leq \alpha ||\mathbf{m}||_q(B). \tag{4.1.9.2}$$

As q is arbitrary in Γ, by Theorem 4.1.4(ii), f is (KL) **m**-integrable in A with values in $\widetilde{X}$ (and hence with values in X by Remark 4.1.5). Thus, if $(KL)\int_B f d\mathbf{m} = z_B \in X$ for $B \in \Sigma_A$, then $\Pi_q(z_B) = (KL)\int_B f d\mathbf{m}_q$ for $q \in \Gamma$ and for $B \in \Sigma_A$. Moreover, by (4.1.9.2) we have

$$q\left((KL) \int_B f d\mathbf{m} \right) = q(z_B) = |\Pi_q(z_B)|_q$$

$$= \left| (KL) \int_B f d\mathbf{m}_q \right|_q \leq \left(\operatorname*{ess\,sup}_{t \in A} |f(t)| \right) \cdot ||\mathbf{m}||_q(B)$$

for $B \in \Sigma_A$. Hence (i)(a) holds.

(i)(b) This is immediate from (i)(a) as $||\mathbf{m}||_q(T) < \infty$ by the hypothesis that $\mathcal{P}$ is a σ-ring and by Proposition 1.1.6 of Chapter 1 and as $f\chi_{T \backslash M \backslash N_q}$ is bounded and $\sigma(\mathcal{P})$-measurable.

(ii) Let $\boldsymbol{\nu}(\cdot) = (KL)\int_{(\cdot)} f d\mathbf{m}$ on $\sigma(\mathcal{P})$. By Theorem 4.1.8(ii), $\boldsymbol{\nu}$ is σ-additive on $\sigma(\mathcal{P})$ and by Theorem 4.1.8(iii)(c), for $q \in \Gamma$, an $\mathbf{m}_q$-null set in $\sigma(\mathcal{P})$ is also $\boldsymbol{\nu}_q$-null. Hence, the **m**-essentially bounded **m**-measurable function φ is also $\boldsymbol{\nu}$-measurable and $\boldsymbol{\nu}$-essentially bounded. Therefore, by (i)(b), φ is (KL) $\boldsymbol{\nu}$-integrable in T with values in X. For $A \in \sigma(\mathcal{P}) \cup \{T\}$, let $(KL)\int_A \varphi d\boldsymbol{\nu} = x_A \in X$. Then, for each $x^* \in X^*$, by Definition 4.1.1 there exists $N_{x^*} \in \sigma(\mathcal{P})$ with $v(x^*\boldsymbol{\nu})(N_{x^*}) = 0$ such that $\varphi\chi_{T \backslash N_{x^*}}$ is $\sigma(\mathcal{P})$-measurable and $x^*(x_A) = \int_A \varphi\chi_{T \backslash N_{x^*}} d(x^*\boldsymbol{\nu})$. As f is (KL) **m**-integrable in T, there exists $M_{x^*} \in \sigma(\mathcal{P})$ with $v(x^*\mathbf{m})(M_{x^*}) = 0$ such that $f\chi_{T \backslash M_{x^*}}$ is $\sigma(\mathcal{P})$-measurable and $x^*\mathbf{m}$-integrable. Then by Theorem

4.1.8(iii)(b) and by Proposition 1.2.13(iv), $0 = v(x^*\boldsymbol{\nu})(N_{x^*}) = ||x^*\boldsymbol{\nu}||(N_{x^*}) = ||\Pi_{q_{x^*}}\boldsymbol{\nu}||(N_{x^*}) = ||\boldsymbol{\nu}||_{q_{x^*}}(N_{x^*}) = \int_{N_{x^*}} |f| dv(x^*\mathbf{m})$ since $E = \{x^*\} \in \mathcal{E}$. Hence either $f = 0$ $x^*\mathbf{m}$-a.e. in N_{x^*} or N_{x^*} is $x^*\mathbf{m}$-null. In either case, $\varphi\chi_{T\backslash N_{x^*}} f\chi_{T\backslash M_{x^*}}$ is $\sigma(\mathcal{P})$-measurable and is equal to φf $x^*\mathbf{m}$-a.e in T. Then

$$x^*(x_A) = \int_A \varphi\chi_{T\backslash N_{x^*}} d(x^*\boldsymbol{\nu}) = \int_A \varphi\chi_{T\backslash N_{x^*}} f\chi_{T\backslash M_{x^*}} d(x^*\mathbf{m})$$

for $A \in \sigma(\mathcal{P}) \cup \{T\}$ and for $x^* \in X^*$, since the last equality holds for $\mathcal{P}$-simple functions φ and then by Proposition 1.1.18 of Chapter 1 and by Theorem 3 (LDCT), §8 of [Din1], it holds for general $\sigma(\mathcal{P})$-measurable $x^*\boldsymbol{\nu}$-integrable $\varphi\chi_{T\backslash N_{x^*}}$. Then by Definition 2.1.1(ii) of Chapter 2, $x^*(x_A) = \int_A \varphi f d(x^*\mathbf{m})$, as $\varphi\chi_{T\backslash N_{x^*}} f\chi_{T\backslash M_{x^*}} = \varphi f$ $(x^*\mathbf{m})$-a.e. in T. Hence $\varphi f \in \mathcal{I}(\mathbf{m})$. The second part follows from the first part by an argument similar to that in the proof of the second part of Theorem 2.1.5(vi) of Chapter 2.

(iii) Let $h(t) = \frac{f(t)}{g(t)}$ for $t \in N(g)$ and $h(t) = 0$ otherwise. Then h is clearly $\mathbf{m}$-measurable, $|h(t)| \leq 1$ $\mathbf{m}$-a.e. in T and $f = gh$ $\mathbf{m}$-a.e. in T. Then by (ii), f is (KL) $\mathbf{m}$-integrable in T with values in X. If $f : T \to \mathbb{K}$ or $[-\infty, \infty]$ is (KL) $\mathbf{m}$-integrable in T, then by Proposition 4.1.3 there exists an $\mathbf{m}$-null set $N \in \sigma(\mathcal{P})$ such that f is finite in $T\backslash N$ and hence $g = f\chi_{T\backslash N} \in \mathcal{I}(\mathbf{m})$ and by the above part, $|g|$ is (KL) $\mathbf{m}$-integrable in T. Then $|f|$ is also (KL) $\mathbf{m}$ integrable in T as $|f| = |g|$ $\mathbf{m}$-a.e. in T. Conversely, if $|f|$ is (KL) $\mathbf{m}$-integrable in T with values in X, then by Theorem 4.1.4(i), $|f|$ is (KL) $\mathbf{m}_q$-integrable in T with values in X_q for each $q \in \Gamma$ and hence by Theorem 2.1.5(vii) of Chapter 2 , f is (KL) $\mathbf{m}_q$-integrable in T with values in $\widetilde{X_q}$ for each $q \in \Gamma$. Consequently, by Theorem 4.1.4(ii) and by Remark 4.1.5, f is (KL) $\mathbf{m}$-integrable in T with values in X. The other parts follow from the first, and from the facts that $\max(f_1, f_2) = \frac{1}{2}(f_1 + f_2 + |f_1 - f_2|)$ and $\min(f_1, f_2) = \frac{1}{2}(f_1 + f_2 - |f_1 - f_2|)$ and that $\mathcal{I}(\mathbf{m})$ is a vector space. $\square$

Remark 4.1.10. The proofs of (ii) and (iii) of the above theorem are respectively similar to those of Lemma II.3.1 and Theorem II.3.1 of [KK]. See Remark 2.1.11 of Chapter 2.

Theorem 4.1.11. (Generalization of Theorem 2.1.7 of Chapter 2 – (LDCT a.e. version)) *Let X be a quasicomplete lcHs and $\mathbf{m} : \mathcal{P} \to X$ be σ-additive. For each $q \in \Gamma$, let $f_n^{(q)}$, $n \in \mathbb{N}$, be $\mathbf{m}_q$-measurable on T with values in $\mathbb{K}$ or in $[-\infty, \infty]$ and let $g^{(q)} : T \to \mathbb{K}$ be $\mathbf{m}_q$-measurable and (KL) $\mathbf{m}_q$-integrable in T with values in $\widetilde{X_q}$ for each $q \in \Gamma$. Suppose $|f_n^{(q)}(t)| \leq |g^{(q)}(t)|$ $\mathbf{m}_q$-a.e. in T for each $q \in \Gamma$ and for each $n \in \mathbb{N}$ and let $f : T \to \mathbb{K}$. If $f_n^{(q)} \to f$ $\mathbf{m}_q$-a.e. in T, then f is $\mathbf{m}_q$-measurable, f, $f_n^{(q)}$, $n \in \mathbb{N}$, are (KL) $\mathbf{m}_q$-integrable in T with values in $\widetilde{X_q}$ and consequently, f is (KL) $\mathbf{m}$-integrable in T with values in X. Moreover, for each $q \in \Gamma$,*

$$\lim_n \left| (KL)\int_A f d\mathbf{m}_q - (KL)\int_A f_n^{(q)} d\mathbf{m}_q \right|_q = 0 \text{ for } A \in \sigma(\mathcal{P}) \cup \{T\} \qquad (4.1.11.1)$$

the limit being uniform with respect to $A \in \sigma(\mathcal{P})$ (for a fixed $q \in \Gamma$) and

$$\lim_n \sup_{x^* \in U_q^o} \int_T |f_n^{(q)} - f| dv(\Psi_{x^*} \mathbf{m}_q) = \lim_n \sup_{x^* \in U_q^o} \int_T |f_n^{(q)} - f| dv(x^* \mathbf{m}) = 0 \quad (4.1.11.2)$$

where Ψ_{x^*} is as in Proposition 1.2.15(ii)(a).

Proof. Let $q \in \Gamma$ be given. By hypothesis there exists an $\mathbf{m}_q$-null set $N^{(q)} \in \sigma(\mathcal{P})$ such that $f_n^{(q)} \to f$ pointwise in $T \backslash N^{(q)}$. As $g^{(q)}$ is (KL) $\mathbf{m}_q$-integrable in T, by Remark 2.1.3 of Chapter 2 there exists an $\mathbf{m}_q$-null set $M(q) \in \sigma(\mathcal{P})$ such that $M(q) \supset \{t \in T : |g^{(q)}(t)| = \infty\}$ so that $g^{(q)}$ is finite in $T \backslash M(q)$. Let $M^{(q)} = M(q) \cup N^{(q)}$. Then $M^{(q)} \in \sigma(\mathcal{P})$ and is $\mathbf{m}_q$-null. As $f_n^{(q)}$, $n \in \mathbb{N}$, are $\mathbf{m}_q$-measurable for each n, by Proposition 1.1.18 of Chapter 1 there exists $M_n^{(q)} \in \sigma(\mathcal{P})$ with $||\mathbf{m}||_q(M_n^{(q)}) = 0$ such that $f_n \chi_{T \backslash M_n^{(q)}}$ is $\sigma(\mathcal{P})$-measurable. Let $M_q = M^{(q)} \bigcup (\bigcup_{n=1}^{\infty} M_n^{(q)})$. Then $M_q \in \sigma(\mathcal{P})$ and $||\mathbf{m}||_q(M_q) = 0$ as $||\mathbf{m}||_q$ is σ-subadditive on $\sigma(\mathcal{P})$. As $f_n^{(q)} \chi_{T \backslash M_q} \to f \chi_{T \backslash M_q}$ pointwise in T and as $f_n^{(q)} \chi_{T \backslash M_q}$, $n \in \mathbb{N}$, are $\sigma(\mathcal{P})$-measurable, it follows that $f \chi_{T \backslash M_q}$ is $\sigma(\mathcal{P})$-measurable so that f is $\mathbf{m}_q$-measurable for each $q \in \Gamma$ and hence, f is $\mathbf{m}$-measurable. Considering $\mathbf{m}_q : \mathcal{P} \to \widetilde{X_q}$, the hypothesis of domination and Theorem 2.1.5(vii) of Chapter 2 imply that f, $f_n^{(q)}$, $n \in \mathbb{N}$, are (KL) $\mathbf{m}_q$-integrable in T with values in $\widetilde{X_q}$. Since q is arbitrary in Γ, by Theorem 4.1.4(ii) and by Remark 4.1.5 we conclude that f is (KL) $\mathbf{m}$-integrable in T with values in X.

Let $q \in \Gamma$. Let $\boldsymbol{\nu}_q(\cdot) = $ (KL) $\int_{(\cdot)} g^{(q)} d\mathbf{m}_q$. Then by Theorem 2.1.5(ii) of Chapter 2, $\boldsymbol{\nu}_q$ is σ-additive on $\sigma(\mathcal{P})$ ($\boldsymbol{\nu}_q$ assumes values in the Banach space $\widetilde{X_q}$). Then by Proposition 1.1.13 of Chapter 1 there exists a control measure $\mu_q : \sigma(\mathcal{P}) \to [0, \infty)$ for $\boldsymbol{\nu}_q$ so that, given $\epsilon > 0$, there exists $\delta^{(q)} > 0$ such that $||\boldsymbol{\nu}_q||(A) < \frac{\epsilon}{3}$ whenever $A \in \sigma(\mathcal{P})$ with $\mu_q(A) < \delta^{(q)}$. Let $F_q = \bigcup_{n=1}^{\infty}(N(f_n^{(q)}) \cap (T \backslash M_q))$. Clearly, $F_q \in \sigma(\mathcal{P})$. Let $A \in \sigma(\mathcal{P}) \cup \{T\}$. Then, arguing as in the proof of Theorem 2.1.7 of Chapter 2 with F_q, $F_k^{(q)}$, M_q, N_q, $||\boldsymbol{\nu}_q||$, μ_q, A, $f_n^{(q)}$ and $||\mathbf{m}||_q$ replacing F, F_k, M, N, $||\boldsymbol{\nu}||$, μ, E, f_n and $||\mathbf{m}||$, respectively, and taking Ψ_{x^*} as in Proposition 1.2.15(ii)(a) and taking n_0 such that $||f_n^{(q)} - f||_{F_{k_0}^{(q)}} \cdot ||\mathbf{m}||_q(F_{k_0}^{(q)}) < \frac{\epsilon}{3}$ for $n \geq n_0$, we have for $x^* \in U_q^o$, $n \geq n_0$ and for $A \in \sigma(\mathcal{P})$ or $A = T$,

$$\int_{A \cap (T \backslash F_{k_0}^{(q)})} |f_n^{(q)} - f| dv(\Psi_{x^*} \mathbf{m}_q)$$

$$\leq 2||\boldsymbol{\nu}_q||(F_q \backslash N_q \backslash F_{k_0}^{(q)}) + 2||\boldsymbol{\nu}_q||(N_q) + 2||\boldsymbol{\nu}_q||(M_q) < \frac{2\epsilon}{3}$$

since $\int_B |f_n^{(q)} - f| dv(\Psi_{x^*} \mathbf{m}_q) \leq 2 \int_B |g^{(q)}| dv(\Psi_{x^*} \mathbf{m}_q) = 2 \int_B |g^{(q)}| dv(x^* \mathbf{m}) \leq 2||\boldsymbol{\nu}_q||(B)$ for $B \in \sigma(\mathcal{P})$ by Proposition 1.2.15(ii) and Theorem 4.1.8(iii)(a).

Moreover,

$$\int_{A\cap F_{k_0}^{(q)}} |f_n^{(q)} - f|\, dv(\Psi_{x^*}\mathbf{m}_q) < \frac{\epsilon}{3}$$

since $v(\Psi_{x^*}\mathbf{m}_q)(F_{k_0}^{(q)}) = v(x^*\mathbf{m})(F_{k_0}^{(q)}) \le ||\mathbf{m}||_q(F_{k_0}^{(q)})$ by Proposition 1.2.15(ii)(c) and since $||f_n^{(q)} - f||_{F_{k_0}^{(q)}} \cdot ||\mathbf{m}||_q(F_{k_0}^{(q)}) < \frac{\epsilon}{3}$ for $n \ge n_0$. Consequently, by the above inequalities and by the fact that f is (KL) $\mathbf{m}$-integrable in T with values in X we have $\Pi_q((KL)\int_A f d\mathbf{m}) = (KL)\int_A f d\mathbf{m}_q$ and by Proposition 1.2.15 we also have

$$\left| (KL)\int_A f_n^{(q)} d\mathbf{m}_q - (KL)\int_A f d\mathbf{m}_q \right|_q$$
$$= \left| (KL)\int_A (f_n^{(q)} - f) d\mathbf{m}_q \right|_q$$
$$\le \sup_{x^*\in U_q^o}\int_A |f_n^{(q)} - f|\, dv(\Psi_{x^*}\mathbf{m}_q) \le \epsilon \qquad (4.1.11.3)$$

for $n \ge n_0$ and for all $A \in \sigma(\mathcal{P}) \cup \{T\}$. Then (4.1.11.3) implies (4.1.11.2) and hence (4.1.11.1) where the limit in (4.1.11.1) is uniform with respect to $A \in \sigma(\mathcal{P})$ (for a fixed $q \in \Gamma$). □

Corollary 4.1.12. (Generalization of Corollary 2.1.8 of Chapter 2-(LBCT a.e.-version)) *Let X be a quasicomplete lcHs and $\mathbf{m} : \mathcal{S} \to X$ be σ-additive, where $\mathcal{S}$ is a σ-ring. Suppose $f, f_n^{(q)} : T \to \mathbb{K}$ or $[-\infty, \infty]$ for $n \in \mathbb{N}$ and let $0 < K^{(q)} < \infty$ for $q \in \Gamma$. If $f_n^{(q)}$, $n \in \mathbb{N}$, are $\mathbf{m}_q$-measurable, $f_n^{(q)} \to f$ $\mathbf{m}_q$-a.e. in T and if $|f_n^{(q)}| \le K^{(q)}$ $\mathbf{m}_q$-a.e. for all n, then $f, f_n^{(q)}$, $n \in \mathbb{N}$, are (KL) $\mathbf{m}_q$-integrable in T with values in $\widetilde{X_q}$ and consequently, f is (KL) $\mathbf{m}$-integrable in T with values in X. Then, for each $q \in \Gamma$,*

$$\lim_n \left| (KL)\int_A f d\mathbf{m}_q - (KL)\int_A f_n^{(q)} d\mathbf{m}_q \right|_q = 0$$

for $A \in \mathcal{S} \cup \{T\}$, the limit being uniform with respect to $A \in \mathcal{S}$ (for a fixed $q \in \Gamma$). Moreover, for each $q \in \Gamma$,

$$\lim_n \sup_{x^*\in U_q^o}\int_T |f_n^{(q)} - f|\, dv(\Psi_{x^*}\mathbf{m}_q) = 0$$

where Ψ_{x^*} is as in Proposition 1.2.15(ii)(a).

Proof. The corollary is immediate from the above theorem and the second part of Theorem 2.1.5(v) of Chapter 2 as $\mathcal{S}$ is a σ-ring. □

The following theorem plays a key role in Section 4.3.

Theorem 4.1.13. (Generalization of Corollary 2.1.9 of Chapter 2) *Let X be a quasicomplete lcHs and* $\mathbf{m} : \mathcal{P} \to X$ *be σ-additive. If f is an* **m***-measurable (KL)* $\mathbf{m}_q$*-integrable scalar function on T with values in $\widetilde{X}_q$ for each $q \in \Gamma$, then f is (KL)* **m***-integrable in T with values in $\tilde{X}$. (In fact, (KL) $\int_A f d\mathbf{m} \in X$ by Remark 4.1.5.) Moreover, for each $q \in \Gamma$, there exist a set $N_q \in \sigma(\mathcal{P})$ with $||\mathbf{m}||_q(N_q) = 0$ and a sequence $(s_n^{(q)})_{n=1}^\infty \subset \mathcal{I}_s$ such that $s_n^{(q)} \to f$ and $|s_n^{(q)}| \nearrow |f|$ pointwise in $T \backslash N_q$. Then for any such sequence $(s_n^{(q)})$, by 1.2.16,*

$$\lim_n q \left(\int_A s_n^{(q)} d\mathbf{m} - (KL) \int_A f d\mathbf{m} \right)$$
$$= \lim_n \left| \int_A s_n^{(q)} d\mathbf{m}_q - (KL) \int_A f d\mathbf{m}_q \right|_q = 0$$

for $A \in \sigma(\mathcal{P}) \cup \{T\}$, the limit being uniform with respect to $A \in \sigma(\mathcal{P})$ (for a fixed $q \in \Gamma$).

Consequently, for $A \in \sigma(\mathcal{P})$ and $q \in \Gamma$,

$$||\mathbf{m}||_q(A) = \sup \left\{ \left| (KL) \int_A h d\mathbf{m}_q \right|_q : h \in \mathcal{I}(\mathbf{m}_q), |h| \leq \chi_A \ \mathbf{m}_q\text{-}a.e. \ in \ T \right\}.$$

Proof. Let $q \in \Gamma$. By hypothesis, f is (KL) $\mathbf{m}_q$-integrable in T with values in $\widetilde{X}_q$. Then by Corollary 2.1.9 of Chapter 2 for $\mathbf{m}_q$, there exists $(s_n^{(q)}) \subset \mathcal{I}_s$ such that $s_n^{(q)} \to f$ and $|s_n^{(q)}| \nearrow |f|$ $\mathbf{m}_q$-a.e. in T so that by the same corollary we have

$$\lim_n \left| (KL) \int_A f d\mathbf{m}_q - \int_A s_n^{(q)} d\mathbf{m}_q \right|_q = 0, \ A \in \sigma(\mathcal{P}) \cup \{T\}, \tag{4.1.13.1}$$

the limit being uniform with respect to $A \in \sigma(\mathcal{P})$. Thus

$$\left| (KL) \int_A f d\mathbf{m}_q \right|_q = \lim_n \left| \int_A s_n^{(q)} d\mathbf{m}_q \right|_q \ \text{for } A \in \sigma(\mathcal{P}) \cup \{T\}. \tag{4.1.13.2}$$

Since f is (KL) $\mathbf{m}_q$-integrable in T for each $q \in \Gamma$, by Theorem 4.1.4(ii) f is (KL) **m**-integrable in T with values in $\tilde{X}$ and by (4.1.13.1) we have

$$\lim_n q \left((KL) \int_A f d\mathbf{m} - \int_A s_n^{(q)} d\mathbf{m} \right) = 0$$

where we use Notation 1.2.16 and where the limit is uniform with respect to $A \in \sigma(\mathcal{P})$ for a fixed $q \in \Gamma$.

Consequently, by Theorem 4.1.8(i), by (4.1.13.2) and by Proposition 1.2.15(ii) we have

$$||\mathbf{m}||_q(A) = \sup\left\{\left|\int_A sd\mathbf{m}_q\right|_q : s \in \mathcal{I}_s, |s| \le \chi_A\right\}$$

$$= \sup\left\{\left|\int_A sd\mathbf{m}_q\right|_q : s \in \mathcal{I}_s, |s| \le \chi_A\ \mathbf{m}_q\text{-a.e. in } T\right\}$$

$$\le \sup\left\{(KL)\left|\int_A hd\mathbf{m}_q\right|_q : h \in \mathcal{I}(\mathbf{m}_q), |h| \le \chi_A\ \mathbf{m}_q\text{-a.e. in } T\right\}$$

$$= \sup\left\{\sup_{x^* \in U_q^o}\left|\int_A hd(\Psi_{x^*}\mathbf{m}_q)\right| : h \in \mathcal{I}(\mathbf{m}_q), |h| \le \chi_A\ \mathbf{m}_q\text{-a.e. in } T\right\}$$

$$\le \sup_{x^* \in U_q^o} v(\Psi_{x^*}\mathbf{m}_q)(A) = \sup_{x^* \in U_q^o} v(x^*\mathbf{m})(A)$$

$$= ||\mathbf{m}||_q(A)$$

for $A \in \sigma(\mathcal{P})$ and for $q \in \Gamma$.   □

Corollary 4.1.14. (Simple function approximation) *Let X be a quasicomplete lcHs and $\mathbf{m} : \mathcal{P} \to X$ be σ-additive. If $f : T \to \mathbb{K}$ is $\mathbf{m}$-measurable and (KL) $\mathbf{m}_q$-integrable in T with values in $\widetilde{X}_q$ for each $q \in \Gamma$, then f is (KL) $\mathbf{m}$-integrable in T with values in $\tilde{X}$ (by Remark 4.1.5 with values in X). Moreover, for each $q \in \Gamma$, there exists $(s_n^{(q)})_{n=1}^\infty \subset \mathcal{I}_s$ such that $s_n^{(q)} \to f$ $\mathbf{m}_q$-a.e. in T with $|s_n^{(q)}| \nearrow |f|$ $\mathbf{m}_q$-a.e. in T and such that*

$$\lim_n \sup_{x^* \in U_q^o} \int_T |f - s_n^{(q)}|dv(\Psi_{x^*}\mathbf{m}_q) = \lim_n \sup_{x^* \in U_q^o} \int_T |f - s_n^{(q)}|dv(x^*\mathbf{m}) = 0.$$

Proof. This follows from Theorems 4.1.9 and 4.1.11.   □

In the following remark we include the analogues of Theorems 4.1.9, 4.1.11 and 4.1.13 and those of Corollaries 4.1.12 and 4.1.14 when $\mathbf{m}$ takes values in a sequentially complete lcHs and when the functions considered are $\sigma(\mathcal{P})$-measurable.

Remark 4.1.15. ((KL)-integrability with respect to a sequentially complete lcHs-valued $\mathbf{m}$) Let X be an lcHs, $\mathbf{m} : \mathcal{P} \to X$ be σ-additive and $\mathcal{I}(\sigma(\mathcal{P}), \mathbf{m})$ be the collection of all (KL) $\mathbf{m}$-integrable $\sigma(\mathcal{P})$-measurable scalar functions with values in X.

Theorem 4.1.9′. This is the same as Theorem 4.1.9 excepting that X is a sequentially complete lcHs and all the functions considered are $\sigma(\mathcal{P})$-measurable or belong to $\mathcal{I}(\sigma(\mathcal{P}), \mathbf{m})$.

(*Proof.* Let $A \in \mathcal{P}$ and let $\alpha = \operatorname{ess\,sup}_{t\in A} |f(t)| < \infty$. Then there exists an $\mathbf{m}$-null set $M \in \sigma(\mathcal{P}) \cap A$ such that $\sup_{t\in A\setminus M} |f(t)| = \alpha$. Let $\Sigma_A = \mathcal{P} \cap A$. Then

Σ_A is a σ-algebra of subsets of A. As $f\chi_{A\setminus M}$ is $\sigma(\mathcal{P})$-measurable, it is also Σ_A-measurable and is moreover bounded. Hence there exists a sequence $(s_n)_1^\infty$ of Σ_A-simple functions such that $s_n \to f\chi_{A\setminus M}$ uniformly in A and $|s_n| \nearrow |f\chi_{A\setminus M}|$ in T. Let $q \in \Gamma$. Arguing as in the proof of Theorem 2.1.5(v) of Chapter 2, replacing $|\cdot|$ by q and $||\mathbf{m}||$ by $||\mathbf{m}||_q$, we can show that $q(\int_B s_n d\mathbf{m} - \int_B s_r d\mathbf{m}) \to 0$ as $n, r \to \infty$ for each $B \in \Sigma_A$. Hence $(\int_B s_n d\mathbf{m})$ is Cauchy in X and as X is sequentially complete, there exists $x_B \in X$ such that $\lim_n \int_B s_n d\mathbf{m} = x_B$ and consequently, by LDCT for scalar measures

$$x^*(x_B) = \lim_n \int_B s_n d(x^*\mathbf{m}) = \int_B f d(x^*\mathbf{m}).$$

Hence f is (KL) **m**-integrable in A, (KL)$\int_B f d\mathbf{m} = x_B$ and $|$(KL) $\int_B f d\mathbf{m}|_q = |\lim_n \int_B s_n d\mathbf{m}|_q \leq \alpha ||\mathbf{m}||_q(B)$ for $B \in \sigma(\mathcal{P}) \cap A$ and for $q \in \Gamma$. Hence (i)(a) of Theorem 4.1.9$'$ holds. Theorem 4.1.9$'$(i)(b) is immediate from Theorem 4.1.9$'$(i)(a). The proof of Theorem 4.1.9$'$(ii) is similar to that of Theorem 4.1.9(ii). If $|f|$ is (KL) **m**-integrable in T with values in X, then by the last part of Proposition 1.2.10 there exists a sequence $(s_n)_1^\infty \subset \mathcal{I}_s$ such that $s_n \to f$ and $|s_n| \nearrow |f|$ pointwise in T. Then, given $q \in \Gamma$, by Theorem 4.1.4(i), $|f|$ is (KL) $\mathbf{m}_q$-integrable in T with values in $X_q \subset \widetilde{X}_q$ and hence by Theorem 2.1.7 of Chapter 2 for $\mathbf{m}_q$, $(\int_A s_n d\mathbf{m}_q)_1^\infty$ is Cauchy in $\widetilde{X}_q$ (in fact, Cauchy in X_q) for each $q \in \Gamma$ and for each $A \in \sigma(\mathcal{P})$. Thus $(\int_A s_n d\mathbf{m})_1^\infty$ is Cauchy in X and as X is sequentially complete, $\lim_n \int_A s_n d\mathbf{m} \in X$ for each $A \in \sigma(\mathcal{P})$. Thus $f \in (\mathcal{I}(\sigma(\mathcal{P}), \mathbf{m})$. The other parts are proved as in the proof of Theorem 4.1.9(iii).) $\qquad\square$

Theorem 4.1.11$'$. This is the same as Theorem 4.1.11 not only with the changes as in Theorem 4.1.9$'$ but also with $g^{(q)} = g$ for all $q \in \Gamma$, $g \in \mathcal{I}(\sigma(\mathcal{P}), \mathbf{m})$, $f_n^{(q)} = f_n$ for all $q \in \Gamma$ and for all n, f_n being $\sigma(\mathcal{P})$-measurable and $|f_n| \leq |g|$ **m**-a.e. in T for all n and $f_n \to f$ **m**-a.e. in T with f $\sigma(\mathcal{P})$-measurable.

(*Proof.* By Theorem 4.1.9$'$, $(f_n)_1^\infty$ and f are (KL) **m**-integrable in T. By hypothesis and by Proposition 4.1.3, there exists an **m**-null set M in $\sigma(\mathcal{P})$ such that $(f_n)_1^\infty$ and f are finite on $T\setminus M$ and $f_n(t) \to f(t)$ for $t \in T\setminus M$. Then $F = \bigcup_1^\infty (N(f_n) \cap (T\setminus M)) \in \sigma(\mathcal{P})$. Let $q \in \Gamma$ and let $\boldsymbol{\nu}(\cdot) = $ (KL) $\int_{(\cdot)} g d\mathbf{m} \in X$. By Theorem 4.1.8(ii), $\boldsymbol{\nu}$ is σ-additive on $\sigma(\mathcal{P})$ and $\boldsymbol{\nu}_q = \Pi_q \circ \boldsymbol{\nu}$ is σ-additive with values in X_q and hence with values in the Banach space $\widetilde{X}_q$. Then arguing as in the proof of Theorem 4.1.11, there exists a control measure μ_q on $\sigma(\mathcal{P})$ for $\boldsymbol{\nu}_q$ so that given $\epsilon > 0$ there exists $\delta > 0$ such that $||\boldsymbol{\nu}||_q(A) < \frac{\epsilon}{3}$ if $\mu_q(A) < \delta$. Moreover, by the Egoroff-Lusin theorem, there exist $N \in \sigma(\mathcal{P}) \cap F$ with $\mu_q(N) = 0$ and a sequence $(F_k^{(q)})_1^\infty \subset \mathcal{P}$ with $F_k^{(q)} \nearrow F\setminus N$ such that $f_n \to f$ uniformly in each $F_k^{(q)}$. Then there exists k_0 such that $\mu_q(F\setminus N\setminus F_k) < \delta$ for $k \geq k_0$. Choose n_0 such that $||f_n - f_r||_{F_{k_0}} \cdot ||\mathbf{m}||_q(F_{k_0}) < \frac{\epsilon}{3}$ for $n, r \geq n_0$. Then arguing as in the proof of Theorem 4.1.11 one can show that $q((\text{KL}) \int_A f_r d\mathbf{m} - (\text{KL}) \int_A f_n d\mathbf{m}) = |(\text{KL}) \int_A (f_n - f_r) d\mathbf{m}_q|_q \leq \sup_{x^* \in U_q^0} \int_A |f_n - f_r| dv(\Psi_{x^*} \mathbf{m}_q) = \sup_{x^* \in U_q^0} \int_A |f_n -$

$f_r|dv(x^*\mathbf{m}) < \epsilon$ (4.1.15.1) for $n, r \geq n_0$ and for $A \in \sigma(\mathcal{P})$. Since q is arbitrary in Γ, it follows that $((\mathrm{KL}) \int_A f_n d\mathbf{m})_1^\infty$ is Cauchy in X for each $A \in \sigma(\mathcal{P})$ and hence by the hypothesis that X is sequentially complete, there exists $x_A \in X$ such that $\lim_n (\mathrm{KL}) \int_A f_n d\mathbf{m} = x_A$ for $A \in \sigma(\mathcal{P}) \cup \{T\}$. Then, for $x^* \in X^*$, we have $x^*(x_A) = \lim_n \int_A f_n d(x^*\mathbf{m}) = \int_A f d(x^*\mathbf{m})$ by LDCT for scalar measures and hence f is (KL) $\mathbf{m}$-integrable in T with values in X. Then $x_A = (\mathrm{KL}) \int_A f d\mathbf{m} = \lim_n (\mathrm{KL}) \int_A f_n d\mathbf{m}$ and $q\left((\mathrm{KL}) \int_A f d\mathbf{m} - (\mathrm{KL}) \int_A f_n d\mathbf{m}\right) \leq \sup_{x^* \in U_q^0} |\int_A |f_n - f| dv(\Psi_{x^*} \circ \mathbf{m}_q) = \sup_{x^* \in U_q^0} \lim_{r \to \infty} \int_A |f_r - f_n| dv(\Psi_{x^*} \circ \mathbf{m}_q) < \epsilon$ for $n \geq n_0$ and for $A \in \sigma(\mathcal{P}) \cup \{T\}$ by (4.1.15.1). Hence Theorem 4.1.11$'$ holds. $\square$

(Note that Theorem 4.1.11$'$ is essentially the version of LDCT given in Theorem 3.3 of [L2], which remains unestablished as observed in Remark 2.1.12 of Chapter 2 .))

Corollary 4.1.12$'$. This is the corresponding version of Corollary 4.1.12 (LBCT) deduced from Theorem 4.1.11$'$. If f is $\sigma(\mathcal{P})$-measurable, then by Proposition 1.2.10 there exists a sequence $(s_n) \subset \mathcal{I}_s$ such that $s_n \to f$ and $|s_n| \nearrow |f|$ pointwise in T. Thus, if **Theorem 4.1.13$'$** and **Corollary 4.1.14$'$** are the analogues of Theorem 4.1.13 and Corollary 4.1.14, respectively, with X sequentially complete, f $\sigma(\mathcal{P})$-measurable and $(s_n^{(q)})_{n=1}^\infty$ being replaced by $(s_n)_1^\infty$ for all $q \in \Gamma$, then they hold by Theorem 4.1.11$'$.

Remark 4.1.16. LDCT for (KL) $\mathbf{m}$-integrals with respect to a σ-additive vector measure defined on a δ-ring τ with values in a sequentially complete lcHs is given in [L2] under the hypothesis that the dominated sequence converges pointwise, but, as observed in Remark 2.1.12 of Chapter 2, its proof is incorrect and is corrected in the said Remark. For the case of σ-additive vector measures defined on σ-algebras with values in a sequentially complete lcHs, Theorems 4.1.8, 4.1.9$'$ and 4.1.11$'$ are obtained in [KK] (for real lcHs) and [OR4] (for complex lcHs) by different methods. Theorem 3.5 of [L2], whose incorrect proof is corrected in Remark 2.1.12 of Chapter 2, is easily deducible from Corollary 4.1.14$'$.

4.2   (BDS) m-integrability (m lcHs-valued)

Let X be a quasicomplete lcHs and $\mathbf{m} : \mathcal{P} \to X$ be σ-additive. For an $\mathbf{m}$-measurable function f we define (BDS) $\mathbf{m}$-integrability and the (BDS) $\mathbf{m}$-integral of f with values in $\tilde{X}$ and show that f is (BDS) $\mathbf{m}$-integrable in T if and only if it is (KL) $\mathbf{m}$-integrable in T and that (BDS)$\int_A f d\mathbf{m} = (\mathrm{KL}) \int_A f d\mathbf{m} \in X$, $A \in \sigma(\mathcal{P}) \cup \{T\}$ (Theorems 4.2.2 and 4.2.3). Also we generalize Theorems 2.2.4, 2.2.5 and 2.2.8 and Corollary 2.2.11 of Chapter 2. We define (BDS) $\mathbf{m}$-integrability for $\sigma(\mathcal{P})$-measurable functions in Remark 4.2.12 when $\mathbf{m}$ assumes values in a sequentially complete lcHs and Theorem 4.2.2$'$ in Remark 4.2.12 gives an analogue of Theorem 4.2.2 for such $\mathbf{m}$.

Definition 4.2.1. Let X be a quasicomplete lcHs and $\mathbf{m} : \mathcal{P} \to X$ be σ-additive. An $\mathbf{m}$-measurable function $f : T \to \mathbb{K}$ or $[-\infty, \infty]$ is said to be $\mathbf{m}$-integrable in T in the sense of Bartle-Dunford-Schwartz or (BDS) $\mathbf{m}$-integrable in T, if f is $\mathbf{m}_q$-integrable in T with values in $\widetilde{X}_q$ (considering $\mathbf{m}_q : \mathcal{P} \to X \subset \widetilde{X}_q$) for each $q \in \Gamma$ (see Definition 2.2.1 and Theorem 2.2.2 of Chapter 2). In that case, using Notation 1.2.17, we define

$$\text{(BDS)} \int_A f d\mathbf{m} = \lim_{\longleftarrow} \int_A f d\mathbf{m}_q, \quad A \in \sigma(\mathcal{P})$$

and

$$\text{(BDS)} \int_T f d\mathbf{m} = \lim_{\longleftarrow} \int_T f d\mathbf{m}_q = \lim_{\longleftarrow} \int_{N(f) \backslash N_q} f d\mathbf{m}_q$$

with values in $\tilde{X}$, where $N_q \in \sigma(\mathcal{P})$ with $||\mathbf{m}||_q(N_q) = 0$ such that $f\chi_{T \backslash N_q}$ is $\sigma(\mathcal{P})$-measurable. (In the light of Theorem 4.2.3 below, these integrals indeed assume values in X as X is quasicomplete.)

Theorem 4.2.2. *Let X be a quasicomplete lcHs and $\mathbf{m} : \mathcal{P} \to X$ be σ-additive. Let $f : T \to \mathbb{K}$ or $[-\infty, \infty]$ be $\mathbf{m}$-measurable. Then:*

(i) *If f is (KL) $\mathbf{m}$-integrable in T with values in X, then f is (BDS) $\mathbf{m}$-integrable in T and (BDS) $\int_A f d\mathbf{m} = (KL) \int_A f d\mathbf{m} \in X$ for $A \in \sigma(\mathcal{P}) \cup \{T\}$.*

(ii) *If f is (BDS) $\mathbf{m}$-integrable in T, then f is (KL) $\mathbf{m}$-integrable in T and (KL) $\int_A f d\mathbf{m} = (BDS) \int_A f d\mathbf{m} \in \tilde{X}$ for $A \in \sigma(\mathcal{P}) \cup \{T\}$.*

(*Moreover, as X is quasicomplete, by Theorem 4.2.3 below* (BDS) $\int_A f d\mathbf{m} \in X$ *for $A \in \sigma(\mathcal{P}) \cup \{T\}$ whenever f is (BDS) $\mathbf{m}$-integrable in T.) **Hereafter we shall denote either of the integrals of f over $A \in \sigma(\mathcal{P}) \cup \{T\}$ by $\int_A f d\mathbf{m}$ and say that f is m-integrable in T.***

Proof. (i) As f is (KL) $\mathbf{m}$-integrable in T with values in X, then, for $A \in \sigma(\mathcal{P}) \cup \{T\}$, there exists $x_A \in X$ such that $x^*(x_A) = \int_A f d(x^*\mathbf{m})$ for $x^* \in X^*$ so that $(KL) \int_A f d\mathbf{m} = x_A$. Then by Theorem 4.1.4(i), for each $q \in \Gamma$, we have

$$\Pi_q(x_A) = \Pi_q((KL) \int_A f d\mathbf{m}) = (KL) \int_A f d(\Pi_q \circ \mathbf{m}) \quad \text{(by Theorem 4.1.8(v))}$$

$$= (KL) \int_A f d\mathbf{m}_q = (BDS) \int_A f d\mathbf{m}_q$$

by Theorem 2.2.2 of Chapter 2 (with X in Theorem 2.2.2 of Chapter 2 being replaced by $\widetilde{X}_q$ and $\mathbf{m}$ by $\mathbf{m}_q$). Then by Definition 4.2.1, f is (BDS) $\mathbf{m}$-integrable in T and

$$\text{(BDS)} \int_A f d\mathbf{m} = \lim_{\longleftarrow} \text{(BDS)} \int_A f d\mathbf{m}_q = \lim_{\longleftarrow}(KL) \int_A f d\mathbf{m}_q$$

$$= \lim_{\longleftarrow}\Pi_q(x_A) = x_A = (KL) \int_A f d\mathbf{m}$$

belongs to X for $A \in \sigma(\mathcal{P}) \cup \{T\}$.

Conversely, let f be (BDS) $\mathbf{m}$-integrable in T. Then, for each $q \in \Gamma$, by Definition 4.2.1 and by Definition 2.2.1 of Chapter 2 there exist $(s_n^{(q)})_{n=1}^\infty \subset \mathcal{I}_s$ and an $\mathbf{m}_q$-null set $N_q \in \sigma(\mathcal{P})$ such that $s_n^{(q)} \to f$ pointwise in $T \backslash N_q$ and such that $\lim_n \int_A s_n^{(q)} d\mathbf{m}_q = x_A^{(q)}$ (say) exists in $\widetilde{X_q}$ for each $A \in \sigma(\mathcal{P}) \cup \{T\}$. Then by Theorem 2.2.2 of Chapter 2 with respect to $\mathbf{m}_q$ and $\widetilde{X_q}$, f is (KL) $\mathbf{m}_q$-integrable in T and (KL)$\int_A f d\mathbf{m}_q =$ (BDS)$\int_A f d\mathbf{m}_q = x_A^{(q)} \in \widetilde{X_q}$ for $A \in \sigma(\mathcal{P}) \cup \{T\}$. Consequently, by Theorem 4.1.4(ii) and by Definition 4.2.1,

$$(\text{KL}) \int_A f d\mathbf{m} = \varprojlim x_A^{(q)} = \varprojlim (\text{KL}) \int_A f d\mathbf{m}_q$$

$$= \varprojlim (\text{BDS}) \int_A f d\mathbf{m}_q = (\text{BDS}) \int_A f d\mathbf{m} \in \tilde{X}$$

for $A \in \sigma(\mathcal{P}) \cup \{T\}$. □

Theorem 4.2.3. *Let X be a quasicomplete lcHs and $\mathbf{m} : \sigma(\mathcal{P}) \to X$ be σ-additive. If $f : T \to \mathbb{K}$ or $[-\infty, \infty]$ is $\mathbf{m}$-measurable and $\mathbf{m}$-integrable in T, then $\int_A f d\mathbf{m} \in X$ for $A \in \sigma(\mathcal{P}) \cup \{T\}$.*

Proof. Let $A \in \sigma(\mathcal{P}) \cup \{T\}$. Let $\Phi = \{s \in \mathcal{I}_s : |s| \leq |f| \text{ in } T\}$ and let $G_A = \{\int_A s d\mathbf{m} : s \in \Phi\}$. Let $x^* \in X^*$. Then by Theorem 4.2.2(ii), f is (KL) $\mathbf{m}$-integrable in T and hence f is $x^* \mathbf{m}$-integrable in T and therefore,

$$\sup_{s \in \Phi} \left| \int_A s d(x^* \mathbf{m}) \right| \leq \int_A |f| dv(x^* \mathbf{m}) = M_{x^*} (\text{say}) < \infty.$$

Thus G_A is weakly bounded. Then by Theorem 3.18 of [Ru2], G_A is τ-bounded in X.

Let $q \in \Gamma$. Taking $\omega_n^{(q)} = s_n^{(q)} \chi_{T \backslash N_q}$ in Proposition 1.2.10, we have $(\omega_n^{(q)})_{n=1}^\infty \subset \mathcal{I}_s$, $|\omega_n^{(q)}| \leq |f|$ in T for $n \in \mathbb{N}$, $\omega_n^{(q)} \to f$ and $|\omega_n^{(q)}| \nearrow |f|$ pointwise in $T \backslash N_q$. Then by Theorems 4.1.13 and 4.2.2(ii) and by Notation and Convention 1.2.16 we have

$$\lim_n q \left(\int_A \omega_n^{(q)} d\mathbf{m} - (\text{KL}) \int_A f d\mathbf{m} \right) = \lim_n q \left(\int_A \omega_n^{(q)} d\mathbf{m} - \int_A f d\mathbf{m} \right) = 0.$$

Hence, given $\epsilon > 0$, for each $q \in \Gamma$ there exists $s_q \in \Phi$ such that $q(\int_A s_q d\mathbf{m} - \int_A f d\mathbf{m}) < \epsilon$ and therefore, $\int_A f d\mathbf{m}$ belongs to the $\tilde{\tau}$-closure of G_A (in $\tilde{X}$). Consequently, there exists a net $(x_\alpha) \subset G_A$ such that $x_\alpha \to \int_A f d\mathbf{m}$ in $\tilde{\tau}$. Therefore, (x_α) is $\tilde{\tau}$-Cauchy.

On the other hand, as $G_A \subset X$ is τ-bounded and as X is quasicomplete, the τ-closure of G_A is τ-complete. Since $\tilde{\tau}|_X = \tau$, it follows that (x_α) is also τ-Cauchy. Moreover, as the τ-closure of G_A is τ-complete, there exists x_0 in the τ-closure of G_A (so that $x_0 \in X$) such that $x_\alpha \to x_0$ in τ and hence in $\tilde{\tau}$. Since $\tilde{\tau}$ is Hausdorff, we conclude that $\int_A f d\mathbf{m} = x_0 \in X$. □

Remark 4.2.4. The proof of Theorem 1.35 of [T] is modified in Section 7.1 of Chapter 7 to prove the result for complex functions. See Theorem 7.1.24 of Chapter 7.

Remark 4.2.5. **In Definition 4.2.1 one can take X to be an arbitrary lcHs and in that case, the definitions of the (BDS) integrals with respect to m hold with values of the integrals in $\tilde{X}$. For the validity of Theorem 4.2.3 we need the hypothesis that X be quasicomplete. It is not known if Theorem 4.2.3 still holds if X is only sequentially complete, and not quasicomplete. For an example of such an lcHs, see pp. 295–296, §23 of [KÖ]. Consequently, it is not known whether $\int f d\mathbf{m}$ belongs to X if X is only sequentially complete in Theorems 4.2.4, 4.2.7, 4.2.8, and 4.2.9 and Corollary 4.2.11 below. Thus we are not able to answer the question raised by the referee in this regard.**

Remark 4.2.6. Let X be a quasicomplete lcHs and $\mathbf{m} : \mathcal{P} \to X$ be σ-additive. Then, in the light of Theorems 4.2.2 and 4.2.3, $\mathcal{I}(\mathbf{m})$ in **Notation 4.1.6 is the same as the class of all m-measurable scalar functions on T which are m-integrable in T with values in X. Then Theorems 4.2.7, 4.2.8, and 4.2.9 and Corollary 4.2.11 hold for functions $f \in \mathcal{I}(\mathbf{m})$.**

We now generalize Theorems 2.2.4, 2.2.5 and 2.2.8 of Chapter 2 to quasicomplete lcHs-valued **m** in Theorems 4.2.7, 4.2.8 and 4.2.9, respectively.

Theorem 4.2.7. *Let X be a quasicomplete lcHs and $\mathbf{m} : \mathcal{P} \to X$ be σ-additive. Let $f : T \to \mathbb{K}$ or $[-\infty, \infty]$ be $\mathbf{m}$-measurable. For each $q \in \Gamma$, let $(s_n^{(q)})_{n=1}^{\infty} \subset \mathcal{I}_s$ be such that $s_n^{(q)} \to f \, \mathbf{m}_q$-a.e. in T and let $\gamma_n^{(q)} : \sigma(\mathcal{P}) \to X_q \subset \widetilde{X_q}$ be given by $\gamma_n^{(q)}(\cdot) = \int_{(\cdot)} s_n^{(q)} d\mathbf{m}_q, \; n \in \mathbb{N}$. Then:*

(a) *The following statements are equivalent:*

   (i) $\lim_n \gamma_n^{(q)}(A) = \gamma^{(q)}(A)$ *exists in $\widetilde{X_q}$ for each $A \in \sigma(\mathcal{P})$.*

   (ii) $\gamma_n^{(q)}, \; n \in \mathbb{N}, \;$ *are uniformly σ-additive on $\sigma(\mathcal{P})$.*

   (iii) $\lim_n \gamma_n^{(q)}(A)$ *exists in $\widetilde{X_q}$ uniformly with respect to $A \in \sigma(\mathcal{P})$ (for a fixed $q \in \Gamma$).*

(b) *If any one of (i), (ii) or (iii) in (a) holds for some $q \in \Gamma$, then f is $\mathbf{m}_q$ integrable in T with values in $\widetilde{X_q}$ and $\int_A f d\mathbf{m}_q = \gamma^{(q)}(A)$ for $A \in \sigma(\mathcal{P}) \cup \{T\}$. Thus, if (i), (ii) or (iii) in (a) holds for each $q \in \Gamma$, then f is $\mathbf{m}$-integrable in T with values in X and*

$$\int_A f d\mathbf{m} = \varprojlim \gamma^{(q)}(A)$$

for each $A \in \sigma(\mathcal{P}) \cup \{T\}$.

Proof. (a) and the first part of (b) hold by Theorem 2.2.4 of Chapter 2 applied to $\mathbf{m}_q, \; q \in \Gamma$. The last part of (b) holds by Definition 4.2.1 and by Theorem 4.2.3. $\qquad \square$

Theorem 4.2.8. *Let X, $\mathbf{m}$ and f be as in Theorem 4.2.7 and let $\mathcal{P}$ be a σ-ring $\mathcal{S}$. Then f is $\mathbf{m}$-integrable in T if and only if, for each $q \in \Gamma$, there exists a sequence $(f_n^{(q)})_{n=1}^\infty$ of bounded $\mathbf{m}_q$-measurable functions on T such that $f_n^{(q)} \to f$ $\mathbf{m}_q$-a.e. in T and such that $\lim_n \int_A f_n^{(q)} d\mathbf{m}_q$ exists in $\widetilde{X}_q$ for each $A \in \sigma(\mathcal{P})$. In that case, f is $\mathbf{m}_q$-integrable in T for each $q \in \Gamma$ and $\int_A f d\mathbf{m}_q = \lim_n \int_A f_n^{(q)} d\mathbf{m}_q$, $A \in \sigma(\mathcal{P})$ and the limit is uniform with respect to $A \in \sigma(\mathcal{P})$ (for a fixed $q \in \Gamma$). Moreover,*

$$\int_A f d\mathbf{m} = \lim_{\longleftarrow} \int_A f d\mathbf{m}_q \ \in X \tag{4.2.8.1}$$

for $A \in \sigma(\mathcal{P})$ and

$$\int_T f d\mathbf{m} = \lim_{\longleftarrow} \int_{N(f)\backslash N_q} f d\mathbf{m}_q \in X \tag{4.2.8.2}$$

where N_q is as in Definition 4.2.1.

Proof. The condition is necessary by Definitions 2.2.1 of Chapter 2 and by Definition 4.2.1. Let $q \in \Gamma$. As $\mathcal{S}$ is a σ-ring, by Theorem 2.2.2 of Chapter 2 and by the last part of Theorem 2.1.5(v) of Chapter 2, bounded $\mathbf{m}_q$-measurable functions on T are $\mathbf{m}_q$-integrable in T with values in $\widetilde{X}_q$. Let $\lim_n \int_A f_n^{(q)} d\mathbf{m}_q = x_A^{(q)} \in \widetilde{X}_q$ and let $x_A = \lim_{\longleftarrow} x_A^{(q)}$ for $A \in \sigma(\mathcal{P})$ and let $\lim_n \int_{N(f)\backslash N_q} f_n^{(q)} d\mathbf{m}_q = x_T^{(q)}$ and let $x_T = \lim_{\longleftarrow} x_T^{(q)}$, where $N_q \in \sigma(\mathcal{P})$ with $\|\mathbf{m}\|_q(N_q) = 0$ such that $f \chi_{T\backslash N_q}$ is $\sigma(\mathcal{P})$-measurable. Let $A \in \sigma(\mathcal{P}) \cup \{T\}$. Then $x_A \in \check{X}$. By Theorem 2.2.5 of Chapter 2, f is $\mathbf{m}_q$-integrable in T with values in $\widetilde{X}_q$ and $x_A^{(q)} = \int_A f d\mathbf{m}_q = \lim_n \int_A f_n^{(q)} d\mathbf{m}_q$, $A \in \sigma(\mathcal{P}) \cup \{T\}$ and the limit is uniform with respect to $A \in \sigma(\mathcal{P}) \cup \{T\}$ (for a fixed $q \in \Gamma$). Consequently, $x_A = \lim_{\longleftarrow} \int_A f d\mathbf{m}_q$ and hence by Definition 4.2.1, f is $\mathbf{m}$-integrable in T and $\int_A f d\mathbf{m} = x_A \in \check{X}$ for $A \in \sigma(\mathcal{P}) \cup \{T\}$. But by Theorem 4.2.3, $\int_A f d\mathbf{m} \in X$ for $A \in \sigma(\mathcal{P}) \cup \{T\}$ and hence (4.2.8.1) and (4.2.8.2) hold. $\qquad\square$

Theorem 4.2.9. (Closure theorem) *Let X and $\mathbf{m}$, be as in Theorem 4.2.7 and let $f : T \to \mathbb{K}$. For each $q \in \Gamma$, let $(f_n^{(q)})_{n=1}^\infty \subset \mathcal{I}(\mathbf{m}_q)$. (Note that $\mathbf{m}_q : \mathcal{P} \to X_q \subset \widetilde{X}_q$). If $f_n^{(q)} \to f$ $\mathbf{m}_q$-a.e. in T for each $q \in \Gamma$, then f is $\mathbf{m}$-measurable. Let $\gamma_n^{(q)}(A) = \int_A f_n^{(q)} d\mathbf{m}_q$ for $A \in \sigma(\mathcal{P})$. Then, for each $q \in \Gamma$, the following statements are equivalent:*

(i) $\lim_n \gamma_n^{(q)}(A) = \gamma^{(q)}(A)$ *exists in $\widetilde{X}_q$ for each $A \in \sigma(\mathcal{P})$.*

(ii) $\gamma_n^{(q)}$, $n \in \mathbb{N}$, *(q fixed), are uniformly σ-additive on $\sigma(\mathcal{P})$.*

(iii) $\lim_n \gamma_n^{(q)}(A) = \gamma(A) \in \widetilde{X}_q$ *exists uniformly with respect to $A \in \sigma(\mathcal{P})$ (for a fixed $q \in \Gamma$).*

If any one of (i), (ii) *or* (iii) *holds for each* $q \in \Gamma$, *then* f *is* $\mathbf{m}_q$-*integrable in* T *with values in* $\widetilde{X_q}$ *for each* $q \in \Gamma$ *and*

$$\int_A f d\mathbf{m}_q = \lim_n \int_A f_n^{(q)} d\mathbf{m}_q \text{ for } A \in \sigma(\mathcal{P}), \tag{4.2.9.1}$$

the limit being uniform with respect to $A \in \sigma(\mathcal{P})$ *(for a fixed* $q \in \Gamma$). *Moreover,* f *is* **m**-*integrable in* T *with values in* X *and*

$$\int_A f d\mathbf{m} = \lim_{\longleftarrow} \int_A f d\mathbf{m}_q \text{ for } A \in \sigma(\mathcal{P}) \cup \{T\}. \tag{4.2.9.2}$$

Proof. Clearly, f is $\mathbf{m}_q$-measurable for each $q \in \Gamma$ and hence f is **m**-measurable. By hypothesis and by Theorem 2.2.8 of Chapter 2, (i)$\Leftrightarrow$(ii)$\Leftrightarrow$(iii) for $q \in \Gamma$ and if any one of (i),(ii) or (iii) holds for each $q \in \Gamma$, then by the last part of the said theorem, f is $\mathbf{m}_q$-integrable in T with values in $\widetilde{X_q}$ and (4.2.9.1) (with the limit being uniform with respect to $A \in \sigma(\mathcal{P})$) holds for $q \in \Gamma$. Consequently, by Definition 4.2.1 and by Theorem 4.2.3, f is **m**-integrable in T with values in X and (4.2.9.2) holds. $\qquad\square$

Remark 4.2.10. If we replace the simple functions in Definition 2.2.1 of Chapter 2 by functions $\mathbf{m}_q$-integrable in T, then Theorem 4.2.9 above says that the process described in Definition 4.2.1 yields functions which are already in $\mathcal{I}(\mathbf{m})$ and no new functions are rendered **m**-integrable in T. Hence Theorem 4.2.9 is called the closure theorem (compare Theorem 9 of [DP2]).

The following result generalizes Corollary 2.2.11 of Chapter 2 to quasicomplete lcHs-valued vector measures.

Corollary 4.2.11. *Let* X *be a quasicomplete lcHs,* $\mathcal{P}$ *be a* σ-*ring* $\mathcal{S}$ *and* $\mathbf{m} : \mathcal{S} \to X$ *be* σ-*additive. If* X, $f_n^{(q)}$, $n \in \mathbb{N}$, q *and* f *are as in Theorem 4.2.9 and if* $f_n^{(q)} \to f$ $\mathbf{m}_q$-*a.e. in* T *for each* $q \in \Gamma$ *and if, for each* $q \in \Gamma$,

$$\lim_{\|\mathbf{m}\|_q(A) \to 0} \int_A f_n^{(q)} d\mathbf{m}_q = 0, \ A \in \mathcal{S}$$

uniformly for $n \in \mathbb{N}$, *then* f *is* $\mathbf{m}_q$-*integrable in* T *with values in* $\widetilde{X_q}$ *and*

$$\int_A f d\mathbf{m}_q = \lim_n \int_A f_n^{(q)} d\mathbf{m}_q, \ A \in \mathcal{S}, \tag{4.2.11.1}$$

the limit being uniform with respect to $A \in \mathcal{S}$ *(for a fixed* $q \in \Gamma$). *Consequently,* f *is* **m**-*integrable in* T *with values in* X *and*

$$\int_A f d\mathbf{m} = \lim_{\longleftarrow} \int_A f d\mathbf{m}_q, \ A \in \mathcal{S} \cup \{T\}. \tag{4.2.11.2}$$

Proof. Let $\gamma_n^{(q)}(\cdot) = \int_{(\cdot)} f_n^{(q)} d\mathbf{m}_q$, $n \in \mathbb{N}$. Then by hypothesis and by Corollary 2.2.11 of Chapter 2, f is $\mathbf{m}_q$-integrable in T and (4.2.11.1) holds (with the limit being uniform with respect to $A \in \mathcal{S}$). Then by Definition 4.2.1 and by Theorem 4.2.3, f is $\mathbf{m}$-integrable in T with values in X and (4.2.11.2) holds. $\qquad\square$

Remark 4.2.12. (**m-integrability of $\sigma(\mathcal{P})$-measurable functions with respect to a sequentially complete lcHs-valued m**).

Definition 4.2.1$'$. Let X be a sequentially complete lcHs and let $\mathbf{m} : \mathcal{P} \to X$ be σ-additive. A $\sigma(\mathcal{P})$-measurable function $f : T \to \mathbb{K}$ or $[-\infty, \infty]$ is said to be (BDS) $\mathbf{m}$-integrable in T if there exists a sequence $(s_n) \subset \mathcal{I}_s$ such that $s_n(t) \to f(t)$ pointwise in T and such that $\lim_n \int_A s_n d\mathbf{m}$ exists in X for each $A \in \sigma(\mathcal{P})$. In that case, we define (BDS)$\int_A f d\mathbf{m} = \lim_n \int_A s_n d\mathbf{m}$, $A \in \sigma(\mathcal{P})$ and (BDS)$\int_T f d\mathbf{m} = \lim_n \int_{N(f)} s_n d\mathbf{m}$.

Theorem 4.2.2$'$. If X is a sequentially complete lcHs and $\mathbf{m} : \mathcal{P} \to X$ is σ-additive, then a $\sigma(\mathcal{P})$-measurable function $f : T \to \mathbb{K}$ or $[-\infty, \infty]$ is (BDS) $\mathbf{m}$-integrable in T (with values in X) if and only if it is (KL) $\mathbf{m}$-integrable in T (with values in X) and in that case, (BDS)$\int_A f d\mathbf{m} = $ (KL)$\int_A f d\mathbf{m} \in X$ for $A \in \sigma(\mathcal{P}) \cup \{T\}$. This shows that $\int_A f d\mathbf{m}$ is well defined for $A \in \sigma(\mathcal{P}) \cup \{T\}$.

Proof. If f is (BDS) $\mathbf{m}$-integrable in T, then by hypothesis and by Proposition 1.1.21 of Chapter 1, it follows that f is (KL) $\mathbf{m}$-integrable in T and that (BDS)$\int_A f d\mathbf{m} = $ (KL)$\int_A f d\mathbf{m}$ for $A \in \sigma(\mathcal{P}) \cup \{T\}$. Conversely, if f is (KL) $\mathbf{m}$-integrable in T, take $(s_n) \subset \mathcal{I}_s$ as in the last part of Proposition 1.2.10. Then given $\epsilon > 0$ and $q \in \Gamma$, by Theorem 4.1.11$'$ in Remark 4.1.15, $q(\int_A s_n d\mathbf{m} - \int_A s_r d\mathbf{m}) \leq q(\int_A s_n d\mathbf{m} - $ (KL)$\int_A f d\mathbf{m}) + q((KL)\int_A f d\mathbf{m} - \int_A s_r d\mathbf{m}) < \epsilon$ for n, r sufficiently large. Thus $(\int_A s_n d\mathbf{m})_1^\infty$ is Cauchy in X and as X is sequentially complete, there exists $x_A \in X$ such that $x_A = \lim_n \int_A s_n d\mathbf{m}$ so that f is (BDS) $\mathbf{m}$-integrable in T and (BDS)$\int_A f d\mathbf{m} = x_A$. Then by LDCT for scalar measures, $x^*(x_A) = \int_A f d(x^* \mathbf{m})$ for $x^* \in X^*$ so that f is (KL) $\mathbf{m}$-integrable in T and (KL)$\int_A f d\mathbf{m} = x_A = $ (BDS)$\int_A f d\mathbf{m}$. Hence Theorem 4.2.2$'$ holds. $\qquad\square$

Consequently, when X is a sequentially complete lcHs, $\mathcal{I}(\sigma(\mathcal{P}), \mathbf{m})$ in Remark 4.1.15 is the same as the family of all $\sigma(\mathcal{P})$-measurable (BDS) $\mathbf{m}$-integrable scalar functions on T with values in X and, for $f \in \mathcal{I}(\sigma(\mathcal{P}), \mathbf{m})$, (KL)$\int_A f d\mathbf{m} = $ (BDS)$\int_A f d\mathbf{m}$ for $A \in \sigma(\mathcal{P}) \cup \{T\}$. Hence, **hereafter, when X is sequentially complete, we shall denote either of the integrals by $\int_A f d\mathbf{m}$ for $f \in \mathcal{I}(\sigma(\mathcal{P}), \mathbf{m})$ and for $A \in \sigma(\mathcal{P}) \cup \{T\}$.** Then in that case, Remark 4.2.6$'$ is the same as Remark 4.2.6 in which $\mathcal{I}(\mathbf{m})$ is replaced by $\mathcal{I}(\sigma(\mathcal{P}), \mathbf{m})$ and X is sequentially complete, reference to Theorems 4.1.9, 4.1.11 and 4.1.13 is changed to Theorems 4.1.9$'$, 4.1.11$'$ and 4.1.13$'$ in Remark 4.1.15, respectively (in which no reference is made to Theorem 4.1.4 and reference to Theorem 4.1.8 remains unchanged) and reference to Corollaries 4.1.12 and 4.1.14 is changed to Corollaries 4.1.12$'$ and 4.1.14$'$ in Remark 4.1.15, respectively.

Remark 4.2.13. If $\mathbf{m} : \mathcal{P} \to X$ is σ-additive and X is quasicomplete, by Theorem 4.1.4(ii), an $\mathbf{m}$-measurable function which is (KL) $\mathbf{m}_q$-integrable in T for each $q \in \Gamma$, is (KL) $\mathbf{m}$-integrable in T with values in $\tilde{X}$, but in the light of Theorems 4.2.2 and 4.2.3, f is also (BDS) $\mathbf{m}$-integrable in T and (KL)$\int_A f d\mathbf{m} = $ (BDS) $\int_A f d\mathbf{m} \in X$ for $A \in \sigma(\mathcal{P}) \cup \{T\}$. **Hence the concept of (BDS) m-integrability of m-measurable functions is needed here to show that (KL) $\int_A f d\mathbf{m}$ belongs to X for $A \in \sigma(\mathcal{P}) \cup \{T\}$ whenever X is quasicomplete.**

Remark 4.2.14. If X is quasicomplete and $\mathbf{m} : \mathcal{P} \to X$ is σ-additive, then the definitions of $\mathbf{m}$-integrability and the $\mathbf{m}$-integral for a $\sigma(\mathcal{P})$-measurable function f as given in Definitions 4.2.1 and 4.2.1$'$ coincide. In fact, if X is quasicomplete, it is also sequentially complete. If f is $\sigma(\mathcal{P})$-measurable, there exists $(s_n)_1^\infty \subset \mathcal{I}_s$ such that $s_n \to f$ and $|s_n| \nearrow |f|$ pointwise in T. Then for each $q \in \Gamma$, by LDCT for $\mathbf{m}_q$ (Theorem 2.1.7 of Chapter 2) $q(\int_A s_n d\mathbf{m} - \int_A s_r d\mathbf{m}) = |\int_A s_n d\mathbf{m}_q - \int_A s_r d\mathbf{m}_q|_q \to 0$ as $n \to \infty$ and hence $(\int_A s_n d\mathbf{m})_1^\infty$ is Cauchy in X. Since X is also sequentially complete, there exists $x_A \in X$ such that $\int_A s_n d\mathbf{m} \to x_A$ and hence by Definition 4.2.1$'$, $\int_A f d\mathbf{m} = x_A$ for $A \in \sigma(\mathcal{P}) \cup \{T\}$. On the other hand,

$$\int_A f d\mathbf{m}_q = \Pi_q \left(\int_A f d\mathbf{m} \right) = \Pi_q(x_A) \in X_q$$

and

$$x_A = \underleftarrow{\lim}\Pi_q(x_A) = \underleftarrow{\lim} \int_A f d\mathbf{m}_q = \int_A f d\mathbf{m}$$

in the sense of Definition 4.2.1 for $A \in \sigma(\mathcal{P}) \cup \{T\}$. Hence the remark holds.

4.3 The locally convex spaces $\mathcal{L}_p\mathcal{M}(\mathbf{m})$, $\mathcal{L}_p\mathcal{M}(\sigma(\mathcal{P}),\mathbf{m})$, $\mathcal{L}_p\mathcal{I}(\mathbf{m})$ and $\mathcal{L}_p\mathcal{I}(\sigma(\mathcal{P}), \mathbf{m})$, $1 \leq p < \infty$

We generalize the results in Section 3.1 of Chapter 3 to an lcHs-valued σ-additive vector measure on $\mathcal{P}$ and this section plays a key role in the study of the $\mathcal{L}_p$-spaces when X is quasicomplete (resp. sequentially complete).

Let X be an lcHs and $\mathbf{m} : \mathcal{P} \to X$ be σ-additive. Then $\mathbf{m}_q = \Pi_q \circ \mathbf{m} : \mathcal{P} \to X_q \subset \widetilde{X_q}$ is σ-additive for $q \in \Gamma$.

Definition 4.3.1. Let X be an lcHs and $\mathbf{m} : \mathcal{P} \to X$ be σ-additive. Let $g : T \to \mathbb{K}$ or $[-\infty, \infty]$ be $\mathbf{m}$-measurable, $1 \leq p < \infty$ and $A \in \sigma(\mathcal{P})$. For $q \in \Gamma$, we define

$$(\mathbf{m}_q)_p^\bullet(g, A) = \sup \left\{ \left| \int_A s d\mathbf{m}_q \right|_q^{\frac{1}{p}} : s \in \mathcal{I}_s, |s| \leq |g|^p \ \mathbf{m}_q\text{-a.e. in } A \right\}$$

and

$$(\mathbf{m}_q)_p^\bullet(g, T) = \sup_{A \in \sigma(\mathcal{P})} (\mathbf{m}_q)_p^\bullet(g, A).$$

By Definition 3.1.4 of Chapter 3, for $A \in \widetilde{\sigma(\mathcal{P})}_q$ with $A = B_q \cup N_q$, $B_q \in \sigma(\mathcal{P})$, $N_q \subset M_q \in \sigma(\mathcal{P})$ with $\|\mathbf{m}\|_q(M_q) = 0$, we define

$$(\mathbf{m}_q)_p^\bullet(g, A) = (\mathbf{m}_q)_p^\bullet(g, B_q)$$

and it can be shown to be well defined by an argument as in the paragraph preceding Remark 3.1.5 of Chapter 3.

Theorem 4.3.2. *Let g, p and q be as in Definition 4.3.1. For $A \in \sigma(\mathcal{P})$,*

$$(\mathbf{m}_q)_p^\bullet(g, A) = \sup_{x^* \in U_q^o} \left(\int_A |g|^p dv(x^*\mathbf{m}) \right)^{\frac{1}{p}}$$

$$= \sup_{x^* \in U_q^o} (\Psi_{x^*}\mathbf{m}_q)_p^\bullet(g, A)$$

$$= \sup \left\{ \left| (KL) \int_A f d\mathbf{m}_q \right|_q^{\frac{1}{p}} : f \in \mathcal{I}(\sigma(\mathcal{P}), \mathbf{m}_q), |f| \leq |g|^p \, \mathbf{m}_q\text{-a.e. in } A \right\}$$

$$= \sup \left\{ \left| (KL) \int_A f d\mathbf{m}_q \right|_q^{\frac{1}{p}} : f \in \mathcal{I}(\mathbf{m}_q), |f| \leq |g|^p \, \mathbf{m}_q\text{-a.e. in } A \right\}$$

where $\mathcal{I}(\mathbf{m}_q)$ is as in Notation 2.1.2 of Chapter 2 (with values of the integrals in $\widetilde{X_q}$), $\mathcal{I}(\sigma(\mathcal{P}), \mathbf{m}_q)$ is as in Remark 4.1.15 for $\mathbf{m}_q$ (with values in $\widetilde{X_q}$) and Ψ_{x^*} is as in Proposition 1.2.15. Consequently, for $A \in \sigma(\mathcal{P})$,

$$\left| (KL) \int_A |f|^p d\mathbf{m}_q \right|_q^{\frac{1}{p}} \leq (\mathbf{m}_q)_p^\bullet(f, A)$$

if f is an $\mathbf{m}_q$-measurable scalar function with $|f|^p \in I(\mathbf{m}_q)$. Moreover, for $A \in \sigma(\mathcal{P})$,

$$\left| (KL) \int_A f d\mathbf{m}_q \right|_q \leq (\mathbf{m}_q)_1^\bullet(f, A) \qquad (4.3.2.1)$$

if $f \in \mathcal{I}(\mathbf{m}_q)$.

Proof. By Proposition 1.2.15(ii)(b), $\{\Psi_{x^*} : x^* \in U_q^o\}$ is a norm determining subset of the closed unit ball of $(X_q)^*$ and for $x^* \in U_q^o$, $x^*(\Pi_q \circ \mathbf{m}) = \Psi_{x^*}\mathbf{m}_q = x^*\mathbf{m}$ by (ii)(a) of the said proposition. Then by Lemma 3.1.2(ii) of Chapter 3 for $\mathbf{m}_q$ we have

$$(\Psi_{x^*}\mathbf{m}_q)_p^\bullet(g, A) = \left(\int_A |g|^p dv(\Psi_{x^*}\mathbf{m}_q) \right)^{\frac{1}{p}} = \left(\int_A |g|^p dv(x^*\mathbf{m}) \right)^{\frac{1}{p}} \qquad (4.3.2.2)$$

for $x^* \in U_q^0$. Arguing as in the proof of Theorem 3.1.3 of Chapter 3, using Lemma 3.1.2(i) of Chapter 3 for $\mathbf{m}_q$, and using Proposition 1.2.15(ii)(b), by (4.3.2.2) we have

$$(\mathbf{m}_q)^\bullet_p(g, A) = \sup\left\{\left(\sup_{x^*\in U^o_q}\left|\int_A sd(\Psi_{x^*}\mathbf{m}_q)\right|^{\frac{1}{p}}_q\right) : s \in \mathcal{I}_s,\ |s| \le |g|^p\ \mathbf{m}_q\text{-a.e.in } A\right\}$$

$$= \sup_{x^*\in U^o_q}(\Psi_{x^*}\mathbf{m}_q)^\bullet_p(g, A) = \sup_{x^*\in U^o_q}\left(\int_A |g|^p dv(x^*\mathbf{m})\right)^{\frac{1}{p}}$$

for $A \in \sigma(\mathcal{P})$.

Let $q \in \Gamma$. Note that $\mathcal{I}_s \subset \mathcal{I}(\sigma(\mathcal{P}), \mathbf{m}_q) \subset \mathcal{I}(\mathbf{m}_q)$. Given $f \in \mathcal{I}(\mathbf{m}_q)$, by Proposition 1.1.18 of Chapter 1 there exists a sequence $(s^{(q)}_n) \subset \mathcal{I}_s$ such that $s^{(q)}_n \to f\ \mathbf{m}_q$-a.e. in T and $|s^{(q)}_n| \nearrow |f|\ \mathbf{m}_q$-a.e. in T and hence by Corollary 2.1.9 of Chapter 2 applied to $\mathbf{m}_q : \mathcal{P} \to X_q$ we have $|\,(KL)\int_A f d\mathbf{m}_q|_q = \lim_n |\int_A s^{(q)}_n d\mathbf{m}_q|_q$. This proves the other equalities in the first part. The second part is evident from the first. □

Remark 4.3.3. The above proof is more general and elementary than that of Lemma II.2.2 of [KK]. Also compare Remark 3.1.5 of Chapter 3.

The following theorem generalizes Theorem 3.1.6 of Chapter 3 to lcHs-valued σ-additive vector measures on $\mathcal{P}$.

Theorem 4.3.4. *Let X be an lcHs, $\mathbf{m} : \mathcal{P} \to X$ be σ-additive and $1 \le p < \infty$. Let $f : T \to \mathbb{K}$ or $[-\infty, \infty]$ be $\mathbf{m}$-measurable and let $|f|^p$ be (KL) $\mathbf{m}$-integrable in T (with values in X). Let $\gamma(\cdot) = (KL)\int_{(\cdot)} |f|^p d\mathbf{m}$. Then $\gamma : \sigma(\mathcal{P}) \to X$ is σ-additive. For $q \in \Gamma$, $(\mathbf{m}_q)^\bullet_p(f, A) = (\|\gamma\|_q(A))^{\frac{1}{p}}$ for $A \in \sigma(\mathcal{P})$. Consequently, $(\mathbf{m}_q)^\bullet_p(f, T) = (\|\gamma\|_q(T))^{\frac{1}{p}} < \infty$ and $(\mathbf{m}_q)^\bullet_p(f, \cdot)$ is continuous on $\sigma(\mathcal{P})$ for each $q \in \Gamma$.*

Proof. By (ii) of Theorem 4.1.8, γ is σ-additive on $\sigma(\mathcal{P})$ and hence, $\gamma_q = \Pi_q \circ \gamma : \sigma(\mathcal{P}) \to X_q \subset \widetilde{X_q}$ is σ-additive. Then by Theorems 4.1.8(iii)(a) and 4.3.2, $(\mathbf{m}_q)^\bullet_p(f, A) = (\|\gamma\|_q(A))^{\frac{1}{p}}$, $A \in \sigma(\mathcal{P})$. By Proposition 1.1.5 of Chapter 1 for $\mathbf{m}_q$ other results hold. □

Theorem 4.3.5. *Let X be a quasicomplete lcHs, $\mathbf{m} : \mathcal{P} \to X$ be σ-additive, $1 \le p < \infty$ and $f : T \to \mathbb{K}$ or $[-\infty, \infty]$. Then:*

(i) *If f is $\mathbf{m}$-measurable and $c_0 \not\subset \widetilde{X_q}$ for each $q \in \Gamma$, then $|f|^p$ is $\mathbf{m}$-integrable in T with values in X if and only if $(\mathbf{m}_q)^\bullet_p(f, T) < \infty$ for each $q \in \Gamma$. Moreover, if $|f|^p$ is $(x^*\mathbf{m})$-integrable for each $x^* \in X^*$, then $|f|^p$ is $\mathbf{m}$-integrable in T with values in X and hence $(\mathbf{m}_q)^\bullet_p(f, T) < \infty$ for each $q \in \Gamma$.*

(ii) *If f is $\sigma(\mathcal{P})$-measurable and $c_0 \not\subset X$, then $|f|^p$ is $\mathbf{m}$-integrable in T (and hence (KL) $\mathbf{m}$-integrable in T) with values in X if and only if $(\mathbf{m}_q)^\bullet_p(f, T) < \infty$ for each $q \in \Gamma$. Moreover, if $|f|^p$ is is $(x^*\mathbf{m})$-integrable for each $x^* \in X^*$, then*

$|f|^p$ *is* **m**-*integrable in* T *with values in* X *and hence* $(\mathbf{m}_q)_p^{\bullet}(f, T) < \infty$ *for each* $q \in \Gamma$.

Proof. (i) The condition is necessary by Theorems 4.3.4 and 4.2.2 (note that this part holds for any quasicomplete lcHs). Conversely, let $(\mathbf{m}_q)_p^{\bullet}(f, T) < \infty$ for each $q \in \Gamma$ and let $c_0 \not\subset \widetilde{X_q}$ for each $q \in \Gamma$. For $x^* \in X^*$, let $q_{x^*}(x) = |x^*(x)|$, $x \in X$. Then by Theorem 4.3.2 and by hypothesis,

$$\left(\int_A |f|^p dv(x^* \mathbf{m}) \right)^{\frac{1}{p}} \leq (\mathbf{m}_{q_{x^*}})_p^{\bullet}(f, T) < \infty \qquad (4.3.5.1)$$

for each $A \in \sigma(\mathcal{P})$.

Let $q \in \Gamma$. Then by Proposition 1.2.9 there exists a sequence $(s_n^{(q)}) \subset \mathcal{I}_s$ such that $0 \leq s_n^{(q)} \nearrow |f|^p$ $\mathbf{m}_q$-a.e. in T. If $u_n^{(q)} = s_n^{(q)} - s_{n-1}^{(q)}$ for $n \geq 1$, where $s_0^{(q)} = 0$, then as in the proof of Theorem 3.1.8 of Chapter 3 we have $\sum_1^{\infty} \int_A |u_n^{(q)}| dv(y^* \mathbf{m}_q) = \int_A |f|^p dv(y^* \Pi_q \mathbf{m}) < \infty$ by (4.3.5.1) for $y^* \in (X_q)^*$, since $y^* \Pi_q \in X^*$. As $c_0 \not\subset \widetilde{X_q}$ by hypothesis, arguing as in the proof of Theorem 3.1.8 of Chapter 3 we observe that there exists a vector $x_A \in \widetilde{X_q}$ such that $x_A = \lim_n \int_A s_n^{(q)} d\mathbf{m}_q$ and this holds for each $A \in \sigma(\mathcal{P})$. Then by Definition 2.2.1 of Chapter 2, $|f|^p$ is $\mathbf{m}_q$-integrable in T with values in $\widetilde{X_q}$. Since q is arbitrary in Γ and since X is quasicomplete by hypothesis, by Definition 4.2.1 and by Theorem 4.2.3, $|f|^p$ is **m**-integrable in T with values in X.

Now let $|f|^p$ be $(x^* \mathbf{m})$-integrable for each $x^* \in X^*$. Let $q \in \Gamma$ and let $y^* \in X_q^*$. Then $y^* \Pi_q \in X^*$ and hence by hypothesis, $|f|^p$ is $(y^* \Pi_q \mathbf{m})$-integrable; i.e., $|f|^p$ is $y^* \mathbf{m}_q$-integrable. As $c_0 \not\subset \widetilde{X_q}$, by the last part of Theorem 3.1.8 of Chapter 3, $|f|^p$ is $\mathbf{m}_q$-integrable in T with values in $\widetilde{X_q}$. As q is arbitrary in Γ, by Definition 4.2.1 and by Theorem 4.2.3, $|f|^p$ is **m**-integrable in T with values in X and moreover, $(\mathbf{m}_q)_p^{\bullet}(f, T) < \infty$ for each $q \in \Gamma$ by Theorems 3.3.2 and 4.3.4.

(ii) The condition is necessary by Theorems 4.3.4 and 4.2.2 (this part holds for any quasicomplete lcHs X). Conversely, let f be $\sigma(\mathcal{P})$-measurable, let $c_0 \not\subset X$ and let $(\mathbf{m}_q)_p^{\bullet}(f, T) < \infty$ for each $q \in \Gamma$. Then particularly (4.3.5.1) holds for $x^* \in X^*$. Now by the last part of Proposition 1.2.9 there exists $(s_n)_1^{\infty} \subset \mathcal{I}_s$ such that $0 \leq s_n \nearrow |f|^p$ pointwise in T. Then arguing as in the proof of Theorem 3.1.8 of Chapter 3 and using (4.3.5.1), we have $\sum_1^{\infty} |x^*(\int_A u_n d\mathbf{m})| < \infty$ for $A \in \sigma(\mathcal{P})$ and for each $x^* \in X^*$, where $u_n = s_n - s_{n-1}$ for $n \geq 1$ and $s_0 = 0$. Then, as $c_0 \not\subset X$, by Theorem 4 of [Tu] there exists a vector $x_A \in X$ such that $x_A = \sum_1^{\infty} \int_A u_n d\mathbf{m} = \lim_n \int_A s_n d\mathbf{m}$ for $A \in \sigma(\mathcal{P})$ and hence by Definition 4.2.1' in Remark 4.2.12 and by Remark 4.2.14, $|f|^p$ is **m**-integrable in T with values in X. Then by Theorem 4.2.2' in the said remark, $|f|^p$ is also (KL) **m**-integrable in T with values in X. When $|f|^p$ is $(x^* \mathbf{m})$-integrable for each $x^* \in X^*$, then $\int_T |f|^p dv(x^* \mathbf{m}) < \infty$ so that (4.3.5.1) holds for $x^* \in X^*$ and hence the above argument holds to prove that $|f|^p$ is **m**-integrable in T with values in X and hence by Theorem 4.3.4, $(\mathbf{m}_q)_p^\bullet(f,T) < \infty$ for each $q \in \Gamma$. $\square$

Remark 4.3.6. Let X be a sequentially complete lcHs. By Theorem 4.3.4, $(\mathbf{m}_q)_p^\bullet(f,T) < \infty$ if f is $\sigma(\mathcal{P})$-measurable, and $|f|^p$ is (KL) **m**-integrable in T with values in X. Conversely, if f is $\sigma(\mathcal{P})$-measurable and if $c_0 \not\subset X$ and if $(\mathbf{m}_q)_p^\bullet(f,T) < \infty$ for each $q \in \Gamma$, then the proof of Theorem 4.3.5(ii) holds here verbatim as Theorem 4 of [Tu] holds for weakly sequentially complete lcHs. Thus **this result generalizes Theorem 1, Chapter II.5 of [KK] to δ-rings and Theorem 5.1 of [L2] to sequentially complete lcHs. Moreover, the question of the referee about the validity of Theorem 4.3.5(ii) for sequentially complete lcHs is answered in the affirmative.**

Definition 4.3.7. Let X be an lcHs, $\mathbf{m} : \mathcal{P} \to X$ be σ-additive and $1 \le p < \infty$. We define $\mathcal{L}_p\mathcal{M}(\mathbf{m}) = \{f : T \to \mathbb{K}, f \text{ **m**-measurable with } (\mathbf{m}_q)_p^\bullet(f,T) < \infty \text{ for each } q \in \Gamma\}$; $\mathcal{I}_p(\mathbf{m}) = \{f : T \to \mathbb{K}, f \text{ **m**-measurable and } |f|^p(\text{KL}) \text{ **m**-integrable in } T \text{ with values in } X\}$ and $\mathcal{L}_p\mathcal{I}(\mathbf{m}) = \mathcal{L}_p\mathcal{M}(\mathbf{m}) \bigcap \mathcal{I}_p(\mathbf{m})$. Let $\mathcal{M}(\sigma(\mathcal{P})) = \{f : T \to \mathbb{K}, f \,\sigma(\mathcal{P})\text{-measurable}\}$. Then we define:

$$\mathcal{L}_p\mathcal{M}(\sigma(\mathcal{P}), \mathbf{m}) = \mathcal{L}_p\mathcal{M}(\mathbf{m}) \bigcap \mathcal{M}(\sigma(\mathcal{P}));$$

$$\mathcal{I}_p(\sigma(\mathcal{P}), \mathbf{m}) = \mathcal{I}_p(\mathbf{m}) \bigcap \mathcal{M}(\sigma(\mathcal{P}));$$

$$\mathcal{L}_p\mathcal{I}(\sigma(\mathcal{P}), \mathbf{m}) = \mathcal{L}_p\mathcal{I}(\mathbf{m}) \bigcap \mathcal{M}(\sigma(\mathcal{P})).$$

Theorem 3.1.10 of Chapter 3 is generalized to lcHs-valued **m** as follows.

Theorem 4.3.8. *Let X be an lcHs, $\mathbf{m} : \mathcal{P} \to X$ be σ-additive and $1 \le p < \infty$. Then:*

(i) $\mathcal{L}_p\mathcal{I}(\mathbf{m}) = \mathcal{I}_p(\mathbf{m}) \subset \mathcal{L}_p\mathcal{M}(\mathbf{m})$ *and* $\mathcal{L}_p\mathcal{I}(\sigma(\mathcal{P}), \mathbf{m}) = \mathcal{I}_p(\sigma(\mathcal{P}), \mathbf{m}) \subset \mathcal{L}_p\mathcal{M}(\sigma(\mathcal{P}), \mathbf{m})$.

(ii) *If X is quasicomplete, then $\mathcal{I}_1(\mathbf{m}) = \mathcal{I}(\mathbf{m})$ (see Notation 4.1.6).*

(iii) *If X is sequentially complete, then $\mathcal{I}_1(\sigma(\mathcal{P}), \mathbf{m}) = \mathcal{I}(\sigma(\mathcal{P}), \mathbf{m})$. (See Remark 4.1.15.)*

(iv) *If X is quasicomplete and $c_0 \not\subset \widetilde{X_q}$ for each $q \in \Gamma$, then $\mathcal{L}_p\mathcal{I}(\mathbf{m}) = \mathcal{I}_p(\mathbf{m}) = \mathcal{L}_p\mathcal{M}(\mathbf{m})$.*

(v) *If X is a quasicomplete lcHs and if $c_0 \not\subset X$, then $\mathcal{L}_p\mathcal{I}(\sigma(\mathcal{P}), \mathbf{m}) = \mathcal{I}_p(\sigma(\mathcal{P}), \mathbf{m}) = \mathcal{L}_p\mathcal{M}(\sigma(\mathcal{P}), \mathbf{m})$.*

Proof. (i) holds by Theorem 4.3.4. (ii) (resp. (iii)) is due to Theorem 4.1.9(iii) (resp. due to Theorem 4.2.9' in Remark 4.1.15. (iv) is due to (i) and Theorem 4.3.5(i) while (v) is due to (i) and Theorem 4.3.5(ii). $\square$

If X is an lcHs and $\mathbf{m} : \mathcal{P} \to X$ is σ-additive, then by Definition 4.3.1 and by Theorem 4.3.2, it is clear that **Theorems 3.1.11–3.1.13 of Chapter 3 hold for**

m-measurable functions on T and hence for $\sigma(\mathcal{P})$-measurable functions on T with values in $\mathbb{K}$ or in $[-\infty,\infty]$ for each $q \in \Gamma$, if we replace $\mathbf{m}_p^\bullet(g,\cdot)$ by $(\mathbf{m}_q)_p^\bullet(g,\cdot)$, $||\mathbf{m}||$ by $||\mathbf{m}||_q$, $\mathbf{m}_p^\bullet(ag,\cdot)$ by $(\mathbf{m}_q)_p^\bullet(ag,\cdot)$ and $\mathbf{m}_p^\bullet(f+g,\cdot)$ by $(\mathbf{m}_q)_p^\bullet(f+g,\cdot)$. Hence, when these results are used, we simply make references analogous to those for Theorem 3.1.11 of Chapter 3 for $\mathbf{m}_q$, etc.

(i)–(iv) of the following theorem generalize Theorem 3.1.14 of Chapter 3 to an lcHs-valued $\mathbf{m}$ on $\mathcal{P}$ with Notation 3.1.15 of Chapter 3 being suitably interpreted here.

Theorem 4.3.9. *Let X be an lcHs, $\mathbf{m} : \mathcal{P} \to X$ be σ-additive, $1 \le p < \infty$ and $\xi_\Gamma^{(p)} = \{(\mathbf{m}_q)_p^\bullet(\cdot,T) : q \in \Gamma\}$. Then:*

(i) *$\mathcal{L}_p\mathcal{M}(\mathbf{m})$ (resp. $\mathcal{L}_p\mathcal{M}(\sigma(\mathcal{P}),\mathbf{m})$) is a vector space over $\mathbb{K}$.*

(ii) *$\xi_\Gamma^{(p)}$ is a family of seminorms on $\mathcal{L}_p\mathcal{M}(\mathbf{m})$. If $\tau_\mathbf{m}^{(p)}$ is the locally convex topology generated by the family $\xi_\Gamma^{(p)}$ on $\mathcal{L}_p\mathcal{M}(\mathbf{m})$, then by $\mathcal{L}_p\mathcal{M}(\mathbf{m})$ we mean the locally convex space $(\mathcal{L}_p\mathcal{M}(\mathbf{m}), \tau_\mathbf{m}^{(p)})$.*

(iii) *If X is quasicomplete, then $\mathcal{L}_p\mathcal{I}(\mathbf{m})$ is a linear subspace of $\mathcal{L}_p\mathcal{M}(\mathbf{m})$. In that case, by $\mathcal{L}_p\mathcal{I}(\mathbf{m})$ we mean the locally convex space $(\mathcal{L}_p\mathcal{I}(\mathbf{m}), \tau_\mathbf{m}^{(p)}|_{\mathcal{L}_p\mathcal{I}(\mathbf{m})})$.*

(iv) *If X is sequentially complete, then $\mathcal{L}_p\mathcal{I}(\sigma(\mathcal{P}),\mathbf{m})$ is a linear subspace of $\mathcal{L}_p\mathcal{M}(\sigma(\mathcal{P}),\mathbf{m})$. In that case, by $\mathcal{L}_p\mathcal{I}(\sigma(\mathcal{P}),\mathbf{m})$ we mean the locally convex space $(\mathcal{L}_p\mathcal{I}(\sigma(\mathcal{P}),\mathbf{m}), \tau_\mathbf{m}^{(p)}|_{\mathcal{L}_p\mathcal{I}(\sigma(\mathcal{P}),\mathbf{m})})$.*

(v) *For $f,g \in \mathcal{L}_p\mathcal{M}(\mathbf{m})$, we write $f \sim g$ if $f = g$ $\mathbf{m}$-a.e. in T (see Definition 1.2.4). Then '$\sim$' is an equivalence relation and we denote $\mathcal{L}_p\mathcal{M}(\mathbf{m})/\sim$ by $L_p\mathcal{M}(\mathbf{m})$; when X is quasicomplete, $\mathcal{L}_p\mathcal{I}(\mathbf{m})/\sim$ by $L_p\mathcal{I}(\mathbf{m})$ and when X is sequentially complete, $\mathcal{L}_p\mathcal{I}(\sigma(\mathcal{P}),\mathbf{m})/\sim$ by $L_p\mathcal{I}(\sigma(\mathcal{P}),\mathbf{m})$. Then $f \sim g$ if and only if $(\mathbf{m}_q)_p^\bullet(f-g,T) = 0$ for all $q \in \Gamma$. If $\mathcal{F} = \mathcal{L}_p\mathcal{M}(\mathbf{m})$ and if $N_\mathcal{F} = \{f \in \mathcal{F} : f \sim 0\}$, then $N_\mathcal{F}$ is a closed linear subspace of $\mathcal{F}$. If X is quasicomplete and $\mathcal{F} = \mathcal{L}_p\mathcal{I}(\mathbf{m})$, then $N_\mathcal{F} = \{f \in \mathcal{F} : f \sim 0\}$ is a closed linear subspace of $\mathcal{L}_p\mathcal{I}(\mathbf{m})$. If X is sequentially complete and $\mathcal{F} = \mathcal{L}_p\mathcal{M}(\sigma(\mathcal{P}),\mathbf{m})$ (or $=\mathcal{L}_p\mathcal{I}(\sigma(\mathcal{P},\mathbf{m}))$, then $N_\mathcal{F} = \{f \in \mathcal{F} : f \sim 0\}$ is a closed linear subspace of $\mathcal{F}$. Consequently, $L_p\mathcal{M}(\mathbf{m})$, $L_p\mathcal{I}(\mathbf{m})$, $L_p\mathcal{M}(\sigma(P),\mathbf{m})$ and $L_p\mathcal{I}(\sigma(\mathcal{P}),\mathbf{m})$ are lcHs.*

Proof. By Theorems 3.1.11(ii) and 3.1.13(ii) of Chapter 3 for $\mathbf{m}_q$, $q \in \Gamma$ and by Definition 4.3.7, (i) and (ii) hold. Arguing as in the proof of Theorem 3.1.14 of Chapter 3, using Theorems 4.1.8(iv) and 4.1.9(iii) (resp. and 4.1.9'(iii) in Remark 4.1.15) for $\mathbf{m}_q$, $q \in \Gamma$ in place of (iv) and (vii) of Theorem 2.1.5 of Chapter 2 and invoking Theorem 4.3.8(i), one can show that (iii) (resp. (iv)) holds.

(v) Clearly, '$\sim$' is an equivalence relation. By Definition 1.2.4, $f \sim g$ if and only if there exists $M \in \sigma(\mathcal{P})$ with $||\mathbf{m}||_q(M) = 0$ for all $q \in \Gamma$ such that $f(t) - g(t) = 0$ for $t \in T\backslash M$. Let $f \sim g$ with M as above and let $q \in \Gamma$. Then by

Theorem 4.3.2,

$$((\mathbf{m}_q)^{\bullet}_p(f-g,T))^p = \sup_{x^* \in U^o_q} \int_T |f-g|^p dv(x^*\mathbf{m}) = \sup_{x^* \in U^o_q} \int_M |f-g|^p dv(x^*\mathbf{m}) = 0$$

since by Theorem 1.2.15(ii)(c), $v(x^*\mathbf{m})(M) \leq ||\mathbf{m}||_q(M) = 0$ for $x^* \in U^o_q$. Hence $(\mathbf{m}_q)^{\bullet}_p(f-g,T) = 0$ for all $q \in \Gamma$. Conversely, if $(\mathbf{m}_q)^{\bullet}_p(f-g,T) = 0$ for all $q \in \Gamma$, let $A = \{t \in T : f(t) - g(t) \neq 0\}$. For $q \in \Gamma$, A is of the form $A = B_q \cup N_q$, where $B_q \in \sigma(\mathcal{P})$, $N_q \subset M_q \in \sigma(\mathcal{P})$ with $||\mathbf{m}||_q(M_q) = 0$ if f, g are $\mathbf{m}$-measurable and $A = B_q \in \sigma(\mathcal{P})$ if f, g are $\sigma(\mathcal{P})$-measurable. Then by hypothesis and by Theorem 4.3.2, $((\mathbf{m}_q)^{\bullet}_p(f-g,B_q))^p = \sup_{x^* \in U^o_q} \int_{B_q} |f-g|^p dv(x^*\mathbf{m}) = 0$ and hence $v(x^*\mathbf{m})(B_q) = 0$ for $x^* \in U^0_q$. Then by Proposition 1.2.15(ii)(c) we have $||\mathbf{m}||_q(A) = ||\mathbf{m}||_q(B_q) = 0$ (in both cases). Since $q \in \Gamma$ is arbitrary, it follows that A is $\mathbf{m}$-null. Using this characterization, it is easy to check that $N_{\mathcal{F}}$ is a closed linear subspace of $\mathcal{F}$ and hence the last part of (v) holds. $\qquad\square$

Definition 4.3.10. Definition 3.1.16 of Chapter 3 is suitably modified for $\mathbf{m}$-measurable (resp. $\sigma(\mathcal{P})$-measurable) scalar functions on T to define (i) convergence in measure (resp. Cauchy in measure) in T or in $A \in \sigma(\mathcal{P})$ with respect to $\mathbf{m}_q$ for $q \in \Gamma$; (ii) almost uniform convergence (resp. Cauchy for almost uniform convergence) in T or in $A \in \sigma(\mathcal{P})$ with respect to $\mathbf{m}_q$ for $q \in \Gamma$; and (iii) convergence to f in (mean^p) with respect to $\mathbf{m}_q$ for $q \in \Gamma$ and for $1 \leq p < \infty$.

Then for functions in $\mathcal{L}_p\mathcal{M}(\mathbf{m})$ (resp. for $\mathbf{m}$-measurable scalar functions on T) (iv)–(viii) of Theorem 3.1.18 (resp. Theorem 3.1.19) of Chapter 3 hold (resp. holds) with $\mathbf{m}$ being replaced by $\mathbf{m}_q$, $q \in \Gamma$. Hence when these results are used, we simply make references analogous to those for (vii) of Theorem 3.1.18 of Chapter 3 for $\mathbf{m}_q$, etc.

Remark 4.3.11. If the topology τ of X is generated by another family Γ_1 of seminorms on X, then it is easy to show that the topology $\tau^{(p)}_{\mathbf{m}}$ in Theorem 4.3.9(ii) is the same as the locally convex topology generated by $\xi^{(p)}_{\Gamma_1}$ on the space $\mathcal{L}_p\mathcal{M}(\mathbf{m})$ for $1 \leq p < \infty$.

4.4 Completeness of $\mathcal{L}_p\mathcal{M}(\mathbf{m}), \mathcal{L}_p\mathcal{I}(\mathbf{m})$, $\mathcal{L}_p\mathcal{M}(\sigma(\mathcal{P}), \mathbf{m})$ and $\mathcal{L}_p\mathcal{I}(\sigma(\mathcal{P}), \mathbf{m})$, for suitable X

If X is a metrizable lcHs and $\mathbf{m} : \mathcal{P} \to X$ is σ-additive, then we show that $\mathcal{L}_p\mathcal{M}(\mathbf{m})$ and $\mathcal{L}_p\mathcal{M}(\sigma(\mathcal{P}), \mathbf{m})$ are complete pseudo-metrizable locally convex spaces so that $L_p\mathcal{M}(\mathbf{m})$ and $L_p\mathcal{M}(\sigma(\mathcal{P}), \mathbf{m})$ are Fréchet spaces for $1 \leq p < \infty$. Similar to the case of Banach space-valued measures in Chapter 3, we introduce the locally convex spaces $\mathcal{L}_p(\mathbf{m})$ (resp. $\mathcal{L}_p(\sigma(\mathcal{P}), \mathbf{m})$) for $1 \leq p < \infty$, when X is quasicomplete (resp. sequentially complete) and show that (i) $\mathcal{L}_p(\mathbf{m}) = \mathcal{L}_p\mathcal{I}(\mathbf{m})$ (resp. (i') $\mathcal{L}_p(\sigma(\mathcal{P}), \mathbf{m}) = \mathcal{L}_p\mathcal{I}(\sigma(\mathcal{P}), \mathbf{m})$) and (ii) $\mathcal{L}_p(\mathbf{m})$ is closed in $\mathcal{L}_p\mathcal{M}(\mathbf{m})$

(resp. (ii$'$) $\mathcal{L}_p(\sigma(\mathcal{P}), \mathbf{m})$ is closed in $\mathcal{L}_p\mathcal{M}(\sigma(\mathcal{P}), \mathbf{m})$). Consequently, $L_p(\mathbf{m})$ (resp. $L_p(\sigma(\mathcal{P}), \mathbf{m})$) is a Fréchet space for $1 \leq p < \infty$ whenever X is a Fréchet space. See Remark 4.4.10 for the existing literature on the completeness of $L_1\mathcal{I}(\sigma(\mathcal{P}), \mathbf{m})$ and Remark 4.4.15 for comments on the recent developments.

The following theorem is obtained by adapting the proof of Theorem 3.2.1 of Chapter 3 with the use of Theorem 4.3.2 in place of Theorem 3.1.3 of Chapter 3. Details are left to the reader.

Theorem 4.4.1. *Let X be an lcHs, $\mathbf{m} : \mathcal{P} \to X$ be σ-additive and $1 \leq p < \infty$. Let $f, f_n, n \in \mathbb{N}$, be $\mathbf{m}$-measurable on T with values in $\mathbb{K}$ or in $[-\infty, \infty]$ and let $q \in \Gamma$. Then:*

(i) **(The Fatou property of $(\mathbf{m}_q)_p^{\bullet}(\cdot, A)$).** *If $|f_n| \nearrow |f|$ $\mathbf{m}_q$-a.e. in $A \in \widetilde{\sigma(\mathcal{P})}_q$, then*
$$(\mathbf{m}_q)_p^{\bullet}(f, A) = \sup_n (\mathbf{m}_q)_p^{\bullet}(f_n, A) = \lim_n (\mathbf{m}_q)_p^{\bullet}(f_n, A).$$

(ii) **(Generalizad Fatou's lemma).** *For $A \in \widetilde{\sigma(\mathcal{P})}_q$,*
$$(\mathbf{m}_q)_p^{\bullet}(\liminf_{n \to \infty} |f_n|, A) \leq \liminf_{n \to \infty} (\mathbf{m}_q)_p^{\bullet}(f_n, A).$$

Theorem 4.4.2. *Let X be a metrizable lcHs with its topology generated by the seminorms $(q_n)_1^{\infty}$, $\mathbf{m} : \mathcal{P} \to X$ be σ-additive and $1 \leq p < \infty$. Then:*

(i) *$\mathcal{L}_p\mathcal{M}(\mathbf{m})$ (resp. $\mathcal{L}_p\mathcal{M}(\sigma(\mathcal{P}), \mathbf{m})$) is a complete pseudo-metrizable locally convex space so that $L_p\mathcal{M}(\mathbf{m})$ (resp. $L_p\mathcal{M}(\sigma(\mathcal{P}), \mathbf{m})$) is a Fréchet space.*

(ii) *Let X be a Fréchet space. If $c_0 \not\subset \widetilde{X_q}$ for each $q \in \Gamma$ (resp. if $c_0 \not\subset X$), then $\mathcal{L}_p\mathcal{I}(\mathbf{m})$ (resp. $\mathcal{L}_p\mathcal{I}(\sigma(\mathcal{P}), \mathbf{m})$) is a complete pseudo-metrizable locally convex space so that $L_p\mathcal{I}(\mathbf{m})$ (resp. $L_p\mathcal{I}(\sigma(\mathcal{P}), \mathbf{m})$) is a Fréchet space.*

Proof. (i) By Theorem 4.3.9 and Remark 4.3.11, $\mathcal{L}_p\mathcal{M}(\mathbf{m})$ (resp. $\mathcal{L}_p\mathcal{M}(\sigma(\mathcal{P}), \mathbf{m})$) is pseudo-metrizable with respect to $\tau_{\mathbf{m}}^{(p)}$ which is generated by $((\mathbf{m}_{q_n})_p^{\bullet}(\cdot, T))_{n=1}^{\infty}$. Suppose $(f_r)_1^{\infty} \subset \mathcal{L}_p\mathcal{M}(\mathbf{m})$ (resp. $(f_r)_1^{\infty} \subset \mathcal{L}_p\mathcal{M}(\sigma(\mathcal{P}), \mathbf{m})$) is Cauchy in $\tau_{\mathbf{m}}^{(p)}$. Proceeding as in the proof of Theorem 3.2.3(i) of Chapter 3, we can choose a subsequence $(f_{1,r})_{r=1}^{\infty}$ of $(f_r)_1^{\infty}$ such that $(\mathbf{m}_{q_1})^{\bullet}(f_{1,k+1} - f_{1,k}, T) < \frac{1}{2^k}$, $k \in \mathbb{N}$. Let
$$g_{1,k} = \sum_{r=1}^{k} |f_{1,r+1} - f_{1,r}| \quad \text{and} \quad g_1 = \sum_{r=1}^{\infty} |f_{1,r+1} - f_{1,r}|.$$

Then $g_{1,k}, k \in \mathbb{N}$ and g_1 are $\mathbf{m}$-measurable (resp. $\sigma(\mathcal{P})$-measurable) and $g_{1,k} \nearrow g_1$. Moreover, by Theorem 3.1.13(ii) of Chapter 3 for $\mathbf{m}_{q_1}$, $(\mathbf{m}_{q_1})_p^{\bullet}(g_{1,k}, T) < 1$ for all $k \in \mathbb{N}$ so that by Theorem 4.4.1(i), $(\mathbf{m}_{q_1})_p^{\bullet}(g_1, T) = \sup_k (\mathbf{m}_{q_1})_p^{\bullet}(g_{1,k}, T) \leq 1$. Then by Theorem 3.1.12(ii) of Chapter 3 for $\mathbf{m}_{q_1}$, g_1 is finite $\mathbf{m}_{q_1}$-a.e. in T and hence there exists $N_1 \in \sigma(\mathcal{P})$ with $||\mathbf{m}||_{q_1}(N_1) = 0$ such that g_1 is finite in $T \backslash N_1$.

Arguing similarly with $(f_{1,r})_{r=1}^{\infty}$ and $\mathbf{m}_{q_2}$, there exist a subsequence $(f_{2,r})_{r=1}^{\infty}$ of $(f_{1,r})_{r=1}^{\infty}$ and a set $N_2 \in \sigma(\mathcal{P})$ with $||\mathbf{m}||_{q_2}(N_2) = 0$ such that

$$g_2 = \sum_{r=1}^{\infty} |f_{2,r+1} - f_{2,r}|$$

is $\mathbf{m}$-measurable (resp. $\sigma(\mathcal{P})$-measurable), $(\mathbf{m}_{q_2})_p^{\bullet}(g_2, T) \leq 1$ and g_2 is finite in $T\backslash N_2$. Proceeding successively, in the $(k+1)^{th}$-step we shall have a subsequence $(f_{k+1,r})_{r=1}^{\infty}$ of $(f_{k,r})_{r=1}^{\infty}$ and a set $N_{k+1} \in \sigma(\mathcal{P})$ with $||\mathbf{m}||_{q_{k+1}}(N_{k+1}) = 0$ such that

$$g_{k+1} = \sum_{r=1}^{\infty} |f_{k+1,r+1} - f_{k+1,r}|$$

is $\mathbf{m}$-measurable (resp. $\sigma(\mathcal{P})$-measurable), $(\mathbf{m}_{q_{k+1}})_p^{\bullet}(g_{k+1}, T) \leq 1$ and g_{k+1} is finite in $T\backslash N_{k+1}$. Then the diagonal sequence $(f_{k,k})_{k=1}^{\infty}$ is a subsequence of each subsequence $(f_{k,r})_{r=1}^{\infty}$ starting with the term $f_{k,k}$; and if

$$g = \sum_{k=1}^{\infty} |f_{k+1,k+1} - f_{k,k}|,$$

then g is $\mathbf{m}$-measurable (resp. $\sigma(\mathcal{P})$-measurable). Moreover, $(\mathbf{m}_{q_n})_p^{\bullet}(g, T) < \infty$ for all n since, for a fixed n, $(\mathbf{m}_{q_n})_p^{\bullet}(g_n, T) \leq 1$ and $(\mathbf{m}_{q_n})_p^{\bullet}(f_{i,i}, T) < \infty$ for $i = 1, 2, \ldots, n$.

Let $N = \bigcap_1^{\infty} N_n$. Then $N \in \sigma(\mathcal{P})$ and $||\mathbf{m}||_{q_n}(N) = 0$ for all n. Given $t \in T\backslash N$, there exists n_0 such that $t \in T\backslash N_{n_0}$ so that $g_{n_0}(t)$ is finite. As each f_r is finite-valued in T and as all but a finite number of terms of $(f_{k,k})_1^{\infty}$ belong to $(f_{n_0,r})_{r=1}^{\infty}$, it follows that $g(t)$ is finite. Hence g is finite in $T\backslash N$. By Remark 1.2.5 of Chapter 1 and by the hypothesis that $(q_n)_1^{\infty}$ generate the topology of X, it follows that N is $\mathbf{m}$-null and hence g is finite $\mathbf{m}$-a.e. in T.

Then the series

$$\sum_{k=1}^{\infty} (f_{k+1,k+1} - f_{k,k})$$

is absolutely convergent in $T\backslash N$. As $(f_{k,k})_1^{\infty}$ is a subsequence of $(f_r)_1^{\infty}$, let $f_{n_k} = f_{k,k}$. Let $h_k = f_{n_k}$ for $k \geq 1$ and $h_0 = 0$. Define

$$f(t) = \begin{cases} \sum_{k=0}^{\infty}(h_{k+1}(t) - h_k(t)) = \lim_k h_k(t), & \text{for } t \in T\backslash N, \\ 0 & \text{otherwise.} \end{cases}$$

Then f is $\mathbb{K}$-valued in T, is $\mathbf{m}$-measurable (resp. $\sigma(\mathcal{P})$-measurable) and is the $\mathbf{m}$-a.e. pointwise limit of $(h_k)_1^{\infty}$. Let $\epsilon > 0$ and let n_0 be given. By hypothesis, there exists r_0 such that $(\mathbf{m}_{q_{n_0}})_p^{\bullet}(f_r - f_\ell, T) < \epsilon$ for $r, \ell \geq r_0$ so that $(\mathbf{m}_{q_{n_0}})_p^{\bullet}(h_k - h_\ell, T) < \epsilon$ for $n_k, n_\ell \geq r_0$. Let $F = \bigcup_1^{\infty} N(h_k)\backslash N$. Then $F \in \widetilde{\sigma(\mathcal{P})}_{q_{n_0}}$ (resp. $F \in \sigma(\mathcal{P})$) and $N(f) \subset F$. Then, arguing as in the proof of Theorem 3.2.3(i) of Chapter

3 with $\mathbf{m}_{q_{n_0}}$ in place of $\mathbf{m}$, we have $(\mathbf{m}_{q_{n_0}})_p^\bullet(f - h_k, T) < \epsilon$ for $n_k \geq r_0$. Then by the triangle inequality, $(\mathbf{m}_{q_{n_0}})_p^\bullet(f, T) < \infty$. As n_0 is arbitrary in $\mathbb{N}$, we conclude that $f \in \mathcal{L}_p\mathcal{M}(\mathbf{m})$ (resp. $f \in \mathcal{L}_p\mathcal{M}(\sigma(\mathcal{P}), \mathbf{m})$). Moreover, $\lim_k (\mathbf{m}_{q_{n_0}})_p^\bullet(f - h_k, T) = 0$. Consequently,

$$(\mathbf{m}_{q_{n_0}})_p^\bullet(f - f_r, T) \leq (\mathbf{m}_{q_{n_0}})_p^\bullet(f_r - f_{n_k}, T) + (\mathbf{m}_{q_{n_0}})_p^\bullet(f - h_k, T) \to 0$$

as $n_k, r \to \infty$ and hence $\lim_r (\mathbf{m}_{q_{n_0}})_p^\bullet(f_r - f, T) = 0$. As n_0 is arbitrary in $\mathbb{N}$, it follows that $f_r \to f$ in $\tau_\mathbf{m}^{(p)}$. Hence $\mathcal{L}_p\mathcal{M}(\mathbf{m})$ (resp. $\mathcal{L}_p\mathcal{M}(\sigma(\mathcal{P}), \mathbf{m})$) is complete. Consequently, by the last part of Theorem 4.3.9(v) and by Theorem 5.7 of Chapter 2 and Lemma 11.3 of Chapter 3 of [KN], $L_p\mathcal{M}(\mathbf{m})$ (resp. $L_p\mathcal{M}(\sigma(\mathcal{P}), \mathbf{m})$) is a Fréchet space.

(ii) is immediate from (i) and from (iv) (resp. (v)) of Theorem 4.3.8.   □

The following corollary is immediate from the proof of the last part of Theorem 4.4.2(i) and it holds even if X is not metrizable.

Corollary 4.4.3. *Let X be an lcHs, $\mathbf{m} : \mathcal{P} \to X$ be σ-additive and $1 \leq p < \infty$. If $(f_n)_1^\infty \subset \mathcal{L}_p\mathcal{M}(\mathbf{m})$ (resp. $\mathcal{L}_p\mathcal{M}(\sigma(\mathcal{P}), \mathbf{m})$) is Cauchy in $(\mathbf{m}_q)_p^\bullet(\cdot, T)$ for each $q \in \Gamma$, and if there exist a scalar function f on T, a subsequence (f_{n_k}) of (f_n) and an $\mathbf{m}$-null set $N \in \sigma(\mathcal{P})$ such that $f_{n_k} \to f$ in $T \backslash N$ (resp. and such that $f(t) = 0$ for $t \in N$), then f is $\mathbf{m}$-measurable (resp. $\sigma(\mathcal{P})$-measurable), $f \in \mathcal{L}_p\mathcal{M}(\mathbf{m})$ (resp. $f \in \mathcal{L}_p\mathcal{M}(\sigma(\mathcal{P}), \mathbf{m})$) and $\lim_n (\mathbf{m}_q)_p^\bullet(f_n - f, T) = 0$ for each $q \in \Gamma$.*

Similar to the second part of Definition 3.2.5 of Chapter 3 we give the following

Definition 4.4.4. *Let X be an lcHs, $\mathbf{m} : \mathcal{P} \to X$ be σ-additive and $1 \leq p < \infty$. Let $\mathcal{L}_p(\mathbf{m}) = \{f \in \mathcal{L}_p\mathcal{M}(\mathbf{m}) : (\mathbf{m}_q)_p^\bullet(f, \cdot)$ is continuous on $\sigma(\mathcal{P})$ for each $q \in \Gamma\}$ (resp. $\mathcal{L}_p(\sigma(\mathcal{P}), \mathbf{m}) = \mathcal{L}_p(\mathbf{m}) \bigcap \mathcal{M}(\sigma(\mathcal{P}))$) (see Definition 4.3.7) be provided with the relative topology induced by $\tau_\mathbf{m}^{(p)}$. If $' \sim '$ is given as in Theorem 4.3.9(v), then $\mathcal{L}_p(\mathbf{m})/\sim$ is denoted by $L_p(\mathbf{m})$ and $\mathcal{L}_p(\sigma(\mathcal{P}), \mathbf{m})/\sim$ by $L_p(\sigma(\mathcal{P}), \mathbf{m})$.*

Since $(\mathbf{m}_q)_p^\bullet(f, \cdot)$ is subadditive and positively homogeneous, $\mathcal{L}_p(\mathbf{m})$ (resp. $\mathcal{L}_p(\sigma(\mathcal{P}), \mathbf{m})$) is a linear subspace of $\mathcal{L}_p\mathcal{M}(\mathbf{m})$ (resp. of $\mathcal{L}_p\mathcal{M}(\sigma(\mathcal{P}), \mathbf{m})$) so that $\mathcal{L}_p(\mathbf{m})$ (resp. $\mathcal{L}_p(\sigma(\mathcal{P}), \mathbf{m})$) is a locally convex space and $L_p(\mathbf{m})$ (resp. $L_p(\sigma(\mathcal{P}), \mathbf{m})$) is an lcHs when X is quasicomplete (resp. sequentially complete). (See Theorem 4.3.9(v).)

The following result generalizes Theorem 3.2.7 of Chapter 3 to quasicomplete lcHs-valued $\mathbf{m}$.

Theorem 4.4.5. *Let X be a quasicomplete lcHs, $\mathbf{m} : \mathcal{P} \to X$ be σ-additive and $1 \leq p < \infty$. Let f be an $\mathbf{m}$-measurable scalar function on T such that $(\mathbf{m}_q)_p^\bullet(f, \cdot)$ is continuous on $\sigma(\mathcal{P})$ for each $q \in \Gamma$. Then $f \in \mathcal{L}_p\mathcal{M}(\mathbf{m})$, and hence $f \in \mathcal{L}_p(\mathbf{m})$. Moreover, $\mathcal{L}_p\mathcal{I}(\mathbf{m}) = \mathcal{L}_p(\mathbf{m})$.*

Proof. Let $(\mathbf{m}_q)_p^\bullet(f, \cdot)$ be continuous on $\sigma(\mathcal{P})$ for each $q \in \Gamma$. Then by Theorem 3.2.7 of Chapter 3 for $\mathbf{m}_q$, $(\mathbf{m}_q)_p^\bullet(f, T) < \infty$ and $|f|^p$ is $\mathbf{m}_q$-integrable in T with values in $\widetilde{X_q}$ for each $q \in \Gamma$ (where $\mathbf{m}_q$ has values in $X_q \subset \widetilde{X_q}$). Then, in particular, $f \in \mathcal{L}_p\mathcal{M}(\mathbf{m})$ and moreover, by Theorems 2.2.2 of Chapter 2 and by Theorems 4.1.4(ii), 4.2.2 and 4.2.3, $|f|^p$ is (KL) $\mathbf{m}$-integrable in T with values in X so that $f \in \mathcal{I}_p(\mathbf{m})$. Consequently, by Theorem 4.3.8(i), $f \in \mathcal{L}_p\mathcal{I}(\mathbf{m})$. This also proves that $\mathcal{L}_p(\mathbf{m}) \subset \mathcal{L}_p\mathcal{I}(\mathbf{m})$.

Conversely, let $f \in \mathcal{L}_p\mathcal{I}(\mathbf{m}) \, (= \mathcal{I}_p(\mathbf{m})$ by Theorem 4.3.8(i)). Then, by Theorem 4.2.2 and Remark 4.2.6, $|f|^p$ is (KL) $\mathbf{m}$-integrable in T with values in X and hence by Theorem 4.3.4, $f \in \mathcal{L}_p\mathcal{M}(\mathbf{m})$ and $(\mathbf{m}_q)_p^\bullet(f, \cdot)$ is continuous on $\sigma(\mathcal{P})$ for each $q \in \Gamma$. Hence $f \in \mathcal{L}_p(\mathbf{m})$. Therefore, $\mathcal{L}_p\mathcal{I}(\mathbf{m}) = \mathcal{L}_p(\mathbf{m})$. □

The following theorem is an analogue of Theorem 2.2.8 of Chapter 2 for sequentially complete lcHs and plays a key role in the study of $\mathcal{L}_p$-spaces of a sequentially complete lcHs-valued $\mathbf{m}$ (see Theorem 4.4.7 below). Also one can compare with Theorem 4.2.7.

Theorem 4.4.6. *Let X be a sequentially complete lcHs and $\mathbf{m} : \mathcal{P} \to X$ be σ-additive. Let $(f_n)_1^\infty \subset \mathcal{I}(\sigma(\mathcal{P}), \mathbf{m})$ (see Remark 4.1.15) and let f be a $\sigma(\mathcal{P})$-measurable scalar function on T. If $f_n \to f$ $\mathbf{m}$-a.e. in T and if*

$$\gamma_n(\cdot) = (KL) \int_{(\cdot)} f_n d\mathbf{m}, \, n \in \mathbb{N},$$

then the following are equivalent:

(i) $\lim_n \Pi_q \circ \gamma_n(A)$ *exists in* $\widetilde{X_q}$ *for each* $A \in \sigma(\mathcal{P})$ *and for each* $q \in \Gamma$.

(ii) $\Pi_q \circ \gamma_n$, $n \in \mathbb{N}$, *are uniformly σ-additive on* $\sigma(\mathcal{P})$ *for each q fixed in* Γ.

If any one of (i) or (ii) holds, then $\lim_n \gamma_n(A) = \gamma(A)$ (say) exists in X for $A \in \sigma(\mathcal{P})$, $\lim_n \Pi_q \circ \gamma_n(A) = \Pi_q \circ \gamma(A) \in X_q$, $\gamma : \sigma(\mathcal{P}) \to X$ is σ-additive, f is $\mathbf{m}$-integrable in T with values in X and

$$\gamma(A) = \int_A f d\mathbf{m} = (KL) \int_A f d\mathbf{m} = \lim_n (KL) \int_A f_n d\mathbf{m}, \, A \in \sigma(\mathcal{P}).$$

Proof. Suppose (i) holds. Then by Theorem 4.1.8(ii), γ_n, $n \in \mathbb{N}$, are σ-additive on $\sigma(\mathcal{P})$ and hence $\Pi_q \circ \gamma_n$, $n \in \mathbb{N}$, are σ-additive on $\sigma(\mathcal{P})$ for each $q \in \Gamma$. Then, as $\lim_n \Pi_q \circ \gamma_n(A) \in \widetilde{X_q}$, by VHSN (Proposition 1.1.8 of Chapter 1 applied to $(\Pi_q \circ \gamma_n)_{n=1}^\infty$ as $\Pi_q \circ \gamma_n$ are $\widetilde{X_q}$-valued for each $q \in \Gamma$), (i) implies (ii).

Conversely, let (ii) hold. Then by Theorem 2.2.8 of Chapter 2 for $\mathbf{m}_q$,

$$\lim_n \int_A f_n d\mathbf{m}_q = \lim_n \Pi_q \circ \gamma_n(A) = \widetilde{x_{A,q}} \in \widetilde{X_q}$$

exists uniformly with respect to $A \in \sigma(\mathcal{P})$ (for a fixed $q \in \Gamma$). Therefore, given $\epsilon > 0$, there exists $n_0(q) \in \mathbb{N}$ such that $|\Pi_q \circ \gamma_n(A) - \Pi_q \circ \gamma_r(A)|_q = q(\gamma_n(A) -$

$\gamma_r(A)) < \epsilon$ for $n, r \geq n_0(q)$ and for all $A \in \sigma(\mathcal{P})$. Since q is arbitrary in Γ, it follows that $(\gamma_n(A))_{n=1}^{\infty}$ is τ-Cauchy in X (in fact, uniformly τ-Cauchy in X)) for each $A \in \sigma(\mathcal{P})$. Consequently, as X is sequentially complete, there exists $x_A \in X$ such that $\gamma_n(A) \to x_A$ in τ. Then $\Pi_q(x_A) = \lim_n \Pi_q \circ \gamma_n(A) = \widetilde{x_{A,q}}$ and hence $\widetilde{x_{A,q}} = \Pi_q(x_A) \in X_q$. Thus, particularly, (i) holds.

Let (i) or (ii) hold. Then (ii) holds and as shown above $\lim_n \gamma_n(A) = x_A \in X$ exists in τ for each $A \in \sigma(\mathcal{P})$. Hence $\gamma(A) = x_A$.

Now, for $x^* \in X^*$ and for $A \in \sigma(\mathcal{P})$, we have

$$x^*(x_A) = x^* \gamma(A) = \lim_n x^* \gamma_n(A) = \lim_n x^* ((KL) \int_A f_n d\mathbf{m})$$

$$= \lim_n \int_A f_n d(x^* \mathbf{m}) \qquad \text{(by Theorem 4.1.8(v))}$$

$$= \lim_n \int_A f d(x^* \mathbf{m}) \qquad \text{(by Proposition 1.1.21 of Chapter 1)}$$

and hence f is (KL) $\mathbf{m}$-integrable in T with (KL) $\int_A f d\mathbf{m} = x_A \in X$ so that f is $\mathbf{m}$-integrable in T (i.e., (BDS)-integrable) with values in X by Theorem 4.2.2′ in Remark 4.2.12 and $\gamma(A) = (KL) \int_A f d\mathbf{m} = \int_A f d\mathbf{m} \in X$ for $A \in \sigma(\mathcal{P})$. Then by Theorem 4.1.8(ii), $\gamma : \sigma(\mathcal{P}) \to X$ is σ-additive. $\qquad \square$

For a sequentially complete lcHs-valued $\mathbf{m}$, (i) (resp. (ii)) in the following theorem is an analogue of Theorem 4.4.5 (resp. of Theorem 4.1.4 combined with Remark 4.1.5) for $\sigma(\mathcal{P})$-measurable scalar functions. The following theorem plays a key role in Section 4.5 for generalizing the results in Sections 3.3, 3.4 and 3.5 of Chapter 3 to σ-additive vector measures with values in such spaces.

Theorem 4.4.7. *Let X be a sequentially complete lcHs, $\mathbf{m} : \mathcal{P} \to X$ be σ-additive and $1 \leq p < \infty$. Let f be a $\sigma(\mathcal{P})$-measurable scalar function on T. Then:*

(i) *If $(\mathbf{m}_q)_p^{\bullet}(f, \cdot)$ is continuous on $\sigma(\mathcal{P})$ for each $q \in \Gamma$, then $f \in \mathcal{L}_p(\sigma(\mathcal{P}), \mathbf{m})$ as well as $|f|^p$ is $\mathbf{m}$-integrable in T with values in X. Moreover, $\mathcal{L}_p \mathcal{I}(\sigma(\mathcal{P}), \mathbf{m}) = \mathcal{L}_p(\sigma(\mathcal{P}), \mathbf{m})$.*

(ii) *$f \in \mathcal{L}_p(\sigma(\mathcal{P}), \mathbf{m})$ if and only if $|f|^p$ is $\mathbf{m}_q$-integrable in T with values in $\widetilde{X_q}$ for each $q \in \Gamma$.*

Proof. (i) Let $(\mathbf{m}_q)_p^{\bullet}(f, \cdot)$ be continuous on $\sigma(\mathcal{P})$ for each $q \in \Gamma$. Then by Theorem 3.2.7 of Chapter 3, $(\mathbf{m}_q)_p^{\bullet}(f, T) < \infty$ for each $q \in \Gamma$ and hence $f \in \mathcal{L}_p \mathcal{M}(\sigma(\mathcal{P}), \mathbf{m}_q)$ for each $q \in \Gamma$. Then by hypothesis and by Definition 4.4.4, $f \in \mathcal{L}_p(\sigma(\mathcal{P}), \mathbf{m})$. As $|f|^p$ is $\sigma(\mathcal{P})$-measurable, by the last part of Proposition 1.2.10 there exists a sequence $(s_n)_1^{\infty} \subset \mathcal{I}_s$ such that $0 \leq s_n \nearrow |f|^p$ pointwise in T. Let $q \in \Gamma$. Then by hypothesis and by Theorem 3.2.5 of Chapter 3 for $\mathbf{m}_q$, $|f|^p$ is (KL) $\mathbf{m}_q$-integrable in T with values in $\widetilde{X_q}$. Therefore, by Theorem 2.1.7 of Chapter 2 for $\mathbf{m}_q$, $(\gamma_n^{(q)}(A))_1^{\infty}$ is Cauchy in X_q where $\gamma_n^{(q)}(A) = \int_A s_n d\mathbf{m}_q$ for $A \in \sigma(\mathcal{P})$. Let $\gamma_n(A) = \int_A s_n d\mathbf{m}$

for $A \in \sigma(\mathcal{P})$. Since $\Pi_q(\boldsymbol{\gamma}_n(A)) = \boldsymbol{\gamma}_n^{(q)}(A)$ and since $(\boldsymbol{\gamma}_n^{(q)}(A))_1^\infty$ is convergent in $\widetilde{X}_q$ for each $q \in \Gamma$, (i) of Theorem 4.4.6 holds and hence by Theorem 4.4.6 $|f|^p$ is (KL)$\mathbf{m}$-integrable in T with values in X. Therefore, $f \in \mathcal{I}_p(\sigma(\mathcal{P}), \mathbf{m})$. Then by Theorem 4.3.8(i), $f \in \mathcal{L}_p\mathcal{I}(\sigma(\mathcal{P}), \mathbf{m})$. This also proves that $\mathcal{L}_p(\sigma(\mathcal{P}), \mathbf{m}) \subset \mathcal{L}_p\mathcal{I}(\sigma(\mathcal{P}), \mathbf{m})$.

Conversely, if $f \in \mathcal{L}_p\mathcal{I}(\sigma(\mathcal{P}), \mathbf{m})$ $(= \mathcal{I}_p(\sigma(\mathcal{P}), \mathbf{m})$ by Theorem 4.3.8(i)), then $|f|^p$ is (KL) $\mathbf{m}$-integrable in T with values in X and hence $f \in \mathcal{L}_p(\sigma(\mathcal{P}), \mathbf{m})$ by Theorem 4.3.4 and Definition 4.4.4. Therefore, $\mathcal{L}_p\mathcal{I}(\sigma(\mathcal{P}), \mathbf{m}) = \mathcal{L}_p(\sigma(\mathcal{P}), \mathbf{m})$.

(ii) If $f \in \mathcal{L}_p(\sigma(\mathcal{P}), \mathbf{m})$, then by (i), $f \in \mathcal{L}_p\mathcal{I}(\sigma(\mathcal{P}), \mathbf{m})$ and hence by Definition 4.3.7, $|f|^p$ is (KL) $\mathbf{m}$-integrable in T with values in X. Then by Theorem 4.1.8(v), $|f|^p$ is (KL) $\mathbf{m}_q$-integrable in T with values in $\widetilde{X}_q$ for each $q \in \Gamma$ since $\mathbf{m}_q = \Pi_q \circ \mathbf{m}$ has range in $X_q \subset \widetilde{X}_q$. Then the condition is necessary by Theorem 2.2.2 of Chapter 2. Conversely, if $|f|^p$ is $\mathbf{m}_q$-integrable in T with values in $\widetilde{X}_q$ for each $q \in \Gamma$, then by Theorem 3.3.6 of Chapter 3, $(\mathbf{m}_q)_p^\bullet(f, \cdot)$ is continuous on $\sigma(\mathcal{P})$ for each $q \in \Gamma$ and hence by (i), $f \in \mathcal{L}_p(\sigma(\mathcal{P}), \mathbf{m})$. Thus (ii) holds. $\qquad\square$

Theorem 4.4.8. *Let X be an lcHs, $\mathbf{m} : \mathcal{P} \to X$ be σ-additive and $1 \le p < \infty$. Then:*

(i) *$\mathcal{L}_p(\mathbf{m})$ is closed in $\mathcal{L}_p\mathcal{M}(\mathbf{m})$. Consequently, when X is quasicomplete, $\mathcal{L}_p\mathcal{I}(\mathbf{m})$ is closed in $\mathcal{L}_p\mathcal{M}(\mathbf{m})$.*

(ii) *If X is a Fréchet space, then $\mathcal{L}_p\mathcal{I}(\mathbf{m})$ is a complete pseudo-metrizable locally convex space. Consequently, $\mathcal{L}_p\mathcal{I}(\mathbf{m})$ is a Fréchet space.*

(iii) *$\mathcal{L}_p(\sigma(\mathcal{P}), \mathbf{m})$ is closed in $\mathcal{L}_p\mathcal{M}(\sigma(\mathcal{P}), \mathbf{m})$. Consequently, if X is sequentially complete, then $\mathcal{L}_p\mathcal{I}(\sigma(\mathcal{P}), \mathbf{m})$ is closed in $\mathcal{L}_p\mathcal{M}(\sigma(\mathcal{P}), \mathbf{m})$.*

(iv) *If X is a Fréchet space, then $\mathcal{L}_p\mathcal{I}(\sigma(\mathcal{P}), \mathbf{m})$ is a complete pseudo-metrizable locally convex space. Consequently, $L_p\mathcal{I}(\sigma(\mathcal{P}), \mathbf{m})$ is a Fréchet space.*

Proof. (i) (resp. (iii)) Let f be an element in the closure of $\mathcal{L}_p(\mathbf{m})$ in $\mathcal{L}_p\mathcal{M}(\mathbf{m})$ (resp. in the closure of $\mathcal{L}_p(\sigma(\mathcal{P}), \mathbf{m})$ in $\mathcal{L}_p\mathcal{M}(\sigma(\mathcal{P}), \mathbf{m}))$. Let $q \in \Gamma$. Then, for each $n \in \mathbb{N}$, there exists $f_n^{(q)} \in \mathcal{L}_p(\mathbf{m})$ (resp. $f_n^{(q)} \in \mathcal{L}_p(\sigma(\mathcal{P}), \mathbf{m})$) such that $(\mathbf{m}_q)_p^\bullet(f - f_n^{(q)}, T) < \frac{1}{n}$. Given $\epsilon > 0$, choose n_0 such that $\frac{2}{n_0} < \epsilon$. Then $(\mathbf{m}_q)_p^\bullet(f - f_n^{(q)}, T) < \frac{\epsilon}{2}$ for $n \ge n_0$. Let $(E_k)_1^\infty \subset \sigma(\mathcal{P})$ such that $E_k \searrow \emptyset$. Since $f_{n_0}^{(q)} \in \mathcal{L}_p(\mathbf{m})$ (resp. $\in \mathcal{L}_p(\sigma(\mathcal{P}), \mathbf{m})$), there exists k_0 such that $(\mathbf{m}_q)_p^\bullet(f_{n_0}^{(q)}, E_k) < \frac{\epsilon}{2}$ for $k \ge k_0$. Then

$$(\mathbf{m}_q)_p^\bullet(f, E_k) \le (\mathbf{m}_q)_p^\bullet(f - f_{n_0}^{(q)}, E_k)$$
$$+ (\mathbf{m}_q)_p^\bullet(f_{n_0}^{(q)}, E_k) \le (\mathbf{m}_q)_p^\bullet(f - f_{n_0}^{(q)}, T) + (\mathbf{m}_q)_p^\bullet(f_{n_0}^{(q)}, E_k) < \epsilon$$

for $k \ge k_0$. Hence $(\mathbf{m}_q)_p^\bullet(f, \cdot)$ is continuous on $\sigma(\mathcal{P})$. As $q \in \Gamma$ is arbitrary, we conclude that $f \in \mathcal{L}_p(\mathbf{m})$ (resp. $f \in \mathcal{L}_p(\sigma(\mathcal{P}), \mathbf{m})$) and hence the first part of (i) (resp. of (iii)) holds. The last part of (i) (resp. of (iii)) holds by the first part and by Theorem 4.4.5 (resp. and by Theorem 4.4.7(i)).

(ii) (resp. (iv)) The first part is immediate from (i) (resp. from (iii)) and from the first part of Theorem 4.4.2(i). Then $L_p\mathcal{I}(\mathbf{m})$ (resp. $L_p\mathcal{I}(\sigma(\mathcal{P}),\mathbf{m})$) is a Fréchet space by Theorem 5.7, Chapter 2 and Lemma 11.3, Chapter 3 of [KN] and by the fact that $L_p\mathcal{I}(\mathbf{m})$ (resp. $L_p\mathcal{I}(\sigma(\mathcal{P}),\mathbf{m})$) is an LcHs by Theorem 4.3.9(v). □

Remark 4.4.9. One can directly prove that $\mathcal{L}_p(\mathbf{m})$ (resp. $\mathcal{L}_p(\sigma(\mathcal{P}),\mathbf{m})$) is a complete pseudo-metrizable space for $1 \leq p < \infty$ whenever X is a Fréchet space. In fact, let $(f_r)_1^\infty \subset \mathcal{L}_p(\mathbf{m})$ (resp. $\subset \mathcal{L}_p(\sigma(\mathcal{P}),\mathbf{m})$) be Cauchy in $(\mathbf{m}_{q_n})_p^\bullet(\cdot,T)$ for each $n \in \mathbb{N}$, where q_n, $n \in \mathbb{N}$, generate the topology of X. Arguing as in the proof of Theorem 4.4.2(i), we obtain a subsequence $(f_{n_k})_{k=1}^\infty$ of $(f_r)_1^\infty$ and an $\mathbf{m}$-measurable (resp. $\sigma(\mathcal{P})$-measurable) scalar function f on T such that $f_{n_k} \to f$ $\mathbf{m}$-a.e. in T and such that $\lim_k (\mathbf{m}_{q_n})_p^\bullet(f_{n_k} - f, T) = 0$, for each $n \in \mathbb{N}$. Then, given $n \in \mathbb{N}$ and $\epsilon > 0$, there exists k_0 such that $(\mathbf{m}_{q_n})_p^\bullet(f_{n_k} - f, T) < \frac{\epsilon}{2}$ for $k \geq k_0$. Let $E_\ell \searrow \emptyset$ in $\sigma(\mathcal{P})$. Then there exists an ℓ_0 such that $(\mathbf{m}_{q_n})_p^\bullet(f_{n_{k_0}}, E_\ell) < \frac{\epsilon}{2}$ for $\ell \geq \ell_0$, since $f_{n_{k_0}} \in \mathcal{L}_p(\mathbf{m})$. Then $(\mathbf{m}_{q_n})_p^\bullet(f, E_\ell) \leq (\mathbf{m}q_n)_p^\bullet(f - f_{n_{k_0}}, T) + (\mathbf{m}_{q_n})_p^\bullet(f_{n_{k_0}}, E_\ell) < \epsilon$ for $\ell \geq \ell_0$ and hence $(\mathbf{m}_{q_n})_p^\bullet(f, \cdot)$ is continuous on $\sigma(\mathcal{P})$. Since n is arbitrary in $\mathbb{N}$, $f \in \mathcal{L}_p(\mathbf{m})$ (resp. $\mathcal{L}_p(\sigma(\mathcal{P}),\mathbf{m})$) by Theorem 4.4.5 (resp. by Theorem 4.4.7) . Moreover, by an argument similar to that in the last part of the proof of Theorem 4.4.2(i), $\lim_r(\mathbf{m}_{q_n})_p^\bullet(f_r - f, T) = 0$. Hence $\mathcal{L}_p(\mathbf{m})$ (resp. $\mathcal{L}_p(\sigma(\mathcal{P}),\mathbf{m})$) is complete and consequently, $L_p(\mathbf{m})$ (resp. $L_p(\sigma(\mathcal{P}),\mathbf{m})$) is a Fréchet space. Then by Theorem 4.4.5 (resp. by Theorem 4.4.7(i)), $\mathcal{L}_p\mathcal{I}(\mathbf{m})$ (resp. $\mathcal{L}_p\mathcal{I}(\sigma(\mathcal{P}),\mathbf{m})$) is complete and consequently, $L_p\mathcal{I}(\mathbf{m})$ (resp. $L_p\mathcal{I}(\sigma(\mathcal{P}),\mathbf{m})$) is a Fréchet space.

Remark 4.4.10. When $p = 1$ and when the domain of $\mathbf{m}$ is a σ-algebra, the second part of Theorem 4.4.8(iv) is obtained in Theorem 4.1, Chapter IV of [KK] for a real Fréchet space X, using the concept of closed vector measures. Later, when X is a complex Fréchet space admitting a continuous norm, a simple direct proof (without using closed measures) is given for the said result in [Ri4]. Recently, for an arbitrary complex Fréchet space X, a direct proof of the above result is given in [FNR] and the proof in [FNR] uses the complex version of Theorem 4.1, Chapter II of [KK] and the diagonal sequence argument. As the reader can observe, our proof is also direct and moreover, is much stronger than the proofs in the literature, since not only the domain of the vector measure $\mathbf{m}$ is assumed to be just a δ-ring but also p is arbitrary in $[1,\infty)$. As for the problem of completeness of $\mathcal{L}_p$-spaces for vector measures defined on δ-rings, the reader may note that the concept of closed vector measures is not useful. In [SP1] the completeness is proved by arguments very special to real functions and real Banach spaces when the vector measure is defined on a σ-algebra and hence cannot be used here.

Let us generalize Definition 2.2.12 and Examples 2.2.13.–2.2.16 of the case of lcHs as follows.

Definition 4.4.11. In Definition 2.2.12, let X be an lcHs. Then a collection $(x_\alpha)_{\alpha \in A} \subset X$ is said to be summable in X if the net $(\sum_{\alpha \in A} x_\alpha)$ formed by sums over finite sets of indices and directed by inclusion, is convergent in X. The collection $(x_\alpha)_{\alpha \in A}$ is said to be absolutely summable in X if it is summable in X and if for each seminorm $q \in \Gamma$, the family $\{q(x_\alpha) : \alpha \in A\}$ is summable in $\mathbb{R}$. Finally, a sequence $(x_i)_1^\infty \subset X$ is said to be subseries convergent if $\sum_{k \le n} \chi_A(k) x_k$ exists in X for each $A \subset \mathbb{N}$.

Example 4.4.12. Let X be an lcHs and let $(x_n)_1^\infty$ be a summable sequence in X. Let $\mathbf{m} : \mathcal{P}(\mathbb{N}) \to X$ be defined by $\mathbf{m}(E) = \sum_{n \in E} x_n$. Then $\mathbf{m}$ is an X-valued σ-additive measure on $\mathcal{P}(\mathbb{N})$.

By Theorem 2.2 of [L2] we have

Example 4.4.13. Let X be an lcHs. Any X-valued σ-additive measure $\mathbf{m}$ defined on a δ-ring $\mathcal{P}$ of sets is of finite variation if and only if $(\mathbf{m}(A_n)_1^\infty)$ is absolutely summable to $\mathbf{m}(\bigcup_1^\infty A_n)$ for each pairwise disjoint sequence $(A_n)_1^\infty \subset \mathcal{P}$ with $\bigcup_1^\infty A_i \in \mathcal{P}$ (i.e., if and only if $q(\bigcup_1^\infty A_n) = \sum_1^\infty q(A_n)$ for each $q \in \Gamma$).

By Corollary 4.3 of [L2] we have

Example 4.4.14. Let X be an lcHs. If $\mathbf{m}$ is an X-valued σ-additive measure defined on a δ-ring $\mathcal{P}$ of subsets of T, and if every summable sequence (x_n) in X is absolutely summable, then every $\mathbf{m}$-integrable function in T is $v_q(\mathbf{m})$-integrable in T for each $q \in \Gamma$.

Remark 4.4.15. Let us recall that a real Fréchet lattice E is an AL-space if its topology can be defined by a family of lattice seminorms q that are additive in the positive cone, that is, such that

$$q(x + y) = q(x) + q(y), \text{for all } x, y \in E^+.$$

A real Fréchet lattice E is an AM-space if its topology can be defined by a family of lattice seminorms q such that

$$q(\sup\{x, y\}) = \sup\{q(x), q(y)\} \text{ for all } x, y \in E^+.$$

In [FN2] the authors characterize when $L_1(\mathbf{m})$ is an AL-space and when it is an AM-space, whenever $\mathbf{m} : \Sigma \to X$ is σ-additive on a σ-algebra Σ and X is a real Fréchet space. For Banach spaces, such a study was carried out by Curbera in [C2]. Later in [OR6], Okada and Ricker obtained the following results:

(i) Let X be a complex Fréchet space. Then X is nuclear if and only if $L_1(\mathbf{m}, \Sigma)$ is a Fréchet AL-lattice for each X-valued σ-additive vector measure $\mathbf{m}$ defined on a σ-algebra Σ.

(ii) Let X be a complex Fréchet Montel space and $\mathbf{m}$ be an X-valued σ-additive vector measure defined on a σ-algebra Σ of subsets of a non-void set Ω. Then the following statements are equivalent:

(a) The Fréchet lattice $L_1(\mathbf{m}, \Sigma)$ is isomorphic (orderwise and topologically) to a Banach AL-lattice.

(b) There exists $r \in \mathbb{N}$ such that the Fréchet lattice $L_1(\mathbf{m}, \Sigma)$ is isomorphic to the Banach lattice $L_1(|\mathbf{m}_r|, \Sigma)$ where $|\mathbf{m}_k|$ is as on p. 306 of [OR6].

(c) The Fréchet space $L_1(\mathbf{m}, \Sigma)$ is normable.

(d) The integration operator $I_\mathbf{m} : L_1(\mathbf{m}, \Sigma) \to X$ is compact, where $I_\mathbf{m}(f) = \int_\Omega f \, d\mathbf{m}$.

(iii) In [OR5], necessary and sufficient conditions are given for the integration map $I_\mathbf{m} : L_1(\mathbf{m}, \Sigma) \to X$ to be compact, where Σ is a σ-algebra of subsets of a non-void set Ω and $\mathbf{m}$ is a σ-additive complex Fréchet space-valued measure on Σ.

In [FMNP] the following interesting results are obtained.

Let X be a complex Fréchet space, Σ a σ-algebra of subsets of a non-void set Ω, $f : \Omega \to X$ be measurable in the sense of p. 224 of [FMNP] and μ a probability measure on Σ. Then the following hold:

(i) If X is weakly sequentially complete, then $L_p(\mu, X)$ is also weakly sequentially complete for $1 \le p < \infty$.

(ii) If X is reflexive, then $L_1(\mu, X)$ does not contain a copy of c_0.

(iii) Let X be a real Fréchet space and $\mathbf{m} : \Sigma \to X$ be σ-additive. Then:

(a) If X does not contain a copy of c_0, then $L_1(\mathbf{m}, \Sigma)$ is weakly sequentially complete.

(b) If X is weakly sequentially complete, then $L_1(\mathbf{m}, \Sigma)$ is weakly sequentially complete.

(c) If X does not contain any copy of c_0, then $L_1(\mathbf{m}, \Sigma)$ also does not contain any copy of c_0.

(iv) For a real Fréchet lattice X, the following conditions are equivalent:

(a) X is weakly sequentially complete.

(b) X does not contain a copy of c_0.

(c) X does not contain any lattice copy of c_0.

(v) For a real Fréchet lattice X, the following conditions are equivalent:

(α) X does not contain any copy of c_0.

(β) $L_1(\mathbf{m}, \Sigma)$ does not contain any copy of c_0 where $\mathbf{m} : \Sigma \to X$ is σ-additive.

4.5 Characterizations of $\mathcal{L}_p$-spaces, convergence theorems and relations between $\mathcal{L}_p$-spaces

In this section, using Theorem 4.4.5 (resp. Theorem 4.4.7) we generalize the results in Sections 3.3, 3.4 and 3.5 of Chapter 3 to a quasicomplete (resp. sequentially complete) lcHs-valued σ-additive vector measure $\mathbf{m}$ on $\mathcal{P}$. Similar to that in Definition 3.2.5 of Chapter 3, we introduce the space $\mathcal{L}_p\mathcal{I}_s(\mathbf{m})$ (resp. $\mathcal{L}_p\mathcal{I}_s(\sigma(\mathcal{P}),\mathbf{m})$) and show that

$$\mathcal{L}_p(\mathbf{m}) = \mathcal{L}_p\mathcal{I}(\mathbf{m}) = \mathcal{L}_p\mathcal{I}_s(\mathbf{m})$$

(resp.

$$\mathcal{L}_p(\sigma(\mathcal{P}),\mathbf{m}) = \mathcal{L}_p\mathcal{I}(\sigma(\mathcal{P}),\mathbf{m}) = \mathcal{L}_p\mathcal{I}_s(\sigma(\mathcal{P}),\mathbf{m}))$$

for $1 \le p < \infty$.

Theorem 4.5.1. (Generalizations of Theorem 3.3.1 of Chapter 3) *Let X be a quasicomplete (resp. sequentially complete) lcHs, $\mathbf{m} : \mathcal{P} \to X$ be σ-additive and $1 \le p < \infty$. Let $f : T \to \mathbb{K}$ (resp. and be $\sigma(\mathcal{P})$-measurable) and let $(f_n^{(q)})_1^\infty \subset \mathcal{L}_p(\mathbf{m}_q)$ (resp. $\subset \mathcal{L}_p(\sigma(\mathcal{P}),\mathbf{m}_q)$) for $q \in \Gamma$. Suppose $f_n^{(q)} \to f$ $\mathbf{m}_q$-a.e. in T for $q \in \Gamma$. Then $(\mathbf{m}_q)_p^\bullet(f_n^{(q)} - f, T) \to 0$ as $n \to \infty$ for $q \in \Gamma$ if and only if $(\mathbf{m}_q)_p^\bullet(f_n^{(q)}, \cdot)$, $n \in \mathbb{N}$, are uniformly continuous on $\sigma(\mathcal{P})$ for each $q \in \Gamma$. In that case, $f \in \mathcal{L}_p(\mathbf{m}_q)$ (resp. $f \in \mathcal{L}_p(\sigma(\mathcal{P}),\mathbf{m}_q)$) for each $q \in \Gamma$ and hence $f \in \mathcal{L}_p(\mathbf{m})$ (resp. $f \in \mathcal{L}_p(\sigma(\mathcal{P}),\mathbf{m})$). Moreover, in that case, for $p = 1$,*

$$\lim_n \left| q\left(\int_A f\,d\mathbf{m}\right) - \left|\left|\int_A f_n^{(q)}\,d\mathbf{m}_q\right|\right|_q \right| \le \lim_n \left|\left|\int_A f\,d\mathbf{m}_q - \int_A f_n^{(q)}\,d\mathbf{m}_q\right|\right|_q = 0 \quad (4.5.1.1)$$

for $A \in \sigma(\mathcal{P}) \cup \{T\}$ and for $q \in \Gamma$. Moreover, for a fixed $q \in \Gamma$, the limit in (4.5.1.1) is uniform with respect to $A \in \sigma(\mathcal{P})$.

Proof. The convergence hypothesis implies that f is $\mathbf{m}_q$-measurable for each $q \in \Gamma$ and hence f is $\mathbf{m}$-measurable (resp. by hypothesis, f is $\sigma(\mathcal{P})$-measurable). As $\mathbf{m}_q = \Pi_q \circ \mathbf{m} : \mathcal{P} \to \widetilde{X_q}$ is σ-additive, the first part is immediate from Theorem 3.3.1 of Chapter 3. Moreover, in that case, by the said theorem, $f \in \mathcal{L}_p(\mathbf{m}_q)$ (resp. $f \in \mathcal{L}_p(\sigma(\mathcal{P}),\mathbf{m}_q)$) for each q and hence, by Theorem 3.3.5 of Chapter 3, $|f|^p$ is $\mathbf{m}_q$-integrable in T with values in $\widetilde{X_q}$ for each $q \in \Gamma$. Then by Definition 4.2.1 and by Theorem 4.2.3, $|f|^p$ is $\mathbf{m}$-integrable in T with values in X and hence $f \in \mathcal{I}_p(\mathbf{m}) = \mathcal{L}_p(\mathbf{m})$ by Theorem 4.2.2, by (i) and (ii) of Theorem 4.3.8 and by Theorem 4.4.5 when X is quasicomplete and by Theorem 4.4.7(ii), $f \in \mathcal{L}_p(\sigma(\mathcal{P}),\mathbf{m})$ when X is sequentially complete.

Now let $p = 1$. Given $q \in \Gamma$ and $\epsilon > 0$, by hypothesis there exists n_0 such that $(\mathbf{m}_q)_1^\bullet(f - f_n^{(q)}, T) < \epsilon$ for $n \ge n_0$ and by the first part, $f \in \mathcal{L}_1(\mathbf{m})$ (resp. $f \in \mathcal{L}_1(\sigma(\mathcal{P}),\mathbf{m})$). Then by Definition 4.4.4 and by Theorems 4.4.5 and 4.1.9(iii), f is (KL) $\mathbf{m}$-integrable in T and then by Theorems 4.2.2 and 4.2.3, f is $\mathbf{m}$-integrable in

T with values in X (resp. and then by Theorem 4.4.7 and by Theorem 4.1.9'(iii) in Remark 4.1.15, f is (KL) $\mathbf{m}$-integrable in T with values in X and hence by Theorem 4.2.2' in Remark 4.2.12 f is $\mathbf{m}$-integrable in T with values in X). Moreover, by Theorem 4.1.8(v) and by Theorem 2.2.2 of Chapter 2, f is (KL) $\mathbf{m}_q$-integrable in T for each $q \in \Gamma$ and by (4.3.2.1) we have

$$\left| q\left(\int_A f d\mathbf{m} \right) - \left| \int_A f_n^{(q)} d\mathbf{m}_q \right|_q \right|$$

$$\leq \left| (\mathrm{KL}) \int_A f d\mathbf{m}_q - (\mathrm{KL}) \int_A f_n^{(q)} d\mathbf{m}_q \right|_q$$

$$= \left| (\mathrm{KL}) \int_A (f - f_n^{(q)}) d\mathbf{m}_q \right|_q \leq (\mathbf{m}_q)_1^\bullet (f - f_n^{(q)}, T) < \epsilon \qquad (4.5.1.2)$$

for $n \geq n_0$ and for all $A \in \sigma(\mathcal{P}) \cup \{T\}$. Then (4.5.1.2) implies (4.5.1.1) and that the limit in (4.5.1.1) is uniform with respect to $A \in \sigma(\mathcal{P})$ for a fixed $q \in \Gamma$. □

Theorem 4.5.2. (Generalizations of Theorem 3.3.2 of Chapter 3) *Let X be a quasicomplete (resp. sequentially complete) lcHs, $\mathbf{m} : \mathcal{P} \to X$ be σ-additive and $1 \leq p < \infty$. Let $f : T \to \mathbb{K}$ be $\mathbf{m}$-measurable (resp. $\sigma(\mathcal{P})$-measurable) and let $(f_n^{(q)})_1^\infty \subset \mathcal{L}_p(\mathbf{m}_q)$ (resp. $\subset \mathcal{L}_p(\sigma(\mathcal{P}), \mathbf{m}_q)$) for $q \in \Gamma$. Then the sequence $(f_n^{(q)})_1^\infty$ converges to f in (mean^p) with respect to $\mathbf{m}_q$ for $q \in \Gamma$ if and only if $(f_n^{(q)})$ converges to f in measure in T with respect to $\mathbf{m}_q$ for $q \in \Gamma$, $f \in \mathcal{L}_p \mathcal{M}(\mathbf{m})$ and $(\mathbf{m}_q)_p^\bullet (f_n^{(q)}, \cdot)$, $n \in \mathbb{N}$, are uniformly continuous on $\sigma(\mathcal{P})$ for each $q \in \Gamma$. In that case, $f \in \mathcal{L}_p(\mathbf{m})$ (resp. $f \in \mathcal{L}_p(\sigma(\mathcal{P}), \mathbf{m})$). In such case, for $p = 1$, results similar to those in the last part of Theorem 4.5.1 hold.*

Proof. The first part is immediate from Theorem 3.3.2 of Chapter 3. Then by the said theorem, $f \in \mathcal{L}_p(\mathbf{m}_q)$ (resp. $f \in \mathcal{L}_p(\sigma(\mathcal{P}), \mathbf{m}_q)$) for each $q \in \Gamma$. Then as shown in the proof of Theorem 4.5.1, we conclude that $f \in \mathcal{L}_p(\mathbf{m})$ (resp. $f \in \mathcal{L}_p(\sigma(\mathcal{P}), \mathbf{m})$) and the results asserted for $p = 1$ hold. □

The a.e. convergence version in the following theorem generalizes Theorem 4.1.11 and Corollary 4.1.12 when X is quasicomplete (resp. Theorem 4.1.11' and Corollary 4.1.12' in Remark 4.1.15 when X is sequentially complete) for general $p \in [1, \infty)$.

Theorem 4.5.3. *Let X be an lcHs, $\mathbf{m} : \mathcal{P} \to X$ be σ-additive and $1 \leq p < \infty$. Then:*

(i) **(LDCT and LBCT for $\mathcal{L}_p(\mathbf{m})$, X quasicomplete)**. *Let X be quasicomplete and let $f_n^{(q)}$, $n \in \mathbb{N}$, be $\mathbf{m}_q$-measurable scalar functions on T for $q \in \Gamma$. Let $g^{(q)} \in \mathcal{L}_p(\mathbf{m}_q)$ for $q \in \Gamma$ and let $|f_n^{(q)}| \leq |g^{(q)}|$ $\mathbf{m}_q$-a.e. in T for $n \in \mathbb{N}$ and for $q \in \Gamma$ (resp. let $\mathcal{P}$ be a σ-ring $\mathcal{S}$ and let $K^{(q)}$ be a finite constant such that $|f_n^{(q)}| \leq K^{(q)}$ $\mathbf{m}_q$-a.e. in T for $n \in \mathbb{N}$ and for $q \in \Gamma$). If $f_n^{(q)} \to f$ $\mathbf{m}_q$-a.e.*

in T where f is a scalar function on T or if f is an $\mathbf{m}$-measurable scalar function on T and if $f_n^{(q)} \to f$ in measure in T with respect to $\mathbf{m}_q$ for $q \in \Gamma$, then $f, f_n^{(q)}, n \in \mathbb{N}$, belong to $\mathcal{L}_p(\mathbf{m}_q)$ and $\lim_n (\mathbf{m}_q)_p^\bullet (f_n^{(q)} - f, T) = 0$ for $q \in \Gamma$. Consequently, $f \in \mathcal{L}_p(\mathbf{m})$. When $p = 1$, f is $\mathbf{m}$-integrable in T with values in X and

$$\lim_n \left| q\left(\int_A f \, d\mathbf{m} \right) - \left| \left| \int_A f_n^{(q)} \, d\mathbf{m}_q \right| \right|_q \right| \leq \lim_n \left| \left| \int_A f \, d\mathbf{m}_q - \int_A f_n^{(q)} \, d\mathbf{m}_q \right| \right|_q = 0$$

(4.5.3.1)

for $A \in \sigma(\mathcal{P}) \cup \{T\}$ (resp. for $A \in \mathcal{S} \cup \{T\}$) for $q \in \Gamma$ and, for a fixed $q \in \Gamma$, the limit in (4.5.3.1) is uniform with respect to $A \in \sigma(\mathcal{P})$ (resp. $A \in \mathcal{S}$).

(ii) **(LDCT and LBCT for $\mathcal{L}_p(\sigma(\mathcal{P}), \mathbf{m})$, X sequentially complete).** *Let X be sequentially complete and for each $q \in \Gamma$, let $(f_n^{(q)})_1^\infty$, $g^{(q)}$ and f be $\sigma(\mathcal{P})$-measurable scalar functions (resp. let $\mathcal{P}$ be a σ-ring $\mathcal{S}$ and $K^{(q)}$ be finite constants) satisfying the other hypothesis in (i). Then $f, f_n^{(q)}, n \in \mathbb{N}$, belong to $\mathcal{L}_p(\sigma(\mathcal{P}), \mathbf{m}_q)$ and $\lim_n (\mathbf{m}_q)_p^\bullet (f_n^{(q)} - f, T) = 0$ for $q \in \Gamma$ and $f \in \mathcal{L}_p(\sigma(\mathcal{P}), \mathbf{m})$. Then, for $p = 1$, the remaining assertions in (i) hold here verbatim.*

Proof. By hypothesis, (i) is immediate from Theorems 4.5.1 and 4.5.2, in the light of Theorems 3.1.11(iv) of Chapter 3 (for $\mathbf{m}_q$) and of Theorems 4.1.9(i)(b), 4.1.9(iii), 4.4.5 and 4.2.2 and of Remark 4.2.6. By hypothesis, (ii) follows from Theorems 4.5.1 and 4.5.2 in view of Theorem 3.1.11(iv) of Chapter 3 (for $\mathbf{m}_q$) and of Theorems 4.1.9'(i)(b) and 4.1.9'(iii) in Remark 4.1.15 and of Theorems 4.4.7(i) and 4.1.12' (in Remark 4.2.12). $\qquad\square$

The following is motivated by the first part of Definition 3.2.5 in Chapter 3.

Definition 4.5.4. Let X be a quasicomplete (resp. sequentially complete) lcHs and let $\mathbf{m} : \mathcal{P} \to X$ be σ-additive. For $1 \leq p < \infty$, let $\mathcal{L}_p \mathcal{I}_s(\mathbf{m}) =$ closure of $\mathcal{I}_s$ in the locally convex space $(\mathcal{L}_p \mathcal{M}(\mathbf{m}), \tau_{\mathbf{m}}^{(p)})$ (resp. $\mathcal{L}_p \mathcal{I}_s(\sigma(\mathcal{P}), \mathbf{m}) =$ closure of $\mathcal{I}_s$ in the locally convex space $(\mathcal{L}_p \mathcal{M}(\sigma(\mathcal{P}), \mathbf{m}), \tau_{\mathbf{m}}^{(p)})$). Let $L_p \mathcal{I}_s(\mathbf{m}) = \mathcal{L}_p \mathcal{I}_s(\mathbf{m})/\sim$ and $L_p \mathcal{I}_s(\sigma(\mathcal{P}), \mathbf{m}) = \mathcal{L}_p \mathcal{I}_s(\sigma(\mathcal{P}), \mathbf{m})/\sim$. By $\mathcal{L}_p \mathcal{I}_s(\mathbf{m})$ we mean the locally convex space $(\mathcal{L}_p \mathcal{I}_s(\mathbf{m}), \tau_{\mathbf{m}}^{(p)}|_{\mathcal{L}_p \mathcal{I}_s(\mathbf{m})})$ and $L_p \mathcal{I}_s(\mathbf{m})$ is the lcHs with the corresponding quotient topology.

The following result generalizes Theorem 3.3.6 of Chapter 3 to a quasicomplete (resp. to a sequentially complete) lcHs-valued σ-additive vector measure.

Theorem 4.5.5. (Characterizations of $\mathcal{L}_p \mathcal{I}(\mathbf{m})$) (resp. $\mathcal{L}_p \mathcal{I}(\sigma(\mathcal{P}), \mathbf{m})$) *Let X be a quasicomplete (resp. sequentially complete) lcHs, $\mathbf{m} : \mathcal{P} \to X$ be σ-additive and $1 \leq p < \infty$. Let $f : T \to \mathbb{K}$ be $\mathbf{m}$-measurable (resp. $\sigma(\mathcal{P})$-measurable). Then the following statements are equivalent:*

(i) $f \in \mathcal{I}_p(\mathbf{m})$ *(resp.* (i') $f \in \mathcal{I}_p(\sigma(\mathcal{P}), \mathbf{m})$*).*

(ii) $(\mathbf{m}_q)_p^\bullet(f, \cdot)$ *is continuous on $\sigma(\mathcal{P})$ for each $q \in \Gamma$.*

(iii) **(Simple function approximation)**. *For each $q \in \Gamma$, there exists a sequence $(s_n^{(q)})_1^\infty \subset \mathcal{I}_s$ such that $s_n^{(q)} \to f$ $\mathbf{m}_q$-a.e. in T and such that $\lim_n (\mathbf{m}_q)_p^\bullet(s_n^{(q)} - f, T) = 0$ (resp. (iii′) there exists a sequence $(s_n)_1^\infty \subset \mathcal{I}_s$ such that $s_n \to f$ pointwise in T, and such that $\lim_n (\mathbf{m}_q)_p^\bullet(s_n - f, T) = 0$ for each $q \in \Gamma$).*

Consequently,

$$\mathcal{L}_p\mathcal{I}(\mathbf{m}) = \mathcal{I}_p(\mathbf{m}) = \mathcal{L}_p\mathcal{I}_s(\mathbf{m}) = \mathcal{L}_p(\mathbf{m}) \tag{4.5.5.1}$$

(resp.

$$\mathcal{L}_p\mathcal{I}(\sigma(\mathcal{P}), \mathbf{m}) = \mathcal{I}_p(\sigma(\mathcal{P}), \mathbf{m}) = \mathcal{L}_p\mathcal{I}_s(\sigma(\mathcal{P}), \mathbf{m}) = \mathcal{L}_p(\sigma(\mathcal{P}), \mathbf{m}). \tag{4.5.5.1′}$$

If $c_0 \not\subset \tilde{X}_q$ for each $q \in \Gamma$, then

$$\mathcal{L}_p\mathcal{M}(\mathbf{m}) = \mathcal{L}_p\mathcal{I}(\mathbf{m}) = \mathcal{I}_p(\mathbf{m}) = \mathcal{L}_p\mathcal{I}_s(\mathbf{m}) = \mathcal{L}_p(\mathbf{m}) \tag{4.5.5.2}$$

(resp. if $c_0 \not\subset X$, then

$$\mathcal{L}_p\mathcal{M}(\sigma(\mathcal{P}), \mathbf{m}) = \mathcal{L}_p\mathcal{I}(\sigma(\mathcal{P}), \mathbf{m}) = \mathcal{I}_p(\sigma(\mathcal{P}), \mathbf{m})$$
$$= \mathcal{L}_p\mathcal{I}_s(\sigma(\mathcal{P}), \mathbf{m}) = \mathcal{L}_p(\sigma(\mathcal{P}), \mathbf{m}). \tag{4.5.5.2′}$$

Proof. (i)⇔(ii) by Theorems 4.3.8(i) and 4.4.5 (resp. (i′)⇔(ii) by Theorems 4.3.8(i) and 4.4.7(i)).

(ii)⇒(iii) by Proposition 1.2.9 and by Theorems 4.4.7 and 4.5.3(i) (resp. (ii)⇒(iii′) by the last part of Proposition 1.2.10 and by Theorems 4.4.7 and 4.5.3(ii)).

(iii)⇒(ii) (resp. (iii′)⇒(ii)) Let $\epsilon > 0$ and $q \in \Gamma$. Let $A_k \searrow \emptyset$ in $\sigma(\mathcal{P})$. Arguing as in the proof of (iii)⇒(ii) of Theorem 3.3.6 Chapter 3 with $\mathbf{m}_q$ in place of $\mathbf{m}$, we can show that there exists $k_0(q)$ such that $(\mathbf{m}_q)_p^\bullet(f, A_k) < \epsilon$ for $k \geq k_0(q)$ and hence (ii) holds.

Thus (i)⇔(ii)⇔(iii) (resp. (i′)⇔(ii)⇔(iii′)).

Since $\mathcal{I}_s \subset \mathcal{I}_p(\mathbf{m}) = \mathcal{L}_p\mathcal{I}(\mathbf{m})$ by Theorem 4.5.7(i), $\mathcal{L}_p\mathcal{I}_s(\mathbf{m}) \subset$ closure of $\mathcal{L}_p\mathcal{I}(\mathbf{m})$ in $\mathcal{L}_p\mathcal{M}(\mathbf{m})=$ closure of $\mathcal{L}_p(\mathbf{m})$ in $\mathcal{L}_p\mathcal{M}(\mathbf{m}) =\mathcal{L}_p(\mathbf{m})$ by Theorems 4.4.5 and 4.4.8(i). On the other hand, $\mathcal{I}_s(\mathbf{m})$ is dense in $\mathcal{I}_p(\mathbf{m}) = \mathcal{L}_p\mathcal{I}(\mathbf{m}) = \mathcal{L}_p(\mathbf{m})$ by (iii), by Theorem 4.3.8(i) and by Theorem 4.4.5 and hence $\mathcal{L}_p(\mathbf{m}) \subset \mathcal{L}_p\mathcal{I}_s(\mathbf{m})$. Therefore, $\mathcal{L}_p\mathcal{I}_s(\mathbf{m}) = \mathcal{L}_p(\mathbf{m})$. Consequently, (4.5.5.1) holds by Theorems 4.3.8(i) and 4.4.5. A similar argument invoking (iii′) and Theorems 4.3.8(i), 4.4.7(i) and 4.4.8(iii) proves (4.5.5.1′). (4.5.5.2) (resp. (4.5.5.2′)) holds by (4.5.5.1) and by Theorem 4.3.8(iv) (resp. by (4.5.5.1′) and by Theorem 4.3.8(v)). $\square$

The following result which generalizes Theorem 3.3.8 of Chapter 3 is immediate from Theorem 4.5.5 and from the last part of Proposition 1.2.10.

Theorem 4.5.6. *Let X be a quasicomplete (resp. sequentially complete) lcHs, $\mathbf{m} : \mathcal{P} \to X$ be σ-additive and $1 \leq p < \infty$. Then $\mathcal{I}_s$ is dense in $\mathcal{L}_p(\mathbf{m})$ (resp. $\mathcal{I}_s$ is dense in $\mathcal{L}_p(\sigma(\mathcal{P}), \mathbf{m})$ and moreover, given $f \in \mathcal{L}_p(\sigma(\mathcal{P}), \mathbf{m})$ there exists a sequence $(s_n)_1^\infty \subset \mathcal{I}_s$ such that $s_n \to f$ in $\tau_\mathbf{m}^{(p)}$ with $|s_n| \nearrow |f|$ pointwise in T).*

Similar to Definition 3.2.10 of Chapter 3 we can introduce $\mathcal{L}_\infty(\mathbf{m})$ and $\mathcal{L}_\infty(\sigma(\mathcal{P}), \mathbf{m})$ for an lcHs-valued σ-additive vector measure $\mathbf{m}$ on $\mathcal{P}$.

Definition 4.5.7. Let X be a quasicomplete lcHs and $\mathbf{m} : \mathcal{P} \to X$ be σ-additive. Then we define

$$\mathcal{L}_\infty(\mathbf{m}) = \{f : T \to \mathbb{K}, \ f \ \mathbf{m}\text{-essentially bounded } \mathbf{m}\text{-measurable function}\}$$

and

$$\|f\|_\infty = \operatorname*{ess\,sup}_{t \in T} |f(t)| \quad \text{for} \quad f \in \mathcal{L}_\infty(\mathbf{m}).$$

Then $L_\infty(\mathbf{m}) = \mathcal{L}_\infty(\mathbf{m})/\sim$. In the above, if X is sequentially complete, then we define $\mathcal{L}_\infty(\sigma(\mathcal{P}), \mathbf{m}) = \{f \in \mathcal{M}(\sigma(\mathcal{P}) : f\mathbf{m}\text{-essentially bounded}\}$, $\|f\|_\infty = \operatorname{ess\,sup}_{t \in T} |f(t)|$ for $f \in \mathcal{L}_\infty(\sigma(\mathcal{P}), \mathbf{m})$. Then $L_\infty(\sigma(\mathcal{P}), \mathbf{m}) = \mathcal{L}_\infty(\sigma(\mathcal{P}), \mathbf{m})/\sim$. As in Convention 3.3.12 of Chapter 3 , the members of $L_\infty(\mathbf{m})$ and $L_\infty(\sigma(\mathcal{P}), \mathbf{m})$ are treated as functions in which two functions which are equal $\mathbf{m}$-a.e. in T are identified.

In the light of (3.1.3.1) and the σ-subadditivity of $\|\mathbf{m}\|_q$ on $\sigma(\mathcal{P})$ for $q \in \Gamma$, the proof of Theorem 3.3.13 of Chapter 3 holds here to obtain

Theorem 4.5.8. *Let X be a quasicomplete (resp. sequentially complete) lcHs and $\mathbf{m} : \mathcal{P} \to X$ be σ-additive. Then $\mathcal{L}_\infty(\mathbf{m}, \|\cdot\|_\infty)$ (resp. $(\mathcal{L}_\infty(\sigma(\mathcal{P}), \mathbf{m}), \|\cdot\|_\infty))$ is a complete seminormed space so that $L_\infty(\mathbf{m})$ (resp. $L_\infty(\sigma(\mathcal{P}), \mathbf{m})$) is a Banach space.*

Notation 4.5.9. In the light of Theorem 4.5.5 **we shall hereafter use the symbol $\mathcal{L}_p(\mathbf{m})$ (resp. $\mathcal{L}_p(\sigma(\mathcal{P}), \mathbf{m})$) to denote not only the space given in Definition 4.4.4 but also any one of the spaces $\mathcal{L}_p\mathcal{I}_s(\mathbf{m}), \mathcal{L}_p\mathcal{I}(\mathbf{m})$ or $\mathcal{I}_p(\mathbf{m})$ (resp. $\mathcal{L}_p\mathcal{I}_s(\sigma(\mathcal{P}), \mathbf{m})$, $\mathcal{L}_p\mathcal{I}(\sigma(\mathcal{P}), \mathbf{m})$ or $\mathcal{I}_p(\sigma(\mathcal{P}), \mathbf{m})$). The quotient $\mathcal{L}_p(\mathbf{m})/\sim$ is denoted by $L_p(\mathbf{m})$, (resp. $\mathcal{L}_p(\sigma(\mathcal{P}), \mathbf{m})/\sim$) is denoted by $L_p(\sigma(\mathcal{P}), \mathbf{m}))$ and as in Convention 2.3.10, the members of the latter are treated as functions in which two functions which are equal m-a.e. in T are identified.**

The following result generalizes Theorem 3.3.13 of Chapter 3 and is proved by an argument similar to that in the proof of the said theorem, in which one has to use (iv) and (vi) of Theorem 3.1.18 and Theorem 3.1.19 of Chapter 3 with respect to $\mathbf{m}_q$, Remark 4.2.6, Theorems 4.1.4, 4.5.5 and 4.4.8(i), Corollary 4.4.3 and inequality (4.3.2.1). Details are left to the reader.

Theorem 4.5.10. *Let X be a quasicomplete lcHs, $\mathbf{m} : \mathcal{P} \to X$ be σ-additive and $1 \leq p < \infty$. An $\mathbf{m}$-measurable scalar function f on T belongs to $\mathcal{L}_p(\mathbf{m})$ if and only if, for each $q \in \Gamma$, there exists a sequence $(s_n^{(q)})_1^\infty \subset \mathcal{I}_s$ (resp. $(f_n^{(q)})_1^\infty \subset \mathcal{L}_p(\mathbf{m})$) such that $s_n^{(q)} \to f$ (resp. $f_n^{(q)} \to f$) in measure in T with respect to $\mathbf{m}_q$ and $(s_n^{(q)})$ (resp. $(f_n^{(q)})$) is Cauchy in $(mean^p)$ with respect to $\mathbf{m}_q$. In that case, for $p = 1$, $f \in \mathcal{L}_1(\mathbf{m})$ and*

$$\lim_n \left| \int_A (f - s_n^{(q)}) d\mathbf{m}_q \right|_q = 0 \ (resp. \ \lim_n \left| \int_A (f - f_n^{(q)}) d\mathbf{m}_q \right|_q = 0)$$

for $A \in \sigma(\mathcal{P}) \cup \{T\}$ and for $q \in \Gamma$, the limit being uniform with respect to $A \in \sigma(\mathcal{P})$ for a fixed $q \in \Gamma$.

The result (iii) in the following theorem generalizes Theorem 3.2.11 of Chapter 3 to Fréchet space-valued σ-additive vector measures on $\mathcal{P}$.

Theorem 4.5.11. *Let X be a Fréchet space with its topology generated by the seminorms $(q_n)_1^\infty$ and let $\mathbf{m} : \mathcal{P} \to X$ be σ-additive. For $1 \leq p < \infty$, let $\mathcal{L}_p^r(\mathbf{m}) = \{f \in \mathcal{L}_p(\mathbf{m}) : f \text{ real-valued}\}$ (resp. $\mathcal{L}_p^r(\sigma(\mathcal{P}), \mathbf{m}) = \{f \in \mathcal{L}_p(\sigma(\mathcal{P}), \mathbf{m}) : f \text{ real-valued}\}$). Let $|f|_{\mathbf{m}}^{(p)} = \sum_1^\infty \frac{1}{2^n} \frac{(\mathbf{m}_{q_n})_p^\bullet(f,T)}{1+(\mathbf{m}_{q_n})_p^\bullet(f,T)}$ for*

$$f \in L_p(\mathbf{m}) \ (\text{resp. } f \in L_p(\sigma(\mathcal{P}), \mathbf{m})) \ (4.5.11.1). \ \text{Then:}$$

(i) *$| \cdot |_{\mathbf{m}}^{(p)}$ is a complete quasinorm on $L_p(\mathbf{m})$ (resp. on $L_p(\sigma(\mathcal{P}), \mathbf{m})$) in the sense of Definition 2, Section 2, Chapter I of [Y] and generates the quotient topology induced by $\tau_{\mathbf{m}}^{(p)}$ on $L_p(\mathbf{m})$ so that $(L_p(\mathbf{m}), | \cdot |_{\mathbf{m}}^{(p)})$ (resp. $(L_p(\sigma(\mathcal{P}), \mathbf{m}), | \cdot |_{\mathbf{m}}^{(p)})$) is the Fréchet space $L_p(\mathbf{m})$ (resp. $L_p(\sigma(\mathcal{P}), \mathbf{m})$).*

(ii) *$(L_p^r(\mathbf{m}), | \cdot |_{\mathbf{m}}^{(p)})$ (resp. $(L_p^r(\sigma(\mathcal{P}), \mathbf{m}), | \cdot |_{\mathbf{m}}^{(p)})$) is a Fréchet lattice (i.e., an F-lattice in the sense of the definition in Section 3, Chapter XII and Definition 1 in Section 9, Chapter I of [Y]) under the partial order $f \leq g$ if and only if $f(t) \leq g(t)$ $\mathbf{m}$-a.e. in T for $f, g \in \mathcal{L}_p^r(\mathbf{m})$ (resp. for $f, g \in \mathcal{L}_p^r(\sigma(\mathcal{P}), \mathbf{m})$).*

(iii) *$L_\infty^r(\mathbf{m})$ (resp. $L_\infty^r(\sigma(\mathcal{P}), \mathbf{m})$) is a Banach lattice.*

Proof. We shall prove the results for $L_p(\mathbf{m})$ and $L_p^r(\mathbf{m})$ only. In the light of Theorem 4.4.8(iv) and Theorem $4.1.9'$(iii) in Remark 4.1.15, the proof of the other case is similar.

(i) $L_p(\mathbf{m})$ is a Fréchet space by Theorem 4.4.8(ii) and by Notation 4.5.9. Consequently, $L_p^r(\mathbf{m})$ is a Fréchet space over $\mathbb{R}$. Arguing as in the proof of the first part of Proposition 2, §6, Chapter 2 of [Ho] and using the fact that $(\mathbf{m}_{q_n})_p^\bullet(\cdot, T)$, $n \in \mathbb{N}$, are seminorms, one can show that $| \cdot |_{\mathbf{m}}^{(p)}$ given in (4.5.11.1) is a quasinorm on $L_p(\mathbf{m})$. Let the sequence $f_r \to f$ in $L_p(\mathbf{m})$. Given $\epsilon > 0$, choose $k_0 > 1$ such that $\frac{1}{2^{k_0}} < \frac{\epsilon}{2}$. By hypothesis, there exists r_0 such that $(\mathbf{m}_{q_n})_p^\bullet(f_r - f, T) < \frac{\epsilon}{2}$ for $r \geq r_0$ and for $n = 1, 2, \ldots, k_0$. Then

$$\begin{aligned}
|f_r - f|_{\mathbf{m}}^{(p)} &= \sum_{n=1}^\infty \frac{1}{2^n} \frac{(\mathbf{m}_{q_n})_p^\bullet(f_r - f, T)}{1+(\mathbf{m}_{q_n})_p^\bullet(f_r - f, T)} \\
&< \sum_{n=1}^{k_0} \frac{1}{2^n} \frac{(\mathbf{m}_{q_n})_p^\bullet(f_r - f, T)}{1+(\mathbf{m}_{q_n})_p^\bullet(f_r - f, T)} + \frac{\epsilon}{2} \\
&< \sum_{n=1}^{k_0} \frac{1}{2^n} \cdot \frac{\epsilon}{2} + \frac{\epsilon}{2} \\
&< \epsilon
\end{aligned}$$

for $r \geq r_0$. Hence $\lim_r |f_r - f|_{\mathbf{m}}^{(p)} = 0$. Conversely, let $|f_r - f|_{\mathbf{m}}^{(p)} \to 0$. Let $n \in \mathbb{N}$ be given and let $\frac{1}{2} > \epsilon > 0$. Then there exists r_0 such that $|f_r - f|_{\mathbf{m}}^{(p)} < \frac{\epsilon}{2^{n+1}}$ for $r \geq r_0$. Then $(\mathbf{m}_{q_n})_p^{\bullet}(f_r - f, T) < \frac{\epsilon}{2}(1 + (\mathbf{m}_{q_n})_p^{\bullet}(f_r - f, T))$ so that $(\mathbf{m}_{q_n})_p^{\bullet}(f_r - f, T) < \epsilon$ for $r \geq r_0$. Hence $(\mathbf{m}_{q_n})_p^{\bullet}(f_r - f, T) \to 0$ as $r \to 0$. Since n is arbitrary, it follows that $f_r \to f$ in the topology of $L_p(\mathbf{m})$ (see Remark 4.3.11). Hence the topology of $L_p(\mathbf{m})$ is generated by $|\cdot|_{\mathbf{m}}^{(p)}$.

If $(f_r)_1^{\infty}$ is Cauchy in $|\cdot|_{\mathbf{m}}^{(p)}$, then replacing $f_r - f$ by $f_r - f_k$ in the above argument (in the converse part), it follows that (f_r) is Cauchy in $(\mathbf{m}_{q_n})_p^{\bullet}(\cdot, T)$ for each n and hence by Theorem 4.4.8(ii), there exists $f \in L_p(\mathbf{m})$ such that $f_r \to f$ in $L_p(\mathbf{m})$ and hence in $|\cdot|_{\mathbf{m}}^{(p)}$. This completes the proof of (i). (The compatibility of the topology generated by $|\cdot|_{\mathbf{m}}^{(p)}$ with that of $L_p(\mathbf{m})$ follows from Remarks 1.38(c) of [Ru2], but not the completeness of $|\cdot|_{\mathbf{m}}^{(p)}$.)

(ii) If $|f| \leq |g|$ in $L_p^r(\mathbf{m})$, then by Theorem 4.3.2 $(\mathbf{m}_{q_n})_p^{\bullet}(f, T) \leq (\mathbf{m}_{q_n})_p^{\bullet}(g, T)$ for each n and consequently, $|f|_{\mathbf{m}}^{(p)} \leq |g|_{\mathbf{m}}^{(p)}$. As $L_p^r(\mathbf{m})$ is a vector space and as $L_1^r(\mathbf{m})$ is a lattice by Theorem 4.1.9'(iii), it follows that $L_p^r(\mathbf{m})$ is a lattice. Clearly, $f \leq g$ implies $f + h \leq g + h$ and $\alpha f \leq \alpha g$ for $f, g, h \in L_p^r(\mathbf{m})$ and $\alpha \geq 0$. Hence $L_p^r(\mathbf{m})$ is a vector lattice. Consequently, by (i), $L_p^r(\mathbf{m})$ is a Fréchet lattice with respect to $|\cdot|_{\mathbf{m}}^{(p)}$.

(iii) In the light of Theorem 4.5.8, the result is obvious. $\qquad\square$

The following theorem generalizes Theorems 3.4.5, 3.4.6, 3.4.7 and 3.4.10 and Corollary 3.4.11 of Chapter 3 to a quasicomplete (resp. to a sequentially complete) lcHs-valued vector measure.

Theorem 4.5.12. *Let X be an lcHs, $\mathbf{m} : \mathcal{P} \to X$ be σ-additive and $1 \leq p < \infty$. Let X be quasicomplete (resp. sequentially complete). Then:*

(i) **(Generalizations of Theorem 3.4.5 of Chapter 3).** *Let $(f_n^{(q)})_{n=1}^{\infty} \subset \mathcal{L}_p(\mathbf{m}_q)$ (resp. $\subset \mathcal{L}_p(\sigma(\mathcal{P}), \mathbf{m}_q)$) for $q \in \Gamma$ and let $f : T \to \mathbb{K}$ be $\mathbf{m}$-measurable (resp. $\sigma(\mathcal{P})$-measurable). Then $f \in \mathcal{L}_p(\mathbf{m})$ (resp. $f \in \mathcal{L}_p(\sigma(\mathcal{P}), \mathbf{m})$) and $\lim_n (\mathbf{m}_q)_p^{\bullet}(f_n^{(q)} - f, T) = 0$ for $q \in \Gamma$ if and only if the following conditions hold:*

(a) *$f_n^{(q)} \to f$ in measure in each $E \in \mathcal{P}$ with respect to $\mathbf{m}_q$ for $q \in \Gamma$.*

(b) *$(\mathbf{m}_q)_p^{\bullet}(f_n^{(q)}, \cdot)$, $n \in \mathbb{N}$, are uniformly $\mathbf{m}_q$-continuous on $\sigma(\mathcal{P})$ for $q \in \Gamma$.*

(c) *For each $\epsilon > 0$ and $q \in \Gamma$, there exists $A_\epsilon^{(q)} \in \mathcal{P}$ such that*

$$(\mathbf{m}_q)_p^{\bullet}(f_n^{(q)}, T \backslash A_\epsilon^{(q)}) < \epsilon \quad \text{for all} \quad n \in \mathbb{N}.$$

(See Notation 3.4.1 of Chapter 3.)

In such case, for $p = 1$, $\int_A f d\mathbf{m}_q = \lim_n \int_A f_n^{(q)} d\mathbf{m}_q$, $A \in \sigma(\mathcal{P})$ and the limit is uniform with respect to $A \in \sigma(\mathcal{P})$ for a fixed $q \in \Gamma$.

(ii) **(Generalizations of Theorem 3.4.6 of Chapter 3).** *Let* $(f_n^{(q)})_{n=1}^\infty \subset \mathcal{L}_p(\mathbf{m}_q)$ *(resp.* $\subset \mathcal{L}_p(\sigma(\mathcal{P}), \mathbf{m}_q))$ *for* $q \in \Gamma$ *and let* $f : T \to \mathbb{K}$ *(resp. and be* $\sigma(\mathcal{P})$-*measurable). Suppose* $f_n^{(q)} \to f$ $\mathbf{m}_q$-*a.e. in* T *for* $q \in \Gamma$. *Then* $f \in \mathcal{L}_p(\mathbf{m})$ *(resp.* $f \in \mathcal{L}_p(\sigma(\mathcal{P}), \mathbf{m}))$ *and* $\lim_n (\mathbf{m}_q)_p^\bullet(f_n^{(q)} - f, T) = 0$ *for* $q \in \Gamma$ *if and only if the following conditions are satisfied:*

(a) $(\mathbf{m}_q)_p^\bullet(f_n^{(q)}, \cdot)$, $n \in \mathbb{N}$, *are uniformly* $\mathbf{m}_q$-*continuous on* $\sigma(\mathcal{P})$ *for* $q \in \Gamma$.

(b) *For each* $\epsilon > 0$ *and* $q \in \Gamma$, *there exists* $A_\epsilon^{(q)} \in \mathcal{P}$ *such that*

$$(\mathbf{m}_q)_p^\bullet(f_n^{(q)}, T \backslash A_\epsilon^{(q)}) < \epsilon \quad \text{for all} \quad n.$$

In such case, for $p = 1$, $\int_A f d\mathbf{m}_q = \lim_n \int_A f_n^{(q)} d\mathbf{m}_q$ *for* $A \in \sigma(\mathcal{P})$ *and the limit is uniform with respect to* $A \in \sigma(\mathcal{P})$ *for a fixed* $q \in \Gamma$.

(iii) *LDCT and LBCT as given in* (i) *and* (ii) *of Theorem 4.5.3 are deducible from* (i) *and* (ii) *above.*

(iv) **(Generalizations of Theorem** 3.4.10 **of Chapter 3).** *For each* $q \in \Gamma$, *let* $(f_\alpha^{(q)})_{\alpha \in (D^{(q)}, \geq_q)}$, *be a net of* $\mathbf{m}_q$-*measurable (resp.* $\sigma(\mathcal{P})$-*measurable) scalar functions on* T *and let* $f : T \to \mathbb{K}$ *be* $\mathbf{m}$-*measurable (resp.* $\sigma(\mathcal{P})$-*measurable). Let* $g^{(q)} \in \mathcal{L}_p(\mathbf{m}_q)$ *(resp.* $\in \mathcal{L}_p(\sigma(\mathcal{P}), \mathbf{m}_q))$ *and let* $|f_\alpha^{(q)}| \leq |g^{(q)}|$ $\mathbf{m}_q$-*a.e. in* T *for each* $\alpha \in (D^{(q)}, \geq_q)$ *and for* $q \in \Gamma$. *Then* $f_\alpha^{(q)} \to f$ *in measure in* T *with respect to* $\mathbf{m}_q$, $q \in \Gamma$, *if and only if* $f \in \mathcal{L}_p(\mathbf{m})$ *(resp.* $f \in \mathcal{L}_p(\sigma(\mathcal{P}), \mathbf{m}))$ *and* $\lim_{\alpha \in (D^{(q)}, \geq_q)}(\mathbf{m}_q)_p^\bullet(f_\alpha^{(q)} - f, T) = 0$ *for* $q \in \Gamma$. *In such case, for* $p = 1$, $\int_A f d\mathbf{m}_q = \lim_\alpha \int_A f_\alpha^{(q)} d\mathbf{m}_q$ *for* $A \in \sigma(\mathcal{P})$ *where the limit is uniform with respect to* $A \in \sigma(\mathcal{P})$ *for a fixed* $q \in \Gamma$.

(v) **(Generalizations of Corollary** 3.4.11 **of Chapter 3).** *Let* $\mathcal{P}$ *be a* σ-*ring* $\mathcal{S}$ *and let* $0 < K_q < \infty$ *for each* $q \in \Gamma$. *If* $(f_\alpha^{(q)})$ *is a net as in* (iv), *if* $|f_\alpha^{(q)}| \leq K_q$ $\mathbf{m}_q$-*a.e. in* T *for each* $\alpha \in (D^{(q)}, \geq_q)$ *and for* $q \in \Gamma$ *and if* $f : T \to \mathbb{K}$ *is* $\mathbf{m}$-*measurable (resp.* $\sigma(\mathcal{P})$-*measurable), then* $f_\alpha^{(q)} \to f$ *in measure in* T *with respect to* $\mathbf{m}_q$ *for* $q \in \Gamma$ *if and only if* $f \in \mathcal{L}_p(\mathbf{m})$ *(resp.* $f \in \mathcal{L}_p(\sigma(\mathcal{P}), \mathbf{m}))$ *and* $\lim_\alpha(\mathbf{m}_q)_p^\bullet(f_\alpha^{(q)} - f, T) = 0$ *for* $q \in \Gamma$. *In such case, for* $p = 1$, $\int_A f d\mathbf{m}_q = \lim_\alpha \int_A f_\alpha^{(q)} d\mathbf{m}_q$ *for* $A \in \sigma(\mathcal{P})$ *where the limit is uniform with respect to* $A \in \sigma(\mathcal{P})$ *for a fixed* $q \in \Gamma$.

Proof. By (i) of the following theorem and by the theorems in Chapter 3 which they generalize, the results hold. $\qquad\square$

The following theorem generalizes the results in Section 3.5 of Chapter 3 to a quasicomplete (resp. sequentially complete) lcHs-valued σ-additive vector measure.

Theorem 4.5.13. *Let* X *be a quasicomplete (resp. sequentially complete) lcHs,* $\mathbf{m} : \mathcal{P} \to X$ *be* σ-*additive and* $1 \leq p < \infty$. *Then the following statements hold:*

(i) $\mathcal{L}_p(\mathbf{m}) = \bigcap_{q \in \Gamma} \mathcal{L}_p(\mathbf{m}_q)$ (resp. (i') $\mathcal{L}_p(\sigma(\mathcal{P}),\mathbf{m}) = \bigcap_{q \in \Gamma} \mathcal{L}_p(\sigma(\mathcal{P}),\mathbf{m}_q)$).

(ii) If $1 \leq r < p < s \leq \infty$, then $\mathcal{L}_r(\mathbf{m}) \bigcap \mathcal{L}_s(\mathbf{m}) \subset \mathcal{L}_p(\mathbf{m})$.
(resp. (ii') $\mathcal{L}_r(\sigma(\mathcal{P}),\mathbf{m}) \bigcap \mathcal{L}_s(\sigma(\mathcal{P}),\mathbf{m}) \subset \mathcal{L}_p(\sigma(\mathcal{P}),\mathbf{m})$).

(iii) If $f : T \to \mathbb{K}$ is $\mathbf{m}$-measurable (resp. (iii') is $\sigma(\mathcal{P})$-measurable), then $\mathcal{I}_f = \{p : 1 \leq p < \infty, f \in \mathcal{L}_p(\mathbf{m})\}$ (resp. $\mathcal{I}_f(\sigma(\mathcal{P})) = \{f : 1 \leq p < \infty, f \in \mathcal{L}_p(\sigma(\mathcal{P}),\mathbf{m})\}$) is either void or an interval, where singletons are considered as intervals.

(iv) Let $A \in \widetilde{\sigma(\mathcal{P})}$ (see Definition 3.1.3) (resp. (iv') $A \in \sigma(\mathcal{P})$) such that χ_A is $\mathbf{m}$-integrable in T. Then the set $\mathcal{I}_f(A) = \{p : 1 \leq p < \infty, f\chi_A \in \mathcal{L}_p(\mathbf{m})\}$ (resp. $\mathcal{I}_f(\sigma(\mathcal{P}), A) = \{p : 1 \leq p < \infty, f\chi_A \in \mathcal{L}_p(\sigma(\mathcal{P}),\mathbf{m})\}$) is either void or an interval containing 1, $\mathcal{I}_f(A) = \{1\}$ (resp. $\mathcal{I}_f(\sigma(\mathcal{P}), A)) = \{1\}$) being permitted.

(v) Let $\mathcal{P}$ be a σ-ring $\mathcal{S}$. Then:

    (a) If $1 \leq r < s < \infty$, then $\mathcal{L}_s(\mathbf{m}) \subset \mathcal{L}_r(\mathbf{m})$ (resp. (a') $\mathcal{L}_s(\sigma(\mathcal{P}),\mathbf{m}) \subset \mathcal{L}_r(\sigma(\mathcal{P}),\mathbf{m})$) and the topology of $\mathcal{L}_s(\mathbf{m})$ (resp. of $\mathcal{L}_s(\sigma(\mathcal{P}),\mathbf{m})$) is finer than that of $\mathcal{L}_r(\mathbf{m})$ (resp. of $\mathcal{L}_r(\sigma(\mathcal{P}),\mathbf{m})$).

    (b) If $f : T \to \mathbb{K}$ is $\mathbf{m}$-measurable (resp. (b') $\sigma(\mathcal{P})$-measurable), then the set $\mathcal{I}_f$ (resp. $\mathcal{I}_f(\sigma(\mathcal{P}))$ is either void or an interval containing 1, ($\mathcal{I}_f = \{1\}$ (resp. $\mathcal{I}_f(\sigma(\mathcal{P}))=\{1\}$) being permitted).

Proof. (i) holds by Remark 4.2.6 and by (i) and (ii) of Theorem 4.2.2 (resp. (i') holds by Theorem 4.4.7(ii)). (ii) and (iii) hold by (i) and by Theorem 3.5.2 of Chapter 3. (ii') and (iii') hold by (i'), by Theorem 3.5.2 of Chapter 3 and by the fact that $\mathcal{L}_p(\sigma(\mathcal{P}),\mathbf{m}) = \mathcal{L}_p(\mathbf{m}) \bigcap \mathcal{M}(\mathcal{P})$. (iv) is due to (i) and Theorem 3.5.3 of Chapter 3. (iv') is due to (i'), Theorem 3.5.3 of Chapter 3 and the definition of $\mathcal{L}_p(\sigma(\mathcal{P}),\mathbf{m})$. (v)(a) (resp. (v)(a')) is due to Corollary 3.5.4 of Chapter 3 and the fact that $(\mathbf{m}_q)_r^\bullet(f,T) \leq (\mathbf{m}_q)_s^\bullet(f,T) \cdot (\|\mathbf{m}\|_q(N(f))^{\frac{1}{r}-\frac{1}{s}}$ for each $q \in \Gamma$ and for $f \in \mathcal{L}_s(\mathbf{m})$ (resp. and for $f \in \mathcal{L}_s(\sigma(\mathcal{P}),\mathbf{m})$) by Theorem 3.5.3 of Chapter 3. (v)(b) (resp. (v)(b')) follows from (v)(a) and (iii) (resp. (v)(a') and (iii')). $\qquad\square$

4.6 Separability of $\mathcal{L}_p(\mathbf{m})$ and $\mathcal{L}_p(\sigma(\mathcal{P}),\mathbf{m}), 1 \leq p < \infty$, $\mathbf{m}$ lcHs-valued

In this section we give some sufficient conditions for the separability of $\mathcal{L}_p(\mathbf{m})$ (resp. of $\mathcal{L}_p(\sigma(\mathcal{P}),\mathbf{m})$) for $1 \leq p < \infty$, when $\mathbf{m} : \mathcal{P} \to X$ is σ-additive and X is quasicomplete (resp. sequentially complete). For such p and for a σ-ring $\mathcal{S}$, we also include a generalization of Propositions 2 and 3(ii) of [Ri3] to $\mathcal{L}_p(\mathbf{m})$ and to $L_p(\sigma(\mathcal{S}),\mathbf{m})=L_p(\mathcal{S},\mathbf{m})$.

Definition 4.6.1. Let X be a quasicomplete (resp. sequentially complete) lcHs, $\mathbf{m} : \mathcal{P} \to X$ be σ-additive and $1 \leq p < \infty$. We identify $\mathcal{P}$ with the subset

$\mathcal{F} = \{\chi_A : A \in \mathcal{P}\}$ of $\mathcal{L}_p(\mathbf{m})$ (resp. of $\mathcal{L}_p(\sigma(\mathcal{P}), \mathbf{m})$) and endow $\mathcal{P}$ with the relative topology $\tau_{\mathbf{m}}^{(p)}|_{\mathcal{P}}$. Then we write $(\mathcal{P}, \tau_{\mathbf{m}}^{(p)})$.

For $A, B \in \mathcal{P}$, we define $\rho(\mathbf{m})_q^{(p)}(A, B) = (\mathbf{m}_q)_p^{\bullet}(\chi_A - \chi_B, T)$ for $q \in \Gamma$. By Theorem 4.3.2 and Proposition 1.2.15(ii)(c) we have

$$\rho(\mathbf{m})_q^{(p)}(A, B) = \sup_{x^* \in U_q^o} \left(\int_T \chi_{A \triangle B} dv(x^*\mathbf{m}) \right)^{\frac{1}{p}}$$
$$= \sup_{x^* \in U_q^o} (v(x^*\mathbf{m})(A \triangle B))^{\frac{1}{p}} = (\|\mathbf{m}\|_q(A \triangle B))^{\frac{1}{p}}$$

for $A, B \in \mathcal{P}$. Moreover, by Theorem 3.1.13(ii) of Chapter 3 for $\mathbf{m}_q$, $\rho(\mathbf{m})_q^{(p)}$ is a pseudo-metric on $\mathcal{P}$. Thus, the topology $\tau_{\mathbf{m}}^{(p)}|_{\mathcal{P}}$ is generated by the pseudo-metrics $\{\rho(\mathbf{m})_q^{(p)}, q \in \Gamma\}$.

Theorem 4.6.2. *Let X be a quasicomplete (resp. sequentially complete) lcHs, $\mathbf{m} : \mathcal{P} \to X$ be σ-additive and $1 \le p < \infty$. Then:*

(i) *For $1 \le r < \infty$, $\tau_{\mathbf{m}}^{(r)}|_{\mathcal{P}} = \tau_{\mathbf{m}}^{(p)}|_{\mathcal{P}}$.*

(ii) *If $(\mathcal{P}, \tau_{\mathbf{m}}^{(p)})$ is separable, then $\mathcal{L}_p(\mathbf{m})$ and $L_p(\mathbf{m})$ (resp. $\mathcal{L}_p(\sigma(\mathcal{P}), \mathbf{m})$ and $L_p(\sigma(\mathcal{P}), \mathbf{m})$) are separable. Consequently, for $1 \le r < \infty$, $\tau_{\mathbf{m}}^{(r)}|_{\mathcal{P}}$ is separable whenever $(\mathcal{P}, \tau_{\mathbf{m}}^{(p)})$ is separable and in that case, $\mathcal{L}_r(\mathbf{m})$ and $L_r(\mathbf{m})$ (resp. $\mathcal{L}_r(\sigma(\mathcal{P}), \mathbf{m})$ and $L_r(\sigma(\mathcal{P}), \mathbf{m})$) are also separable for each $r \in [1, \infty)$.*

(iii) *If X is further metrizable, then $\mathcal{L}_p(\mathbf{m})$ (resp. $\mathcal{L}_p(\sigma(\mathcal{P}), \mathbf{m})$) is separable if and only if $(\mathcal{P}, \tau_{\mathbf{m}}^{(p)})$ is separable.*

Proof. (i) Let $1 \le r < \infty$ and let $q \in \Gamma$. Since $\rho(\mathbf{m})_q^{(r)}(A, B) = (\|\mathbf{m}\|_q(A \triangle B))^{\frac{1}{r}}$, it follows that $\rho(\mathbf{m})_q^{(r)}(A, B_\alpha) \to 0$ if and only if $\rho(\mathbf{m})_q^{(p)}(A, B_\alpha) \to 0$ for $A \in \mathcal{P}$ and for a net $(B_\alpha) \subset \mathcal{P}$. Hence (i) holds.

(ii) Let $(\mathcal{P}, \tau_{\mathbf{m}}^{(p)})$ be separable. Then by hypothesis there exists a countable subset D of $\mathcal{P}$ such that D is $\tau_{\mathbf{m}}^{(p)}$-dense in $\mathcal{P}$. Let $W = \{\sum_{j=1}^r \alpha_j \chi_{F_j} : \alpha_j = a_j + ib_j, a_j, b_j$ real rational, $(F_j)_1^r \subset D, r \in \mathbb{N}\}$. Then W is countable. Let $s = \sum_{j=1}^k \beta_j \chi_{A_j} \in \mathcal{I}_s$, $\beta_j \ne 0$ for all j. Then $c = \sum_{j=1}^k |\beta_j| > 0$. Let U be a $\tau_{\mathbf{m}}^{(p)}$-neighborhood of 0 in $\mathcal{L}_p(\mathbf{m})$ (resp. in $\mathcal{L}_p(\sigma(\mathcal{P}), \mathbf{m})$). Then there exist an $\epsilon \in (0, 1)$ and $q_1, q_2, \ldots, q_n$ in Γ such that $\{f \in \mathcal{L}_p(\mathbf{m})(\text{resp.} f \in \mathcal{L}_p(\sigma(\mathcal{P}), \mathbf{m})) : (\mathbf{m}_{q_i})_p^{\bullet}(f, T) < \epsilon, i = 1, 2, \ldots, n\} \subset U$. Let $M = 1 + \sup_{1 \le i \le n, 1 \le j \le k}(\|\mathbf{m}\|_{q_i}(A_j))^{\frac{1}{p}}$. Since D is $\tau_{\mathbf{m}}^{(p)}$-dense in $\mathcal{P}$, there exist $(F_j)_1^k \subset D$ such that

$$(\mathbf{m}_{q_i})_p^{\bullet}(\chi_{F_j} - \chi_{A_j}, T) = (\|\mathbf{m}\|_{q_i}(A_j \triangle F_j))^{\frac{1}{p}} < \frac{\epsilon}{3c} \qquad (4.6.2.1)$$

for $i = 1, 2, \ldots, n$ and $j = 1, 2, \ldots, k$. Choose $\alpha_j = a_j + ib_j, a_j, b_j$ real rational for $j = 1, 2, \ldots, k$ such that $\sum_{j=1}^k |\beta_j - \alpha_j| < \frac{\epsilon c}{3M}$. Let $\omega = \sum_{j=1}^k \alpha_j \chi_{F_j}$. Then

$\omega \in W$. Now by Theorem 3.1.13(i) of Chapter 3 for $\mathbf{m}_q$ and by (4.6.2.1) we have

$$
\begin{aligned}
(||\mathbf{m}||_{q_i}(F_j))^{\frac{1}{p}} &\leq (||\mathbf{m}||_{q_i}(F_j \setminus A_j))^{\frac{1}{p}} + (||\mathbf{m}||_{q_i}(F_j \cap A_j))^{\frac{1}{p}} \\
&\leq (||\mathbf{m}||_{q_i}(F_j \Delta A_j))^{\frac{1}{p}} + (||\mathbf{m}||_{q_i}(A_j))^{\frac{1}{p}} \\
&< \frac{\epsilon}{3c} + (||\mathbf{m}||_{q_i}(A_j))^{\frac{1}{p}}
\end{aligned}
\tag{4.6.2.2}
$$

for $j = 1, 2, \ldots, k$ and $i = 1, 2, \ldots, n$.

Again by Theorem 3.1.13(i) of Chapter 3 for $\mathbf{m}_q$ and by (4.6.2.1) and (4.6.2.2) we have

$$
\begin{aligned}
(\mathbf{m}_{q_i})_p^\bullet(s - \omega, T) &\leq \sum_{j=1}^{k} (\mathbf{m}_{q_i})_p^\bullet(\beta_j \chi_{A_j} - \alpha_j \chi_{F_j}, T) \\
&\leq \sum_{j=1}^{k} |\beta_j - \alpha_j|(\mathbf{m}_{q_i})_p^\bullet(\chi_{F_j}, T) + \sum_{j=1}^{k} |\beta_j|(\mathbf{m}_{q_i})_p^\bullet(\chi_{A_j} - \chi_{F_j}, T) \\
&= \sum_{j=1}^{k} |\beta_j - \alpha_j|(||\mathbf{m}||_{q_i}(F_j))^{\frac{1}{p}} + \sum_{j=1}^{k} |\beta_j| \left(\frac{\epsilon}{3c} \right) \\
&< \sum_{j=1}^{k} \left(|\beta_j - \alpha_j| \frac{\epsilon}{3c} \right) + \sum_{j=1}^{k} |\beta_j - \alpha_j|(||\mathbf{m}_{q_i}||(A_j))^{\frac{1}{p}} + \frac{\epsilon}{3} \\
&< \frac{\epsilon c}{3c} + \frac{\epsilon}{3M} M + \frac{\epsilon}{3} = \epsilon
\end{aligned}
$$

for $i = 1, 2, \ldots, n$ since $M > 1$ and $0 < \epsilon < 1$ and hence $\omega \in s + U$. This shows that W is $\tau_{\mathbf{m}}^{(p)}$-dense in $\mathcal{I}_s$. Since $\mathcal{I}_s$ is $\tau_{\mathbf{m}}^{(p)}$-dense in $\mathcal{L}_p(\mathbf{m})$ (resp. in $\mathcal{L}_p(\sigma(\mathcal{P}), \mathbf{m})$) by Theorem 3.3.8, it follows that W is dense in $\mathcal{L}_p(\mathbf{m})$ (resp. in $\mathcal{L}_p(\sigma(\mathcal{P}), \mathbf{m})$) and hence $\mathcal{L}_p(\mathbf{m})$ (resp. $\mathcal{L}_p(\sigma(\mathcal{P}), \mathbf{m})$) is separable. Then $L_p(\mathbf{m})$ (resp. $L_p(\sigma(\mathcal{P}), \mathbf{m})$ is separable by Problem H, Chapter 4 of [Ke].

(iii) If X is metrizable, then $\mathcal{L}_p(\mathbf{m})$ (resp. $\mathcal{L}_p(\sigma(\mathcal{P}), \mathbf{m})$) is pseudo-metrizable by Theorem 4.4.2(i) and hence, if $\mathcal{L}_p(\mathbf{m})$ (resp. $\mathcal{L}_p(\sigma(\mathcal{P}), \mathbf{m})$) is separable, then by Theorem 11, Chapter 4 of [Ke], $(\mathcal{P}, \tau_{\mathbf{m}}^{(p)})$ is separable. The converse holds by (ii). $\square$

The following theorem gives a sufficient condition for the separability of $(\mathcal{P}, \tau_{\mathbf{m}}^{(p)})$.

Theorem 4.6.3. *Let X be a quasicomplete (resp. sequentially complete) lcHs and let $\mathcal{S}$ be a σ-ring of subsets of T. Suppose $\mathbf{m} : \mathcal{P} \to X$ (resp. $\mathbf{n} : \mathcal{S} \to X$) is σ-additive. If there exists a countable family $\mathcal{R}$ such that (i) $\mathcal{P}$ is the δ-ring (resp. (ii) $\mathcal{S}$ is the σ-ring) generated by $\mathcal{R}$, then $(\mathcal{P}, \tau_{\mathbf{m}}^{(p)})$ (resp. $(\mathcal{S}, \tau_{\mathbf{n}}^{(p)})$) is separable for $1 \leq p < \infty$.*

Proof. In the light of Theorem 5C of [H], without loss of generality we shall assume that $\mathcal{R}$ is a countable subring of $\mathcal{P}$ (resp. of $\mathcal{S}$). Let $1 \leq p < \infty$, p fixed.

(i) Let $A \in \mathcal{P}$ and let U be a $\tau_{\mathbf{m}}^{(p)}|_{\mathcal{P}}$-neighborhood of A in $\mathcal{P}$. Then there exist an $\epsilon > 0$ and $q_1, q_2, \ldots, q_k$ in Γ such that $\{B \in \mathcal{P} : \rho(\mathbf{m})_{q_j}^{(p)}(A, B) < \epsilon, j = 1, 2, \ldots, k\} \subset U$. By hypothesis and by a corollary to Proposition 10, §1 of [Din1], there exists $F \in \mathcal{R}$ such that $A \subset F$. Then $F \cap \mathcal{P}$ is a σ-algebra of subsets of F and hence by Proposition 1.1.13 of Chapter 1 there exist control measures $\mu_F^{(j)} : F \cap \mathcal{P} \to [0, \infty)$ for $\mathbf{m}_{q_j} : F \cap \mathcal{P} \to \widetilde{X_{q_j}}$, $j = 1, 2, \ldots, k$. Hence there exists $\delta > 0$ such that $\mu_F^{(j)}(B) < \delta$ implies $||\mathbf{m}||_{q_j}(B) < \epsilon^p$ for $j = 1, 2, \ldots, k$. Since $F \cap \mathcal{P}$ is the σ-ring generated by the ring $F \cap \mathcal{R} (\subset \mathcal{R})$ (to prove this use Theorem 5E of [H]), by ex.13.8 of [H] there exists $B \in F \cap \mathcal{R}$ such that $\sum_{j=1}^{k-1} \mu_F^{(j)}(A \Delta B) < \delta$ and $\mu_F^{(k)}(A \Delta B) < \delta$ so that $\mu_F^{(j)}(A \Delta B) < \delta$ for $j = 1, 2, \ldots, k$. Consequently, $\rho(\mathbf{m})_{q_j}^{(p)}(A, B) = (\mathbf{m}_{q_j})_p^{\bullet}(\chi_A - \chi_B, T) = (||\mathbf{m}||_{q_j}(A \Delta B))^{\frac{1}{p}} < \epsilon$ for $j = 1, 2, \ldots, k$ and hence $B \in U$. This shows that the countable subring $\mathcal{R}$ is $\tau_{\mathbf{m}}^{(p)}$-dense in $\mathcal{P}$ and hence $(\mathcal{P}, \tau_{\mathbf{m}}^{(p)})$ is separable for $p \in [1, \infty)$.

(ii) Let $A \in \mathcal{S}$ and let U be a $\tau_{\mathbf{n}}^{(p)}$-neighborhood of A in $\mathcal{S}$. Then there exist an $\epsilon > 0$ and $q_1, q_2, \ldots, q_k$ in Γ such that $\{B \in \mathcal{S} : \rho(\mathbf{n})_{q_j}^{(p)}(A, B) < \epsilon, j = 1, 2, \ldots, k\} \subset U$. Then by Proposition 1.1.13 of Chapter 1 there exist control measures $\mu_j : \mathcal{S} \to [0, \infty)$ for $\mathbf{n}_{q_j} : \mathcal{S} \to \widetilde{X_{q_j}}$, $j = 1, 2, \ldots, k$. Now using the hypothesis that $\mathcal{S}$ is the σ-ring generated by the countable subring $\mathcal{R}$ and arguing as in the proof of (i) we conclude that $(\mathcal{S}, \tau_{\mathbf{n}}^{(p)})$ is separable for $p \in [1, \infty)$.   □

Definition 4.6.4. Let T be a locally compact Hausdorff space. Let $\mathcal{U}, \mathcal{C}$ and $\mathcal{C}_0$ be respectively the families of all open sets, all compact sets and all compact G_δ sets in T. For a family $\mathcal{F}$ of subsets of T, let $\sigma(\mathcal{F})$ be the σ-ring generated by $\mathcal{F}$. Then $\mathcal{B}(T)$ (resp. $\mathcal{B}_c(T)$, $\mathcal{B}_0(T)$) denotes $\sigma(\mathcal{U})$ (resp. $\sigma(\mathcal{C})$, $\sigma(\mathcal{C}_0)$). The members of $\mathcal{B}(T)$ (resp. $\mathcal{B}_c(T)$, $\mathcal{B}_0(T)$) are called Borel (resp. σ-Borel, Baire) sets in T.

Theorem 4.6.5. *Let T be a locally compact Hausdorff space with a countable base of open sets and let X be a quasicomplete (resp. sequentially complete) lcHs. Let $\mathbf{m} : \mathcal{S} \to X$ be σ-additive, where $\mathcal{S} = \mathcal{B}(T)$ or $\mathcal{B}_c(T)$ or $\mathcal{B}_0(T)$. Then $\mathcal{L}_p(\mathbf{m})$ and $L_p(\mathbf{m})$*

$$(\text{resp. } \mathcal{L}_p(\sigma(\mathcal{S}), \mathbf{m}) = \mathcal{L}_p(\mathcal{S}, \mathbf{m}) \quad \text{and} \quad L_p(\sigma(\mathcal{S}), \mathbf{m}) = L_p(\mathcal{S}, \mathbf{m}))$$

are separable for $1 \leq p < \infty$.

Proof. By hypothesis, $\mathcal{B}(T)$ is a countably generated σ-ring; by hypothesis and by Corollary to Proposition 2 (resp. and by Corollary to Proposition 16) in §14 of [Din1], $\mathcal{B}_c(T)$ (resp. $\mathcal{B}_0(T)$) is a countably generated σ-ring. Hence the results hold by Theorems 4.6.3(ii) and 4.6.2(ii).   □

To obtain some useful sufficient conditions for the separability of $\mathcal{L}_p(\mathbf{m})$ and $L_p(\mathbf{m})$ (resp. $\mathcal{L}_p(\sigma(\mathcal{P}),\mathbf{m})$ and $L_p(\sigma(\mathcal{P}),\mathbf{m})$) for $\mathcal{P} = \delta(\mathcal{C})$ or $\delta(\mathcal{C}_0)$ (i.e., the δ-rings generated by $\mathcal{C}$ and $\mathcal{C}_0$, respectively) in a locally compact Hausdorff space T we give the following

Theorem 4.6.6. *Let X be a Fréchet space, $\mathbf{m} : \mathcal{P} \to X$ be σ-additive and $1 \le p < \infty$. Suppose that $\mathbf{m}$ admits a σ-additive X-valued extension $\tilde{\mathbf{m}}$ on $\sigma(\mathcal{P})$ and that $\sigma(\mathcal{P})$ is countably generated. Then $\mathcal{L}_p(\mathbf{m})$, $\mathcal{L}_p(\sigma(\mathcal{P}),\mathbf{m})$, $L_p(\mathbf{m})$ and $L_p(\sigma(\mathcal{P}),\mathbf{m})$ are separable.*

Proof. By hypothesis and by Theorems 4.6.3 and 4.6.2(ii), $\mathcal{L}_p(\tilde{\mathbf{m}})$ and $L_p(\tilde{\mathbf{m}})$ are separable. If $f \in \mathcal{L}_p(\mathbf{m})$, then f is $\mathbf{m}$-measurable and hence is $\tilde{\mathbf{m}}$-measurable. As $\tilde{\mathbf{m}}|_\mathcal{P} = \mathbf{m}$, $\int_A s d\tilde{\mathbf{m}} = \int_A s d\mathbf{m}$ for $A \in \sigma(\mathcal{P})$ and for $s \in \mathcal{I}_s$. Then by Definition 4.2.1, by Theorem 4.2.3 and by the fact $|f|^p$ is $\mathbf{m}$-integrable in T (see Notation 4.5.9), $|f|^p$ is $\tilde{\mathbf{m}}$-integrable in T with values in X so that $f \in \mathcal{L}_p(\tilde{\mathbf{m}})$ and clearly, $\int_A |f|^p d\tilde{\mathbf{m}} = \int_A |f|^p d\mathbf{m}$. Moreover, by Theorem 4.3.2 we have

$$(\tilde{\mathbf{m}}_q)_p^\bullet(f,T) = \sup_{A\in\sigma(\mathcal{P}),x^*\in U_q^\circ} (\int_A |f|^p dv(x^*\tilde{\mathbf{m}}))^{\frac{1}{p}}$$

$$= \sup_{A\in\sigma(\mathcal{P}),x^*\in U_q^\circ} (\int_A |f|^p dv(x^*\mathbf{m}))^{\frac{1}{p}} = (\mathbf{m}_q)_p^\bullet(f,T)$$

for $q \in \Gamma$. Hence $\mathcal{L}_p(\mathbf{m})$ is a subspace of $\mathcal{L}_p(\tilde{\mathbf{m}})$ with $\tau_{\tilde{\mathbf{m}}}^{(p)}|_{\mathcal{L}_p(\mathbf{m})} = \tau_{\mathbf{m}}^{(p)}$. As X is metrizable, $\mathcal{L}_p(\tilde{\mathbf{m}})$ is pseudo-metrizable by Theorem 4.4.2(i) and consequently, by Theorem 11, Chapter 4, of [Ke], $\mathcal{L}_p(\mathbf{m})$ and $\mathcal{L}_p(\sigma(\mathcal{P}),\mathbf{m})$ are separable and then by Problem H, Chapter 4 of [Ke], $L_p(\mathbf{m})$ and $L_p(\sigma(\mathcal{P}),\mathbf{m})$ are separable. $\square$

Let us recall the following definition from [P9].

Definition 4.6.7. Let X be an lcHs and let T be a locally compact Hausdorff space. Suppose $\mathcal{R}$ is a ring of sets in T such that $\mathcal{R} \supset \delta(\mathcal{C})$ or $\delta(\mathcal{C}_0)$. An X-valued σ-additive vector measure $\mathbf{m}$ on $\mathcal{R}$ is said to be $\mathcal{R}$-regular if, given $\epsilon > 0$, $A \in \mathcal{R}$ and $q \in \Gamma$, there exist an open set $U \in \mathcal{R}$ and a compact $K \in \mathcal{R}$ such that $K \subset A \subset U$ and such that $||\mathbf{m}||_q(B) < \epsilon$ for each $B \in \mathcal{R}$ with $B \subset U\backslash K$. (In the light of Proposition 1.1.6 of Chapter 1, this is equivalent to the $\mathcal{R}$-regularity given in Definition 5 of [P9].)

Theorem 4.6.8. *Let T be a locally compact Hausdorff space and let X be a Fréchet space. Let $\mathbf{m} : \delta(\mathcal{C}_0) \to X$ be σ-additive. If the range of $\mathbf{m}$ is relatively weakly compact, then $\mathbf{m}$ has a unique σ-additive X-valued $\mathcal{B}_0(T)$-regular (resp. $\mathcal{B}_c(T)$-regular) extension $\mathbf{m}_0$ on $\mathcal{B}_0(T)$ (resp. $\mathbf{m}_c$ on $\mathcal{B}_c(T)$). Let $\mathbf{m}_c' = \mathbf{m}_c|_{\delta(\mathcal{C})}$. If T has a countable base of open sets, then $\mathcal{L}_p(\mathbf{m})$, $\mathcal{L}_p(\sigma(\mathcal{C}_0),\mathbf{m})$, $L_p(\mathbf{m})$ and $L_p(\sigma(\mathcal{C}_0),\mathbf{m})$ (resp. $\mathcal{L}_p(\mathbf{m}_c')$, $\mathcal{L}_p(\sigma(\mathcal{C}),\mathbf{m}_c')$, $L_p(\mathbf{m}_c')$ and $L_p(\sigma(\mathcal{C}),\mathbf{m}_c')$) are separable for $1 \le p < \infty$.*

Proof. By the hypothesis on the range of $\mathbf{m}$ and by the Theorem on Extension of [K3] or by Corollary 2 of [P7] (where a self-contained short proof of the said theorem of [K3] is given), $\mathbf{m}$ admits an X-valued σ-additive extension $\mathbf{m}_0$ on $\mathcal{B}_0(T)$, which is unique by the Hahn-Banach theorem. Then by Theorem 10 of [P9] or by Theorem 1 of [DP1] (where a simple proof of the said theorem of [P9] is given), $\mathbf{m}_0$ is $\mathcal{B}_0(T)$-regular and has a unique X-valued $\mathcal{B}_c(T)$-regular σ-additive extension $\mathbf{m}_c$ on $\mathcal{B}_c(T)$. Then the conclusions follow from Theorem 4.6.6 since $\mathcal{B}_c(T)$ (resp. $\mathcal{B}_0(T)$) is countably generated by the hypothesis on T and by Corollary to Proposition 2 (resp. and by Corollary to Proposition 16) of §14 of [Din1]. □

Following [Ri3] we give the following

Definition 4.6.9. Let X be a quasicomplete (resp. sequentially complete) lcHs and let $\mathbf{m} : \mathcal{P} \to X$ be σ-additive. Let $\mathbf{m}(\mathcal{P}) = \{\mathbf{m}(A) : A \in \mathcal{P}\}$. For $A, B \in \mathcal{P}$, we write $A \sim B$ if $\chi_A = \chi_B$ $\mathbf{m}$-a.e. in T. Then we write $\mathcal{P}(\mathbf{m}) = \mathbf{m}(\mathcal{P})/ \sim$. (Note that '$\sim$' is an equivalence relation.)

Remark 4.6.10. If $A, B \in \mathcal{P}$ and $A \sim B$, then $||\mathbf{m}||_q(A\Delta B) = 0$ for each $q \in \Gamma$ and hence $\rho(\mathbf{m})_q^{(p)}(A, B) = 0$ for $1 \le p < \infty$. Consequently, $\mathcal{P}(\mathbf{m}) \subset L_p(\mathbf{m})$ for all $p \in [1, \infty)$ and by Theorem 4.6.2(i), $\tau_\mathbf{m}^{(p)}|_{\mathcal{P}(\mathbf{m})} = \tau_\mathbf{m}^{(r)}|_\mathcal{P}(\mathbf{m})$ for all $p, r \in [1, \infty)$. (Here by an abuse of notation we also denote by $\tau_\mathbf{m}^{(p)}$ the quotient topology induced on $L_p(\mathbf{m})$.)

Notation 4.6.11. In the light of Remark 4.6.10, when we write $\mathcal{P}(\mathbf{m})$, we consider it as a subset of some $L_r(\mathbf{m})$, $1 \le r < \infty$, with the relative topology from $L_r(\mathbf{m})$.

Theorem 4.6.12. *Let X and $\mathbf{m}$ be as in Definition* 4.6.9. *Then:*

(i) *If for some $p \in [1, \infty)$, $(\mathcal{P}(\mathbf{m}), \tau_\mathbf{m}^{(p)})$ is separable, then $L_r(\mathbf{m})$ (resp. $L_r(\sigma(\mathcal{P}), \mathbf{m}))$ is separable for all $r \in [1, \infty)$. In that case, in the light of Notation* 4.6.11 *we simply say that $\mathcal{P}(\mathbf{m})$ is separable.*

(ii) *If X is further metrizable, then $(\mathcal{P}(\mathbf{m}), \tau_\mathbf{m}^{(r)})$ is separable for all $r \in [1, \infty)$ whenever $L_p(\mathbf{m})$ (resp. $L_p(\sigma(\mathcal{P}), \mathbf{m}))$ is separable for some $p \in [1, \infty)$.*

Proof. (i) By hypothesis there exists a countable $\tau_\mathbf{m}^{(p)}$-dense set D in $\mathcal{P}(\mathbf{m})$. Then by Remark 4.6.10, D is also $\tau_\mathbf{m}^{(r)}$-dense in $\mathcal{P}(\mathbf{m})$ for any $r \in [1, \infty)$. Now taking W as in the proof of Theorem 4.6.2(ii) and using the fact that $\mathcal{I}_s/ \sim$ is dense in $L_r(\mathbf{m})$ (resp. in $L_r(\sigma(\mathcal{P}), \mathbf{m})$) by Theorem 4.5.6, we conclude that W is dense in $L_r(\mathbf{m})$ (resp. $L_r(\sigma(\mathcal{P}), \mathbf{m})$) and hence (i) holds.

(ii) If X is further metrizable, then by Theorem 4.4.2(i) $L_p(\mathbf{m})$ (resp. $L_p(\sigma(\mathcal{P}), \mathbf{m}))$ is metrizable. If $L_p(\mathbf{m})$ (resp. $L_p(\sigma(\mathcal{P}), \mathbf{m}))$ is separable, then by Theorem 11, Chapter 4 of [Ke], $\mathcal{P}(\mathbf{m})$ is separable for the relative topology from $L_p(\mathbf{m})$ (resp. from $L_p(\sigma(\mathcal{P}), \mathbf{m}))$. Then by (i), $L_r(\mathbf{m})$ (resp. $L_r(\sigma(\mathcal{P}), \mathbf{m}))$ is separable for all $r \in [1, \infty)$. □

To give a characterization of the separability of $\mathcal{S}(\mathbf{n})$ similar to Propositions 2 and 3(ii) of [Ri3] when $\mathcal{S}$ is a σ-ring of sets and $\mathbf{n} : \mathcal{S} \to X$ is σ-additive, we give the following concept. See also ([KK], pp. 32–33).

Definition 4.6.13. Let X be a quasicomplete (resp. sequentially complete) lcHs and let $\mathcal{S}$ be a σ-ring of subsets of T. Let $\mathbf{m} : \mathcal{P} \to X$ (resp. $\mathbf{n} : \mathcal{S} \to X$) be σ-additive. If there exists a countably generated δ-ring $\mathcal{P}_0 \subset \mathcal{P}$ (resp. σ-ring $\mathcal{S}_0 \subset \mathcal{S}$) such that $\mathcal{P}_0(\mathbf{m}) = \mathcal{P}(\mathbf{m})$ (resp. $\mathcal{S}_0(\mathbf{n}) = \mathcal{S}(\mathbf{n})$)(see Definition 4.6.9), then $\mathcal{P}$ (resp. $\mathcal{S}$) is said to be $\mathbf{m}$-essentially (resp. $\mathbf{n}$-essentially) countably generated.

Theorem 4.6.14. *Let X ba a quasicomplete (resp. sequentially complete) lcHs and let $\mathcal{S}$ be a σ-ring of subsets of T. Let $\mathbf{m} : \mathcal{P} \to X$ (resp. $\mathbf{n} : \mathcal{S} \to X$) be σ-additive. Then:*

(i) *If $\mathcal{P}$ is $\mathbf{m}$-essentially (resp. $\mathcal{S}$ is $\mathbf{n}$-essentially) countably generated, then $\mathcal{P}(\mathbf{m})$ (resp. $\mathcal{S}(\mathbf{n})$) is separable (see Notation 4.6.11) and hence, for $1 \le p < \infty$, $\mathcal{L}_p(\mathbf{m})$, $L_p(\mathbf{m})$, $\mathcal{L}_p(\sigma(\mathcal{P}), \mathbf{m})$ and $L_p(\sigma(\mathcal{P}), \mathbf{m})$ are separable; (resp. $\mathcal{L}_p(\mathbf{n})$, $L_p(\mathbf{n})$, $\mathcal{L}_p(\sigma(\mathcal{S}), \mathbf{n})$ $(= \mathcal{L}_p(\mathcal{S}, \mathbf{n}))$ and $L_p(\sigma(\mathcal{S}), \mathbf{n})$ $(= L_p(\mathcal{S}, \mathbf{n}))$ are separable).*

(ii) *If $\mathcal{S}(\mathbf{n})$ is separable and if the range $\mathbf{n}(\mathcal{S})$ is metrizable for the relative topology from X, then $\mathcal{S}$ is $\mathbf{n}$-essentially countably generated. (Consequently, by (i), $\mathcal{L}_p(\mathbf{n})$, $L_p(\mathbf{n})$ $\mathcal{L}_p(\sigma(\mathcal{S}), \mathbf{n})$ $(= \mathcal{L}_p(\mathcal{S}, \mathbf{n}))$ and $L_p(\sigma(\mathcal{S}), \mathbf{n})$ $(= L_p(\mathcal{S}, \mathbf{n}))$ are separable for all $p \in [1, \infty)$.)*

Proof. (i) We shall prove for the case of $\mathcal{P}(\mathbf{m})$. The proof for $\mathcal{S}(\mathbf{n})$ is similar. Let $\mathcal{P}_0$ be a countably generated δ-ring such that $\mathcal{P}_0(\mathbf{m}) = \mathcal{P}(\mathbf{m})$. Then by Theorem 4.6.3, for all $p \in [1, \infty)$, $\mathcal{P}_0$ is separable for the topology $\tau_{\mathbf{m}}^{(p)}$. Let $1 \le p < \infty$, p fixed. Then in the light of Theorem 5C of [H], there exists a $\tau_{\mathbf{m}}^{(p)}$-dense countable subring $\mathcal{R}_0$ of $\mathcal{P}_0$. Let $\mathcal{I}_{\mathcal{R}_0}$ be the set of all $\mathcal{R}_0$-simple functions and let $W = \{\sum_1^r \alpha_j \chi_{F_j} : \alpha_j = a_j + ib_j, a_j, b_j \text{ real rational}, F_j \in \mathcal{R}_0, j = 1, 2, \ldots, r, r \in \mathbb{N}\}$. Then W is countable and by an argument similar to that in the proof of Theorem 4.6.2(ii), W is $\tau_{\mathbf{m}}^{(p)}$-dense in $\mathcal{I}_{\mathcal{R}_0}$. Let $f \in \mathcal{L}_p(\mathbf{m})$ (resp. $f \in \mathcal{L}_p(\sigma(\mathcal{P}), \mathbf{m})$) and let V be a neighborhood of f in $\mathcal{L}_p(\mathbf{m})$ (resp. in $\mathcal{L}_p(\sigma(\mathcal{P}), \mathbf{m})$). Then there exist an $\epsilon > 0$ and $q_1, q_2, \ldots, q_n$ in Γ such that $V \supset \{g \in \mathcal{L}_p(\mathbf{m}) \text{ (resp. } g \in \mathcal{L}_p(\sigma(\mathcal{P}), \mathbf{m})) : (\mathbf{m}_{q_i})_p^{\bullet}(f - g, T) < \epsilon, i = 1, 2, \ldots, n\}$. By Theorem 4.5.6, there exists $s \in \mathcal{I}_s$ such that $(\mathbf{m}_{q_i})_p^{\bullet}(f - s, T) < \frac{\epsilon}{2}$ for $i = 1, 2, \ldots, n$ and as $\mathcal{P}_0(\mathbf{m}) = \mathcal{P}(\mathbf{m})$, there exists $s' \in \mathcal{I}_{\mathcal{R}_0}$ such that $s = s'$ $\mathbf{m}$-a.e. in T. Moreover, as W is dense in $\mathcal{I}_{\mathcal{R}_0}$, there exists $\omega \in W$ such that $(\mathbf{m}_{q_i})_p^{\bullet}(s' - \omega, T) < \frac{\epsilon}{2}$ for $i = 1, 2, \ldots, n$. Then it follows that $\omega \in V$ and hence $\mathcal{L}_p(\mathbf{m})$ (resp. $\mathcal{L}_p(\sigma(\mathcal{P}), \mathbf{m})$) is separable. Then by Problem H, Chapter 4 of [Ke], $L_p(\mathbf{m})$ (resp. $L_p(\sigma(\mathcal{P}), \mathbf{m})$) is separable. Hence (i) holds.

(ii) Let $\Omega = \mathbf{n}(\mathcal{S})$. By hypothesis, Ω and hence $\Omega - \Omega$ is metrizable for the relative topology from X. Hence there exists a sequence $(V_n)_1^{\infty}$ of τ-closed absolutely convex τ-neighborhoods of 0 in X such that

(i) $V_{n+1} + V_{n+1} \subset V_n$ for $n \ge 1$; and

(ii) $V_n \bigcap (\Omega - \Omega), n \in \mathbb{N}$, form a neighborhood base of 0 in $(\Omega - \Omega, \tau | (\Omega - \Omega))$.

Let τ_1 be the locally convex topology on X for which $(V_n)_1^{\infty}$ is a neighborhood base of 0 in X. To show that $\tau_1|_{\Omega - \Omega} = \tau|_{\Omega - \Omega}$, we argue as in the proof of Theorem 2.1

of [S]. Let $x \in \Omega$ and let V be a τ-neighborhood of 0 in X. Then there exists n such that $V_n \cap (\Omega - \Omega) \subset V \cap (\Omega - \Omega)$ as $V \cap (\Omega - \Omega)$ is a neighborhood of 0 in $\tau|_{\Omega - \Omega}$. Let $y \in (x + V_n) \cap \Omega$ so that there exist $v_n \in V_n$ and $\omega \in \Omega$ such that $y = x + v_n = \omega$. Then $y - x = v_n = \omega - x$ and hence $y - x \in V_n \cap (\Omega - \Omega) \subset V \cap (\Omega - \Omega)$. Therefore, $(x + V_n) \cap \Omega \subset (x + V) \cap \Omega$. Thus $\tau|_\Omega \leq \tau_1|_\Omega$. Clearly, $\tau_1|_\Omega \leq \tau|_\Omega$ and hence $\tau_1|_\Omega = \tau|_\Omega$. If q_n is the Minkowski functional of V_n, then $(q_n|_\Omega)_1^\infty$ generate the topology $\tau_1|_\Omega$ and hence $\tau_{\mathbf{n}}^{(p)}|_{\mathbf{n}(\mathcal{S})}$ is generated by $(\rho(\mathbf{n})_{q_k}^{(p)}|_{\mathbf{n}(\mathcal{S})})_{k=1}^\infty$ for $p \in [1, \infty)$.

By hypothesis, $\mathcal{S}(\mathbf{n}) = \mathbf{n}(\mathcal{S})/ \sim$ is separable in the topology induced by $\tau_{\mathbf{n}}^{(p)}|_{\mathcal{S}(\mathbf{n})} = \tau_0$ (say) (by an abuse of notation (see Remark 4.6.10)). Hence there exists a countable τ_0-dense family $\mathcal{F} \subset \mathcal{S}(\mathbf{n})$. Let $\mathcal{S}_0 = \sigma(\mathcal{F})$. Then, obviously, $\mathcal{S}_0(\mathbf{n}) \subset \mathcal{S}(\mathbf{n})$. Now let $x \in \mathcal{S}(\mathbf{n})$. Then there exists $A \in \mathcal{S}$ such that $\mathbf{n}(A) = x$. As τ_0 is pseudo-metrizable and as $\mathcal{F}$ is τ_0-dense in $\mathcal{S}(\mathbf{n})$, there exists a sequence $(F_n)_1^\infty \subset \mathcal{F}$ such that $\lim_n (\mathbf{n}_{q_k})_p^\bullet(\chi_A - \chi_{F_n}, T) = 0$ for each $k \in \mathbb{N}$. Then by Theorems 3.1.18(vi) and 3.1.19 of Chapter 3 for $\mathbf{n}_{q_k}$, there exists a subsequence $(F_{n,1})_{n=1}^\infty$ of $(F_n)_1^\infty$ such that $\chi_{F_{n,1}} \to \chi_A \; \mathbf{n}_{q_1}$-a.e. in T. As $(\mathbf{n}_{q_2})_p^\bullet(\chi_A - \chi_{F_{n,1}}, T) \to 0$, by the said theorems there exists a subsequence $(F_{n,2})_{n=1}^\infty$ of $(F_{n,1})_{n=1}^\infty$ such that $\chi_{F_{n,2}} \to \chi_A \; \mathbf{n}_{q_2}$-a.e. in T. Continuing this process indefinitely, we note that the diagonal sequence $(\chi_{F_{n,n}})_{n=1}^\infty$ is some subsequence $(\chi_{F_{n_k}})_{k=1}^\infty$ of $(\chi_{F_n})_1^\infty$ and $\chi_{F_{n_k}} \to \chi_A \; \mathbf{n}$-a.e. in T. Let $N \in \mathcal{S}$ with $\|\mathbf{m}\|_q(N) = 0$ for all $q \in \Gamma$ such that $\chi_{F_{n_k}}(t) \to \chi_A(t)$ for $t \in T \backslash N$. Then, for each $t \in A \backslash N$, $\chi_{F_{n_k}}(t) \to 1$ and hence $\liminf_k F_{n_k} \supset A \backslash N$. Similarly, for each $t \in T \backslash (A \backslash N)$, $\chi_{F_{n_k}}(t) \to 0$ so that $\liminf_k (T \backslash F_{n_k}) \supset T \backslash (A \backslash N)$ or equivalently, $\limsup_k F_{n_k} \subset A \backslash N$. Hence $\chi_{A \backslash N} = \lim_k F_{n_k}$ and hence $A \backslash N \in \mathcal{S}_0$. As N is $\mathbf{n}$-null, it follows that $x = \mathbf{n}(A)$ and $A \sim (A \backslash N) \in \mathcal{S}_0$. Therefore, $x \in \mathcal{S}_0(\mathbf{n})$ and hence $\mathcal{S}(\mathbf{n}) = \mathcal{S}_0(\mathbf{n})$. Thus $\mathcal{S}$ is $\mathbf{n}$-essentially countably generated. $\qquad \square$

Remark 4.6.15. The above theorem for $\mathbf{n}$ and for $p = 1$ subsumes Propositions 2 and 3(ii) of [Ri3] given for a σ-additive vector measure $\mathbf{n}$ defined on a σ-algebra Σ of subsets of T with values in a sequentially complete lcHs. There the proof is based on the facts that $\mathbf{n}$ is a closed vector measure as shown in [Ri1] and that the σ-algebra generated by an algebra of sets is the same as its sequential closure. But our proof is different and dispenses with the concept of closed vector measures.

Chapter 5

Applications to Integration in Locally Compact Hausdorff Spaces – Part I

5.1 Generalizations of the Vitali-Carathéodory Integrability Criterion Theorem

The results of the present section are needed in Section 6.5 of Chapter 6 to describe the duals of $L_1(\mathbf{m})$ and $\mathcal{L}_1(\mathbf{n})$, where $\mathbf{m} : \mathcal{B}(T) \to X$ is σ-additive and $\mathcal{B}(T)$-regular and $\mathbf{n} : \delta(\mathcal{C}) \to X$ is σ-additive and $\delta(\mathcal{C})$-regular and X is a Banach space.

In the sequel, T denotes a locally compact Hausdorff space and $\mathcal{U}$, $\mathcal{C}$, and $\mathcal{C}_0$ are as in Definition 4.6.4 of Chapter 4; i.e., $\mathcal{U}$ is the family of open sets in T, $\mathcal{C}$ that of compact sets in T and $\mathcal{C}_0$ that of compact G_δ sets in T. Then $\mathcal{B}(T) = \sigma(\mathcal{U})$, the σ-algebra of the Borel sets in T; $\mathcal{B}_c(T) = \sigma(\mathcal{C})$, the σ-ring of the σ-Borel sets in T and $\mathcal{B}_0(T) = \sigma(\mathcal{C}_0)$, the σ-ring of the Baire sets in T. $\delta(\mathcal{C})$ and $\delta(\mathcal{C}_0)$ are the δ-rings generated by $\mathcal{C}$ and $\mathcal{C}_0$, respectively.

As in Chapters 1–4, X denotes a Banach space or an lcHs over $\mathbb{K}$ (= $\mathbb{R}$ or $\mathbb{C}$) with Γ, the family of continuous seminorms on X, unless otherwise stated and it will be explicitly specified whether X is a Banach space or an lcHs. Let $\mathcal{R} = \mathcal{B}(T)$ or $\delta(\mathcal{C})$; a σ-additive set function $\mathbf{m} : \mathcal{R} \to X$ is said to be $\mathcal{R}$-regular if it satisfies the conditions in Definition 4.6.7 of Chapter 4. Recall the definition of $\mathcal{L}_p(\sigma(\mathcal{P}), \mathbf{m})$ given in Theorem 3.2.8 of Chapter 3.

Lemma 5.1.1. *Let X be a quasicomplete (resp. sequentially complete) lcHs, $\mathbf{m} : \mathcal{B}(T) \to X$ be σ-additive and Borel regular and $f : T \to [0, \infty)$ (resp. and be $\mathcal{B}(T)$-measurable). Then $f \in \mathcal{L}_1(\mathbf{m})$ (resp. $f \in \mathcal{L}_1(\mathcal{B}(T), \mathbf{m})$) if and only if, given*

$\epsilon > 0$ and $q \in \Gamma$, there exist functions $u^{(q)}$ and $v^{(q)}$ on T such that $u^{(q)} \leq f \leq v^{(q)}$ $\mathbf{m}_q$-a.e. in T, $u^{(q)}$ is upper semicontinuous in T with compact support and with $u^{(q)}(T) \subset [0,\infty)$, $v^{(q)}$ is lower semicontinuous and $\mathbf{m}_q$-integrable in T, and $(\mathbf{m}_q)_1^\bullet(v^{(q)} - u^{(q)}, T) < \epsilon$.

Proof. In the light of Theorem 4.5.13(i) of Chapter 4 which states that $\mathcal{L}_1(\mathbf{m}) = \bigcap_{q \in \Gamma} \mathcal{L}_1(\mathbf{m}_q)$ (resp. $\mathcal{L}_1(\mathcal{B}(T), \mathbf{m}) = \bigcap_{q \in \Gamma} \mathcal{L}_1(\mathcal{B}(T), \mathbf{m}_q)$), it suffices to prove the result for Banach spaces. So we shall assume X to be a Banach space. Suppose the conditions are satisfied for each $\epsilon > 0$. If f is not $\mathcal{B}(T)$-measurable, we first show that it is $\mathbf{m}$-measurable. By hypothesis, for each n, there exist such functions u_n and v_n with $0 \leq u_n \leq f \leq v_n$ $\mathbf{m}$-a.e. in T and with $\mathbf{m}_1^\bullet(v_n - u_n, T) < \frac{1}{n}$. Since v_n is $\mathbf{m}$-integrable in T and since u_n is Borel measurable, by the domination principle (see Theorem 2.1.5(vii) and Remark 2.3.3 of Chapter 2) u_n is also $\mathbf{m}$-integrable in T. Let $g_n = \max_{1 \leq i \leq n} u_i$ and $h_n = \min_{1 \leq i \leq n} v_i$. Then by Theorems 3.3 and 3.5, §3, Chapter III of [MB], g_n is upper semicontinuous and h_n is lower semicontinuous for each n. Moreover, $g_n \nearrow$ and $h_n \searrow$ and by hypothesis and by Theorem 2.1.5(vii) and Remark 2.2.3 of Chapter 2, g_n and h_n are $\mathbf{m}$-integrable in T for each n. Let $g = \sup_n g_n$ and $h = \inf_n h_n$. Then $0 \leq g \leq f \leq h$ $\mathbf{m}$-a.e. in T and g and h are $\mathcal{B}(T)$-measurable. Moreover, $0 \leq h_n - g_n \leq v_1 \in \mathcal{L}_1(\mathbf{m})$ and $h_n - g_n \to h - g$ pointwise in T. Hence by LDCT (see Theorem 2.1.7 and Remark 2.2.3 of Chapter 2), $h - g \in \mathcal{L}_1(\mathbf{m})$ and $\mathbf{m}_1^\bullet((h-g)-(h_n - g_n), T) \to 0$. Moreover, as $0 \leq h_n - g_n \leq v_n - u_n$, by Theorem 3.1.11(i) of Chapter 3, $\mathbf{m}_1^\bullet(h_n - g_n, T) \leq \mathbf{m}_1^\bullet(v_n - u_n, T) < \frac{1}{n}$ for $n \in \mathbb{N}$. Consequently, by Theorem 3.1.13(ii) of Chapter 3, $\mathbf{m}_1^\bullet(h-g, T) \leq \mathbf{m}_1^\bullet((h-g)-(h_n-g_n), T) + \mathbf{m}_1^\bullet(h_n - g_n, T) \to 0$ as $n \to \infty$. Hence by Theorem 3.1.18(ii) of Chapter 3, $h = g$ $\mathbf{m}$-a.e. in T. Then $f = g = h$ $\mathbf{m}$-a.e. in T and hence f is $\mathbf{m}$-measurable. Moreover, as $\mathbf{m}_1^\bullet(v_n - f, T) \leq \mathbf{m}_1^\bullet(v_n - u_n, T) < \frac{1}{n}$ and as $\mathcal{L}_1(\mathbf{m})$ is closed in $\mathcal{L}_1\mathcal{M}(\mathbf{m})$ by Theorem 3.2.8 of Chapter 3, we conclude that f is $\mathbf{m}$-integrable in T. If f is $\mathcal{B}(T)$-measurable, then by the above argument $\lim_n \mathbf{m}_1^\bullet(v_n - f, T) = 0$ and hence by the second part of Theorem 3.2.8 of Chapter 3, $f \in \mathcal{L}_1(\mathcal{B}(T), \mathbf{m})$.

To prove the converse, let us assume that f is not identically zero, $f \geq 0$ and $f \in \mathcal{L}_1(\mathbf{m})$ (resp. $f \in \mathcal{L}_1(\mathcal{B}(T), \mathbf{m})$). By Definition 2.1.1 of Chapter 2, there exists a $\mathcal{B}(T)$-measurable function $\hat{f} : T \to [0,\infty)$ such that $f = \hat{f}$ $\mathbf{m}$-a.e. in T or $\hat{f} = f$ if f is $\mathcal{B}(T)$-measurable. Arguing as in the first paragraph on p. 51 of [Ru1], we have

$$\hat{f}(t) = \sum_{i=1}^\infty c_i \chi_{E_i}(t), \quad t \in T$$

where $c_i > 0$ and $E_i \in \mathcal{B}(T)$ for all i. Let

$$f_n = \sum_{i=1}^n c_i \chi_{E_i}, \quad n \in \mathbb{N}.$$

Then $0 \leq f_n \nearrow \hat{f}$ and hence by LDCT (see Theorem 2.1.7 and Remark 2.2.3 of Chapter 2) $\lim_n \mathbf{m}_1^\bullet(\hat{f} - f_n, T) = 0$. Thus, given $\epsilon > 0$, there exists n_0 such that

$\mathbf{m}_1^\bullet(\hat{f} - f_{n_0}, T) < \frac{\epsilon}{2}$. That is,

$$\mathbf{m}_1^\bullet\left(\sum_{n_0+1}^\infty c_i \chi_{E_i}, T\right) < \frac{\epsilon}{2}. \tag{5.1.1.1}$$

By the Borel regularity of $\mathbf{m}$ there exist compact sets $(K_i)_1^\infty$ and open sets $(V_i)_1^\infty$ such that $K_i \subset E_i \subset V_i$ and such that

$$c_i||\mathbf{m}||(V_i \backslash K_i) < \frac{\epsilon}{2^{i+1}} \tag{5.1.1.2}$$

for $i \in \mathbb{N}$. Let $v = \sum_{i=1}^\infty c_i \chi_{V_i}$ and $u = \sum_{i=1}^{n_0} c_i \chi_{K_i}$. Then by Theorems 3.3, 3.4 and 3.5 and Example 2 of §3, Chapter III of [MB], v is lower semicontinuous and u is upper semicontinuous in T, $u(T) \subset [0, \infty)$, supp u is compact and $u \leq \hat{f} \leq v$ in T. If $v_n = \sum_{i=1}^n c_i \chi_{V_i}$, then $(v_n)_1^\infty \subset \mathcal{L}_1(\mathbf{m})$ as v_n are $\mathcal{B}(T)$-simple functions. For $A \in \mathcal{B}(T)$, by Theorem 3.1.3 of Chapter 3 we have

$$\mathbf{m}_1^\bullet(\chi_A, T) = \sup_{|x^*| \leq 1} \int_T \chi_A dv(x^*\mathbf{m}) = \sup_{|x^*| \leq 1} v(x^*\mathbf{m})(A) = ||\mathbf{m}||(A). \tag{5.1.1.3}$$

Now, by (5.1.1.1), (5.1.1.2) and (5.1.1.3) and by Theorem 3.1.13(ii) of Chapter 3 we have $\mathbf{m}_1^\bullet(v - v_n, T) \leq \mathbf{m}_1^\bullet(\sum_{n+1}^\infty c_i \chi_{E_i}, T) + \mathbf{m}_1^\bullet(\sum_{n+1}^\infty c_i \chi_{V_i \backslash E_i}, T) < \frac{\epsilon}{2} + \sum_{n+1}^\infty c_i||\mathbf{m}||(V_i \backslash E_i) < \epsilon$ for $n \geq n_0$. Hence $\lim_n \mathbf{m}_1^\bullet(v - v_n, T) = 0$ and consequently, by Theorem 3.2.8 of Chapter 3, v is $\mathbf{m}$-integrable in T.

Finally, by Theorems 3.1.11 and 3.1.13(ii) of Chapter 3 and by (5.1.1.3) we have

$$\mathbf{m}_1^\bullet(v - u, T) \leq \mathbf{m}_1^\bullet\left(\sum_1^{n_0} c_i \chi_{V_i \backslash K_i}, T\right) + \mathbf{m}_1^\bullet\left(\sum_{n_0+1}^\infty c_i \chi_{V_i}, T\right)$$

$$\leq \sum_1^{n_0} c_i||\mathbf{m}||(V_i \backslash K_i) + \mathbf{m}_1^\bullet\left(\sum_{n_0+1}^\infty c_i \chi_{E_i}, T\right) + \mathbf{m}_1^\bullet\left(\sum_{n_0+1}^\infty c_i \chi_{V_i \backslash K_i}, T\right)$$

$$\leq \sum_1^\infty c_i||\mathbf{m}||(V_i \backslash K_i) + \mathbf{m}_1^\bullet\left(\sum_{n_0+1}^\infty c_i \chi_{E_i}, T\right) < \epsilon.$$

Hence the lemma holds. $\qquad\square$

Theorem 5.1.2. (Generalization of the Vitali-Carathéodory integrability criterion theorem for Borel regular m). *Let X be a quasicomplete (resp. sequentially complete) lcHs, $\mathbf{m} : \mathcal{B}(T) \to X$ be σ-additive and Borel regular and $f : T \to \mathbb{R}$ (resp. and be $\mathcal{B}(T)$-measurable). Then $f \in \mathcal{L}_1(\mathbf{m})$ (resp. $f \in \mathcal{L}_1(\mathcal{B}(T), \mathbf{m})$) if and only if, given $\epsilon > 0$ and $q \in \Gamma$, there exist functions $u^{(q)}$ and $v^{(q)}$ on T such that $u^{(q)} \leq f \leq v^{(q)}$ $\mathbf{m}_q$-a.e. in T, $u^{(q)}$ is upper semicontinuous, bounded above and $\mathbf{m}_q$-integrable in T, $v^{(q)}$ is lower semicontinuous, bounded below and $\mathbf{m}_q$-integrable in T and $(\mathbf{m}_q)_1^\bullet(v^{(q)} - u^{(q)}, T) < \epsilon$.*

Proof. In the light of Theorem 4.5.13(i) of Chapter 4, it suffices to prove the result for a Banach space X and hence let X be a Banach space. Suppose the conditions hold for each $\epsilon > 0$. If f is not $\mathcal{B}(T)$-measurable, we first show that f is $\mathbf{m}$-measurable. For each n, there exist such functions u_n and v_n with $u_n \leq f \leq v_n$ $\mathbf{m}$-a.e. in T and $\mathbf{m}_1^{\bullet}(v_n - u_n, T) < \frac{1}{n}$. Let $g_n = \max_{1 \leq i \leq n} u_i$ and $h_n = \min_{1 \leq i \leq n} v_i$, $g = \sup_n g_n$ and $h = \inf_n h_n$. Then by Theorems 3.3 and 3.5, §3, Chapter III of [MB], $(g_n)_1^{\infty}$ are upper semicontinuous and $(h_n)_1^{\infty}$ are lower semicontinuous. Moreover, $g_n \nearrow g$ and $h_n \searrow h$. Hence g and h are $\mathcal{B}(T)$-measurable and $g \leq f \leq h$ $\mathbf{m}$-a.e. in T. Now, $0 \leq h_n - g_n \leq v_n - u_n$ and by hypothesis, $h_n - g_n$ and $v_n - u_n$ are well defined on T. By hypothesis, v_n and u_n are $\mathbf{m}$-integrable in T and hence by Theorem 3.1.12(ii) of Chapter 3, v_n and u_n are finite $\mathbf{m}$-a.e. in T. Hence $v_n - u_n$ is $\mathbf{m}$-integrable in T. Consequently, by Theorem 2.2.5(vii) and Remark 2.2.3 of Chapter 2, $h_n - g_n$ is $\mathbf{m}$-integrable in T for $n \in \mathbb{N}$ and as $0 \leq h_n - g_n \leq v_1 - u_1$, $h_n - g_n \to h - g$ in T and $v_1 - u_1$ is $\mathbf{m}$-integrable in T, by LDCT (see Theorem 2.1.7 and Remark 2.2.3 of Chapter 2), $\lim_n \mathbf{m}_1^{\bullet}((h_n - g_n) - (h - g), T) = 0$. Moreover, $\mathbf{m}_1^{\bullet}(h_n - g_n, T) \leq \mathbf{m}_1^{\bullet}(v_n - u_n, T) < \frac{1}{n} \to 0$ as $n \to \infty$. Hence $\mathbf{m}_1^{\bullet}(h - g, T) = 0$ so that by Theorem 3.1.18(ii) of Chapter 3, $h = g$ $\mathbf{m}$-a.e. in T. Consequently, $f = h = g$ $\mathbf{m}$-a.e. in T and hence f is $\mathbf{m}$-measurable.

As $\mathbf{m}_1^{\bullet}(v_n - f, T) \leq \mathbf{m}_1^{\bullet}(v_n - u_n, T) < \frac{1}{n} \to 0$ as $n \to \infty$, by Theorem 3.2.8 of Chapter 3 the function $f \in \mathcal{L}_1(\mathbf{m})$ (resp. $f \in \mathcal{L}_1(\mathcal{B}(T), \mathbf{m})$).

Conversely, let $f \in \mathcal{L}_1(\mathbf{m})$ (resp. $f \in \mathcal{L}_1(\mathcal{B}(T), \mathbf{m})$). By Theorem 2.1.5(vii) and Remark 2.2.3 of Chapter 2, $f^+, f^- \in \mathcal{L}_1(\mathbf{m})$ (resp. $f^+, f^- \in \mathcal{L}_1(\mathcal{B}(T), \mathbf{m})$). Using Lemma 5.1.1 above and Theorem 2.1.5(vii) and Remark 2.2.3 of Chapter 2 and arguing as in the general case in the proof of Theorem 2.24 of [Ru1], one can prove the converse. Details are left to the reader. $\qquad\square$

Theorem 5.1.3. (Generalization of the Vitali-Carathéodory integrability criterion theorem for $\delta(\mathcal{C})$-regular n). *Let X be a quasicomplete (resp. sequentially complete) lcHs, $\mathbf{n} : \delta(\mathcal{C}) \to X$ be σ-additive and $\delta(\mathcal{C})$-regular and $f : T \to \mathbb{R}$ (resp. and be $\mathcal{B}_c(T)$-measurable). Then $f \in \mathcal{L}_1(\mathbf{n})$ (resp. $f \in \mathcal{L}_1(\mathcal{B}_c(T), \mathbf{n})$) if and only if, given $q \in \Gamma$ and $\epsilon > 0$, there exist functions $u^{(q)}$ and $v^{(q)}$ on T such that $u^{(q)} \leq f \leq v^{(q)}$ $\mathbf{n}_q$-a.e. in T, $v^{(q)}$ is lower semicontinuous, bounded below and $\mathbf{n}_q$-integrable in T, $u^{(q)}$ is upper semicontinuous, bounded above and $\mathbf{n}_q$-integrable in T and $(\mathbf{n}_q)_1^{\bullet}(v^{(q)} - u^{(q)}, T) < \epsilon$.*

Proof. First we observe that a Borel measurable function h with $N(h) = \{t \in T : h(t) \neq 0\}$ σ-bounded (i.e., contained in a countable union of compact sets) is necessarily $\mathcal{B}_c(T)$-measurable and hence $\mathbf{n}_q$-integrable upper semicontinuous and lower semicontinuous functions are $\mathcal{B}_c(T)$-measurable. Using this observation and an argument quite similar to the proof of the sufficiency part of Theorem 5.1.2 in which $\mathcal{B}(T)$ is replaced by $\mathcal{B}_c(T)$ and $\mathbf{m}$ by $\mathbf{n}$, one can show that the conditions are sufficient.

As seen in the proof of Lemma 5.1.1, we prove the result assuming X to be a Banach space and $f \in \mathcal{L}_1(\mathbf{n})$ (resp. $f \in \mathcal{L}_1(\mathcal{B}_c(T), \mathbf{n})$), $f \geq 0$ and f not identically zero. Then there exists an **n**-null set $N \in \mathcal{B}_c(T)$ such that $\hat{f} = f\chi_{T \setminus N}$ is $\mathcal{B}_c(T)$-measurable and $N = \emptyset$ when f is $\mathcal{B}_c(T)$-measurable. Let $s_0 = 0$ and

$$s_n(t) = \begin{cases} \frac{i-1}{2^n} & \text{if } \frac{i-1}{2^n} \leq \hat{f}(t) < \frac{i}{2^n},\ i = 1, 2, \ldots, 2^n n, \\ n & \text{if } \hat{f}(t) \geq n. \end{cases}$$

Then $s_n \nearrow \hat{f}$ in T and $\hat{f}(t) = \sum_{n=1}^{\infty}(s_n - s_{n-1})(t)$. As $N(\hat{f}) \in \mathcal{B}_c(T)$, there exists $(A_n)_1^{\infty} \subset \delta(\mathcal{C})$ with $A_n \nearrow N(\hat{f})$. Then $\sum_{n=1}^{k}(s_n(t)\chi_{A_n}(t) - s_{n-1}(t)\chi_{A_{n-1}}(t)) = s_k(t)\chi_{A_k}(t)$ and hence $\hat{f}(t) = \lim_k s_k(t) = \lim_k s_k(t)\chi_{A_k}(t)$. Moreover, it is easy to check that $\sum_1^{\infty}(s_n\chi_{A_n} - s_{n-1}\chi_{A_{n-1}})$ is of the form $\sum_1^{\infty} c_k\chi_{E_k}$ with $c_k > 0$ for each k and $(E_k)_1^{\infty} \subset \delta(\mathcal{C})$ since $A \cap B \in \delta(\mathcal{C})$ for $A \in \delta(\mathcal{C})$ and $B \in \mathcal{B}_c(T)$. Thus

$$\hat{f}(t) = \sum_{k=1}^{\infty} c_k\chi_{E_k}, \ c_k > 0 \text{ for each } k \text{ and } (E_k)_1^{\infty} \subset \delta(\mathcal{C}).$$

As **n** is $\delta(\mathcal{C})$-regular, given $\epsilon > 0$, there exist an open set $V_n \in \delta(\mathcal{C})$ and a compact K_n such that $K_n \subset E_n \subset V_n$ and such that $||\mathbf{n}||(V_n \setminus K_n) < \frac{\epsilon}{2^{n+1}}$ for $n \in \mathbb{N}$. Then arguing as in the proof of the converse part of Lemma 5.1.1, one can show the existence of u and v as in the said lemma with u and v $\mathcal{B}_c(T)$-measurable. Then arguing as in the proof of the necessity part of Theorem 5.1.2 the theorem is proved for real-valued f. $\square$

Remark 5.1.4. Arguing as in the proof of Lemma 5.1.1 and Theorem 5.1.2, a result generalizing the corollary of Theorem 3, no. 4, §4, Chapter IV of [B] can be obtained for $f \in \mathcal{L}_1(\mathbf{m})$ where **m** is as in Theorem 5.1.2. Similarly, an analogous result is true for $f \in \mathcal{L}_1(\mathbf{n})$ where $\mathbf{n} : \delta(\mathcal{C}) \to X$ is σ-additive and $\delta(\mathcal{C})$-regular.

5.2 The Baire version of the Dieudonné-Grothendieck theorem and its vector-valued generalizations

We show that the boundedness hypothesis in Corollary 1 of [P8] is redundant and thereby we obtain the Baire version of the Dieudonné-Grothendieck theorem in Theorem 5.2.6 below. Also see Remark 5.2.7 below. Then using the ideas in the proofs of Proposition 2.11 and Theorem 2.12 of [T], we generalize Theorem 5.2.6 to σ-additive Borel regular vector measures. (See Theorems 5.2.21 and 5.2.23.)

Notation 5.2.1. $C_c(T) = \{f : T \to \mathbb{K}, f \text{ continuous with compact support}\}$; $C_0(T) = \{f : T \to \mathbb{K}, f \text{ continuous and vanishes at infinity in } T\}$, both the spaces being provided with norm $|| \cdot ||_T$. $M(T)$ denotes the dual of $(C_0(T), || \cdot ||_T)$ and each member of $M(T)$ is considered as a σ-additive Borel regular scalar measure on $\mathcal{B}(T)$. We write $|\mu|(\cdot) = v(\mu, \mathcal{B}(T))(\cdot)$ for $\mu \in M(T)$. Then $||\mu|| = |\mu|(T)$, for $\mu \in M(T)$. $\mathcal{V}$ is the family of relatively compact open sets in T. $C_c(T)$ endowed with the inductive limit locally convex topology as in §1, Chapter III of [B] is denoted by $\mathcal{K}(T)$.

Lemma 5.2.2. $\delta(\mathcal{C}) = \{A \in \mathcal{B}(T) : \bar{A} \in \mathcal{C}\}$ and $\delta(\mathcal{C}_0) = \{A \in \mathcal{B}_0(T) : \bar{A} \in \mathcal{C}\}$ where $\bar{A}$ denotes the closure of A in T.

Proof. Let $\mathcal{F}$ be the family of closed sets in T. Since $\delta(\mathcal{C}) \subset \mathcal{B}_c(T)$ and $\delta(\mathcal{C}_0) \subset \mathcal{B}_0(T)$, and since each member of $\delta(\mathcal{C}) \cup \delta(\mathcal{C}_0)$ is relatively compact, it suffices to show that $\{A \in \mathcal{B}(T) : \bar{A} \in \mathcal{C}\} \subset \delta(\mathcal{C})$ and $\{A \in \mathcal{B}_0(T) : \bar{A} \in \mathcal{C}\} \subset \delta(\mathcal{C}_0)$. Let $A \in \mathcal{B}(T)$ (resp. $A \in \mathcal{B}_0(T)$) with $\bar{A} \in \mathcal{C}$. Then by Theorem 50.D of [H] there exists $C_0 \in \mathcal{C}_0$ such that $\bar{A} \subset C_0$ and hence by Theorem 5.E of [H] we have $A = A \cap C_0 \in \sigma(\mathcal{F}) \cap C_0 = \sigma(\mathcal{F} \cap C_0) = \sigma(\mathcal{C} \cap C_0) = \delta(\mathcal{C} \cap C_0) \subset \delta(\mathcal{C})$ (resp. $A = A \cap C_0 \in \sigma(\mathcal{C}_0) \cap C_0 = \sigma(\mathcal{C}_0 \cap C_0) = \delta(\mathcal{C}_0 \cap C_0) \subset \delta(\mathcal{C}_0)$). Hence the lemma holds. $\square$

Lemma 5.2.3. *A σ-compact open set in T is a Baire set. Conversely, every open Baire set in T is σ-compact.*

Proof. Let U be open in T and let $U = \bigcup_1^\infty K_n$, $(K_n)_1^\infty \subset \mathcal{C}$. Then by Theorem 50.D of [H], for each K_n there exists $C_n \in \mathcal{C}_0$ such that $K_n \subset C_n \subset U$ and hence $U = \bigcup_1^\infty C_n \in \mathcal{B}_0(T)$. Conversely, if $U \in \mathcal{B}_0(T)$ is open in T, then U is σ-bounded so that there exists $(K_n)_1^\infty \subset \mathcal{C}$ such that $U \subset \bigcup_1^\infty K_n$. Then by Theorem 50.D of [H] and by the previous part there exist relatively compact open Baire sets $(V_n)_1^\infty$ such that $K_n \subset V_n$ for each n. Then $U = \bigcup_1^\infty (U \cap V_n)$ and by Lemma 5.2.2, each $U \cap V_n \in \delta(\mathcal{C}_0)$. Then U is σ-compact by Proposition 15, §14 of [Din1]. $\square$

Lemma 5.2.4. *Let $(\mu_n)_1^\infty \subset M(T)$ (resp. $\mathbf{m}_n : \mathcal{B}(T) \to X$, $n \in \mathbb{N}$, be σ-additive and Borel regular, where X is an lcHs). Then:*

(i) *For each open set U in T, there exists an open Baire set V_U in T such that $V_U \subset U$ and $|\mu_n|(U \backslash V_U) = 0$ for all n and consequently, $\mu_n(U) = \mu_n(V_U)$ for all n (resp. given $q \in \Gamma$, there exists an open Baire set $V_U^{(q)}$ in T such that $V_U^{(q)} \subset U$ and $||\mathbf{m}_n||_q(U \backslash V_U^{(q)}) = 0$ for all n and hence $|\mathbf{m}_n(U) - \mathbf{m}_n(V_U^{(q)})|_q = 0$ for all n).*

(ii) *If, for each open Baire set V in T, $\sup_n |\mu_n(V)| < \infty$, then*

$$\sup_n |\mu_n(U)| < \infty$$

for each open set U in T and consequently, $\sup_n ||\mu_n|| < \infty$.

Proof. Claim 1. Given an open set U in T, (resp. and $q \in \Gamma$), for each $n \in \mathbb{N}$, there exists an open Baire set V_n in T such that $V_n \subset U$ and $|\mu_n|(U \backslash V_n) = 0$ so that $\mu_n(U) = \mu_n(V_n)$ (resp. there exists an open Baire set $V_n^{(q)}$ in T such that $V_n^{(q)} \subset U$ and $||\mathbf{m}_n||_q(U \backslash V_n^{(q)}) = 0$ so that $|\mathbf{m}_n(U) - \mathbf{m}_n(V_n^{(q)})|_q = 0$).

In fact, let $\nu_n = |\mu_n|$ or $||\mathbf{m}_n||_q$ as the case may be. Then, given $\epsilon = \frac{1}{k}$, $k \in \mathbb{N}$, by the Borel regularity of μ_n and of $\mathbf{m}_n$, there exists $K_k^{(n)} \in \mathcal{C}$, $K_k^{(n)} \subset U$ such that $\nu_n(U \backslash K_k^{(n)}) < \frac{1}{k}$. Then by Theorem 50.D of [H] and by Lemma 5.2.3 there exists an open Baire set $V_k^{(n)}$ in T such that $K_k^n \subset V_k^{(n)} \subset U$. Then $\nu_n(U \backslash V_k^{(n)}) < \frac{1}{k}$. Let $V_n = \bigcup_{k=1}^{\infty} V_k^{(n)}$. Then V_n is an open Baire set in T, $V_n \subset U$, and $\nu_n(U \backslash V_n) = 0$.

(i) Let $V_U = \bigcup_{n=1}^{\infty} V_n$ (resp. $V_U^{(q)} = \bigcup_{n=1}^{\infty} V_n^{(q)}$) where V_n (resp. $V_n^{(q)}$) are chosen as in Claim 1 with respect to U (resp. with respect to U and q). Clearly, V_U (resp. $V_U^{(q)}$) is an open Baire set in T, $V_U \subset U$ and $|\mu_n|(U \backslash V_U) = 0$ (resp. $V_U^{(q)} \subset U$ and $\|\mathbf{m}_n\|_q (U \backslash V_U^{(q)}) = 0$) for all n. Hence (i) holds.

(ii) Let U be an open set in T. Then by (i) there exists an open Baire set V_U such that $V_U \subset U$ and such that $|\mu_n|(U \backslash V_U) = 0$ for all n. Then by hypothesis $\sup_n |\mu_n(U)| = \sup_n |\mu_n(V_U)| < \infty$ for each open set U in T. Then by Theorem T_4 in Appendix I of [T], $\sup_n \|\mu_n\| < \infty$. □

The following result is an improvement of the remark under Theorem T_4 in Appendix I of [T].

Corollary 5.2.5. *Let $(\mu_\alpha)_{\alpha \in I} \subset M(T)$. Suppose every sequence from $(\mu_\alpha)_{\alpha \in I}$ is bounded in each open Baire set in T. Then $\sup_{\alpha \in I} \|\mu_\alpha\| < \infty$.*

Proof. Otherwise, for each $n \in \mathbb{N}$, there would exist $\alpha_n \in I$ such that $\|\mu_{\alpha_n}\| > n$. On the other hand, the hypothesis and Lemma 5.2.4(ii) would imply that $\sup_n \|\mu_{\alpha_n}\| < \infty$, a contradiction. □

Theorem 5.2.6. (The Baire version of the Dieudonné-Grothendieck theorem). *A sequence (μ_n) in $M(T)$ is weakly convergent if and only if, for each open Baire set U in T, $\lim_n \mu_n(U)$ exists in $\mathbb{K}$ or equivalently, there exists $\mu \in M(T)$ such that*

$$\lim_n \int_T f d\mu_n = \int_T f d\mu \qquad (5.2.6.1)$$

for each bounded Borel measurable scalar function f on T if and only if $\lim_n \mu_n(U)$ exists in $\mathbb{K}$ for each open Baire set U in T. In that case, μ is unique.

Proof. If (μ_n) converges weakly to $\mu \in M(T)$, then (5.2.6.1) holds and particularly, $\lim_n \mu_n(U) = \mu(U) \in \mathbb{K}$ holds for each open Baire set U in T.

Conversely, if $\lim_n \mu_n(U) \in \mathbb{K}$ for each open Baire set U in T, then $\sup_n |\mu_n(U)| < \infty$ for each open Baire set U in T and hence by Lemma 5.2.4(ii), $\sup_n \|\mu_n\| < \infty$. Consequently, by Corollary 1 of [P8], (μ_n) converges weakly to some $\mu \in M(T)$ and hence particularly (5.2.6.1) holds. Since the weak topology of $M(T)$ is Hausdorff, the weak limit μ is unique. □

Remark 5.2.7. In the light of Lemma 5.2.4(ii), the boundedness hypothesis in Corollary 1 of [P8] is redundant. This has already been noted in Remark 9.18 of [P13]. Also see [SK] for a proof based on the one-point compactification of T.

The following theorem generalizes Lemma 5.2.4(ii) to lcHs-valued σ-additive regular Borel measures on T.

Theorem 5.2.8. *Let X be an lcHs and let $\mathbf{m}_n : \mathcal{B}(T) \to X$ be σ-additive and Borel regular for $n \in \mathbb{N}$. If $(\mathbf{m}_n(V))_1^\infty$ is bounded for each open Baire set V in T, then $\sup_n \|\mathbf{m}_n\|_q(T) < \infty$ for each $q \in \Gamma$.*

Proof. Let $q \in \Gamma$ and let V be an open Baire set in T. Then by hypothesis, $\sup_n q(\mathbf{m}_n(V)) < \infty$. If $U_q = \{x \in X : q(x) \leq 1\}$, then by hypothesis, by Proposition 1.2.15(i) and by Proposition 1.1.6 of Chapter 1, $\sup_n \sup_{x^* \in U_q^\circ} |x^* \circ \mathbf{m}_n|(V) < \infty$. Consequently, by Corollary 5.2.5, $\sup_n \sup_{x^* \in U_q^\circ} |x^* \circ \mathbf{m}_n|(T) < \infty$ and hence by Proposition 1.2.15(ii)(c) of Chapter 1 we have

$$\sup_n \|\mathbf{m}_n\|_q(T) = \sup_n \sup_{x^* \in U_q^\circ} |x^* \circ \mathbf{m}_n|(T) < \infty.$$

Hence the theorem holds. $\qquad\square$

The rest of the section is devoted to generalizing Theorem 5.2.6 to Banach space as well as to sequentially complete lcHs-valued σ-additive regular Borel measures on T.

We start by recalling the following definition from [T].

Definition 5.2.9. Let X be an lcHs with topology τ. A locally convex Hausdorff topology τ' on X is said to possess the Orlicz property when all the formal series $\sum x_n$ of elements in X which are subseries convergent in the topology τ' are unconditionally convergent in τ. A subset H of X^* is said to possess the Orlicz property when the topology $\sigma(X, H)$ is Hausdorff and possesses the Orlicz property.

Notation and Terminology 5.2.10. Let X be an lcHs with topology τ. X^{**} is the bidual of X when X^* is endowed with the strong topology $\beta(X^*, X)$ generated by the seminorms $\{q_B : B \text{ bounded in } X\}$, where

$$q_B(x^*) = \sup_{x \in B} |x^*(x)| \quad \text{for } x^* \in X^*.$$

The topology τ_e on X^{**} of uniform convergence in equicontinuous subsets of X^* is generated by the seminorms $\{q_E : E \in \mathcal{E}\}$ (see Notation 1.2.11 of Chapter 1) where $q_E(x^{**}) = \sup_{x^* \in E} |x^{**}(x^*)|$ for $x^{**} \in X^{**}$. If $u : C_0(T) \to X$ is a continuous linear map, then the adjoint $u^* : (X^*, \beta(X^*, X)) \to M(T)$ and biadjoint $u^{**} : (C_0(T))^{**} \to (X^{**}, \tau_e)$ are continuous and linear and $u^{**}|_{C_0(T)} = u$, where (X, τ) is identified as a subspace of (X^{**}, τ_e). For details see [Ho]. By Theorem 1 of [P9], for each continuous linear mapping $u : C_0(T) \to X$ there exists a unique X^{**}-valued vector measure (i.e., additive set function) $\mathbf{m}$ on $\mathcal{B}(T)$ such that $x^* \circ \mathbf{m} = u^* x^* \in M(T)$ for $x^* \in X^*$, the mapping $x^* \to x^* \circ \mathbf{m}$ of X^* into $M(T)$ is weak*-weak* continuous and $x^* u(\varphi) = \int_T \varphi d(x^* \circ \mathbf{m})$ for each $\varphi \in C_0(T)$ and $x^* \in X^*$. Then $\mathbf{m}(A) = u^{**}(\chi_A)$ for $A \in \mathcal{B}(T)$ and $\{\mathbf{m}(A) : A \in \mathcal{B}(T)\}$ is τ_e-bounded in X^{**}. Such $\mathbf{m}$ is called the representing measure of u (see Definition 4 of [P9]).

Proposition 5.2.11. *Let X be an lcHs and let $u : C_0(T) \to X$ be a continuous linear mapping with the representing measure $\mathbf{m}$. Then each $\varphi \in C_0(T)$ is $\mathbf{m}$-integrable in the sense of Definition 1 of [P7] with respect to the completion of (X^{**}, τ_e) and $\int_T \varphi d\mathbf{m} = u(\varphi) \in X$.*

Proof. By Theorem 1 of [P9], the range of $\mathbf{m}$ is bounded in (X^{**}, τ_e). Let $\varphi \in C_0(T)$. Then φ is a bounded Borel measurable function and hence there exists a sequence (s_n) of $\mathcal{B}(T)$-simple functions converging to φ uniformly in T. Therefore, φ is $\mathbf{m}$-integrable in the sense of Definition 1 of [P7] with $\int_T \varphi d\mathbf{m} \in \widetilde{(X^{**}, \tau_e)}$, the completion of (X^{**}, τ_e). Then, for $x^* \in X^*$, by Theorem 1 of [P9] and by Lemma 6 of [P7] we have, $x^* u(\varphi) = \int_T \varphi d(x^* \circ \mathbf{m}) = x^*(\int_T \varphi d\mathbf{m})$ since x^* is continuous on $\widetilde{(X^{**}, \tau_e)}$. As $u(\varphi) \in X$ and $\int_T \varphi d\mathbf{m} \in \widetilde{(X^{**}, \tau_e)}$, it follows that $q_E(u(\varphi) - \int_T \varphi d\mathbf{m}) = 0$ for each $E \in \mathcal{E}$. Hence $\int_T \varphi d\mathbf{m} = u(\varphi) \in X$. □

Remark 5.2.12. In the light of the above proposition, the hypothesis of quasicompleteness in (vi) of Proposition 5 in [P12] is redundant.

The following lemma is needed in the proof of Theorem 5.2.14.

Lemma 5.2.13. *Let X be a normed space and let H be a norm determining set in X^*. Then $H \subset \{x^* \in X^* : |x^*| \leq 1\}$.*

Proof. Let $x^* \in H$. Then, for $|x| \leq 1$, we have

$$|x^*(x)| \leq \sup_{y^* \in H} |y^*(x)| = |x| \leq 1$$

and hence $|x^*| = \sup_{|x| \leq 1} |x^*(x)| \leq 1$. □

The following theorem is an improvement of Theorem 3(vii) of [P9] and it is motivated by Theorem 2.7 of [T] whose proof is adapted here.

Theorem 5.2.14. *Let X be a Banach space and let $u : C_0(T) \to X$ be a continuous linear mapping with the representing measure $\mathbf{m}$ on $\mathcal{B}(T)$. Let H be a norm determining set in X^* with the Orlicz property. Then u is weakly compact if and only if, for each open Baire set U in T, there exists a vector $x_U \in X$ such that*

$$(x^* \circ \mathbf{m})(U) = x^*(x_U) \tag{5.2.14.1}$$

for $x^* \in H$.

Proof. If u is weakly compact, then by Theorem 2(ii) of [P9], $\mathbf{m}$ has range in X and hence the condition is necessary.

Conversely, let (5.2.14.1) hold. Let (U_n) be a disjoint sequence of open Baire sets in T. For a subsequence P of $\mathbb{N}$, by (5.2.14.1) we have

$$x^*(x_{\bigcup_{n \in P} U_n}) = (x^* \circ \mathbf{m})\left(\bigcup_{n \in P} U_n\right) = \sum_{n \in P}(x^* \circ \mathbf{m})(U_n) = \sum_{n \in P} x^*(x_{U_n}) \tag{5.2.14.2}$$

for each $x^* \in H$ and hence for each $x^* \in \langle H \rangle$, where $\langle H \rangle$ is the linear span of H. Since H is a norm determining set, $\sigma(X, H)$ is Hausdorff. Then by Theorem V.3.9 of [DS1], $(X, \sigma(X, H))^* = \langle H \rangle$ and hence (5.2.14.2) implies that $\sum_{n=1}^{\infty} x_{U_n}$ is subseries convergent in $\sigma(X, H)$. Then, as H has the Orlicz property by hypothesis, $\sum_{1}^{\infty} x_{U_n}$ is unconditionally convergent in the norm topology of X. Therefore, $\lim_n |x_{U_n}| = 0$ so that $\lim_n \sup_{x^* \in H} |x^*(x_{U_n})| = 0$. Consequently, by (5.2.14.1), $\lim_n \sup_{x^* \in H} |(x^* \circ \mathbf{m})(U_n)| = 0$. Then by Lemma 5.2.13 and by the fact that $\{\mathbf{m}(A) : A \in \mathcal{B}(T)\}$ is norm bounded in X^{**} by 5.2.10 (note that τ_e coincides with the norm topology of X^{**}), $\mathbf{m}_H = \{x^* \circ \mathbf{m} : x^* \in H\}$ is bounded in $M(T)$. Consequently, by Theorem 1 of [P8], $\mathbf{m}_H$ is relatively weakly compact in $M(T)$.

Claim 1. That $\mathbf{m}_H$ is relatively weakly compact in $M(T)$ implies that u is weakly compact.

In fact, as $\mathbf{m}_H$ is relatively weakly compact in $M(T)$, by Theorem 1 of [P8], given an open Baire set U in T and $\epsilon > 0$ (resp. for T and $\epsilon > 0$), there exists $K \in \mathcal{C}_0$ such that $K \subset U$ and

$$\sup_{x^* \in H} |x^* \circ \mathbf{m}|(U \backslash K) < \epsilon \tag{5.2.14.3}$$

(and resp.

$$\sup_{x^* \in H} |x^* \circ \mathbf{m}|(T \backslash K) < \epsilon. \tag{5.2.14.3'})$$

Combining (5.2.14.3) and (5.2.14.3') we have

$$\sup_{x^* \in H} |x^* \circ \mathbf{m}|(U \backslash K) < \epsilon \tag{5.2.14.4}$$

where U is the given open Baire set or $U = T$.

For such U, $\chi_{U \backslash K}$ is lower semicontinuous and hence we have

$$\sup_{\varphi \in \mathcal{K}(T), |\varphi| \le \chi_{U \backslash K}} |u(\varphi)| = \sup_{\varphi \in \mathcal{K}(T), |\varphi| \le \chi_{U \backslash K}} \sup_{x^* \in H} |x^* u(\varphi)|$$

$$= \sup_{x^* \in H} \sup_{\varphi \in \mathcal{K}(T), |\varphi| \le \chi_{U \backslash K}} |(u^* x^*)(\varphi)|. \tag{5.2.14.5}$$

On the other hand,

$$\sup_{\varphi \in \mathcal{K}(T), |\varphi| \le \chi_{U \backslash K}} |(u^* x^*)(\varphi)| = \sup_{|\psi| \le |\varphi|, \psi, \varphi \in \mathcal{K}(T), |\varphi| \le \chi_{U \backslash K}} |(u^* x^*)(\psi)|$$

$$= \sup_{\varphi \in \mathcal{K}(T), |\varphi| \le \chi_{U \backslash K}} |u^* x^*|(|\varphi|)$$

$$= |u^* x^*|^*(\chi_{U \backslash K})$$

$$= |u^* x^*|^*(U \backslash K) \tag{5.2.14.6}$$

by (12) on p. 55 of [B] and by Definitions 1 and 2, §1, Chapter IV of [B]. Note that $|u^* x^*|$ is the absolute value of the functional $u^* x^*$ defined as on p. 55 of [B].

By Corollary 3 of Theorem 2, §5, no. 5 of Chapter IV of [B], the Borel sets in T are $|u^*x^*|$-measurable and by an abuse of notation let us denote $|u^*x^*|^*|_{\mathcal{B}(T)}$ also by $|u^*x^*|$. Then by (5.2.14.5) and (5.2.14.6) we have

$$\sup_{\varphi\in\mathcal{K}(T),\,|\phi|\leq\chi_{U\setminus K}} |u(\varphi)| = \sup_{x^*\in H} |(u^*x^*)|(U\setminus K) \qquad (5.2.14.7)$$

where U is the given open Baire set or $U = T$.

By Theorem 4.11 of [P3], by the last part of Theorem 3.3 of [P4] and by Notation 5.2.1 above, we have

$$\mu_{|u^*x^*|} = v(\mu_{u^*x^*}, \mathcal{B}(T)) = |\mu_{u^*x^*}| \qquad (5.2.14.8)$$

where $\mu_{u^*x^*}$ is the complex Radon measure induced by u^*x^* in the sense of Definition 4.3 of [P3]. Note that $\mu_{u^*x^*}$ is the same as $x^* \circ \mathbf{m}$ as $(u^*x^*)(\chi_A) = (x^*u^{**})(\chi_A) = (x^*\circ\mathbf{m})(A)$ for $A \in \mathcal{B}(T)$ (see 5.2.10). Then by (5.2.14.4), (5.2.14.7) and (5.2.14.8) we have

$$
\begin{aligned}
\sup_{\varphi\in\mathcal{K}(T),\,|\varphi|\leq\chi_{U\setminus K}} |u(\varphi)| &= \sup_{x^*\in H} |(u^*x^*)|(U\setminus K) \\
&= \sup_{x^*\in H} \mu_{|u^*x^*|}(U\setminus K) \\
&= \sup_{x^*\in H} |\mu_{u^*x^*}|(U\setminus K) \\
&= \sup_{x^*\in H} |x^* \circ \mathbf{m}|(U\setminus K) \\
&< \epsilon \qquad (5.2.14.9)
\end{aligned}
$$

where $|\mu_{u^*x^*}| = |x^* \circ \mathbf{m}|$ and U is the given open Baire set or $U = T$.

On the other hand, by (5.2.14.4), (5.2.14.6), (5.2.14.8) and (5.2.14.9) and by (12) on p. 55 of [B] we have

$$
\begin{aligned}
\epsilon > \sup_{\varphi\in\mathcal{K}(T),\,|\varphi|\leq\chi_{U\setminus K}} |u(\varphi)| &= \sup_{\varphi\in\mathcal{K}(T),\,|\varphi|\leq\chi_{U\setminus K}} \sup_{|x^*|\leq 1} |(x^*u)(\varphi)| \\
&= \sup_{|x^*|\leq 1} \sup_{\varphi\in\mathcal{K}(T),\,|\varphi|\leq\chi_{U\setminus K}} |(u^*x^*)(\varphi)| \\
&= \sup_{|x^*|\leq 1} |u^*x^*|^*(U\setminus K) \\
&= \sup_{|x^*|\leq 1} |u^*x^*|(U\setminus K) \\
&= \sup_{|x^*|\leq 1} \mu_{|u^*x^*|}(U\setminus K) \\
&= \sup_{|x^*|\leq 1} v(\mu_{u^*x^*}, \mathcal{B}(T))(U\setminus K) \\
&= \sup_{|x^*|\leq 1} v(x^* \circ \mathbf{m}, \mathcal{B}(T))(U\setminus K) \qquad (5.2.14.10)
\end{aligned}
$$

since $U \backslash K$ is $|u^* x^*|$-measurable by Corollary 3 of Theorem 2, §5, no. 5 of Chapter IV of [B] and since $\mu_{u^* x^*} = x^* \circ \mathbf{m}$ as in 5.2.10.

Since $\mathbf{m} : \mathcal{B}(T) \to X^{**}$ is additive and $|x^{**}| = \sup_{|x^*| \le 1} |x^{**}(x^*)|$ for $x^{**} \in X^{**}$, by an argument similar to the proof of Proposition 1.2.13(iii) of Chapter 1 and by (5.2.14.10) we have

$$||\mathbf{m}||(U \backslash K) = \sup_{|x^*| \le 1} v(x^* \circ \mathbf{m}, \mathcal{B}(T))(U \backslash K) < \epsilon$$

where U is the given open Baire set or $U = T$. Therefore, $\mathbf{m}$ is Baire inner regular in each open Baire set U in T and in the set T in the norm topology of X^{**}, which is the same as τ_e for X^{**}. Hence by Theorem 8(xxix) of [P9], u is weakly compact.   □

Corollary 5.2.15. *Let X be a Banach space, H be a norm determining subset of X^* and $u : C_0(T) \to X$ be a continuous linear mapping. Let $\mathcal{K}(T)_b^* = (C_c(T), || \cdot ||_T)^*$ be the set of all bounded linear functionals on $\mathcal{K}(T)$ (see pp. 65 and 69 of [P4]). If $\boldsymbol{\eta} : \mathcal{K}(T)_b^* \to M(T)$ is the isometric isomorphism given in Theorem 5.3 of [P4] ($\boldsymbol{\eta} = \Phi_{\mathcal{B}(T)}^{-1}$ in the notation of Theorem 5.3 of [P4]), then $\boldsymbol{\eta}(\theta) = \mu_\theta|_{\mathcal{B}(T)}$ for $\theta \in \mathcal{K}(T)_b^*$, and hence $\boldsymbol{\eta}(x^* u) = x^* \circ \mathbf{m}$ for $x^* \in X^*$ where $\mathbf{m}$ is the representing measure of u. Moreover, if $\boldsymbol{\eta}\{x^* u : x^* \in H\}$ is relatively weakly compact in $M(T)$, then u is weakly compact.*

Proof. Clearly, $x^* u$ is a bounded linear functional on $C_0(T)$ and hence $x^* u \in \mathcal{K}(T)_b^*$. Moreover, $(x^* u)(\varphi) = (u^* x^*)(\varphi) = (x^* \circ \mathbf{m})(\varphi)$ for $\varphi \in C_0(T)$ (see 5.2.10). Then

$$(x^* u)(\varphi) = \int_T \varphi d(x^* \circ \mathbf{m}), \quad \varphi \in C_0(T).$$

Consequently, by Theorem 5.3 of [P4], $\boldsymbol{\eta}(x^* u) = x^* \circ \mathbf{m}$, $x^* \in X^*$. Then by hypothesis, $\{x^* \circ \mathbf{m} : x^* \in H\}$ is relatively weakly compact in $M(T)$ and hence by Claim 1 in the proof of Theorem 5.2.14, u is weakly compact.   □

The following theorem is motivated by Theorem 2.7 bis in [T] and its proof in [T] is adapted here.

Theorem 5.2.16. *Let X be an lcHs with topology τ and let H be a subset of X^* having the Orlicz property such that the topology τ is identical with the topology of uniform convergence in equicontinuous subsets of H. Let $u : C_0(T) \to X$ be a continuous linear mapping with the representing measure $\mathbf{m}$. Then u is weakly compact if and only if for each open Baire set U in T there exists a vector $x_U \in X$ such that*

$$(x^* \circ \mathbf{m})(U) = x^*(x_U) \tag{5.2.16.1}$$

for $x^* \in H$.

Proof. If u is weakly compact, then by Theorem 2(ii) of [P9] the condition is necessary. Note that the hypothesis of quasicompleteness of X is not needed for (i)$\Rightarrow$(ii) of Theorem 2 of [P9].

Conversely, let (5.2.16.1) hold. Let $H_\mathcal{E} = \{E \subset H : E \text{ is equicontinuous}\}$. If $x^*(x) = 0$ for each $x^* \in H$, then for $E \in H_\mathcal{E}$, $q_E(x) = \sup_{x^* \in E} |x^*(x)| = 0$. Since τ is the same as the locally convex topology generated by $\{q_E : E \in H_\mathcal{E}\}$, it follows that $x = 0$ and hence $\sigma(X, H)$ is Hausdorff.

Let (U_n) be a disjoint sequence of open Baire sets in T. Arguing as in the proof of Theorem 5.2.14, for a subsequence P of $\mathbb{N}$, by (5.2.16.1) we have $\sum_{n \in P} x^*(x_{U_n})$ is subseries convergent for $x^* \in \langle H \rangle$, the linear span of H. Since $(X, \sigma(X, H))^* = \langle H \rangle$ by Theorem V.3.9 of [DS1], $\sum_1^\infty x_{U_n}$ is subseries convergent in $\sigma(X, H)$. By hypothesis, H has the Orlicz property and hence $\sum_1^\infty x_{U_n}$ is unconditionally convergent in τ. By hypothesis, τ is generated by the seminorms $\{q_E : E \in H_\mathcal{E}\}$. Let $E \in H_\mathcal{E}$. Then $\lim_n q_E(x_{U_n}) = 0$ and hence by (5.2.16.1) we have

$$\lim_n \sup_{x^* \in E} (x^* \circ \mathbf{m})(U_n) = 0. \tag{5.2.16.2}$$

Since the range of $\mathbf{m}$ is bounded in τ_e by Theorem 1 of [P9],

$$\sup_{A \in \mathcal{B}(T)} q_E(\mathbf{m}(A)) < \infty$$

and hence $\{x^* \circ \mathbf{m} : x^* \in E\}$ is bounded in $M(T)$. Then by (5.2.16.2) and by Theorem 1 of [P8],

$$\{x^* \circ \mathbf{m} : x^* \in E\} \text{ is relatively weakly compact in } M(T). \tag{5.2.16.3}$$

Let $\widetilde{X}$ be the completion of X. Let $\Pi_{q_E} : \widetilde{X} \to \widetilde{X}_{q_E} \subset \widetilde{X}_{q_E}$ for $E \in H_\mathcal{E}$, where $\widetilde{X}_{q_E}$ is the completion of the normed space $\widetilde{X}_{q_E}$. If Ψ_{x^*} is as in Proposition 1.2.13(i) of Chapter 1 on $\widetilde{X}_{q_E}$, then by Proposition 1.2.13(ii) of Chapter 1 $\{\Psi_{x^*} : x^* \in E\}$ is a norm determining set for $\widetilde{X}_{q_E}$, $E \in H_\mathcal{E}$. Then by Proposition 1.2.13(i) of Chapter 1 and by 5.2.10 we have $(\Psi_{x^*} \circ \Pi_{q_E} \circ u)(\varphi) = (x^* u)(\varphi) = \int_T \varphi d(x^* \circ \mathbf{m})$ for $\varphi \in C_0(T)$ and clearly, $\Psi_{x^*} \circ \Pi_{q_E} \circ u \in \mathcal{K}(T)_b^*$. Therefore, $\boldsymbol{\eta}(\Psi_{x^*} \circ \Pi_{q_E} \circ u) = x^* \circ \mathbf{m}$ where $\boldsymbol{\eta}$ is as in Corollary 5.2.15. Then by (5.2.16.3) and Corollary 5.2.15, $\Pi_{q_E} \circ u$ is weakly compact for each $E \in H_\mathcal{E}$. Consequently, by Lemma 2.21 of [T], which holds for complex lcHs too, we conclude that u is weakly compact. $\square$

Theorem 5.2.17. *Let X be an lcHs with topology τ, H be a subset of X^* having the Orlicz property such that τ is the same as the topology of uniform convergence in equicontinuous subsets of H and $u : C_0(T) \to X$ be a continuous linear mapping with the representing measure $\mathbf{m}$. Suppose u_H is the same as u on $C_0(T)$ with X being provided with the topology $\sigma(X, H)$. Then u is weakly compact if and only if for each open Baire set U in T there exists a vector $x_U \in X$ such that*

$$(x^* \circ u_H^{**})(\chi_U) = x^*(x_U) \tag{5.2.17.1}$$

for $x^* \in H$. *Moreover, condition* (5.2.17.1) *is the same as* $(x^* \circ \mathbf{m})(U) = x^*(x_U)$
for open Baire sets U in T and for $x^* \in H$.

Proof. Suppose (5.2.17.1) holds. As observed in the proof of Theorem 5.2.16, the hypothesis on τ implies that the topology $\sigma(X, H)$ is Hausdorff and hence $(X, \sigma(X, H))$ is an lcHs. As $\sigma(X, H)$ is weaker than τ, $u : C_0(T) \to (X, \sigma(X, H))$ is continuous and hence u_H is a continuous linear map. Therefore, $x^* u_H^{**}(\chi_A) = (u_H^* x^*)(\chi_A)$ for $x^* \in (X, \sigma(X, H))^*$ and for $A \in \mathcal{B}(T)$. Since $(X, \sigma(X, H))^* = \langle H \rangle$, the linear subspace spanned by H by Theorem V.3.9 of [DS1], particularly we have

$$x^* u_H^{**}(\chi_A) = (u_H^* x^*)(\chi_A) \tag{5.2.17.2}$$

for $x^* \in H$. Since

$$(u_H^* x^*)(\varphi) = x^*(u_H \varphi) = x^*(u\varphi) = u^* x^*(\varphi)$$

for $\varphi \in C_0(T)$ and for $x^* \in H$, we have

$$u_H^* x^* = u^* x^* \tag{5.2.17.3}$$

for $x^* \in H$. Consequently, by (5.2.17.2) and (5.2.17.3) we have

$$x^* u_H^{**}(\chi_A) = (u_H^* x^*)(\chi_A) = (u^* x^*)(\chi_A) = x^* u^{**}(\chi_A) = (x^* \circ \mathbf{m})(A) \tag{5.2.17.4}$$

for $x^* \in H$ and for $A \in \mathcal{B}(T)$, since $\mathbf{m}$ is the representing measure of u. Hence the hypothesis is equivalent to saying that

$$(x^* \circ \mathbf{m})(U) = x^* u_H^{**}(\chi_U) = x^*(x_U) \tag{5.2.17.5}$$

for $x^* \in H$ and for open Baire sets U in T. Consequently, by Theorem 5.2.16, u is weakly compact.

Conversely, if u is weakly compact, then by Theorem 2(ii) of [P9] (which holds for an arbitrary lcHs), $\mathbf{m}(U) = x_U$(say) $\in X$ for any open Baire set U in T and hence by (5.2.17.5) we have

$$(x^* u_H^{**})(\chi_U) = (x^* \circ \mathbf{m})(U) = x^*(x_U)$$

for $x^* \in H$ and hence (5.2.17.1) holds. $\qquad \square$

Lemma 5.2.18. *Let X be a sequentially complete lcHs and let $\mathbf{m}_n : \mathcal{B}(T) \to X$ be σ-additive and Borel regular for $n \in \mathbb{N}$. Then $\lim_n \mathbf{m}_n(U) \in X$ for each open Baire set U in T if and only if $\lim_n \mathbf{m}_n(U) \in X$ for each open set U in T.*

Proof. Clearly the condition is sufficient. Conversely, let $\lim_n \mathbf{m}_n(V) \in X$ for each open Baire set V in T. Let U be an open set in T and $q \in \Gamma$. Then by Lemma 5.2.4(i) there exists an open Baire set $V_U^{(q)} \subset U$ such that $|\mathbf{m}_n(U) - \mathbf{m}_n(V_U^{(q)})|_q = 0$

for all n. By hypothesis, $\lim_n \mathbf{m}_n(V_U^{(q)}) = x_q$ (say) exists in X, for each $q \in \Gamma$. Then

$$|\mathbf{m}_n(U) - x_q|_q \le |\mathbf{m}_n(U) - \mathbf{m}_n(V_U^{(q)})|_q + |\mathbf{m}_n(V_U^{(q)}) - x_q|_q \to 0$$

as $n \to \infty$ and hence $|\mathbf{m}_n(U) - \mathbf{m}_k(U)|_q \to 0$ as $n, k \to \infty$. Since q is arbitrary in Γ, this implies that $(\mathbf{m}_n(U))_{n=1}^\infty$ is Cauchy in X. As X is sequentially complete, there exists $x_U \in X$ such that $\lim_n \mathbf{m}_n(U) = x_U$. Hence the lemma holds. □

Lemma 5.2.19. *Let X be a quasicomplete lcHs and $\mathbf{m} : \mathcal{B}(T) \to X$ be σ-additive. Then each $\varphi \in C_0(T)$ is $\mathbf{m}$-integrable in T in the sense of Definition 1 of [P7] as well as $\mathbf{m}$-integrable in the sense of Definitions 4.2.1 and 4.2.1' of Chapter 4 and the corresponding integrals of f coincide. If $u : C_0(T) \to X$ is given by $u(\varphi) = \int_T \varphi d\mathbf{m}$ for $\varphi \in C_0(T)$, then u is a weakly compact operator. If $\mathbf{m}$ is further Borel regular, then $\mathbf{m}$ is the representing measure of u (see Notation and Terminology 5.2.10 or Definition 4 of [P9]).*

Proof. Since $\mathbf{m}$ is σ-additive on the σ-algebra $\mathcal{B}(T)$, $||\mathbf{m}||_q(T) < \infty$ for each $q \in \Gamma$. Since $\varphi \in C_0(T)$ is bounded and Borel measurable, there exists a sequence (s_n) of $\mathcal{B}(T)$-simple functions such that $s_n \to \varphi$ uniformly in T with $|s_n| \nearrow |\varphi|$. Then, for $q \in \Gamma$ and $A \in \mathcal{B}(T)$, by Theorem 4.1.9(i) of Chapter 4,

$$q\left(\int_A s_n d\mathbf{m} - \int_A s_k d\mathbf{m} \right) \le ||s_n - s_k||_A ||\mathbf{m}||_q(T) \le ||s_n - s_k||_T ||\mathbf{m}||_q(T) \to 0$$

as $n, k \to 0$ and hence $(\int_A s_n d\mathbf{m})_1^\infty$ is Cauchy in X for each $A \in \mathcal{B}(T)$. Since X is sequentially complete, φ is $\mathbf{m}$-integrable in the sense of Definition 1 of [P7] and $\int_A \varphi d\mathbf{m} = \lim_n \int_A s_n d\mathbf{m}$ for $A \in \mathcal{B}(T)$. On the other hand, by Definition 4.2.1' in Remark 4.2.12 and by Remark 4.2.14 of Chapter 4, the $\mathcal{B}(T)$-measurable function φ is also $\mathbf{m}$-integrable in T in the sense of Definition 4.2.1 of Chapter 4 with (BDS)$\int_A \varphi d\mathbf{m} = \int_A \varphi d\mathbf{m}$ for $A \in \mathcal{B}(T)$.

Clearly, u is linear. Moreover, by Theorem 4.1.9(ii)(b) of Chapter 4 we have

$$q(u\varphi) = q\left(\int_T \varphi d\mathbf{m} \right) \le ||\varphi||_T \cdot ||\mathbf{m}||_q(T)$$

for each $q \in \Gamma$ and hence u is continuous.

Let $\Sigma(\mathcal{B}(T))$ be the Banach space of all bounded complex functions which are uniform limits of sequences of $\mathcal{B}(T)$-simple functions with norm the supremum norm $||\cdot||_T$. Then $C_0(T)$ is a subspace of $\Sigma(\mathcal{B}(T))$. If $\Phi : \Sigma(\mathcal{B}(T)) \to X$ is given by $\Phi(\varphi) = \int_T \varphi d\mathbf{m}$ with the integral defined in the sense of Definition 1 of [P7], then by Lemma 6 of [P7], Φ is a continuous linear map and $\mathbf{m}$ is the representing measure of Φ in the sense of Definition 2 of [P7]. Since $\mathcal{B}(T)$ is a σ-algebra and $\mathbf{m}$ is σ-additive on $\mathcal{B}(T)$, $\mathbf{m}$ is also strongly additive on $\mathcal{B}(T)$ and hence by Theorem 1 of [P7], Φ is weakly compact. Consequently, $u = \Phi|_{C_0(T)}$ is weakly compact.

Now suppose $\mathbf{m}$ is further Borel regular. Then by Theorem 2(ii) of [P9], the representing measure $\hat{\mathbf{m}}$ (in the sense of Definition 4 of [P9]) of the weakly compact operator u has range in X and by Theorem 1 of [P9], $x^* \circ \hat{\mathbf{m}} \in M(T)$ for $x^* \in X^*$ and $x^* u(\varphi) = \int_T \varphi d(x^* \circ \hat{\mathbf{m}})$ for $\varphi \in C_0(T)$. On the other hand, $u(\varphi) = \Phi(\varphi) = \int_T \varphi d\mathbf{m}$ and hence by Lemma 6(iii) of [P7] we have $\int_T \varphi d(x^* \circ \mathbf{m}) = x^* u(\varphi)$ for $\varphi \in C_0(T)$. Thus we have

$$x^* u(\varphi) = \int_T \varphi d(x^* \circ \mathbf{m}) = \int_T \varphi d(x^* \circ \hat{\mathbf{m}}), \quad \varphi \in C_0(T).$$

Since $x^* \circ \mathbf{m} \in M(T)$ by hypothesis, by the uniqueness part of the Riesz representation theorem, $x^* \circ \hat{\mathbf{m}} = x^* \circ \mathbf{m}$ for $x^* \in X^*$ and consequently, by the Hahn-Banach theorem we have $\mathbf{m} = \hat{\mathbf{m}}$ and hence $\mathbf{m}$ is the representing measure of u (in the sense of 5.2.10). $\qquad\square$

The proof of (i) in the following lemma is motivated by the proof of Theorem 2.12 of [T].

Lemma 5.2.20. *Let X be a sequentially complete lcHs and let $\mathbf{m}_n : \mathcal{B}(T) \to X$ be σ-additive and Borel regular for $n \in \mathbb{N}$. Suppose $\lim_n \mathbf{m}_n(U)$ exists in X for each open Baire set U in T. Let $u_n : C_0(T) \to X$ be given by $u_n(\varphi) = \int_T \varphi d\mathbf{m}_n$ for $\varphi \in C_0(T)$. Then:*

(i) *$\lim_n u_n(\varphi) = u(\varphi)$ (say) exists in X for each $\varphi \in C_0(T)$.*

(ii) *u is an X-valued continuous linear mapping on $C_0(T)$.*

Proof. By hypothesis and by Lemma 5.2.18,

$$\lim_n \mathbf{m}_n(U) = \mathbf{m}(U) \text{ (say)} \qquad (5.2.20.1)$$

exists in X for each open set U in T and moreover, by Theorem 5.2.8,

$$\sup_n ||\mathbf{m}_n||_q(T) = M_q \text{ (say)} < \infty \qquad (5.2.20.2)$$

for each $q \in \Gamma$.

(i) Let $\varphi \in C_0(T)$, $\varphi \geq 0$. Then there exists a sequence (s_n) of $\mathcal{B}(T)$-simple functions such that $s_n \to \varphi$ uniformly in T and

$$s_n = \sum_{i=2}^{n \cdot 2^n} \frac{i-1}{2^n} \chi_{E_{i,n}}$$

where $n \geq ||\varphi||_T$ and $E_{i,n} = \varphi^{-1}([\frac{i-1}{2^n}, \frac{i}{2^n})) = \varphi^{-1}((-n, \frac{i}{2^n})) \backslash \varphi^{-1}((-n, \frac{i-1}{2^n}))$ for $i = 2, 3, \ldots, n \cdot 2^n$. Clearly, $0 \leq s_n \nearrow \varphi$. Then $E_{i,n}$ is the difference of two open sets and hence s_n is a real linear combination of the characteristic functions of open sets in T. Consequently, each $\varphi \in C_0(T)$ is the uniform limit of a sequence (s_n') of $\mathcal{B}(T)$-simple functions with $|s_n'| \nearrow |\varphi|$ and with each s_n' being a complex linear combination of the characteristic functions of open sets in T. Thus, given $\varphi \in C_0(T)$, $q \in \Gamma$ and $\epsilon > 0$, there exists s of the form $s = \sum_{i=1}^{k} \alpha_i \chi_{U_i}$, U_i open in T, $||s||_T \leq ||\varphi||_T$ and

$$||s - \varphi||_T < \frac{\epsilon}{4M_q}. \tag{5.2.20.3}$$

Then by (5.2.20.1) we have

$$\lim_n \int_T s d\mathbf{m}_n = \lim_n \sum_1^k \alpha_i \mathbf{m}_n(U_i) = \sum_1^k \alpha_i \mathbf{m}(U_i) = x \text{ (say)}.$$

Then there exists n_0 such that

$$\left| \int_T s d\mathbf{m}_n - x \right|_q < \frac{\epsilon}{4} \tag{5.2.20.4}$$

for $n \geq n_0$. Then by (5.2.20.3) and (5.2.20.4) and by Theorem 4.1.9(i)(b) and Remark 4.1.5 of Chapter 4 we have

$$|u_n(\varphi) - u_r(\varphi)|_q \leq \left| u_n(\varphi) - \int_T s d\mathbf{m}_n \right|_q + \left| \int_T s d\mathbf{m_n} - \int_T s d\mathbf{m}_r \right|_q$$
$$+ \left| \int_T s d\mathbf{m}_r - u_r(\varphi) \right|_q$$
$$\leq ||\varphi - s||_T ||\mathbf{m}_n||_q(T) + \left| \int_T s d\mathbf{m}_n - x \right|_q + \left| \int_T s d\mathbf{m}_r - x \right|_q$$
$$+ ||s - \varphi||_T ||\mathbf{m}_r||_q(T)$$
$$< ||\varphi - s||_T \cdot (2M_q) + 2\frac{\epsilon}{4} < \epsilon$$

for $n, r \geq n_0$. Since q is arbitrary in Γ, this implies that $(u_n(\varphi))_{n=1}^{\infty}$ is Cauchy in X and as X is sequentially complete, there exists a vector $u(\varphi)$ (say) in X such that $\lim_n u_n(\varphi) = u(\varphi)$ for $\varphi \in C_0(T)$. Hence (i) holds.

(ii) Clearly, $u : C_0(T) \to X$ is linear and u is continuous by (i) and by Theorem 2.8 of [Ru2]. $\square$

The proof of the following theorem is a vector measure adaptation of the proof of Proposition 2.11 of [T].

Theorem 5.2.21. (Generalization of Theorem 5.2.6 to Banach space-valued σ-additive regular Borel measures). Let X be a Banach space and let $\mathbf{m}_n : \mathcal{B}(T) \to X$ be σ-additive and Borel regular for $n \in \mathbb{N}$. Then $\lim_n \mathbf{m}_n(U) \in X$ for each open Baire set U in T if and only if there exists a unique Borel regular X-valued σ-additive measure $\mathbf{m}$ on $\mathcal{B}(T)$ such that

$$\lim_n \int_T f d\mathbf{m}_n = \int_T f d\mathbf{m} (\in X) \tag{5.2.21.1}$$

for each bounded $\mathcal{B}(T)$-measurable scalar function f on T.

Proof. Suppose $\lim_n \mathbf{m}_n(U) \in X$ for each open Baire set U in T. Let $c_X = \{(x_n)_1^\infty \subset X : \lim_n x_n \in X\}$ be provided with norm $\|(x_n)_1^\infty\| = \sup_n |x_n|$. Let $u_n(\varphi) = \int_T \varphi d\mathbf{m}_n$, $\varphi \in C_0(T)$. Then by hypothesis and by Lemma 5.2.19, u_n is an X-valued weakly compact operator on $C_0(T)$ with the representing measure $\mathbf{m}_n$ for each $n \in \mathbb{N}$. Let $\Phi : C_0(T) \to c_X$ be defined by $\Phi(\varphi) = (u_n(\varphi))_1^\infty$ for $\varphi \in C_0(T)$. By Lemma 5.2.20(i), Φ is well defined and clearly, linear. By hypothesis and by Theorem 5.2.8,

$$\sup_n \|\mathbf{m}_n\|(T) = M \text{ (say) } < \infty. \qquad (5.2.21.2)$$

Then by Theorem 2.1.5(v) and Remark 2.2.3 of Chapter 2,

$$\|\Phi(\varphi)\| = \sup_n |u_n(\varphi)| = \sup_n \left| \int_T \varphi d\mathbf{m}_n \right| \leq \|\varphi\|_T \cdot \sup_n \|\mathbf{m}_n\|(T) = M\|\varphi\|_T$$

for $\varphi \in C_0(T)$ and hence Φ is continuous.

Claim 1. Φ is weakly compact.

In fact, let $H = \{I_{n,x^*} : x^* \in X^*, |x^*| \leq 1, n \in \mathbb{N}\}$, where $\langle I_{n,x^*}, (x_k)_1^\infty \rangle = x^*(x_n)$. Clearly, $H \subset (c_X)^*$ is a norm determining set for c_X. The proof of Corollary II.5 in Appendix II of [T] holds for complex spaces too and hence by the complex version of the said corollary, H has the Orlicz property for $(c_X, \|\cdot\|)$. Let $\Phi : C_0(T) \to (c_X, \sigma(c_X, H))$ be designated as Φ_H so that $\Phi_H(\varphi) = (u_n(\varphi))_1^\infty$, $\varphi \in C_0(T)$. Clearly, Φ_H is continuous as $\sigma(c_X, H)$ is weaker than the norm topology of c_X. Moreover,

$$\langle \Phi^* I_{n,x^*}, \varphi \rangle = \langle I_{n,x^*}, \Phi(\varphi) \rangle = \langle I_{n,x^*}, (u_k(\varphi))_1^\infty \rangle = x^* u_n(\varphi) = \langle u_n^* x^*, \varphi \rangle$$

for $\varphi \in C_0(T)$ and hence

$$\Phi^* I_{n,x^*} = u_n^* x^* \qquad (5.2.21.3)$$

for $I_{n,x^*} \in H$.

On the other hand, by Theorem V.3.9 of [DS1], $(c_X, \sigma(c_X, H))^* = \langle H \rangle \subset (c_X)^*$ where $\langle H \rangle$ is the linear span of H, and hence we have

$$\langle \Phi^* I_{n,x^*}, \varphi \rangle = \langle I_{n,x^*}, \Phi(\varphi) \rangle = \langle I_{n,x^*}, \Phi_H(\varphi) \rangle = \langle \Phi_H^* I_{n,x^*}, \varphi \rangle$$

for $\varphi \in C_0(T)$. Therefore, $\Phi^* I_{n,x^*} = \Phi_H^* I_{n,x^*}$ for each $I_{n,x^*} \in H$. Then by (5.2.21.3) we have

$$\Phi_H^* I_{n,x^*} = u_n^* x^* \qquad (5.2.21.4)$$

for $x^* \in X^*$ with $|x^*| \leq 1$. By hypothesis, given an open Baire set U in T there exists a vector $x_U = (\mathbf{m}_n(U))_1^\infty \in c_X$. Then, as u_n is a weakly compact operator with the representing measure $\mathbf{m}_n$ by Lemma 5.2.19, for the open Baire set U in

T with x_U as above , by (5.2.21.4) we have

$$\langle \Phi_H^{**}(\chi_U), I_{n,x^*} \rangle = \langle \chi_U, \Phi_H^* I_{n,x^*} \rangle = \langle \chi_U, u_n^* x^* \rangle = \langle u_n^{**}(\chi_U), x^* \rangle$$
$$= \langle \mathbf{m}_n(U), x^* \rangle = \langle x_U, I_{n,x^*} \rangle.$$

Thus,

$$I_{n,x^*} \circ \Phi_H^{**}(\chi_U) = I_{n,x^*}(x_U) \qquad (5.2.21.5)$$

for $I_{n,x^*} \in H$. Then by (5.2.21.5) and by Theorem 5.2.17 (since H is a norm determining set for c_X, the norm topology of c_X is identical with the topology of uniform convergence in equicontinuous subsets of H), Φ is weakly compact and hence the claim holds.

Let $\hat{\mathbf{m}}$ be the representing measure of Φ. Then by Theorem 2(ii) of [P9], $\hat{\mathbf{m}}$ has range in c_X so that $\hat{\mathbf{m}}(A) = \Phi^{**}(\chi_A) \in c_X$ for $A \in \mathcal{B}(T)$ and we let $\hat{\mathbf{m}}(A) = (x_n)_1^\infty \in c_X$. Then by (5.2.21.3) we have $x^*(x_n) = I_{n,x^*} \hat{\mathbf{m}}(A) = I_{n,x^*} \Phi^{**}(\chi_A) = \langle \Phi^* I_{n,x^*}, \chi_A \rangle = \langle u_n^* x^*, \chi_A \rangle = \langle x^*, u_n^{**}(\chi_A) \rangle = \langle x^*, \mathbf{m}_n(A) \rangle$ for $I_{n,x^*} \in H$ and hence $x^*(x_n) = (x^* \circ \mathbf{m}_n)(A)$ for $x^* \in X^*$ with $|x^*| \le 1$ and for $A \in \mathcal{B}(T)$. Then clearly $x^*(x_n) = (x^* \circ \mathbf{m}_n)(A)$ for $x^* \in X^*$ and hence by the Hahn-Banach theorem, $x_n = \mathbf{m}_n(A)$ for all n and hence

$$(\mathbf{m}_n(A))_1^\infty = (x_n)_1^\infty = \hat{\mathbf{m}}(A) \in c_X. \qquad (5.2.21.6)$$

This implies that $\lim_n \mathbf{m}_n(A) = \mathbf{m}(A)$ (say) exists in X for each $A \in \mathcal{B}(T)$. Then by VHSN (see Proposition 1.1.8 of Chapter 1), $\mathbf{m} : \mathcal{B}(T) \to X$ is σ-additive and hence $||\mathbf{m}||(T) < \infty$.

Let $M_0 = \max(M, ||\mathbf{m}||(T))$ where M is as in (5.2.21.2). Let f be a bounded $\mathcal{B}(T)$-measurable scalar function. Then there exists a sequence (s_n) of Borel simple functions such that $|s_n| \nearrow |f|$ and $||s_n - f||_T \to 0$ as $n \to \infty$. Thus, given $\epsilon > 0$, there exists n_0 such that

$$||s_{n_0} - f||_T < \frac{\epsilon}{3M_0}. \qquad (5.2.21.7)$$

Let $s = s_{n_0} = \sum_1^r \alpha_i \chi_{A_i}$, $(A_i)_1^r \subset \mathcal{B}(T)$. Then by (5.2.21.7) and by Theorem 2.1.5(v) and Remark 2.2.3 of Chapter 2 we have

$$\left| \int_T f d\mathbf{m}_n - \int_T s d\mathbf{m}_n \right| \le ||f - s||_T \cdot ||\mathbf{m}_n||(T) < \frac{\epsilon}{3} \qquad (5.2.21.8)$$

for all n and

$$\left| \int_T f d\mathbf{m} - \int_T s d\mathbf{m} \right| \le ||f - s||_T \cdot ||\mathbf{m}||(T) < \frac{\epsilon}{3}. \qquad (5.2.21.9)$$

As $\lim_n \mathbf{m}_n(A_i) = \mathbf{m}(A_i)$ for $i = 1, 2, \ldots, r$, there exists n_1 such that

$$|\alpha_i| |\mathbf{m}_n(A_i) - \mathbf{m}(A_i)| < \frac{\epsilon}{3r} \qquad (5.2.21.10)$$

for $n \geq n_1$ and for $i = 1, 2, \ldots, r$. Then by (5.2.21.8), (5.2.21.9) and (5.2.21.10) we have

$$\left| \int_T f d\mathbf{m}_n - \int_T f d\mathbf{m} \right| < \epsilon$$

for $n \geq n_1$. Hence $\lim_n \int_T f d\mathbf{m}_n = \int_T f d\mathbf{m} \in X$ and thus (5.2.21.1) holds.

To prove that $\mathbf{m}$ is Borel regular, let $A \in \mathcal{B}(T)$ and $\epsilon > 0$. As $\hat{\mathbf{m}}$ is Borel regular by Theorem 6 of [P9], there exist $K \in \mathcal{C}$ and $U \in \mathcal{U}$ such that

$$K \subset A \subset U \quad \text{with} \quad |\hat{\mathbf{m}}(B)| < \epsilon \qquad (5.2.21.11)$$

for $B \in \mathcal{B}(T)$ such that $B \subset U \backslash K$. Since $\hat{\mathbf{m}}(B) = (\mathbf{m}_n(B))_1^\infty$ by (5.2.21.6), by (5.2.21.11) we have

$$|\mathbf{m}(B)| = \lim_n |\mathbf{m}_n(B)| \leq \sup_n |\mathbf{m}_n(B)| = |\hat{\mathbf{m}}(B)| < \epsilon$$

for $B \in \mathcal{B}(T)$ with $B \subset U \backslash K$ and hence $\mathbf{m}$ is $\mathcal{B}(T)$-regular.

To prove the uniqueness of $\mathbf{m}$, let $\mathbf{m}' : \mathcal{B}(T) \to X$ be another σ-additive Borel regular measure such that

$$\int_T f d\mathbf{m}' = \lim_n \int_T f d\mathbf{m}_n \qquad (5.2.21.12)$$

for each bounded Borel measurable scalar function f on T. Then by Theorem 2.1.5(viii) and Remark 2.2.3 of Chapter 2 and by (5.2.21.1) and (5.2.21.12) we have

$$\int_T f d(x^* \circ \mathbf{m}') = \int_T f d(x^* \circ \mathbf{m})$$

for $x^* \in X^*$ and for $f \in C_0(T)$. Consequently, by the uniqueness part of the Riesz representation theorem, $x^* \circ \mathbf{m} = x^* \circ \mathbf{m}'$ for $x^* \in X^*$ and hence by the Hahn-Banach theorem, $\mathbf{m}' = \mathbf{m}$. Hence $\mathbf{m}$ is unique.

The converse is evident and hence the theorem holds. $\square$

Remark 5.2.22. Unlike in Theorem 5.2.6, evidently we cannot assert the weak convergence of $(\mathbf{m}_n)_1^\infty$ when X is infinite dimensional. Using the above theorem, we give in Theorem 7.2.15 of Chapter 7 an improved version of Theorem 2.12 of [T] (with the open Baire sets in T replacing the open sets in T).

Theorem 5.2.23. (Generalization of Theorem 5.2.6 to sequentially complete lcHs-valued σ-additive regular Borel measures). *Let X be a sequentially complete lcHs and let $\mathbf{m}_n : \mathcal{B}(T) \to X$, $n \in \mathbb{N}$, be σ-additive and Borel regular. Then $\lim_n \mathbf{m}_n(U) \in X$ for each open Baire set U in T if and only if there exists a unique X-valued σ-additive Borel regular measure $\mathbf{m}$ on $\mathcal{B}(T)$ such that*

$$\lim_n \int_T f d\mathbf{m}_n = \int_T f d\mathbf{m} \in X$$

for each bounded $\mathcal{B}(T)$-measurable scalar function f on T.

Proof. For each $q \in \Gamma$, let $(\mathbf{m}_n)_q = \Pi_q \circ \mathbf{m}_n$. Then $(\mathbf{m}_n)_q : \mathcal{B}(T) \to X_q \subset \widetilde{X}_q$ is σ-additive and Borel regular for each $n \in \mathbb{N}$. Suppose there exists $x_U \in X$ such that $\lim_n \mathbf{m}_n(U) = x_U$ for each open Baire set U in T. Then $\lim_n (\mathbf{m}_n)_q(U) = \Pi_q(x_U) \in X_q \subset \widetilde{X}_q$ for each $q \in \Gamma$. Then by Theorem 5.2.21 applied to $(\mathbf{m}_n)_q$, $n \in \mathbb{N}$, there exists a unique σ-additive Borel regular measure $\gamma_q : \mathcal{B}(T) \to \widetilde{X}_q$ such that

$$\lim_n \int_A f d(\mathbf{m}_n)_q = \int_A f d\gamma_q (\in \widetilde{X}_q) \tag{5.2.23.1}$$

for each $A \in \mathcal{B}(T)$ and for each bounded $\mathcal{B}(T)$-measurable scalar function f on T. Then $q \left(\int_A f d\mathbf{m}_n - \int_A f d\mathbf{m}_k \right) \to 0$ as $n, k \to \infty$ for each $A \in \mathcal{B}(T)$ and hence $(\int_A f d\mathbf{m}_n)_1^\infty$ is Cauchy in X, for each $A \in \mathcal{B}(T)$ and for each bounded Borel measurable scalar function f. Consequently, as X is sequentially complete, $\lim_n \mathbf{m}_n(A) = \mathbf{m}(A)$ (say) exists in X, for each $A \in \mathcal{B}(T)$. Clearly, $\mathbf{m} : \mathcal{B}(T) \to X$ is additive. Moreover, $\lim_n (\mathbf{m}_n)_q(A) = (\Pi_q \circ \mathbf{m})(A)$ for $A \in \mathcal{B}(T)$ and for $q \in \Gamma$. But $\lim_n (\mathbf{m}_n)_q(A) = \gamma_q(A)$ by (5.2.23.1) for $q \in \Gamma$. Hence

$$(\Pi_q \circ \mathbf{m})(A) = \gamma_q(A) \tag{5.2.23.2}$$

for $A \in \mathcal{B}(T)$.

Claim 1. $\mathbf{m} : \mathcal{B}(T) \to X$ is σ-additive.

In fact, let $(A_i)_1^\infty \subset \mathcal{B}(T)$ be a disjoint sequence. Given $q \in \Gamma$ and $\epsilon > 0$, there exists $n_0(q)$ such that $|\gamma_q(\bigcup_1^\infty A_i) - \sum_1^n \gamma_q(A_i)|_q < \epsilon$ for $n \geq n_0(q)$, since γ_q is σ-additive on $\mathcal{B}(T)$. Then by (5.2.23.2) we have, $q \left(\mathbf{m}(\bigcup_1^\infty A_i) - \sum_1^n \mathbf{m}(A_i) \right) = |(\Pi_q \circ \mathbf{m})(\bigcup_1^\infty A_i) - \sum_1^n (\Pi_q \circ \mathbf{m})(A_i)|_q = \gamma_q(\bigcup_1^\infty A_i) - \sum_1^n \gamma_q(A_i)|_q < \epsilon$ for $n \geq n_0(q)$. Then, as $q \in \Gamma$ is arbitrary, it follows that $\mathbf{m}(\bigcup_1^\infty (A_i) = \sum_1^\infty \mathbf{m}(A_i)$ and hence the claim holds.

Now, for $q \in \Gamma$, by (5.2.23.1) and (5.2.23.2), by Claim 1 and by Remark 3.3.5′ in Remark 3.3.11 and Theorem 3.3.8(v) of Chapter 3, we have

$$q \left(\int_T f d\mathbf{m}_n - \int_T f d\mathbf{m} \right) = \left| \Pi_q \left(\int_T f d\mathbf{m}_n - \int_T f d\mathbf{m} \right) \right|_q$$

$$= \left| \int_T f d(\Pi_q \circ \mathbf{m}_n) - \int_T f d(\Pi_q \circ \mathbf{m}) \right|_q$$

$$= \left| \int_T f d(\mathbf{m}_n)_q - \int_T f d(\gamma_q) \right|_q \to 0 \quad \text{as} \quad n \to \infty.$$

Hence $\lim_n \int_T f d\mathbf{m}_n = \int_T f d\mathbf{m}$.

To prove that $\mathbf{m}$ is Borel regular, let $A \in \mathcal{B}(T)$, $q \in \Gamma$ and $\epsilon > 0$. Since $\Pi_q \circ \mathbf{m} = \gamma_q$ by (5.2.23.2) and since γ_q is Borel regular for each $q \in \Gamma$, in view of Definition 5 of [P9] it follows that $\mathbf{m}$ is Borel regular. The uniqueness of $\mathbf{m}$ is proved by an argument similar to that in the proof of Theorem 5.2.21, except that we have to appeal to Theorem 3.2.8(v) and Remark 3.3.11 of Chapter 3 instead of Theorem 2.1.5(viii) and Remark 2.2.3 of Chapter 2.

The converse is evident and hence the theorem holds.   □

Remark 5.2.24. Only the Banach space version of Theorem 5.2.17 which is deduced from Theorem 5.2.16 is used in the proof of Theorem 5.2.21. However, Theorem 5.2.16 in its generality is needed in the proof of Theorem 6.3.8 of Chapter 6 which improves Theorem 3.3.2 of Chapter 3 when $\mathcal{P} = \delta(\mathcal{C})$, $\mathbf{m}$ is $\delta(\mathcal{C})$-regular and σ-additive and X is a complete lcHs. Theorem 5.2.16 is used in the proof of Theorem 6.3.4 of Chapter 6 which strengthens Theorem 1.3.2 of Chapter 1 when $\mathcal{P}$ and $\mathbf{m}$ are as above and $\mathbf{m}$ is Banach space-valued.

5.3   Weakly compact bounded Radon operators and prolongable Radon operators

Notation 5.3.1. $C_c(T)$ always denotes the normed space $(C_c(T), ||\cdot||_T)$. For $C \in \mathcal{C}$, let $C_c(T, C) = \{f \in C_c(T) : \operatorname{supp} f \subset C\}$ and let $I_C : C_c(T, C) \hookrightarrow C_c(T)$ be the canonical injection. Let ξ be the inductive limit locally convex topology on $C_c(T)$ induced by the family $\{C_c(T, C), I_C\}$, where $C_c(T, C)$ are provided with the topology τ_u of uniform convergence. Then we denote $(C_c(T), \xi)$ by $\mathcal{K}(T)$. It is well known that $\mathcal{K}(T)$ is an lcHs and $\mathcal{K}(T)^*$ denotes the topological dual of $\mathcal{K}(T)$. Similarly, one defines the locally convex space $\mathcal{K}(T, \mathbb{R})$ corresponding to $C_c^r(T) = \{f \in C_c(T), f \text{real-valued}\}$. See §1, Chapter III of [B].

For the convenience of the reader, let us recall the following notation given in the end of Notation 5.2.1.

Notation 5.3.2. **$\mathcal{V}$ denotes the family of all relatively compact open sets in T.**

Let X be an lcHs with topology τ unless otherwise mentioned. Recall Definition 5 of [P9] for the notions of $\mathcal{R}$-regularity, $\mathcal{R}$-outer regularity and $\mathcal{R}$-inner regularity of a vector measure (i.e., a vector-valued additive set function).

In this section, following Thomas [T], we introduce the notions of weakly compact bounded Radon operators and prolongable Radon operators on $\mathcal{K}(T)$ with values in a quasicomplete lcHs. Using the results of [P9] and those of Section 5.2 above, we give several characterizations of these operators; those for weakly compact bounded Radon operators given here are different from the characterizations obtained in [P9].

Definition 5.3.3. Let X be an lcHs and let $u : \mathcal{K}(T) \to X$ be a continuous linear mapping. This means, for each $C \in \mathcal{C}$ and $q \in \Gamma$, there exists a finite constant $M_{C,q}$ such that $|u(\varphi)|_q \leq M_{C,q}||\varphi||_T$ for all $\varphi \in C_c(T, C)$. Such a mapping u is called an X-valued Radon operator. (Thomas calls it an X-valued Radon measure in [T].)

Theorem 5.3.4. (Integral representation of Radon operators). *Let X be a quasi-complete lcHs and let $u : \mathcal{K}(T) \to X$ be a Radon operator. Then there exists a vector measure (i.e., a vector-valued additive set function) $\mathbf{m} : \delta(\mathcal{C}) \to X^{**}$ such that*

(i) *$x^* \circ \mathbf{m} : \delta(\mathcal{C}) \to \mathbb{K}$ is σ-additive and $\delta(\mathcal{C})$-regular for each $x^* \in X^*$;*

(ii) *$\{\mathbf{m}(A) : A \in \mathcal{B}(V)\}$ is τ_e-bounded in X^{**} (see Notation and Terminology 5.2.10) for each $V \in \mathcal{V}$ and*

(iii) *for each $\varphi \in C_c(T)$, $u(\varphi) = \int_T \varphi d\mathbf{m}$ (in the sense of Definition 1 of [P7]), where X is identified as a subspace of (X^{**}, τ_e). Finally, (i)–(iii) determine $\mathbf{m}$ uniquely.*

Proof. Let $V \in \mathcal{V}$ and let $u_V = u|_{C_c(V)}$. Let $q \in \Gamma$. For $\varphi \in C_c(V)$, $\operatorname{supp} \varphi \subset \bar{V} \in \mathcal{C}$ and hence $|u_V(\varphi)|_q = |u(\varphi)|_q \leq M_{\bar{V},q}\|\varphi\|_T$ so that u_V is continuous. As X is sequentially complete, u_V has a unique continuous linear extension $\widetilde{u_V}$ to the whole of $C_0(V)$ with values in X. Then by Theorem 1 of [P9], $\widetilde{u_V}$ has the representing measure $\mathbf{m}_V$ (as an additive set function) on $\mathcal{B}(V)$ with values in X^{**} and $\mathbf{m}_V(A) = \widetilde{u_V}^{**}(\chi_A) = u_V^{**}(\chi_A)$ for $A \in \mathcal{B}(V)$; $x^* \circ \mathbf{m}_V : \mathcal{B}(V) \to \mathbb{K}$ is σ-additive and $\mathcal{B}(V)$-regular for $x^* \in X^*$; the mapping $x^* \to x^* \circ \mathbf{m}_V$ of X^* into $M(V)$ is weak*-weak* continuous;

$$x^* u_V(\varphi) = \int_T \varphi d(x^* \circ \mathbf{m}_V) \qquad (5.3.4.1)$$

for $\varphi \in C_0(V)$ and for $x^* \in X^*$, and

$$\{\mathbf{m}_V(A) : A \in \mathcal{B}(V)\} \quad \text{is } \tau_e\text{-bounded in } X^{**}. \qquad (*)$$

Let $A \in \delta(\mathcal{C})$. Then there exists $V \in \mathcal{V}$ such that $A \subset V$ so that $A \in \mathcal{B}(V)$. Let $\mathbf{m}(A) = \mathbf{m}_V(A)$.

Claim 1. $\mathbf{m} : \delta(\mathcal{C}) \to X^{**}$ is a well-defined, vector measure (i.e., an additive set function).

In fact, let $A \in \delta(\mathcal{C})$ and let $V_1, V_2 \in \mathcal{V}$ such that $A \subset V_1 \cap V_2$. Then $A \in \mathcal{B}(V_i)$ and the continuous linear mapping $\widetilde{u_{V_i}}$ on $C_0(V_i)$ has the representing measure $\mathbf{m}_{V_i}$ for $i = 1, 2$. Clearly, $A \in \mathcal{B}(V_1 \cap V_2)$ and for $\varphi \in C_c(V_1 \cap V_2)$,

$$x^* u_{V_1}(\varphi) = x^* u_{V_2}(\varphi) = x^* u(\varphi) = x^* u_{V_1 \cap V_2}(\varphi) \text{ for } x^* \in X^* \qquad (5.3.4.1)$$

and hence by (5.3.4.1) we have

$$\int_T \varphi d(x^* \circ \mathbf{m}_{V_1}) = \int_T \varphi d(x^* \circ \mathbf{m}_{V_2}) = \int_T \varphi d(x^* \circ \mathbf{m}_{V_1 \cap V_2}) \qquad (5.3.4.2)$$

for $x^* \in X^*$. As $x^* \circ \mathbf{m}_{V_1 \cap V_2}$, $x^* \circ \mathbf{m}_{V_1}|_{\mathcal{B}(V_1 \cap V_2)}$ and $x^* \circ \mathbf{m}_{V_2}|_{\mathcal{B}(V_1 \cap V_2)}$ belong to $M(V_1 \cap V_2)$, by (5.3.4.2) and by the uniqueness part of the Riesz representation theorem we have $x^* \circ \mathbf{m}_{V_1}(A) = x^* \circ \mathbf{m}_{V_2}(A) = x^* \circ \mathbf{m}_{V_1 \cap V_2}(A)$ for $x^* \in X^*$. As $\mathbf{m}_{V_1}(A)$, $\mathbf{m}_{V_2}(A)$ and $\mathbf{m}_{V_1 \cap V_2}(A)$ belong to X^{**}, we conclude that $\mathbf{m}_{V_1}(A) = \mathbf{m}_{V_2}(A) = \mathbf{m}_{V_1 \cap V_2}(A)$. Hence $\mathbf{m}$ is well defined. Moreover, let $A_1, A_2 \in \delta(\mathcal{C})$ with $A_1 \cap A_2 = \emptyset$. Let $V \in \mathcal{V}$ such that $A_1 \cup A_2 \subset V$. Then, as $\mathbf{m}_V$ is additive on $\mathcal{B}(V)$, we have $\mathbf{m}(A_1 \cup A_2) = \mathbf{m}_V(A_1 \cup A_2) = \mathbf{m}_V(A_1) + \mathbf{m}_V(A_2) = \mathbf{m}(A_1) + \mathbf{m}(A_2)$ and hence $\mathbf{m}$ is additive. Therefore, Claim 1 holds.

Claim 2. $(x^* \circ \mathbf{m})$ is σ-additive on $\delta(\mathcal{C})$ for each $x^* \in X^*$.

In fact, let $(A_i)_1^\infty \subset \delta(\mathcal{C})$, $A_i \cap A_j = \emptyset$ for $i \neq j$ and let $A = \bigcup_1^\infty A_i \in \delta(\mathcal{C})$. Then there exists $V \in \mathcal{V}$ such that $A \subset V$ so that $A, (A_i)_1^\infty \subset \mathcal{B}(V)$. Then $(x^* \circ \mathbf{m})(A) = (x^* \circ \mathbf{m}_V)(A) = \sum_1^\infty (x^* \circ \mathbf{m}_V)(A_i) = \sum_1^\infty (x^* \circ \mathbf{m})(A_i)$ for $x^* \in X^*$. Hence Claim 2 holds.

Claim 3. $(x^* \circ \mathbf{m})$ is $\delta(\mathcal{C})$-regular for $x^* \in X^*$.

In fact, let $A \in \delta(\mathcal{C})$ and $\epsilon > 0$. Choose $V \in \mathcal{V}$ such that $A \subset V$. Then by the $\mathcal{B}(V)$-regularity of $x^* \circ \mathbf{m}_V$ there exist $K \in \mathcal{C}$ and a set U open in V such that $K \subset A \subset U$ and such that $v(x^* \circ \mathbf{m}_V, \mathcal{B}(V))(U \backslash K) < \epsilon$. Then, as V is open in T, U is also open in T. As $\mathbf{m}|_{\mathcal{B}(V)} = \mathbf{m}_V$, and as $v(x^* \circ \mathbf{m}, \delta(\mathcal{C}))(U \backslash K) = v(x^* \circ \mathbf{m}, \mathcal{B}(V))(U \backslash K) = v(x^* \circ \mathbf{m}_V, \mathcal{B}(V))(U \backslash K) < \epsilon$, Claim 3 holds.

By the above claims, $\mathbf{m}$ verifies (i) of the theorem. Since $\{\mathbf{m}(A) : A \in \mathcal{B}(V)\} = \{\mathbf{m}_V(A) : A \in \mathcal{B}(V)\}$, by (*) (ii) of the theorem also holds.

Let $\varphi \in C_c(T)$ and let $\operatorname{supp}\varphi = K$. Then choose $V \in \mathcal{V}$ such that $K \subset V$. As φ is a bounded $\mathcal{B}(V)$-measurable function and as $\mathbf{m}_V$ is an X^{**}-valued τ_e-bounded vector measure on $\mathcal{B}(V)$, by the proof of Proposition 5.2.11 above, φ is not only $\mathbf{m}_V$-integrable in the sense of Definition 1 of [P7], but also $u(\varphi) = u_V(\varphi) = \int_T \varphi d\mathbf{m}_V$, considering X as a subspace of (X^{**}, τ_e). Moreover, as $\mathbf{m}_V = \mathbf{m}|_{\mathcal{B}(V)}$ and as $\varphi \in C_c(V)$, we conclude that $u(\varphi) = \int_T \varphi d\mathbf{m}$, for $\varphi \in C_c(T)$. Thus $\mathbf{m}$ verifies (iii) of the theorem.

To prove the uniqueness of $\mathbf{m}$, if possible let $\mathbf{n}$ be another X^{**}-valued vector measure on $\delta(\mathcal{C})$ such that (i)–(iii) hold for $\mathbf{n}$. Then as $x^* \in X^*$ is continuous on (X^{**}, τ_e), by Lemma 6 of [P7] and by (iii) we have

$$x^* u(\varphi) = \int_T \varphi d(x^* \circ \mathbf{m}) = \int_T \varphi d(x^* \circ \mathbf{n}) \qquad (5.3.4.3)$$

for $\varphi \in C_c(T)$ and for $x^* \in X^*$. Let $V \in \mathcal{V}$. As (5.3.4.3) holds for all $\varphi \in C_c(V)$, by the uniqueness part of the Riesz representation theorem we have $(x^* \circ \mathbf{n})|_{\mathcal{B}(V)} = (x^* \circ \mathbf{m})|_{\mathcal{B}(V)}$ for $x^* \in X^*$. Hence $\mathbf{n}(A) = \mathbf{m}(A)$ for $A \in \mathcal{B}(V)$. Since $\delta(\mathcal{C}) = \bigcup_{V \in \mathcal{V}} \mathcal{B}(V)$, it follows that $\mathbf{n} = \mathbf{m}$. Hence $\mathbf{m}$ is unique. □

The following definition is suggested by Theorem 5.3.4.

Definition 5.3.5. Let X be a quasicomplete lcHs and let $u : \mathcal{K}(T) \to X$ be a Radon operator. The unique X^{**}-valued vector measure $\mathbf{m}$ on $\delta(\mathcal{C})$ satisfying (i)–(iii) of Theorem 5.3.4 is called the representing measure of u.

Following Thomas [T] we give the following definition.

Definition 5.3.6. Let X be a quasicomplete lcHs. A linear mapping $u : \mathcal{K}(T) \to X$ is called a weakly compact bounded Radon operator if u is continuous on $C_c(T)$ for the topology of uniform convergence (i.e., for the topology induced by $|| \cdot ||_T$) and if its continuous extension to $(C_0(T), || \cdot ||_T)$ is weakly compact.

In the light of the above definition, weakly compact bounded Radon operators on $\mathcal{K}(T)$ can be considered as weakly compact operators on $(C_0(T), || \cdot ||_T)$, and [P9] gives 35 characterizations of these operators. An alternative proof based on the Borel extension theorem is given in [P14] to obtain these characterizations. The reader may also refer to [P11] for a simple proof of many of these characterizations where three new characterizations are also given. Part (ii) of the following theorem gives some more characterizations of these operators when the lcHs X satisfies some additional hypothesis and these are suggested by [T]. See also Theorems 5.3.14 and 5.3.15 for further characterizations of these operators.

Theorem 5.3.7. *Let X be a quasicomplete lcHs with topology τ and let $u : C_0(T) \to X$ be a continuous linear mapping. Then:*

(i) *u is weakly compact if and only if , for each uniformly bounded sequence $(\varphi_n)_1^\infty \subset C_0(T)$ with $\varphi_n(t) \to 0$ for $t \in T$, $u(\varphi_n) \to 0$ in X.*

(ii) *Let $H \subset X^*$ have the Orlicz property and let τ be identical with the topology of uniform convergence in equicontinuous subsets of H. Let $\mathbf{m}$ be the representing measure of u in the sense of Definition 4 of [P9]. Then u is weakly compact if and only if any one of the following holds:*

(a) *For each open set U in T there exists a vector $x_U \in X$ such that $(x^* \circ \mathbf{m})(U) = x^*(x_U)$ for each $x^* \in H$.*

(b) *Similar to (a) with U σ-Borel open sets in T.*

(c) *Similar to (a) with U open Baire sets in T.*

(d) *Similar to (a) with U σ-compact open sets in T.*

(e) *Similar to (a) with U open and F_σ in T.*

(f) *For each closed set F in T there exists a vector $x_F \in X$ such that $(x^* \circ \mathbf{m})(F) = x^*(x_F)$ for each $x^* \in H$.*

(g) *Similar to (f) with F closed G_δ's in T.*

Proof. (i) Let $\mathbf{m}$ be the representing measure of u. Then by Theorem 1 of [P9], $u^*x^* = x^* \circ \mathbf{m}$ for $x^* \in X^*$ and $x^*u(\varphi) = \int_T \varphi d(x^* \circ \mathbf{m})$ for $\varphi \in C_0(T)$. Let $(\varphi_n)_1^\infty \subset C_0(T)$ be uniformly bounded and let $\varphi_n(t) \to 0$ for $t \in T$. Then $u(\varphi_n) \to 0$ if and only if $q_E(u(\varphi_n)) = \sup_{x^* \in E} |x^*u(\varphi_n)| = \sup_{x^* \in E} |\int_T \varphi_n d(x^* \circ \mathbf{m})| = \sup_{\mu \in u^*E} |\int_T \varphi_n d\mu| \to 0$ as $n \to \infty$ for each equicontinuous set E in X^* since the topology τ of X is the same as that of uniform convergence in the equicontinuous subsets of X^* by Proposition 7, §4, Chapter 3 of [Ho]. For an equicontinuous set E in X^*, u^*E is bounded by Lemma 2 of [P9] and hence by Theorem 2 of [G] the above condition holds if and only if u^*E is relatively weakly compact in $M(T)$ and hence by Proposition 4 of [P9] or by Corollary 9.3.7 of [E], if and only if u is weakly compact. Hence (i) holds.

(ii) Clearly, (a)$\Rightarrow$(b)$\Rightarrow$(c) and by Theorem 5.2.16, (c) implies that u is weakly compact. Conversely, if u is weakly compact, then by Theorem 2(ii) of [P9], $\mathbf{m}$ has range in X and hence (a), (b) and (c) hold. By Lemma 5.2.3, (c)$\Leftrightarrow$(d). Clearly, (e)$\Rightarrow$(d) and (a)$\Rightarrow$(e). Hence (a)–(e) are equivalent.

(a)$\Rightarrow$(f) In fact, let F be a closed set in T. Let $U = T\backslash F$. Then by (a) there exist vectors $x_U, x_T \in X$ such that $x^*(x_U) = (x^* \circ \mathbf{m})(U)$ and $x^*(x_T) = (x^* \circ \mathbf{m})(T)$ for $x^* \in H$. Then $(x^* \circ \mathbf{m})(F) = x^*(x_T - x_U)$ for $x^* \in H$ and hence (f) holds. Similarly, (f) implies (a) as T is closed and as $F = T\backslash U$ is closed for an open set U in T.

By taking complements, we see that (g) and (e) are equivalent. Hence the theorem holds. $\square$

Following Thomas [T] we give the following definition and its equivalence with Definition 3.1 of [T] will be proved in Theorem 7.3.5 of Chapter 7. (See condition (vi) of the latter theorem.)

Definition 5.3.8. Let X be a quasicomplete lcHs and let $u : \mathcal{K}(T) \to X$ be a Radon operator. Then u is said to be prolongable if, for each $V \in \mathcal{V}$, the continuous linear map $u_V = u|_{C_c(V)}$ is continuous for the topology of uniform convergence on $C_c(V)$ and if its continuous extension $\tilde{u}_V$ to $C_0(V)$ is weakly compact.

The weakly compact bounded Radon operators in Definition 5.3.6 and prolongable Radon operators in Definition 5.3.8 are called respectively weakly compact bounded Radon measures and prolongable Radon measures in [T].

When the Radon operator is prolongable, we can strengthen Theorem 5.3.4 as below.

Theorem 5.3.9. (Integral representation of prolongable Radon operators). *Let X be a quasicomplete lcHs and let $u : \mathcal{K}(T) \to X$ be a prolongable Radon operator. Then the representing measure $\mathbf{m}$ of u as in Definition 5.3.5 is X-valued, σ-additive and $\delta(\mathcal{C})$-regular (considering X as a subspace of (X^{**}, τ_e)) and*

$$u(\varphi) = \int_T \varphi d\mathbf{m}, \ \varphi \in C_c(T) \tag{5.3.9.1}$$

where the integral is a (BDS)-integral.

Conversely, if $\mathbf{m}$ is an X-valued σ-additive $\delta(\mathcal{C})$-regular measure on $\delta(\mathcal{C})$, then the mapping $u : \mathcal{K}(T) \to X$ given by $u(\varphi) = \int_T \varphi d\mathbf{m}$, $\varphi \in \mathcal{K}(T)$ (the integral being a (BDS)-integral), is a prolongable Radon operator. Moreover, the representing measure of u is $\mathbf{m}$. **For an X-valued prolongable or weakly compact bounded Radon operator u on $\mathcal{K}(T)$ (resp. on $\mathcal{K}(T, \mathbb{R})$), let $\mathbf{m}_u$ denote the representing measure of u.**

Proof. Let u be prolongable. Then by Theorem 5.3.4 there exists a unique X^{**}-valued vector measure $\mathbf{m}$ on $\delta(\mathcal{C})$ such that $x^* \circ \mathbf{m}$ is σ-additive and $\delta(\mathcal{C})$-regular for each $x^* \in X^*$; $u(\varphi) = \int_T \varphi d\mathbf{m}$ for $\varphi \in \mathcal{K}(T)$ in the sense of Definition 1 of [P7] and $\{\mathbf{m}(A) : A \in \mathcal{B}(V)\}$ is τ_e-bounded for each $V \in \mathcal{V}$. Let $V \in \mathcal{V}$ and let $\mathbf{m}_V = \mathbf{m}|_{\mathcal{B}(V)}$. Then, from the proof of Theorem 5.3.4 we note that $\mathbf{m}_V$ is the representing measure of the continuous linear map $\widetilde{u_V}$ on $(C_0(V), ||\cdot||_T)$ (in the sense of Definition 4 of [P9]) and by hypothesis, $\widetilde{u_V} : C_0(V) \to X$ is weakly compact. Then by Theorem 2 of [P9], $\mathbf{m}_V$ is σ-additive on $\mathcal{B}(V)$ and has range in X and by Theorem 6 of [P9], $\mathbf{m}_V$ is $\mathcal{B}(V)$-regular. Since V is arbitrary in $\mathcal{V}$ and since $\delta(\mathcal{C}) = \bigcup_{V \in \mathcal{V}} \mathcal{B}(V)$, we conclude that $\mathbf{m}$ is σ-additive on $\delta(\mathcal{C})$, is $\delta(\mathcal{C})$-regular and has range in X. Let $\varphi \in C_c(T)$ with $\operatorname{supp}\varphi \subset K \in \mathcal{C}$. Let $V \in \mathcal{V}$ such that $K \subset V$. Then $\varphi \in C_c(V) \subset C_0(V)$ and $\mathbf{m}_V = \mathbf{m}|_{\mathcal{B}(V)}$ is σ-additive and X-valued. Then by Theorem 5.3.4, $u\varphi = \int_T \varphi d\mathbf{m} = \int_V \varphi d\mathbf{m}_V$ and by Lemma 5.2.19, the integral is a (BDS)-integral.

Conversely, let $\mathbf{m} : \delta(\mathcal{C}) \to X$ be σ-additive and $\delta(\mathcal{C})$-regular. If $V \in \mathcal{V}$, then $\mathbf{m}_V = \mathbf{m}|_{\mathcal{B}(V)}$ is σ-additive, $\mathcal{B}(V)$-regular and X-valued. Let $u : \mathcal{K}(T) \to X$ be given by $u(\varphi) = \int_T \varphi d\mathbf{m}$ for $\varphi \in \mathcal{K}(T)$ where the integral is a (BDS)-integral. Then by Theorem 4.1.9(i)(b) and Remark 4.1.5 of Chapter 4, u is a continuous linear map. Let $U \in \mathcal{V}$. Then by Theorem 50.D of [H], there exists $W \in \mathcal{V}$ such that $\bar{U} \subset W$ and hence $C_0(U) \subset C_c(W)$. Then $\mathbf{m}_W = \mathbf{m}|_{\mathcal{B}(W)}$ is X-valued and σ-additive in τ and hence by Lemma 5.2.19, $\widetilde{u_W} : C_0(W) \to X$ given by $\widetilde{u_W}(\varphi) = \int_W \varphi d\mathbf{m}_W$ is weakly compact and hence $\widetilde{u_U} = \widetilde{u_W}|_{C_0(U)}$ is weakly compact. Hence u is prolongable. Clearly, $\mathbf{m}$ satisfies (i) of Theorem 5.3.4. Since $\mathbf{m}|_{\mathcal{B}(V)}$ is σ-additive, by Proposition 7, §4, Chapter 3 of [Ho], (ii) of Theorem 5.3.4 also holds. By Lemma 5.2.19, the (BDS)-integral $\int_T \varphi d\mathbf{m}$ for $\varphi \in \mathcal{K}(T)$ is the same as the integral in the sense of Definition 1 of [P7], noting that $\tau = \tau_e|_X$ when X is considered as a subspace of (X^{**}, τ_e) by the proposition of [Ho] cited above. Hence $\mathbf{m}$ is the representing measure of u in the sense of Definition 5.3.5. $\qquad\square$

Corollary 5.3.10. *A linear functional θ belongs to $\mathcal{K}(T)^*$ (resp. $\mathcal{K}(T)_b^*$ (see [P4])) if and only if $\theta : \mathcal{K}(T) \to \mathbb{K}$ is a prolongable (resp. weakly compact bounded) Radon operator. In that case, its representing measure $\mathbf{m}_\theta$ is the same as the complex Radon measure μ_θ induced by θ in the sense of Definition 4.3 of [P3].*

Thus prolongable (resp. weakly compact bounded) Radon operators on $\mathcal{K}(T)$ with values in a quasicomplete lcHs generalize complex measures (resp. bounded complex measures) in the sense of [B].

Theorem 5.3.11. *Let X and Y be quasicomplete lcHs over $\mathbb{C}$ (resp. over $\mathbb{R}$). Let $u : \mathcal{K}(T) \to X$ (resp. $u : \mathcal{K}(T, \mathbb{R}) \to X$) be a prolongable Radon operator, let $w : \mathcal{K}(T) \to X$ (resp. $w : \mathcal{K}(T, \mathbb{R}) \to X$) be a weakly compact bounded Radon operator and let $v : X \to Y$ be a continuous linear mapping. Then:*

(i) $v \circ u : \mathcal{K}(T) \to Y$ *(resp.* $v \circ u : \mathcal{K}(T, \mathbb{R}) \to Y$*) is a prolongable Radon operator.*

(ii) $v \circ w : \mathcal{K}(T) \to Y$ (resp. $v \circ w : \mathcal{K}(T, \mathbb{R}) \to Y$) is a weakly compact bounded Radon operator.

(iii) $\mathbf{m}_{v \circ u}(A) = v(\mathbf{m}_u(A))$ for $A \in \delta(\mathcal{C})$, where $\mathbf{m}_u$ and $\mathbf{m}_{v \circ u}$ are the representing measures of u and $v \circ u$, respectively.

(iv) $\mathbf{m}_{v \circ w}(A) = v(\mathbf{m}_w)(A)$ for $A \in \mathcal{B}(T)$, where $\mathbf{m}_w$ and $\mathbf{m}_{v \circ w}$ are the representing measures of w and $v \circ w$, respectively.

(v) If $f \in \mathcal{L}_1(\mathbf{m}_u)$, then $f \in \mathcal{L}_1(\mathbf{m}_{v \circ u})$ and $\int_A f d\mathbf{m}_{v \circ u} = v(\int_A f d\mathbf{m}_u)$ for $A \in \mathcal{B}_c(T)$.

(vi) If $f \in \mathcal{L}_1(\mathbf{m}_w)$, then $f \in \mathcal{L}_1(\mathbf{m}_{v \circ w})$ and $\int_A f d\mathbf{m}_{v \circ w} = v(\int_A f d\mathbf{m}_w)$ for $A \in \mathcal{B}(T)$.

Proof. (i) If Φ is a weakly compact operator with range in X, then it is well known that $v \circ \Phi$ is weakly compact. This result is used to prove (i) and (ii).

(iii) Let u be prolongable and let $V \in \mathcal{V}$. Then $\widetilde{u_V} : C_0(V) \to X$ is weakly compact and its representing measure $(\mathbf{m}_u)_V$ on $\mathcal{B}(V)$ has range in X and is given by $(\mathbf{m}_u)_V = \widetilde{u_V}^{**}|_{\mathcal{B}(V)} = u_V^{**}|_{\mathcal{B}(V)}$. Let $A \in \delta(\mathcal{C})$ and choose $V \in \mathcal{V}$ such that $A \subset V$. Then by 5.2.10 we have $(\mathbf{m}_u)_V(A) = u_V^{**}(\chi_A)$ and $(v \circ u_V)^{**}(\chi_A) = (\mathbf{m}_{v \circ u})_V(A)$. Hence $\mathbf{m}_{v \circ u}(A) = (\mathbf{m}_{v \circ u})_V(A) = (v \circ u_V)^{**}(\chi_A) = v^{**} u_V^{**}(\chi_A) = v^{**}(\mathbf{m}_u)_V(A) = v^{**}\mathbf{m}_u(A) = v \circ \mathbf{m}_u(A)$ as $\mathbf{m}_u(A) \in X$ by Theorem 5.3.9 and as $v^{**}|_X = v$. Hence (iii) holds.

(iv) Similar to the proof of (iii) when w is a weakly compact bounded Radon operator.

(v)(resp. (vi)) Let $u' = u$ or $u' = w$. Let $f \in \mathcal{L}_1(\mathbf{m}_{u'})$. Then by Theorem 4.1.8(v) and Remark 4.1.5 of Chapter 4, $f \in \mathcal{L}_1(v \circ \mathbf{m}_{u'})$ and $\int_A f d(v \circ \mathbf{m}_{u'}) = v(\int_A f d\mathbf{m}_{u'})$ for $A \in \mathcal{B}_c(T)$ (resp. for $A \in \mathcal{B}(T)$) if u' is a prolongable Radon operator (resp. u' is a weakly compact bounded Radon operator). As $\mathbf{m}_{v \circ u'} = v \circ \mathbf{m}_{u'}$ by (iii) (resp. by (iv)), (v) (resp. (vi)) holds. $\square$

The following theorem gives 22 characterizations for a Radon operator to be prolongable and [P9] plays a key role in the proof of the theorem. For the different concepts of regularity used in the following theorem see Definition 5 of [P9].

Theorem 5.3.12. *Let X be a quasicomplete lcHs and let $u : \mathcal{K}(T) \to X$ be a Radon operator. Let $\mathbf{m} : \delta(\mathcal{C}) \to X^{**}$ be the representing measure of u in the sense of Definition 5.3.5 and let $\mathbf{m}_0 = \mathbf{m}|_{\delta(\mathcal{C}_0)}$. By Proposition 7, §4, Chapter 3 of [Ho] we consider X as a subspace of (X^{**}, τ_e). Then the following statements are equivalent:*

(1) *u is prolongable.*

(2) *$\mathbf{m}$ has range in X.*

(3) *$\mathbf{m}$ is σ-additive for the topology τ_e of X^{**}.*

(4) *$\mathbf{m}(V) \in X$ for each $V \in \mathcal{V}$.*

(5) *$\mathbf{m}(V) \in X$ for each $V \in \mathcal{V} \cap \mathcal{B}_0(T)$.*

(6) $\mathbf{m}(K) \in X$ for each $K \in \mathcal{C}$.

(7) $\mathbf{m}(K) \in X$ for each $K \in \mathcal{C}_0$.

(8) For each $U \in \mathcal{V}$ and for each increasing sequence $(f_n)_1^\infty \subset C_0(U)$ with $0 \le f_n \le 1$ in U for all n, (uf_n) converges weakly in X.

(9) Similar to (8) with $U \in \mathcal{V} \cap \mathcal{B}_0(T)$.

(10) $\mathbf{m}_0$ is σ-additive for τ_e.

(11) $\mathbf{m}_0$ has range in X.

(12) $\mathbf{m}$ is $\delta(\mathcal{C})$-regular (for τ_e).

(13) $\mathbf{m}$ is $\delta(\mathcal{C})$-inner regular (for τ_e).

(14) $\mathbf{m}$ is $\delta(\mathcal{C})$-inner regular (for τ_e) in each $V \in \mathcal{V}$.

(15) $\mathbf{m}_0$ is $\delta(\mathcal{C}_0)$-regular (for τ_e).

(16) $\mathbf{m}_0$ is $\delta(\mathcal{C}_0)$-inner regular (for τ_e).

(17) $\mathbf{m}_0$ is $\delta(\mathcal{C}_0)$-inner regular (for τ_e) in each open set $U \in \delta(\mathcal{C}_0)$.

(18) All bounded Borel measurable functions f on T with compact support (equivalently, all bounded σ-Borel measurable functions f on T with compact support) are $\mathbf{m}$-integrable in T (in the sense of Definition 1 of [P7] with respect to the completion of (X^{**}, τ_e)) and $\int_T f \, d\mathbf{m} \in X$.

(19) All bounded Baire measurable functions f on T with compact support are $\mathbf{m}_0$-integrable in T (in the sense of Definition 1 of [P7] as in (18)) and $\int_T f \, d\mathbf{m}_0 \in X$.

(20) All bounded functions f on T belonging to the first Baire class with compact support are $\mathbf{m}$-integrable in T (in the sense of Definition 1 of [P7] as in (18)) and $\int_T f \, d\mathbf{m} \in X$.

(21) $u_V^{**} f \in X$ for all bounded functions f on T belonging to the first Baire class with compact support, the support being contained in $V \in \mathcal{V}$.

(22) For every uniformly bounded sequence (φ_n) of continuous functions vanishing in $T \backslash K$ for some $K \in \mathcal{C}$ (equivalently, by Urysohn's lemma for every sequence (φ_n) of continuous functions dominated by a member of $\mathcal{K}(T)$) with $\lim_n \varphi_n(t) = 0$ for each $t \in T$, $\lim_n u(\varphi_n) = 0$.

Proof. For $V \in \mathcal{V}$, let $u_V = u|_{(C_c(V), \|\cdot\|_T)}$. As u_V is continuous, it has a unique continuous linear extension $\widetilde{u_V}$ to $C_0(V)$. If $\mathbf{m}_V$ is the representing measure of $\widetilde{u_V}$ on $\mathcal{B}(V)$ as in Definition 4 of [P9], then from the proof of Theorem 5.3.4 it is clear that $\mathbf{m}_V = \mathbf{m}|_{\mathcal{B}(V)}$.

(1)$\Leftrightarrow$(2)(resp. (1)$\Leftrightarrow$(3)) By Theorem 2 of [P9], the range of $\mathbf{m}_V$ is contained in X (resp. $\mathbf{m}_V$ is σ-additive on $\mathcal{B}(V)$ for τ_e) if and only if $\widetilde{u_V}$ is weakly compact. Since $\delta(\mathcal{C}) = \bigcup_{V \in \mathcal{V}} \mathcal{B}(V)$ and since $\mathbf{m}_V = \mathbf{m}|_{\mathcal{B}(V)}$, it follows that $\mathbf{m}$ has range in X (resp. $\mathbf{m}$ is σ-additive on $\delta(\mathcal{C})$ for τ_e) if and only if $\mathbf{m}_V(\mathcal{B}(V)) \subset X$ (resp. $\mathbf{m}_V$ is σ-additive on $\mathcal{B}(V)$ for τ_e) for each $V \in \mathcal{V}$ and hence if and only if $\widetilde{u_V}$ is weakly compact for each $V \in \mathcal{V}$. Hence the result holds.

Clearly, $(2)\Rightarrow(4)\Rightarrow(5)$.

$(5)\Rightarrow(1)$

Claim 1. $\mathcal{B}_0(T)$ is the σ-ring generated by all relatively compact open Baire sets in T.

In fact, given $C \in \mathcal{C}_0$, by Theorem 50.D of [H] and by Lemma 5.2.3, there exists a relatively compact open Baire set U in T such that $C \subset U$. Then $C = U \backslash (U \backslash C)$ and hence the claim holds.

Let $V \in \mathcal{V}$. Since V is relatively compact, Then by Lemma 5.2.3 and Claim 1 above, $\mathcal{B}_0(V)$ is the σ-ring generated by $\mathcal{U}_V = \{U : U \text{open in} V, U \in \mathcal{B}_0(V)\} = \{U : U \text{ open in} V, U = \bigcup_1^\infty F_n, F_n \text{ compact in} V\}$. Since V is open in T, $U \in \mathcal{U}_V$ if and only if $U \subset V$ and U is open and σ-compact in T and hence by Lemma 5.2.3, $U \in \mathcal{U}_V$ if and only if $U \subset V$ and U is an open Baire set in T. Hence $\mathcal{U}_V = \{U \in \mathcal{V} \cap \mathcal{B}_0(T) : U \subset V\}$. Then by hypothesis (5), $\mathbf{m}_V(U) \in X$ for $U \in \mathcal{B}_0(V)$ which are open in V and hence by Theorem 3(vii) of [P9], $\widetilde{u_V}$ is weakly compact. Therefore, (1) holds.

$(4)\Rightarrow(6)$ Given $K \in \mathcal{C}$, by Theorem 50.D of [H] there exists an open set $V \in \mathcal{V}$ such that $K \subset V$. Then, as $K = V \backslash (V \backslash K)$, by hypothesis we have $\mathbf{m}(K) = \mathbf{m}(V) - \mathbf{m}(V \backslash K) \in X$.

$(6)\Rightarrow(7)$ Obvious.

$(7)\Rightarrow(5)$ Let $V \in \mathcal{V} \cap \mathcal{B}_0(T)$. Then as V is a relatively compact open Baire set in T, by Theorem 50.D of [H] there exists $K \in \mathcal{C}_0$ such that $V \subset K$. Then again by Theorem 50.D of [H] and by Lemma 5.2.3, there exists a relatively compact open Baire set U in T such that $K \subset U$. Then $V = K \backslash (K \backslash V)$ and $K \backslash V \in \mathcal{C}_0$ by Theorem 51.D of [H]. Then by hypothesis, $\mathbf{m}(V) \in X$.

$(1)\Rightarrow(8)$ Let $U \in \mathcal{V}$. Then by (1), $\widetilde{u_U} : C_0(U) \to X$ is weakly compact, and by hypothesis, $f_n \nearrow$, $0 \le f_n \le 1$ in U and $(f_n)_1^\infty \subset C_0(U)$. Then by Theorem 3(xi) of [P9], $(uf_n) = (\widetilde{u_U} f_n)$ converges weakly in X and hence (8) holds.

$(8)\Rightarrow(9)$ Obvious.

$(9)\Rightarrow(1)$ Let $V \in \mathcal{V} \cap \mathcal{B}_0(T)$. Let (f_n) be an increasing sequence in $C_0(V)$ such that $0 \le f_n \le 1$ in V for $n \in \mathbb{N}$. Then by (9), $(\widetilde{u_V} f_n) = u(f_n)$ converges weakly in X. Then by Theorem 3(xi) of [P9], $\widetilde{u_V}$ is weakly compact. Now let $U \in \mathcal{V}$. Then by Theorem 50.D of [H] and by Lemma 5.2.3 there exists $V \in \mathcal{V} \cap \mathcal{B}_0(T)$ such that $\bar{U} \subset V$ and hence $\widetilde{u_U} = \widetilde{u_V}|_{C_0(U)}$ is weakly compact. Hence (1) holds.

$(3)\Rightarrow(10)$ Obvious.

$(10)\Rightarrow(1)$ Let $\mathcal{U}_0$ be the family of all open Baire sets in T. Let $V \in \mathcal{B}_0(T) \cap \mathcal{V}$. As seen in the proof of '$(5)\Rightarrow(1)$', $\mathcal{B}_0(V)$ is the σ-ring generated by $\{U \in \mathcal{U}_0 : U \subset V\}$ and hence $\mathcal{B}_0(V) = \sigma(\mathcal{U}_0 \cap V) = \sigma(\mathcal{U}_0) \cap V = \mathcal{B}_0(T) \cap V \subset \delta(\mathcal{C}_0)$ by Theorem 5.E of [H] and by Lemma 5.2.2 above. Then by hypothesis, $\mathbf{m}_0|_{\mathcal{B}_0(V)}$ is σ-additive on $\mathcal{B}_0(V)$ in τ_e and hence by Theorem 4(xiii) of [P9], $\widetilde{u_V}$ is weakly compact. Now if $W \in \mathcal{V}$, then by Theorem 50.D of [H] and by Lemma 5.2.3 above, there exists $V \in \mathcal{B}_0(T) \cap \mathcal{V}$ such that $\bar{W} \subset V$ and consequently, $C_c(W) \subset C_0(V)$. Then $\widetilde{u_W} = \widetilde{u_V}|_{C_0(W)}$ is weakly compact and hence (1) holds.

(2)$\Rightarrow$(11) Obvious.

(11)$\Rightarrow$(1) Arguing as in the proof of '(10)$\Rightarrow$ (1)' and using the hypothesis that $\mathbf{m}_0|_{\mathcal{B}_0(V)}$ has range in X, by Theorem 4(xv) of [P9] we observe that $\widetilde{u_V}$ is weakly compact for each $V \in \mathcal{V} \cap \mathcal{B}_0(T)$. Then arguing as in the last part of the proof of '(10)$\Rightarrow$(1)', we can show that $\widetilde{u_W}$ is weakly compact for each $W \in \mathcal{V}$ and hence (1) holds.

(1)$\Rightarrow$(12) Let $A \in \delta(\mathcal{C})$. Then there exists $V \in \mathcal{V}$ such that $A \subset V$. By (1), $\widetilde{u_V} : C_0(V) \to X$ is weakly compact and hence by Theorem 6(xix) of [P9], $\mathbf{m}_V$ is $\mathcal{B}(V)$-regular. As $A \in \mathcal{B}(V)$, given $\epsilon > 0$ and $q \in \Gamma$, by the regularity of $\mathbf{m}_V$ and by Proposition 1.1.6 of Chapter 1 there exist $K \in \mathcal{C}$ and a set U open in V such that $K \subset A \subset U$ and such that $||\mathbf{m}_V||_q(U \backslash K) < \epsilon$. Hence $||\mathbf{m}||_q(U \backslash K) < \epsilon$ since $\mathbf{m}_V = \mathbf{m}|_{\mathcal{B}(V)}$. As V is open in T, U is open in T and as $U \subset V$, U is relatively compact in T. Hence $U \in \delta(\mathcal{C})$ by Lemma 5.2.2. Therefore, (12) holds.

(12)$\Rightarrow$(13)$\Rightarrow$(14) Obvious.

(14)$\Rightarrow$(1) Let $V \in \mathcal{V}$. Then by (14), $\mathbf{m}_V$ is $\mathcal{B}(V)$-inner regular in each open set in V and hence by Theorem 6(xxi) of [P9], $\widetilde{u_V}$ is weakly compact. Hence (1) holds.

(1)$\Rightarrow$(15) Let $A \in \delta(\mathcal{C}_0)$, $q \in \Gamma$ and $\epsilon > 0$. Then by Theorem 50.D of [H] and by Lemma 5.2.3 above there exists $V \in \mathcal{V} \cap \mathcal{B}_0(T)$ such that $A \subset V$. As $\widetilde{u_V}$ is weakly compact, by Theorem 8(xxvii) of [P9] and by Proposition 1.1.6 of Chapter 1 there exist $K \in \mathcal{C}_0$ and an open set U in T belonging to $\mathcal{B}_0(V)$ such that $K \subset A \subset U$ and $||\mathbf{m}_V||_q(U \backslash K) < \epsilon$. From the proof of '(5)$\Rightarrow$(1)', we note that $U \in \mathcal{B}_0(T) \cap \mathcal{V} \subset \delta(\mathcal{C}_0)$ (by Lemma 5.2.2) and hence $\mathbf{m}_0$ is $\delta(\mathcal{C}_0)$-regular.

(15)$\Rightarrow$(16)$\Rightarrow$(17) Obvious.

(17)$\Rightarrow$(1) Let V be a relatively compact open Baire set in T. Then $V \in \mathcal{B}_0(V)$ since it is shown in the proof of '(5)$\Rightarrow$(1)' that the open Baire sets in $\mathcal{B}_0(V)$ are precisely the open Baire sets in T which are contained in V. Then the hypothesis implies that $\mathbf{m}_0$ is $\mathcal{B}_0(V)$-inner regular in each open Baire set in V and hence particularly in V. Therefore, by Theorem 8(xxix) of [P9], $\widetilde{u_V}$ is weakly compact. Then, given $U \in \mathcal{V}$, arguing as in the last part of the proof of '(10)$\Rightarrow$(1)', we can show that $\widetilde{u_U}$ is weakly compact. Hence (1) holds.

(1)$\Rightarrow$(18) Let f be a bounded Borel measurable function on T with support $K \in \mathcal{C}$. Then by Lemma 5.2.2, $N(f) \in \delta(\mathcal{C})$ and $N(f) \cap f^{-1}(U) \in \delta(\mathcal{C}) \subset \mathcal{B}_c(T)$ for open sets U in $\mathbb{K}$. Hence f is $\mathcal{B}_c(T)$-measurable. Clearly, $\mathcal{B}_c(T)$-measurable functions are Borel measurable. Let $V \in \mathcal{V}$ such that $K \subset V$. Let $\mathcal{U}$ be the family of open sets in T. $\mathcal{U} \cap V$ is the family of open sets in V and hence by Theorem 5.E of [H], $\mathcal{B}(V) = \sigma(\mathcal{U} \cap V) = \sigma(\mathcal{U}) \cap V = \mathcal{B}(T) \cap V$. Since f is $\mathcal{B}_c(T)$-measurable, $f^{-1}(U) \cap N(f) \in \mathcal{B}_c(T) \subset \mathcal{B}(T)$ for U open in $\mathbb{K}$ and clearly, $f^{-1}(U) \cap N(f) \subset f^{-1}(U) \cap \overline{N(f)} = f^{-1}(U) \cap K \subset V$. Hence $f^{-1}(U) \cap N(f) \cap V = f^{-1}(U) \cap N(f) \in$

$\mathcal{B}(V)$ and hence f is $\mathcal{B}(V)$-measurable. By (1), $\widetilde{u_V}$ is weakly compact. Then by Theorem 9(xxxi) of [P9], f is $\mathbf{m}_V$-integrable in V and $\int_T f d\mathbf{m} = \int_V f d\mathbf{m}_V \in X$. (See Definition 3 of [P7] and note that $\{\mathbf{m}_V(A) : A \in \mathcal{B}(V)\} = \{\mathbf{m}(A) : A \in \mathcal{B}(V)\}$ is τ_e-bounded in X^{**}.) Hence (18) holds.

(18)$\Rightarrow$(19)$\Rightarrow$(20) Obvious.

(20)$\Rightarrow$(5) Let $U \in \mathcal{V} \cap \mathcal{B}_0(T)$. Then by Lemma 5.2.3 there exists $(K_n)_1^\infty \subset \mathcal{C}$ such that $K_n \nearrow U$. Then by an argument based on Urysohn's lemma there exists $(f_n)_1^\infty \subset C_c(U)$ such that $f_n \nearrow \chi_U$ in T. Hence χ_U belongs to the first Baire class, and clearly has compact support. Then by hypothesis, $\int_T \chi_U d\mathbf{m} = \mathbf{m}(U) \in X$ and hence (5) holds.

(20)$\Rightarrow$(21) Each bounded function f belonging to the first Baire class is Baire measurable. Moreover, let f have compact support K and choose $V \in \mathcal{V}$ such that $K \subset V$. Then by Definition 5.3.3, $u_V : C_c(V) \to X$ is continuous and hence has continuous extension $\widetilde{u_V}$ on $C_0(V)$. Then $\mathbf{m}_0(A) = \mathbf{m}_V(A) = u_V^{**}(\chi_A)$ for $A \in \mathcal{B}_0(V)$ and $(x^* \circ \mathbf{m}_0)_V = x^* u_V^{**} = u_V^* x^*$ for $x^* \in X^*$. Then $x^* \circ u_V \in \mathcal{K}(V)_b^*$ and $\langle \int_T f d(\mathbf{m}_0)_V, x^* \rangle = \int_T f d(x^* \circ \mathbf{m}_0)_V = \int_T f d(u_V^* x^*) = \langle f, u_V^* x^* \rangle = \langle u_V^{**} f, x^* \rangle$ for $x^* \in X^*$. Hence $u_V^{**} f = \int_T f d(\mathbf{m}_0)_V = \int_T f d\mathbf{m} \in X$. Thus (21) holds.

(21)$\Rightarrow$(5) As seen in the proof of '(20)$\Rightarrow$(5)', χ_V belongs to the first Baire class for each $V \in \mathcal{V} \cap \mathcal{B}_0(T)$ and has compact support. Then by (21), $\mathbf{m}(V) = \mathbf{m}_V(V) = u_V^{**}(\chi_V) \in X$ and hence (5) holds.

(1)$\Rightarrow$(22) Let $(\varphi_n)_1^\infty$ satisfy the hypothesis. Then $\sup_n ||\varphi_n||_T = M$ (say) $< \infty$, and there exists $K \in \mathcal{C}$ such that $\varphi_n(t) = 0$ for $t \in T \backslash K$ and for all n and $\lim_n \varphi_n(t) = 0$ for $t \in T$. Let $V \in \mathcal{V}$ with $K \subset V$. Then $(\varphi_n)_1^\infty \subset C_0(V)$ and by (1), $\widetilde{u_V}$ is weakly compact. Then by Theorem 5.3.7(i), $u(\varphi_n) = \widetilde{u_V}(\varphi_n) \to 0$ in X.

(22)$\Rightarrow$(1) Let $V \in \mathcal{V}$. Let $(\varphi_n)_1^\infty \subset C_0(V)$ with $\operatorname{supp} \varphi_n \subset K \subset V$ for some $K \in \mathcal{C}$ for all n, $\lim_n \varphi_n(t) = 0$ for each $t \in V$ and $\sup_n ||\varphi_n||_T = M < \infty$. Then by (22) and by Theorem 5.3.7(i), $\widetilde{u_V}$ is weakly compact and hence (1) holds.

This completes the proof of the theorem. $\qquad \square$

Theorem 5.3.13. *Let X be a quasicomplete lcHs with topology τ and let $u : \mathcal{K}(T) \to X$ be a Radon operator with the representing measure $\mathbf{m}$. Let H be a set in X^* having the Orlicz property such that τ is identical with the topology of uniform convergence in equicontinuous subsets of H. Let $\mu_{x^* u}$ be the complex Radon measure induced by $x^* u$ as in Definition 4.3 of [P3]. Then the following statements are equivalent:*

(i) *u is prolongable.*

(ii) *For each $V \in \mathcal{V}$, there exists $x_V \in X$ such that $x^*(x_V) = (x^* \circ \mathbf{m})(V)$ for each $x^* \in H$.*

(iii) *Similar to (ii) with $V \in \mathcal{B}_0(T) \cap \mathcal{V}$.*

(iv) *For each $K \in \mathcal{C}$, there exists $x_K \in X$ such that $x^*(x_K) = (x^* \circ \mathbf{m})(K)$ for each $x^* \in H$.*

(v) *Similar to* (iv) *with* $K \in \mathcal{C}_0$.

(vi) *For each* $V \in \mathcal{V}$, *there exists* $x_V \in X$ *such that* $x^*(x_V) = \int_T \chi_V d(\mu_{x^*u})$ *for* $x^* \in H$.

(vii) *Similar to* (vi) *with* $V \in \mathcal{B}_0(T) \cap \mathcal{V}$.

(viii) *Similar to* (vi) *with* V *replaced by* $K \in \mathcal{C}$.

(ix) *Similar to* (viii) *with* $K \in \mathcal{C}_0$.

Proof. (i)$\Rightarrow$(ii)$\Rightarrow$(iii) by (2) of Theorem 5.3.12.

(iii)$\Rightarrow$(i) Let $V \in \mathcal{B}_0(T) \cap \mathcal{V}$. Then as observed in the proof of '(5)$\Rightarrow$(1)' in the proof of Theorem 5.3.12, $\mathcal{B}_0(V)$ is the σ-ring generated by the family $\mathcal{U}_V$ of all open Baire sets in T which are contained in V. If $U \in \mathcal{B}_0(V)$ and if U is open in T, then $U \in \mathcal{U}_V \subset V \cap \mathcal{B}_0(T)$ by the proof of '(5)$\Rightarrow$(1)' in the proof of Theorem 5.3.12.

Hence, for each open set $U \in \mathcal{B}_0(V)$ (U being open in V), by hypothesis there exists $x_U \in X$ such that $x^*(x_U) = (x^* \circ \mathbf{m})(U)$ for $x^* \in H$ and hence by Theorem 5.2.16, $\widetilde{u_V}$ is weakly compact. If $W \in \mathcal{V}$, then by Theorem 50.D of [H] and by Lemma 5.2.3, there exists $V \in \mathcal{B}_0(T) \cap \mathcal{V}$ such that $W \subset \bar{W} \subset V$. Then by the above argument $\widetilde{u_V}$ is weakly compact and hence $\widetilde{u_W} = \widetilde{u_V}|_{C_0(W)}$ is weakly compact and hence (i) holds.

By (2) of Theorem 5.3.12, (i)$\Rightarrow$(iv)$\Rightarrow$(v).

(v)$\Rightarrow$(iii) Let $V \in \mathcal{B}_0(T) \cap \mathcal{V}$. Then $\bar{V} \in \mathcal{C}$ and by Theorem 50.D of [H] there exists $K \in \mathcal{C}_0$ such that $\bar{V} \subset K$. Then by hypothesis, there exists $x_K \in X$ such that $x^*(x_K) = (x^* \circ \mathbf{m})(K)$ for $x^* \in H$. As $K \backslash V \in \mathcal{C}_0$ by Theorem 51.D of [H], by hypothesis there exists $x_{K \backslash V} \in X$ such that $x^*(x_{K \backslash V}) = (x^* \circ \mathbf{m})(K \backslash V)$ for $x^* \in H$. Then, as $V = K \backslash (K \backslash V)$, $(x^* \circ \mathbf{m})(V) = x^*(x_K - x_{K \backslash V})$ for $x^* \in H$ and hence (iii) holds.

First we prove the following result.

Claim 1. $(x^*u) \in \mathcal{K}(T)^*$ and $\mu_{x^*u} = x^* \circ \mathbf{m}$ on $\delta(\mathcal{C})$.

In fact, as u is continuous on $\mathcal{K}(T)$, $x^*u \in \mathcal{K}(T)^*$. By Theorem 5.3.4, $u(\varphi) = \int_T \varphi d\mathbf{m}$ for $\varphi \in \mathcal{K}(T)$ and consequently, by Lemma 6(ii) of [P3],

$$\int_T \varphi d(x^* \circ \mathbf{m}) = x^*u(\varphi) = \int_T \varphi d(\mu_{x^*u}) \qquad (5.3.13.1)$$

for $\varphi \in \mathcal{K}(T)$ since $x^* \in X^*$ is a continuous linear form on (X^{**}, τ_e). Choose $V \in \mathcal{V}$ such that $\operatorname{supp} \varphi \subset V$. Since $x^* \circ \mathbf{m}$ and μ_{x^*u} are regular on $\delta(\mathcal{C})$ by Theorem 5.3.4 above and by Theorem 4.4(i) of [P4], respectively, both of them are $\mathcal{B}(V)$-regular on $\mathcal{B}(V)$. Moreover, both of them are σ-additive on $\mathcal{B}(V)$. Since (5.3.13.1) holds for all $\psi \in C_c(V)$, by the uniqueness part of the Riesz representation theorem we have $(x^* \circ \mathbf{m})|_{\mathcal{B}(V)} = (\mu_{x^*u})|_{\mathcal{B}(V)}$. Since $\delta(\mathcal{C}) = \bigcup_{V \in \mathcal{V}} \mathcal{B}(V)$, we have $(x^* \circ \mathbf{m})(A) = \mu_{x^*u}(A)$ for $A \in \delta(\mathcal{C})$. Hence the claim holds.

Let u be prolongable. Let $V \in \mathcal{V}$, $U \in \mathcal{V} \cap \mathcal{B}_0(T)$, $K \in \mathcal{C}$ and $K_0 \in \mathcal{C}_0$. Then by (ii) (resp. (iii), (iv), (v)) there exists $x_V \in X$ (resp. $x_U \in X, x_K \in X, x_{K_0} \in X$) such that $(x^* \circ \mathbf{m})(V) = x^*(x_V)$ (resp. $(x^* \circ \mathbf{m})(U) = x^*(x_U), (x^* \circ \mathbf{m})(K) = x^*(x_K), (x^* \circ \mathbf{m})(K_0) = x^*(x_{K_0})$) for $x^* \in H$. Consequently, by Claim 1, (vi) (resp. (vii), (viii), (ix)) holds.

Finally, by Claim 1, (vi) (resp. (vii), (viii), (ix)) implies (ii) (resp. (iii), (iv), (v)) and hence each one implies that u is prolongable.   □

Theorem 5.3.14. *Let X, u, H, μ_{x^*u} for $x^* \in X^*$ and $\mathbf{m}$ be as in Theorem 5.3.13. Then:*

(a) *The following statements are equivalent:*

   (i) *u is a weakly compact bounded Radon operator.*

  (ii) *For each open set U in T there exists $x_U \in X$ such that $x^*(x_U) = \int_T \chi_U d(\mu_{x^*u})$ for $x^* \in H$.*

 (iii) *Similar to (ii) with U σ-Borel open sets in T.*

 (iv) *Similar to (ii) with U open Baire sets in T.*

  (v) *Similar to (ii) with U F_σ-open sets in T.*

 (vi) *Similar to (ii) with U σ-compact open sets in T.*

(vii) *Similar to (ii) with U replaced by closed sets F in T.*

(viii) *Similar to (vii) with F closed G_δ-sets in T.*

(b) *If u is a weakly compact bounded Radon operator, then u is prolongable and the function χ_T is $\mathbf{m}_u$-integrable in T. (Compare with Proposition 7.3.8 of Chapter 7.)*

Proof. (a) Arguing as in the proof of Claim 1 in the proof of Theorem 5.3.13 we have $\mu_{x^*u} = x^* \circ \mathbf{m}$ on $\mathcal{B}(T)$ if $(x^*u) \in \mathcal{K}(T)_b^*$ and consequently, (a) holds by Theorem 5.3.7(ii).

(b) Let u be a weakly compact bounded Radon operator. Then by Definition 5.3.6, $u : (C_0(T), ||\cdot||_u) \to X$ is weakly compact. Then given $V \in \mathcal{V}$, $\tilde{u}_V = u|_{C_0(V)}$ is weakly compact and hence u is prolongable. As $\mathbf{m}_u$ is X-valued and σ-additive on $\mathcal{B}(T)$ by Theorem 2 of [P9] , χ_T is $\mathbf{m}_u$-integrable in T by Theorem 4.1.9(i) and Remark 4.1.5 of Chapter 4.   □

Theorem 5.3.15. *Let X be a quasicomplete lcHs and let $u : (C_c(T), ||\cdot||_T) \to X$ be a continuous linear mapping. Let $\tilde{u}$ be the continuous linear extension of u to $C_0(T)$ and let $\mathbf{m}$ be the representing measure of $\tilde{u}$ in the sense of Definition 4 of [P9]. Let $\mathbf{m}_c = \mathbf{m}|_{\mathcal{B}_c(T)}$ and $\mathbf{m}_0 = \mathbf{m}|_{\mathcal{B}_0(T)}$. Then the following statements are equivalent:*

 (i) *u is a weakly compact bounded Radon operator.*

(ii) *u is prolongable and given $\epsilon > 0$ and $q \in \Gamma$, there exists $K \in \mathcal{C}$ such that $||\mathbf{m}||_q(T\backslash K) < \epsilon$.*

(iii) *u is prolongable and given $\epsilon > 0$ and $q \in \Gamma$, there exists $K \in \mathcal{C}$ such that* $||\mathbf{m}_c||_q(T\backslash K) < \epsilon$ *where*

$$||\mathbf{m}_c||_q(T\backslash K) = \sup_{A \in \mathcal{B}_c(T), A \subset T\backslash K} ||\mathbf{m}_c||_q(A).$$

(iv) *u is prolongable and given $\epsilon > 0$ and $q \in \Gamma$, there exists $K_0 \in \mathcal{C}_0$ such that* $||\mathbf{m}_0||_q(T\backslash K_0) < \epsilon$ *where*

$$||\mathbf{m}_0||_q(T\backslash K_0) = \sup_{A \in \mathcal{B}_0(T), A \subset T\backslash K_0} ||\mathbf{m}_0||_q(A).$$

Proof. (i)$\Rightarrow$(ii) Let $V \in \mathcal{V}$. Then $\widetilde{u_V} = \tilde{u}|_{C_0(V)}$ is weakly compact and hence u is prolongable. By Theorem 6(xxi) of [P9] and by Proposition 1.1.6 of Chapter 1 the other part of (ii) holds.

Clearly, (ii)$\Rightarrow$(iii)$\Rightarrow$(iv) since for $K \in \mathcal{C}$, by Theorem 50.D of [H] there exists $K_0 \in \mathcal{C}_0$ such that $K \subset K_0$.

(iv)$\Rightarrow$(i) Let $K_0 \in \mathcal{C}_0$. Choose $V \in \mathcal{V}$ such that $K_0 \subset V$. By hypothesis, $\widetilde{u_V}$ is weakly compact and hence by Theorem 8(xxx) of [P9] and by Proposition 1.1.6 of Chapter 1, given $\epsilon > 0$ and $q \in \Gamma$, there exists $U \in \mathcal{B}_0(V)$, U open in V such that $K_0 \subset U$ and $||\mathbf{m}_V||_q(U\backslash K_0) < \epsilon$. As V is open in T, U is open in T and by Lemma 5.2.3, $U \in \mathcal{B}_0(T)$. This proves that $\mathbf{m}_0$ is Baire outer regular in each $K_0 \in \mathcal{C}_0$. The other hypothesis in (iv) implies that $\mathbf{m}_0$ is Baire inner regular in T and hence by Theorem 8(xxx) of [P9], $\tilde{u}$ is weakly compact. Hence (i) holds. $\qquad\square$

Chapter 6

Applications to Integration in Locally Compact Hausdorff Spaces – Part II

6.1 Generalized Lusin's Theorem and its variants

In the sequel, T denotes a locally compact Hausdorff space and $\mathcal{U}$, $\mathcal{C}$, $\mathcal{C}_0$ are as in Definition 4.6.4 of Chapter 4. Then $\mathcal{B}(T) = \sigma(\mathcal{U})$, the σ-algebra of the Borel sets in T; $\mathcal{B}_c(T) = \sigma(\mathcal{C})$, the σ-ring of the σ-Borel sets in T and $\mathcal{B}_0(T) = \sigma(\mathcal{C}_0)$, the σ-ring of the Baire sets in T. $\delta(\mathcal{C})$ and $\delta(\mathcal{C}_0)$ denote the δ-rings generated by $\mathcal{C}$ and $\mathcal{C}_0$.

Notation 6.1.1. $C_c(T) = \{f : T \to \mathbb{K},\ f$ continuous with compact support$\}$; $C_c^r(T) = \{f \in C_c(T) : f$ real$\}$; $C_c^+(T) = \{f \in C_c^r(T) : f \geq 0\}$; $C_0(T) = \{f : T \to \mathbb{K}, f$ is continuous and vanishes at infinity in $T\}$; $C_0^r(T) = \{f \in C_0(T) : f$ real$\}$ and $C_0^+(T) = \{f \in C_0^r(T) : f \geq 0\}$. All these spaces are provided with the supremum norm $\|\cdot\|_T$.

As in Chapters 1, 2, 3 and 4, X denotes a Banach space or an lcHs over $\mathbb{K} = (\mathbb{R}$ or $\mathbb{C})$ with Γ, the family of continuous seminorms on X, unless otherwise stated and it will be explicitly specified whether X is a Banach space or an lcHs. Let $\mathcal{P} = \mathcal{B}(T)$(resp. $\mathcal{B}_c(T)$, $\mathcal{B}_0(T)$, $\delta(\mathcal{C})$, $\delta(\mathcal{C}_0)$) and let $\mathbf{m} : \mathcal{P} \to X$ be σ-additive and $\mathcal{P}$-regular (see Definition 4.6.7 of Chapter 4). In this section we obtain the generalized Lusin's theorem and its variants for $\sigma(\mathcal{P})$-measurable scalar functions on T, with respect to $\mathbf{m}$ when X is a Banach space and when X is an lcHs. Then we deduce that $C_c(T)$ (resp. $C_0(T)$) is dense in $\mathcal{L}_p(\mathbf{m})$ and $\mathcal{L}_p(\sigma(\mathcal{P}), \mathbf{m})$, $1 \leq p < \infty$, for both the cases of X when $\mathcal{P} = \delta(\mathcal{C})$ or $\delta(\mathcal{C}_0)$ (resp. when $\mathcal{P} = \mathcal{B}_0(T)$ or $\mathcal{B}_c(T)$ or $\mathcal{B}(T)$).

Theorem 6.1.2. (Generalized Lusin's theorem for m on $\mathcal{B}(T)$).

(i) **(m normed space-valued).** *Let X be a normed space and let $\mathbf{m} : \mathcal{B}(T) \to X$ be σ-additive and Borel regular. Suppose $f : T \to \mathbb{K}$ is Borel measurable. Then, given $\epsilon > 0$, there exists $g \in C_c(T)$ such that*

$$||\mathbf{m}||(N(f-g)) = ||\mathbf{m}||(\{t \in T : f(t) - g(t) \neq 0\}) < \epsilon \qquad (6.1.2.1)$$

and

$$||g||_T \leq ||f||_T. \qquad (6.1.2.2)$$

(ii) **(m lcHs-valued).** *Let X be an lcHs and let $\mathbf{m} : \mathcal{B}(T) \to X$ be σ-additive and Borel regular and let f be as in (i). Then, given $\epsilon > 0$ and $q \in \Gamma$, there exists $g_q \in C_c(T)$ such that*

$$||\mathbf{m}||_q(N(f-g_q)) < \epsilon \qquad (6.1.2.3)$$

and

$$||g_q||_T \leq ||f||_T. \qquad (6.1.2.4)$$

Proof. (i) Let $\tilde{X}$ be the Banach completion of X. Then $\mathbf{m} : \mathcal{B}(T) \to X \subset \tilde{X}$ and hence $\mathbf{m}$ can be considered as Banach space-valued. As $\mathbf{m}$ is particularly Borel inner regular in T, there exists $K \in \mathcal{C}$ such that $||\mathbf{m}||(T \backslash K) < \frac{\epsilon}{2}$. By hypothesis, $f\chi_K$ is $\mathcal{B}(T)$-measurable and vanishes in $T \backslash K$. If $f\chi_K$ is bounded in T, then the proof of Theorem 2.23 of [Ru1] for the case of bounded Borel functions holds here if we replace μ by $||\mathbf{m}||$, since $||\mathbf{m}||$ is σ-subadditive on $\mathcal{B}(T)$ by Proposition 1.1.12 of Chapter 1. Hence there exists $g \in C_c(T)$ such that $||\mathbf{m}||(N(f\chi_K - g)) < \frac{\epsilon}{2}$. Then $||\mathbf{m}||(N(f-g)) \leq ||\mathbf{m}||(N(f\chi_K - g)) + ||\mathbf{m}||(T \backslash K) < \epsilon$. When $f\chi_K$ is unbounded, the argument in the proof of the said theorem of [Ru1] for the unbounded case also holds here since $||\mathbf{m}||$ is continuous on $\mathcal{B}(T)$ by Proposition 1.1.5 of Chapter 1 and hence there exists $g \in C_c(T)$ such that $||\mathbf{m}||(N(f\chi_K - g)) < \frac{\epsilon}{2}$ so that by the above argument $||\mathbf{m}||(N(f-g)) < \epsilon$. Hence (6.1.2.1) holds.

To prove (6.1.2.2), it suffices to restrict to the case $||f||_T = M < \infty$. We argue as in the last part of the proof of Theorem 2.23 of [Ru1]. Let $g \in C_c(T)$ satisfy (6.1.2.1). Replacing g by $g_1 = \varphi \circ g$, where $\varphi(z) = z$ if $|z| \leq M$ and $\varphi(z) = \frac{Mz}{|z|}$ if $|z| > M$, we deduce that $g_1 \in C_c(T)$, $||\mathbf{m}||(N(f-g_1)) < \epsilon$ and $||g_1||_T \leq ||f||_T$. Hence (6.1.2.1) and (6.1.2.2) hold for g_1.

(ii) Given $q \in \Gamma$, $\mathbf{m}_q = \Pi_q \circ \mathbf{m} : \mathcal{B}(T) \to X_q \subset \widetilde{X_q}$ is σ-additive and Borel regular and hence by (i), (ii) holds. $\square$

To obtain the variants of Theorem 6.1.2 when $\mathbf{m} : \mathcal{R} \to X$ is σ-additive and $\mathcal{R}$-regular, where $\mathcal{R} = \mathcal{B}_c(T)$ (resp. $\mathcal{B}_0(T)$, $\delta(\mathcal{C})$, $\delta(\mathcal{C}_0)$), we give the following lemmas.

Lemma 6.1.3. *Let X be a normed space or an lcHs. Then an X-valued σ-additive measure on $\mathcal{B}_0(T)$ (resp. on $\delta(\mathcal{C}_0)$) is $\mathcal{B}_0(T)$-regular (resp. $\delta(\mathcal{C}_0)$-regular).*

Proof. The result for $\delta(\mathcal{C}_0)$ holds by Theorem in [DL] while that for $\mathcal{B}_0(T)$ holds by Remark on pp. 93–94 of [DL]. □

Lemma 6.1.4.

(i) *Let X be a normed space and let $\mathbf{n}_c : \delta(\mathcal{C}) \to X$ (resp. $\mathbf{n}_0 : \delta(\mathcal{C}_0) \to X$) be σ-additive and let $\mathbf{n}_c$ be $\delta(\mathcal{C})$-regular. If $f : T \to \mathbb{K}$ is $\mathcal{B}_c(T)$-measurable (resp. $\mathcal{B}_0(T)$-measurable) and if A is compact in T such that $f(t) = 0$ for $t \in T \backslash A$, then, given $\epsilon > 0$, there exists $g \in C_c(T)$ such that*

$$||\mathbf{n}_c||(N(f - g)) < \epsilon \qquad (6.1.4.1)$$
$$(resp. \quad ||\mathbf{n}_0||(N(f - g)) < \epsilon \qquad (6.1.4.2))$$

and moreover, we can choose $g \in C_c(T)$ such that

$$||g||_T \leq ||f||_T. \qquad (6.1.4.3)$$

(ii) *If X is an lcHs in (i) and if the remaining hypothesis of $\mathbf{n}_c$ (resp. $\mathbf{n}_0$) and f remain the same, then, given $q \in \Gamma$ and $\epsilon > 0$, there exists $g_q \in C_c(T)$ such that*

$$||\mathbf{n}_c||_q(N(f - g_q)) < \epsilon \qquad (6.1.4.4)$$
$$(resp. \quad ||\mathbf{n}_0||_q(N(f - g_q)) < \epsilon \qquad (6.1.4.5))$$

and moreover, we can choose $g_q \in C_c(T)$ such that

$$||g_q||_T \leq ||f||_T. \qquad (6.1.4.6)$$

Proof. (i) One can adapt the proof of Theorem 2.23 of [Ru1] as follows. Choose a relatively compact open set V such that $A \subset V$. In the construction of the functions on p. 53 of [Ru1], we can observe that $2^n t_n$ (in the notation of [Ru1]) is the characteristic function of some σ-Borel (resp. Baire) set $T_n \subset A$ and

$$f(x) = \sum_1^\infty t_n(x), \ x \in T$$

since f is $\mathcal{B}_c(T)$-measurable (resp. $\mathcal{B}_0(T)$-measurable). By hypothesis, $\mathbf{n}_c$ is $\delta(\mathcal{C})$-regular (resp. by Lemma 6.1.3, $\mathbf{n}_0$ is $\delta(\mathcal{C}_0)$-regular) and hence there exist $K_n \in \mathcal{C}$ (resp. $K_n \in \mathcal{C}_0$) and an open set $V_n \in \delta(\mathcal{C})$ (resp. $V_n \in \delta(\mathcal{C}_0)$) such that $K_n \subset T_n \subset V_n \subset V$ with $||\mathbf{n}_c||(V_n \backslash K_n) < \frac{\epsilon}{2^n}$ (resp. with $||\mathbf{n}_0||(V_n \backslash K_n) < \frac{\epsilon}{2^n}$) for $n \in \mathbb{N}$. Let us suppose that $0 \leq f \leq 1$ in A. Then choosing h_n for all n as on the top of p. 54 of [Ru1] and then defining g as on the latter page of [Ru1] and using the fact that $||\mathbf{n}_c||$ (resp. $||\mathbf{n}_0||$) is σ-subadditive on $\mathcal{B}_c(T)$ (resp. on $\mathcal{B}_0(T)$), we note that $g \in C_c(T)$ and $||\mathbf{n}_c||(N(f - g)) < \epsilon$ (resp. and $||\mathbf{n}_0||(N(f - g)) < \epsilon$) and hence (6.1.4.1) (resp. (6.1.4.2)) holds. From this it follows that these inequalities hold if f is bounded. When f is not bounded, let $B_n = \{x : |f(x)| > n\}$. Then

$B_n \searrow \emptyset$ in $\mathcal{B}_c(T)$ (resp. in $\mathcal{B}_0(T)$) and by hypothesis, B_n is relatively compact for all n. Then by Lemma 5.2.2 of Chapter 5, $(B_n)_1^\infty \subset \delta(\mathcal{C})$ (resp. $(B_n)_1^\infty \subset \delta(\mathcal{C}_0)$). Since $X \subset \tilde{X}$, the Banach completion of X, we can consider $\mathbf{n}_c$ and $\mathbf{n}_0$ as Banach space-valued and hence by Proposition 1.1.5 of Chapter 1, $||\mathbf{n}_c||(B_n) \to 0$ (resp. $||\mathbf{n}_0||(B_n) \to 0$). Then arguing as in the general case of Theorem 2.23 of [Ru1] with $||\mathbf{n}_c||$ (resp. $||\mathbf{n}_0||$) replacing μ there we observe that (6.1.4.1) (resp. (6.1.4.2)) holds for the general case. (6.1.4.3) is proved as in the last part of the proof of Theorem 2.23 of [Ru1].

(ii) This is immediate from (i), since $(\mathbf{n}_c)_q : \delta(\mathcal{C}) \to X_q \subset \widetilde{X_q}$ is σ-additive and $\delta(\mathcal{C})$-regular and $(\mathbf{n}_0)_q : \delta(\mathcal{C}_0) \to X_q \subset \widetilde{X_q}$ is σ-additive for $q \in \Gamma$. $\square$

Lemma 6.1.5. *Let X be an lcHs and let $\mathbf{m}_c : \mathcal{B}_c(T) \to X$ be σ-additive and σ-Borel regular. Then $\boldsymbol{\omega}_c = \mathbf{m}_c|_{\delta(\mathcal{C})}$ is σ-additive and $\delta(\mathcal{C})$-regular.*

Proof. Clearly it suffices to prove the lemma when X is a normed space and hence let X be so. Since $\boldsymbol{\omega}_c$ is σ-additive, it suffices to prove the regularity of $\boldsymbol{\omega}_c$. Let $A \in \delta(\mathcal{C})$ and $\epsilon > 0$. Then by hypothesis, there exist $K \in \mathcal{C}$ and an open set $U \in \mathcal{B}_c(T)$ such that $K \subset A \subset U$ and $||\mathbf{n}_c||(U \backslash K) < \epsilon$. Since A is relatively compact, by Theorem 50.D of [H] there exists a relatively compact open set V such that $\bar{A} \subset V$. Then $W = U \cap V$ is an open set belonging to $\delta(\mathcal{C})$ by Lemma 5.2.2 of Chapter 5, $K \subset A \subset W$ and $||\mathbf{n}_c||(W \backslash K) < \epsilon$. Hence the lemma holds. $\square$

Theorem 6.1.6. (Variants of the generalized Lusin's theorem). *Let X be an lcHs. Let $\mathbf{m}_c : \mathcal{B}_c(T) \to X$ (resp. $\mathbf{n}_c : \delta(\mathcal{C}) \to X$, $\mathbf{m}_0 : \mathcal{B}_0(T) \to X$, $\mathbf{n}_0 : \delta(\mathcal{C}_0) \to X$) be σ-additive and let $\mathbf{m}_c$ be $\mathcal{B}_c(T)$-regular (resp. $\mathbf{n}_c$ be $\delta(\mathcal{C})$-regular). Suppose $f : T \to \mathbb{K}$ is $\mathcal{B}_c(T)$-measurable (resp. Baire measurable). Let $A \in \mathcal{B}_c(T)$ (resp. $A \in \delta(\mathcal{C})$, $A \in \mathcal{B}_0(T)$, $A \in \delta(\mathcal{C}_0)$) such that $f(t) = 0$ for $t \in T \backslash A$ and let $\epsilon > 0$. Then, given $q \in \Gamma$, there exists $g_q \in C_c(T)$ such that*

$$||\boldsymbol{\omega}||_q(N(f - g_q)) < \epsilon \tag{6.1.6.1}$$

where $\boldsymbol{\omega} = \mathbf{m}_c$ or $\mathbf{n}_c$ or $\mathbf{m}_0$ or $\mathbf{n}_0$, as the case may be. Moreover, $g_q \in C_c(T)$ can be chosen such that

$$||g_q||_T \leq ||f||_T. \tag{6.1.6.2}$$

(We say that $\mathbf{m}_c$ (resp. $\mathbf{m}_0$) is σ-Borel (resp. Baire) inner regular in T if, given $q \in \Gamma$ and $\epsilon > 0$, there exists $K \in \mathcal{C}$ (resp. $K \in \mathcal{C}_0$) such that $||\mathbf{m}_c||_q(B) < \epsilon$ for $B \in \mathcal{B}_c(T)$ (resp. $||\mathbf{m}_0||_q(B) < \epsilon$ for $B \in \mathcal{B}_0(T)$) with $B \subset T \backslash K$.) If $\mathbf{m}_c$ (resp. $\mathbf{m}_0$) is further σ-Borel (resp. Baire) inner regular in T, then the above results hold for any $\mathcal{B}_c(T)$-measurable (resp. $\mathcal{B}_0(T)$-measurable) function f on T with values in $\mathbb{K}$.)

Proof. Without loss of generality we shall assume X to be a normed space. Let $\mathcal{R} = \mathcal{B}_c(T)$ and $\boldsymbol{\omega} = \mathbf{m}_c$, or $\mathcal{R} = \delta(\mathcal{C})$ and $\boldsymbol{\omega} = \mathbf{n}_c$ or $\mathcal{R} = \mathcal{B}_0(T)$ and $\boldsymbol{\omega} = \mathbf{m}_0$ or $\mathcal{R} = \delta(\mathcal{C}_0)$ and $\boldsymbol{\omega} = \mathbf{n}_0$. By hypothesis and by Lemma 6.1.3, $\boldsymbol{\omega}$ is $\mathcal{R}$-regular and σ-additive. Then there exists a compact set $K \in \mathcal{R}$ such that $K \subset A$ and

$||\boldsymbol{\omega}||(A\backslash K) < \frac{\epsilon}{2}$. $\mathbf{m}_c$ is σ-additive on $\delta(\mathcal{C})$ and $\mathbf{m}_c|_{\delta(\mathcal{C})}$ is $\delta(\mathcal{C})$-regular by Lemma 6.1.5. $\mathbf{m}_0|_{\delta(\mathcal{C}_0)}$ is σ-additive by hypothesis and $\delta(\mathcal{C}_0)$-regular by Lemma 6.1.3. As $f\chi_K$ satisfies the hypothesis of Lemma 6.1.4(i), by that lemma there exists $g \in C_c(T)$ such that $||\boldsymbol{\omega}||(N(f\chi_K - g)) < \frac{\epsilon}{2}$ with $||g||_T \le ||f\chi_K||_T \le ||f||_T$. By hypothesis and by Theorem 51.B of [H], $f - g$ is $\sigma(\mathcal{R})$-measurable and hence $N(f-g) \in \sigma(\mathcal{R})$. Since $f(t)\chi_K(t) = f(t)$ for $t \in K\cup(T\backslash A)$, $N(f-g) \subset N(f\chi_K - g) \cap (K \cup (T\backslash A)) \cup (A\backslash K)$, and hence $||\boldsymbol{\omega}||(N(f - g)) < \epsilon$. Thus (6.1.6.1) and (6.1.6.2) hold.   □

If $\mathbf{m}_c$ is σ-Borel (resp. $\mathbf{m}_0$ is Baire) inner regular in T, choose $K \in \mathcal{C}$ (resp. $K \in \mathcal{C}_0$) such that $||\mathbf{m}_c||(B) < \frac{\epsilon}{2}$ (resp. $||\mathbf{m}_0||(B) < \frac{\epsilon}{2}$) for $B \in \mathcal{B}_c(T)$ (resp. $B \in \mathcal{B}_0(T)$) with $B \subset T\backslash K$. Let $\boldsymbol{\omega} = \mathbf{m}_c$ or $\mathbf{m}_0$ as the case may be. Then by the above part there exists $g \in C_c(T)$ such that $||\boldsymbol{\omega}||(N(f\chi_K - g)) < \frac{\epsilon}{2}$ with $||g||_T \le ||f\chi_K||_T \le ||f||_T$ and hence $||\boldsymbol{\omega}||(N(f-g)) \le ||\boldsymbol{\omega}||(N(f\chi_K - g)) + ||\boldsymbol{\omega}||(T\backslash K) < \epsilon$.

Corollary 6.1.7. *Let X be an lcHs and $q \in \Gamma$. Suppose $\mathbf{m} : \mathcal{B}(T) \to X$ is σ-additive and Borel regular (resp. $\mathbf{m}_c : \mathcal{B}_c(T) \to X$ is σ-additive and σ-Borel regular and moreover, σ-Borel inner regular in T, $\mathbf{m}_0 : \mathcal{B}_0(T) \to X$ is σ-additive and Baire inner regular in T). Let $f : T \to \mathbb{K}$ be Borel measurable (resp. σ-Borel measurable, Baire measurable). Then given $q \in \Gamma$, there exists a sequence $(g_n^{(q)}) \subset C_c(T)$ such that $\sup_n ||g_n^{(q)}||_T \le ||f||_T$ and $f(t) = \lim_n g_n^{(q)}(t)$ $\mathbf{m}_q$-a.e. in T.*

Proof. Without loss of generality we shall assume X to be a normed space. Let $\mathcal{R} = \mathcal{B}(T)$ (resp. $\mathcal{B}_c(T), \mathcal{B}_0(T)$) and $\boldsymbol{\omega} = \mathbf{m}$ (resp. $\mathbf{m}_c, \mathbf{m}_0$). Then by Theorems 6.1.2 and 6.1.6 there exists $g_n \in C_c(T)$ with $||g_n||_T \le ||f||_T$ such that $||\boldsymbol{\omega}||(N(f - g_n)) < \frac{1}{2^n}$ for $n \in \mathbb{N}$. Let $A_n = N(f - g_n)$ and let $A = \limsup_n A_n$. Clearly, $A \in \mathcal{R}$ and $||\boldsymbol{\omega}||(A) \le ||\boldsymbol{\omega}||(\bigcup_{k\ge n} A_k) < \frac{1}{2^{n-1}} \to 0$ since $||\boldsymbol{\omega}||$ is σ-subadditive on $\mathcal{R}$. Hence $||\boldsymbol{\omega}||(A) = 0$. Clearly, $f(t) = \lim_n g_n(t)$ for $t \in T\backslash A$.   □

Lemma 6.1.8. *Let X be a sequentially complete lcHs, $\mathcal{P} = \delta(\mathcal{C})$ or $\delta(\mathcal{C}_0)$ and $\mathbf{m} : \mathcal{P} \to X$ be σ-additive. Then $C_c(T) \subset \mathcal{L}_p(\sigma(\mathcal{P}), \mathbf{m})$ for $1 \le p < \infty$ (see Definition 4.3.7 of Chapter 4).*

Proof. Let $f \in C_c(T)$ and let $q \in \Gamma$. Then by Theorem 51.B of [H], f is $\sigma(\mathcal{P})$-measurable. Let $\text{supp}\, f = K \in \mathcal{C}$. Then by Theorem 50.D of [H] there exists $C_0 \in \mathcal{C}_0$ such that $K \subset C_0$. As $N(f) \subset C_0$, for a Borel set B in $\mathbb{K}$ we have $f^{-1}(B) \cap N(f) \in \sigma(\mathcal{P}) \cap C_0 = \sigma(\mathcal{P} \cap C_0)$ by Lemma 5.2.2 and by Theorem 5.E of [H]. As $\mathcal{P} \cap C_0$ is a σ-ring, it follows that f is $(\mathcal{P} \cap C_0)$-measurable. Clearly, f is bounded in T. Hence there exists a sequence (s_n) of $(\mathcal{P} \cap C_0)$-simple functions such that $s_n \to f$ and $|s_n| \nearrow |f|$ uniformly in T. Then for $A \in \sigma(\mathcal{P})$, by Theorem 2.1.5(i) of Chapter 2 we have

$$\left| \int_A |s_n|^p d\mathbf{m} - \int_A |s_k|^p d\mathbf{m} \right|_q \le |||s_n|^p - |s_k|^p||_T ||\mathbf{m}||_q(C_0) \to 0$$

as $n, k \to \infty$. As q is arbitrary in Γ and as X is sequentially complete, we conclude that there exists $x_A \in X$ such that $\lim_n \int_A |s_n|^p d\mathbf{m} = x_A$. This holds for each $A \in \sigma(\mathcal{P})$ and consequently, by Definition 4.2.1' in Remark 4.2.12 of Chapter 4, $|f|^p$ is $\mathbf{m}$-integrable in T and hence $f \in \mathcal{L}_p(\sigma(\mathcal{P}), \mathbf{m})$.   □

Lemma 6.1.9. *Let $\mathcal{S} = \mathcal{B}(T)$ or $\mathcal{B}_c(T)$ or $\mathcal{B}_0(T)$, X be a sequentially complete lcHs and $\mathbf{m} : \mathcal{S} \to X$ be σ-additive. Then $C_0(T) \subset \mathcal{L}_p(\sigma(\mathcal{S}), \mathbf{m}) = \mathcal{L}_p(\mathcal{S}, \mathbf{m})$ for $1 \le p < \infty$.*

Proof. Given $f \in C_0(T)$, f is bounded and by Theorem 51.B of [H], f is $\mathcal{S}$-measurable and hence there exists a sequence (s_n) of $\mathcal{S}$-simple functions such that $s_n \to f$ and $|s_n| \nearrow |f|$ uniformly in T. Then arguing as in the last part of the proof of Lemma 6.1.8 we conclude that $f \in \mathcal{L}_p(\mathcal{S}, \mathbf{m})$ for $1 \le p < \infty$.   □

Theorem 6.1.10. *Let X be a sequentially complete (resp. quasicomplete) lcHs and let $1 \le p < \infty$. Suppose $\mathbf{m} : \mathcal{P} \to X$ is σ-additive when $\mathcal{P} = \delta(\mathcal{C}_0)$ or $\mathcal{B}_0(T)$; and $\mathbf{m} : \mathcal{P} \to X$ is σ-additive and $\mathcal{P}$-regular when $\mathcal{P} = \delta(\mathcal{C})$ or $\mathcal{B}_c(T)$ or $\mathcal{B}(T)$. Then $C_c(T)$ is dense in $\mathcal{L}_p(\sigma(\mathcal{P}), \mathbf{m})$ (resp. in $\mathcal{L}_p(\mathbf{m})$). If $\mathcal{P} = \mathcal{B}_0(T)$ or $\mathcal{B}_c(T)$ or $\mathcal{B}(T)$, then $C_0(T)$ is also dense in $\mathcal{L}_p(\sigma(\mathcal{P}), \mathbf{m})$ (resp. in $\mathcal{L}_p(\mathbf{m})$). Moreover, given $f \in \mathcal{L}_p(\mathbf{m})$ (resp. $f \in \mathcal{L}_p(\sigma(\mathcal{P}), \mathbf{m})$), $q \in \Gamma$ and $\epsilon > 0$, there exists $g_q \in C_c(T)$ such that $(\mathbf{m}_q)_p^\bullet(f - g_q, T) < \epsilon$ (resp. and $||g_q||_T \le ||f||_T$).*

Proof. By Lemma 6.1.8, $C_c(T) \subset \mathcal{L}_p(\sigma(\mathcal{P}), \mathbf{m})$ for $\mathcal{P} = \delta(\mathcal{C}_0)$ or $\delta(\mathcal{C})$ and by Lemma 6.1.9, $C_c(T) \subset C_0(T) \subset \mathcal{L}_p(\sigma(\mathcal{P}), \mathbf{m})$ for $\mathcal{P} = \mathcal{B}_0(T)$ or $\mathcal{B}_c(T)$ or $\mathcal{B}(T)$. When X is quasicomplete, $\mathcal{L}_p(\sigma(\mathcal{P}), \mathbf{m}) \subset \mathcal{L}_p(\mathbf{m})$. Let $f \in \mathcal{L}_p(\sigma(\mathcal{P}), \mathbf{m})$ (resp. $f \in \mathcal{L}_p(\mathbf{m})$). Let $q \in \Gamma$ and $\epsilon > 0$. Then by Theorem 4.5.6 of Chapter 4 there exists a $\mathcal{P}$-simple function s such that $(\mathbf{m}_q)_p^\bullet(f - s, T) < \frac{\epsilon}{2}$ and when $f \in \mathcal{L}_p(\sigma(\mathcal{P}), \mathbf{m})$, by the same theorem we can choose s further to satisfy $|s(t)| \le |f(t)|$ for t in T. Then by Theorems 6.1.2(ii) and 6.1.6 there exists $g_q \in C_c(T)$ such that $||\mathbf{m}||_q(N(g_q - s)) < ((\frac{\epsilon}{2})(\frac{1}{2||s||_T}))^p$ and $||g_q||_T \le ||s||_T$. Now by Theorem 4.3.2 and by Proposition 1.2.15(c) of Chapter 1 we have

$$(\mathbf{m}_q)_p^\bullet(s - g_q, T) = (\mathbf{m}_q)_p^\bullet(s - g_q, N(s - g_q))$$

$$= \sup_{x^* \in U_q^0} \left(\int_{N(s - g_q)} |s - g_q|^p dv(x^* \mathbf{m}) \right)^{\frac{1}{p}}$$

$$\le 2||s||_T (||\mathbf{m}||_q(N(s - g_q)))^{\frac{1}{p}} < \frac{\epsilon}{2}$$

and hence, by Theorem 3.1.13(ii) of Chapter 3 we have

$$(\mathbf{m}_q)_p^\bullet(f - g_q, T) \le (\mathbf{m}_q)_p^\bullet(f - s, T) + (\mathbf{m}_q)_p^\bullet(s - g_q, T) < \epsilon.$$

Moreover, for $f \in \mathcal{L}_p(\sigma(\mathcal{P}), \mathbf{m})$, $||g_q||_T \le ||s||_T \le ||f||_T$. Hence the result holds.   □

Remark 6.1.11. Restricting the argument in the proof of Theorem 6.1.10 to real functions, we have similar results for $\mathcal{L}_p^r(\sigma(\mathcal{P}), \mathbf{m})$ and $\mathcal{L}_p^r(\mathbf{m})$ with $C_c(T)$ and $C_0(T)$ being replaced by $C_c^r(T)$ and $C_0^r(T)$, respectively. (See Notation 6.1.1.)

Theorem 6.1.12. *Let X be an lcHs and let $\mathbf{m} : \mathcal{B}(T) \to X$ be σ-additive and Borel regular. Then $\mathbf{m}_c = \mathbf{m}|_{\mathcal{B}_c(T)}$ (resp. $\mathbf{m}_0 = \mathbf{m}|_{\mathcal{B}_0(T)}$) is σ-additive and $\mathcal{B}_c(T)$-regular (resp. and Baire regular). Consequently, $\mathbf{m}|_{\delta(\mathcal{C})}$ (resp. $\mathbf{m}|_{\delta(\mathcal{C}_0)}$) is σ-additive and $\delta(\mathcal{C})$-regular (resp. and $\delta(\mathcal{C}_0)$-regular). (See Notation 6.1.1.)*

Proof. Let $A \in \mathcal{B}_c(T)$. Then there exists a sequence $(K_n) \subset \mathcal{C}$ such that $A \subset \bigcup_1^\infty K_n$. Let $q \in \Gamma$ and $\epsilon > 0$. Then by hypothesis there exists an open set U_n in T such that $A \cap K_n \subset U_n$ with $||\mathbf{m}||_q(U_n \backslash (A \cap K_n)) < \frac{\epsilon}{2^{n+1}}$ for $n \in \mathbb{N}$. By Theorem 50.D of [H] there exists a (σ-Borel) relatively compact open set V_n in T such that $K_n \subset V_n$ so that $A \cap K_n \subset V_n$. Let $W_n = U_n \cap V_n$. Then $W = \bigcup_1^\infty W_n$ is a σ-Borel open set in T and $A \subset W$. By hypothesis, there exists $K \in \mathcal{C}$ such that $K \subset A$ with $||\mathbf{m}||_q(A \backslash K) < \frac{\epsilon}{2}$. Then $K \subset A \subset W$ and $||\mathbf{m}||_q(W \backslash K) < \epsilon$. In fact, $W \backslash A \subset \bigcup_1^\infty (W_n \backslash (A \cap V_n)) \subset \bigcup_1^\infty (W_n \backslash (A \cap K_n)) \subset \bigcup_1^\infty (U_n \backslash (A \cap K_n))$. As $||\mathbf{m}||_q$ is σ-subadditive on $\mathcal{B}(T)$, we have $||\mathbf{m}||_q(W \backslash A) < \frac{\epsilon}{2}$. Consequently, $||\mathbf{m}||_q(W \backslash K) \le ||\mathbf{m}||_q(W \backslash A) + ||\mathbf{m}||_q(A \backslash K) < \epsilon$ and hence $\mathbf{m}_c$ is $\mathcal{B}_c(T)$-regular. Then the other results hold by Lemmas 6.1.3 and 6.1.5. $\square$

Remark 6.1.13. Theorem 6.1.10 will play a key role in the proof of Theorem 7.6.7 in Chapter 7.

6.2 Lusin measurability of functions and sets

If X is an lcHs and $\mathbf{m} : \mathcal{P} \to X$ is σ-additive, let us recall from Definition 1.2.6 of Chapter 1 that for a set A in T, χ_A is $\mathbf{m}$-measurable if $A \in \widetilde{\sigma(\mathcal{P})}_q$, the generalized Lebesgue completion of $\sigma(\mathcal{P})$ with respect to $||\mathbf{m}||_q$, for each $q \in \Gamma$. In that case, we say that A is $\mathbf{m}$-measurable. When $\mathcal{P} = \mathcal{B}(T)$ (resp. $\delta(\mathcal{C})$) and $\mathbf{m}$ is further $\mathcal{P}$-regular, we introduce the concept of Lusin $\mathbf{m}$-measurability and study the interrelations between the concepts of $\mathbf{m}$-measurability and Lusin $\mathbf{m}$-measurability in Theorems 6.2.5 and 6.2.6. The latter theorems play a key role in Section 6.3.

Theorem 6.2.1. *Let X be an lcHs and let $\mathbf{m} : \mathcal{B}(T) \to X$ be σ-additive and Borel regular. For a set A in T the following statements are equivalent:*

(i) *A is $\mathbf{m}$-measurable.*

(ii) *Given $q \in \Gamma$ and $\epsilon > 0$, there exist $K_q \in \mathcal{C}$ and an open set U_q in T such that $K_q \subset A \subset U_q$ and $||\mathbf{m}||_q(U_q \backslash K_q) < \epsilon$.*

(iii) *Given $q \in \Gamma$, there exists a G_δ G_q in T and an F_σ F_q in T such that $F_q \subset A \subset G_q$ and $||\mathbf{m}||_q(G_q \backslash F_q) = 0$.*

(iv) *Given $q \in \Gamma$, there exist a G_δ G_q and an F_σ F_q in T with $F_q = \bigcup_{n=1}^\infty K_n^{(q)}$ where $(K_n^{(q)})_{n=1}^\infty$ is a disjoint sequence in $\mathcal{C}$ such that $F_q \subset A \subset G_q$ and $||\mathbf{m}||_q(G_q \backslash F_q) = 0$.*

(v) *For each $q \in \Gamma$, $A \cap K \in \widetilde{\mathcal{B}(T)}_q$ for each $K \in \mathcal{C}$.*

(vi) *For each $q \in \Gamma$, $A \cap U \in \widetilde{\mathcal{B}(T)}_q$ for each open set U in T.*

Proof. Without loss of generality we shall assume X to be a normed space.

(i)$\Rightarrow$(ii) By the Borel regularity of $\mathbf{m}$ and by the fact that the $\mathbf{m}$-measurable set A is of the form $A = B \cup N$, $N \subset M \in \mathcal{B}(T)$, $B \in \mathcal{B}(T)$ and $||\mathbf{m}||(M) = 0$, (i)$\Rightarrow$(ii).

(ii)$\Rightarrow$(iii) Taking $\epsilon = \frac{1}{n}$ in (ii), $n \in \mathbb{N}$, there exist $K_n \in \mathcal{C}$ and U_n open in T such that $K_n \subset A \subset U_n$ and $||\mathbf{m}||(U_n \backslash K_n) < \frac{1}{n}$. Let $G = \bigcap_1^\infty U_n$ and $F = \bigcup_1^\infty K_n$. Then $F \subset A \subset G$, F is an F_σ, G is a G_δ and $||\mathbf{m}||_q(G\backslash F) \leq ||\mathbf{m}||_q(U_n\backslash K_n) < \frac{1}{n}$ for $n \in \mathbb{N}$. Hence (iii) holds.

(iii)$\Rightarrow$(i) By (iii), there exist a G_δ G and an F_σ F in T such that $F \subset A \subset G$ and $||\mathbf{m}||(G\backslash F) = 0$. Then $A = F \cup (A\backslash F)$ with $A\backslash F \subset G\backslash F \in \mathcal{B}(T)$, $F \in \mathcal{B}(T)$ and $||\mathbf{m}||(G\backslash F) = 0$. Hence A is $\mathbf{m}$-measurable.

Thus (i)$\Rightarrow$(ii)$\Rightarrow$(iii)$\Rightarrow$(i).

(i)$\Rightarrow$(iv) Given $\epsilon > 0$, by (ii) there exist an open set U in T and a compact K such that $K \subset A \subset U$ and $||\mathbf{m}||(U\backslash K) < \epsilon$. As $A \in \widetilde{\mathcal{B}(T)}$ and $A\backslash K \in \widetilde{\mathcal{B}(T)}$ with $A\backslash K \subset U\backslash K$, we have $||\mathbf{m}||(A\backslash K) < \epsilon$. Taking $\epsilon = 1$, there exists $K_1 \in \mathcal{C}$ such that $K_1 \subset A$ with $||\mathbf{m}||(A\backslash K_1) < 1$. Since $A\backslash K_1 \in \widetilde{\mathcal{B}(T)}$, applying the above argument to $A\backslash K_1$ in place of A for $\epsilon = \frac{1}{2}$, we have a compact $K_2 \subset A\backslash K_1$ with $||\mathbf{m}||(A\backslash(K_1 \cup K_2)) < \frac{1}{2}$. Proceeding step by step, in the nth step we would have chosen mutually disjoint compact sets $(K_i)_{1}^{n}$ such that $\bigcup_1^n K_i \subset A$ and $||\mathbf{m}||(A\backslash\bigcup_1^n K_i) < \frac{1}{n}$. Then $F = \bigcup_1^\infty K_i$ is an F_σ, $(K_i)_1^\infty$ is a disjoint sequence in $\mathcal{C}$, $F \subset A$ and $||\mathbf{m}||(A\backslash F) = 0$. By (iii) there exists a G_δ G such that $A \subset G$ and $||\mathbf{m}||(G\backslash A) = 0$. Then $F \subset A \subset G$ with $||\mathbf{m}||(G\backslash F) = 0$ and hence (i)$\Rightarrow$(iv).

Clearly, (iv)$\Rightarrow$(iii) and hence (i)–(iv) are equivalent.

(i)$\Rightarrow$(v) obviously.

(v)$\Rightarrow$(vi) Let U be an open set in T. Then $U \in \mathcal{B}(T)$ and hence (iv) holds for U. Therefore, there exist a sequence $(K_i)_1^\infty \subset \mathcal{C}$ and a G_δ G in T such that $F = \bigcup_1^\infty K_i \subset U \subset G$ and $N = U\backslash F$ is $\mathbf{m}$-null. Then by (v), $A \cap U = \bigcup_1^\infty (A \cap K_n) \bigcup (A \cap N) \in \widetilde{\mathcal{B}(T)}$. Hence (vi) holds.

(vi)$\Rightarrow$(i) by taking $U = T$.

Hence (i)$\Rightarrow$(v)$\Rightarrow$(vi)$\Rightarrow$(i).

This completes the proof of the theorem. $\qquad\square$

Theorem 6.2.2. *Let X be an lcHs and let $\mathbf{n} : \delta(\mathcal{C}) \to X$ be σ-additive and $\delta(\mathcal{C})$-regular. For a set A in T the following statements are equivalent:*

(i) *A is $\mathbf{n}$-measurable.*

(ii) *Given $q \in \Gamma$ and $\epsilon > 0$, there exist a σ-Borel open set U_q in T and a σ-compact F_q in T such that $F_q \subset A \subset U_q$ and $||\mathbf{n}||_q(U_q \backslash F_q) < \epsilon$.*

(iii) *Given $q \in \Gamma$, there exist a G_δ $G_q \in \mathcal{B}_c(T)$ and a σ-compact F_q in T such that $F_q \subset A \subset G_q$ and $||\mathbf{n}||_q(G_q \backslash F_q) = 0$.*

(iv) *A is σ-bounded and $A \cap K \in \widetilde{\mathcal{B}_c(T)}_q$ for each $K \in \mathcal{C}$ and for each $q \in \Gamma$.*

(v) *A is σ-bounded and $A \cap U \in \widetilde{\mathcal{B}_c(T)}_q$ for each open set U in T and for each $q \in \Gamma$.*

Proof. Without loss of generality we shall assume X to be a normed space.

(i)$\Rightarrow$(ii) By hypothesis, A is of the form $A = B \cup N$, $B \in \mathcal{B}_c(T)$, $N \subset M \in \mathcal{B}_c(T)$ and $||\mathbf{n}||(M) = 0$. As $B \cup M \in \mathcal{B}_c(T)$, there exists $(K_n)_1^\infty \subset \mathcal{C}$ such that $B \cup M \subset \bigcup_1^\infty K_n$ so that $B = \bigcup_1^\infty (B \cap K_n)$. Then by Lemma 5.2.2 of Chapter 5, $(B \cap K_n)_1^\infty \subset \delta(\mathcal{C})$. As $\mathbf{n}$ is $\delta(\mathcal{C})$-regular, given $\epsilon > 0$, there exist an open set V_n in T such that $V_n \in \delta(\mathcal{C})$ and a compact C_n such that $C_n \subset B \cap K_n \subset (B \cup M) \cap K_n \subset V_n$ with $||\mathbf{n}||(V_n \backslash C_n) < \frac{\epsilon}{2^n}$. Then $U = \bigcup_1^\infty V_n$ is an open set in T belonging to $\mathcal{B}_c(T)$, $F = \bigcup_1^\infty C_n \subset \bigcup_1^\infty (B \cap K_n) \subset \bigcup_1^\infty ((B \cup M) \cap K_n) \subset \bigcup_1^\infty V_n = U$ and $||\mathbf{n}||(U \backslash F) \leq \sum_1^\infty ||\mathbf{n}||(V_n \backslash C_n) < \epsilon$. Hence (ii) holds.

(ii)$\Rightarrow$(iii) By (ii), for $\epsilon = \frac{1}{n}$, there exist a σ-compact set $F_n \in \mathcal{B}_c(T)$ and an open set $U_n \in \mathcal{B}_c(T)$ such that $F_n \subset A \subset U_n$ with $||\mathbf{n}||(U_n \backslash F_n) < \frac{1}{n}$. Then $F = \bigcup_1^\infty F_n$ is σ-compact, $F \in \mathcal{B}_c(T)$, $G = \bigcap_1^\infty U_n$ is a G_δ belonging to $\mathcal{B}_c(T)$, $F \subset A \subset G$ and $||\mathbf{n}||(G \backslash F) = 0$. Hence (iii) holds.

(iii)$\Rightarrow$(i) Let G be a G_δ in $\mathcal{B}_c(T)$ and F be a σ-compact such that $F \subset A \subset G$ and $||\mathbf{n}||(G \backslash F) = 0$. Then $A = F \cup (A \backslash F)$, $F \in \mathcal{B}_c(T)$, $A \backslash F \subset G \backslash F \in \mathcal{B}_c(T)$ and $||\mathbf{n}||(G \backslash F) = 0$. Hence $A \in \widetilde{\mathcal{B}_c(T)}$ and therefore, (i) holds.

Hence (i)$\Leftrightarrow$(ii)$\Leftrightarrow$(iii).

(i)$\Rightarrow$(iv) and (v) Take B, N and M as in the proof of '(i)$\Rightarrow$(ii)'. Clearly A is σ-bounded. For $K \in \mathcal{C}$, $B \cap K \in \mathcal{B}_c(T)$ and $N \cap K$ is $\mathbf{n}$-null so that $N \cap K \in \widetilde{\mathcal{B}_c(T)}$. Hence $A \cap K \in \widetilde{\mathcal{B}_c(T)}$. For an open set U in T, $U \cap B \in \mathcal{B}_c(T)$ as $U \cap B$ is a σ-bounded Borel set and $U \cap N$ is $\mathbf{n}$-null so that $U \cap N \in \widetilde{\mathcal{B}_c(T)}$. Hence $U \cap A = (U \cap B) \cup (U \cap N) \in \widetilde{\mathcal{B}_c(T)}$.

(iv)$\Rightarrow$(i) As A is σ-bounded, there exists $(K_n)_1^\infty \subset \mathcal{C}$ such that $A \subset \bigcup_1^\infty K_n$ so that by hypothesis, $A = \bigcup_1^\infty (A \cap K_n) \in \widetilde{\mathcal{B}_c(T)}$.

(v)$\Rightarrow$(i) As A is σ-bounded, by Theorem 50.D of [H] there exist relatively compact open sets $(U_n)_1^\infty$ in T such that $A \subset \bigcup_1^\infty U_n$. Then $A = \bigcup_1^\infty A \cap U_n \in \widetilde{\mathcal{B}_c(T)}$ by (v).

Hence (i), (iv) and (v) are equivalent.

This completes the proof of the theorem. □

Definition 6.2.3. Let X be an lcHs and let $\mathbf{m} : \mathcal{B}(T) \to X$ (resp. $\mathbf{n} : \delta(\mathcal{C}) \to X$) be σ-additive and $\mathcal{B}(T)$-regular (resp. and $\delta(\mathcal{C})$-regular). Then a function $f : T \to \mathbb{K}$ is said to be Lusin $\mathbf{m}$-measurable (resp. Lusin $\mathbf{n}$-measurable) if, given $q \in \Gamma$, $\epsilon > 0$ and $K \in \mathcal{C}$, there exists a compact $K_0^{(q)} \subset K$ such that $f|_{K_0^{(q)}}$ is continuous and $||\mathbf{m}||_q(K \backslash K_0^{(q)}) < \epsilon$ (resp. and $||\mathbf{n}||_q(K \backslash K_0^{(q)}) < \epsilon$).

Theorem 6.2.4. *Let X, $\mathbf{m}$, $\mathbf{n}$ and f be as in Definition 6.2.3. Then f is Lusin $\mathbf{m}$-measurable (resp. Lusin $\mathbf{n}$-measurable) if and only if, given $q \in \Gamma$ and $K \in \mathcal{C}$, there exist an $\mathbf{m}_q$-null set (resp. $\mathbf{n}_q$-null set) $N_q \subset K$ and a countable disjoint family $(K_i^{(q)})_1^\infty \subset \mathcal{C}$ such that $K = \bigcup_{i=1}^\infty K_i^{(q)} \cup N_q$ and $f|_{K_i^{(q)}}$ is continuous for each $i \in \mathbb{N}$.*

Proof. Without loss of generality we shall assume X to be a normed space. Let $\mathcal{P} = \mathcal{B}(T)$ and $\boldsymbol{\omega} = \mathbf{m}$ or $\mathcal{P} = \delta(\mathcal{C})$ and $\boldsymbol{\omega} = \mathbf{n}$. Let f be Lusin $\boldsymbol{\omega}$-measurable and $K \in \mathcal{C}$. Then by Definition 6.2.3 there exists $K_1 \in \mathcal{C}$ such that $K_1 \subset K$, $f|_{K_1}$ is continuous and $||\boldsymbol{\omega}||(K \backslash K_1) < 1$. Let $n > 1$ and suppose we have chosen $(K_1)_1^n \subset \mathcal{C}$ mutually disjoint such that $\bigcup_1^n K_i \subset K$, $f|_{K_i}$ is continuous for $1 \leq i \leq n$ and $||\boldsymbol{\omega}||(K \backslash \bigcup_1^n K_i) < \frac{1}{n}$. As $K \backslash \bigcup_1^n K_i \in \mathcal{P}$, by the regularity of $\boldsymbol{\omega}$ there exists a compact $C \subset K \backslash \bigcup_1^n K_i$ such that $||\boldsymbol{\omega}||(K \backslash (\bigcup_1^n K_i \cup C)) < \frac{1}{2(n+1)}$. By hypothesis there exists a compact $K_{n+1} \subset C$ such that $f|_{K_{n+1}}$ is continuous and $||\boldsymbol{\omega}||(C \backslash K_{n+1}) < \frac{1}{2(n+1)}$. Then $(K_i)_1^{n+1} \subset \mathcal{C}$ are mutually disjoint, $\bigcup_1^{n+1} K_i \subset K$ and $||\boldsymbol{\omega}||(K \backslash \bigcup_1^{n+1} K_i) < \frac{1}{n+1}$. Therefore, by induction there exists a disjoint sequence $(K_i)_1^\infty \subset \mathcal{C}$ such that $f|_{K_i}$ is continuous for all i and $||\boldsymbol{\omega}||(K \backslash \bigcup_1^n K_i) < \frac{1}{n}$ for all n. Then $N = K \backslash \bigcup_1^\infty K_i$ is $\boldsymbol{\omega}$-null and $f|_{K_i}$ is continuous for all i.

Conversely, suppose $K = \bigcup_1^\infty K_i \cup N$, where $(K_i)_1^\infty \subset \mathcal{C}$, $K_i \cap K_j = \emptyset$ for $i \neq j$, $f|_{K_i}$ is continuous for each i and $||\boldsymbol{\omega}||(N) = 0$. Let $\epsilon > 0$. As $K \backslash \bigcup_1^n K_i \in \mathcal{P}$ for all n, $K \backslash \bigcup_1^n K_i \searrow N \in \delta(\mathcal{C})$ and as $||\boldsymbol{\omega}||$ is continuous on $\mathcal{P}$ by Proposition 1.1.5 of Chapter 1, there exists n_0 such that $||\boldsymbol{\omega}||(K \backslash \bigcup_1^{n_0} K_i) < \epsilon$. Clearly, $K_0 = \bigcup_1^{n_0} K_i \in \mathcal{C}$, $K_0 \subset K$ and $f|_{K_0}$ is continuous since the K_i are mutually disjoint. Hence f is Lusin $\boldsymbol{\omega}$-measurable. $\qquad \square$

Theorem 6.2.5. (resp. Theorem 6.2.6) gives the relation between m-measurability and Lusin m-measurability (resp. n-measurability and Lusin n-measurability).

Theorem 6.2.5. *Let X be an lcHs, $\mathbf{m} : \mathcal{B}(T) \to X$ be σ-additive and Borel regular and $f : T \to \mathbb{K}$. Then f is Lusin $\mathbf{m}$-measurable if and only if it is $\mathbf{m}$-measurable.*

Proof. Without loss of generality we shall assume X to be a normed space. Let f be $\mathbf{m}$-measurable, $K \in \mathcal{C}$ and $\epsilon > 0$. Then $f\chi_K$ is $\mathbf{m}$-measurable and hence by Proposition 1.1.18 of Chapter 1 there exists $N \in \mathcal{B}(T)$ with $||\mathbf{m}||(N) = 0$ such that $h = f\chi_{K \backslash N}$ is $\mathcal{B}(T)$-measurable. Then by Theorem 6.1.2(i) there exists $g \in C_c(T)$ such that $||\mathbf{m}||(N(h - g)) < \frac{\epsilon}{2}$ and $||g||_T \leq ||h||_T$. Let $A = N(h - g)$. Then $A \in \mathcal{B}(T)$ and hence by the Borel regularity of $\mathbf{m}$ there exists a compact

$K_0 \subset (K \backslash N) \backslash A$ such that $||\mathbf{m}||(K \backslash N \backslash A \backslash K_0) < \frac{\epsilon}{2}$. Then $h|_{K_0} = f|_{K_0} = g|_{K_0}$ is continuous and $||\mathbf{m}||(K \backslash K_0) < \epsilon$. Hence f is Lusin $\mathbf{m}$-measurable.

Conversely, let f be Lusin $\mathbf{m}$-measurable. Given $K \in \mathcal{C}$, by Theorem 6.2.4 there exist a disjoint sequence $(K_i)_1^\infty \subset \mathcal{C}$ and an $\mathbf{m}$-null set N disjoint with $\bigcup_1^\infty K_i$ such that $K = \bigcup_1^\infty K_i \cup N$ and such that $f|_{K_i}$ is continuous for each i. Let U be an open set in $\mathbb{K}$. Then $f^{-1}(U) \cap K = \bigcup_1^\infty (f^{-1}(U) \cap K_i) \cup (f^{-1}(U) \cap N)$. As $f|_{K_i}$ is continuous, there exists an open set V_i in T such that $f^{-1}(U) \cap K_i = V_i \cap K_i$ and hence $f^{-1}(U) \cap K = \bigcup_1^\infty (V_i \cap K_i) \cup (f^{-1}(U) \cap N) \in \widehat{\mathcal{B}(T)}$. Then by Theorem 6.2.1(v), $f^{-1}(U) \in \widehat{\mathcal{B}(T)}$ and hence f is $\mathbf{m}$-measurable. $\qquad \square$

Theorem 6.2.6. *Let X be an lcHs, $\mathbf{n} : \delta(\mathcal{C}) \to X$ be σ-additive and $\delta(\mathcal{C})$-regular and $f : T \to \mathbb{K}$. Then f is $\mathbf{n}$-measurable if and only if $N(f)$ is σ-bounded and f is Lusin $\mathbf{n}$-measurable.*

Proof. Without loss of generality we shall assume X to be a normed space. Let f be $\mathbf{n}$-measurable. Then $N(f) \in \widehat{\mathcal{B}_c(T)}$ and hence $N(f)$ is σ-bounded. Let $K \in \mathcal{C}$ and $\epsilon > 0$. Then by Proposition 1.1.18 of Chapter 1 there exists $N \in \mathcal{B}_c(T)$ with $||\mathbf{n}||(N) = 0$ such that $f \chi_{T \backslash N}$ is $\mathcal{B}_c(T)$-measurable. Then $f \chi_{K \backslash N}$ is $\mathcal{B}_c(T)$-measurable. Hence by Theorem 6.1.6 there exists $g \in C_c(T)$ such that $||\mathbf{n}||(N(f \chi_{K \backslash N} - g)) < \frac{\epsilon}{2}$. Let $A = N(g - f \chi_{K \backslash N})$. Then $A \in \mathcal{B}_c(T)$ and hence $K \backslash A \in \delta(\mathcal{C})$ by Lemma 5.2.2 of Chapter 5. Then by the $\delta(\mathcal{C})$-regularity of $\mathbf{n}$ there exists a compact $K_0 \subset K \backslash N \backslash A$ such that $||\mathbf{n}||(K \backslash N \backslash A \backslash K_0) < \frac{\epsilon}{2}$. Then $f|_{K_0} = g|_{K_0}$ is continuous and $||\mathbf{n}||(K \backslash N \backslash K_0) < \epsilon$. Hence f is Lusin $\mathbf{n}$-measurable.

Conversely, let f be Lusin $\mathbf{n}$-measurable and let $N(f)$ be σ-bounded. Let $K \in \mathcal{C}$. Then by Theorem 6.2.4 there exist a disjoint countable family $(K_i)_1^\infty \subset \mathcal{C}$ and an $\mathbf{n}$-null set N disjoint with $\bigcup_1^\infty K_i$ such that $K = \bigcup_1^\infty K_i \cup N$ and $f|_{K_i}$ is continuous for each i. Let U be an open set in $\mathbb{K}$. If $f_i = f|_{K_i}$, then by the continuity of f_i we have $f^{-1}(U \backslash \{0\}) \cap K_i = f_i^{-1}(U \backslash \{0\}) \in \mathcal{B}(K_i)$ and hence $N(f) \cap f^{-1}(U) \cap K = \bigcup_1^\infty (f^{-1}(U \backslash \{0\}) \cap K_n) \cup (N \cap f^{-1}(U \backslash \{0\})) \in \widehat{\mathcal{B}_c(T)}$. As $N(f)$ is σ-bounded by hypothesis, it follows by Theorem 6.2.2(iv) that $N(f) \cap f^{-1}(U) \in \widehat{\mathcal{B}_c(T)}$ and hence f is $\mathbf{n}$-measurable. $\qquad \square$

Corollary 6.2.7. *Let X be an lcHs and let $\mathbf{m} : \mathcal{B}(T) \to X$ (resp. $\mathbf{n} : \delta(\mathcal{C}) \to X$) be σ-additive and Borel regular (resp. and $\delta(\mathcal{C})$-regular). Then a Borel measurable scalar function f on T is Lusin $\mathbf{m}$-measurable (resp. Lusin $\mathbf{n}$-measurable).*

Proof. If f is Borel measurable, then f is $\mathbf{m}$-measurable and hence is Lusin $\mathbf{m}$-measurable by Theorem 6.2.5. Let $K \in \mathcal{C}$. Then by Lemma 5.2.2 of Chapter 5, $f \chi_K$ is $\mathcal{B}_c(T)$-measurable and hence $\mathbf{n}$-measurable. Then by Theorem 6.2.6, $f \chi_K$ is Lusin $\mathbf{n}$-measurable. As K is arbitrary in $\mathcal{C}$, it follows that f is Lusin $\mathbf{n}$-measurable. $\qquad \square$

Definition 6.2.8. Let X be an lcHs and let $\mathbf{m} : \mathcal{B}(T) \to X$ (resp. $\mathbf{n} : \delta(\mathcal{C}) \to X$) be σ-additive and Borel regular (resp. and $\delta(\mathcal{C})$-regular). Then a set A in T is said to be Lusin $\mathbf{m}$-measurable (resp. Lusin $\mathbf{n}$-measurable) if χ_A is so.

The following theorem is immediate from Definition 6.2.8 and Theorem 6.2.4.

Theorem 6.2.9. *Let X, $\mathbf{m}$ and $\mathbf{n}$ be as in Definition 6.2.8. Let $A \subset T$. Then A is Lusin $\mathbf{m}$-measurable (resp. Lusin $\mathbf{n}$-measurable) if for each $q \in \Gamma$ and $K \in \mathcal{C}$, there exist a disjoint sequence $(K_i^{(q)})_1^\infty \subset \mathcal{C}$ and an $\mathbf{m}_q$-null set (resp. and an $\mathbf{n}_q$-null set) N_q disjoint with $\bigcup_1^\infty K_i^{(q)}$ such that $K = \bigcup_1^\infty K_i^{(q)} \cup N_q$ and such that, for each i, $K_i^{(q)} \subset A$ or $K_i^{(q)} \subset T \backslash A$.*

Using Theorem 6.2.1(iv) and the Borel regularity of $\mathbf{m}$, the proof of Proposition 4, no.2, §5, Chapter IV of [B] can be adapted to prove the following

Theorem 6.2.10. (Localization principle). *Let X be an lcHs and let $\mathbf{m} : \mathcal{B}(T) \to X$ be σ-additive and Borel regular. Let $f : T \to \mathbb{K}$ and suppose for each $t \in T$ and $q \in \Gamma$, there exist an open neighborhood $V_t^{(q)}$ of t and a Lusin $\mathbf{m}_q$-measurable scalar function $g_t^{(q)}$ such that $f(t') = g_t^{(q)}(t')$ $\mathbf{m}_q$-a.e. in $V_t^{(q)}$. Then f is Lusin $\mathbf{m}$-measurable.*

As in the classical case of [B], the above theorem motivates the following

Definition 6.2.11. Let X and $\mathbf{m}$ be as in Theorem 6.2.10. A set A in T is said to be locally $\mathbf{m}$-null (briefly, loc. $\mathbf{m}$-null) if, for each $t \in T$, there exists an open neighborhood V_t of t such that $A \cap V_t$ is $\mathbf{m}$-null. (See Definition 1.2.3 of Chapter 1.)

The proof of the following theorem is similar to those on pp. 172–173 of [B] and is based on Theorems 6.2.1(iv), 6.2.5 and 6.2.10 and hence is omitted.

Theorem 6.2.12. *Let X be an lcHs and let $\mathbf{m} : \mathcal{B}(T) \to X$ be σ-additive and Borel regular. Then:*

(i) *Locally $\mathbf{m}$-null sets are $\mathbf{m}$-measurable.*

(ii) *If A is loc. $\mathbf{m}$-null, then all the subsets of A are also loc. $\mathbf{m}$-null.*

(iii) *A is loc. $\mathbf{m}$-null if and only if $A \cap K$ is $\mathbf{m}$-null for each $K \in \mathcal{C}$.*

(iv) *If A_i, $i \in \mathbb{N}$, are loc. $\mathbf{m}$-null, then $\bigcup_1^\infty A_i$ is also loc. $\mathbf{m}$-null.*

(v) *A is loc. $\mathbf{m}$-null if and only if A is $\mathbf{m}$-null. (Use (i) and Theorem 6.2.1(iv).)*

(vi) *$f : T \to \mathbb{K}$ and $N = \{t \in T : f$ is discontinuous in $t\}$ is loc. $\mathbf{m}$-null, then f is $\mathbf{m}$-measurable.*

6.3 Theorems of integrability criteria

The aim of the present section is to improve Theorem 2.2.2 of Chapter 2 and Theorem 4.2.2 of Chapter 4 for $\delta(\mathcal{C})$-regular σ-additive vector measures on $\delta(\mathcal{C})$. The said improvement of Theorem 2.2.2 of Chapter 2 is given in the last part of Theorem 6.3.4 which gives much stronger results and Theorem 6.3.5 improves Theorem 4.2.2 of Chapter 4. We also generalize Theorem 6.3.4 to complete lcHs-valued vector measures. The proofs of Lemmas 3.10 and 3.14, Propositions 2.17, 2.20 and 3.7 and Theorems 3.5, 3.13 and 3.20 of [T] are adapted here in the set-up of vector measures.

Recall from Notation 5.2.1 of Chapter 5 that $\mathcal{V}$ denotes the family of relatively compact open sets in T.

Lemma 6.3.1. *Let X be a Banach space and let H be a norm determining set in X^*. Let $\mathcal{P}$ be a δ-ring of subsets of a set $\Omega(\neq \emptyset)$ and let $\mathbf{m} : \mathcal{P} \to X$ be additive. Then:*

(i) $||\mathbf{m}||(A) = \sup_{x^* \in H} v(x^* \circ \mathbf{m})(A)$, $A \in \sigma(\mathcal{P})$.

(ii) *Suppose $\mathbf{m}$ is σ-additive and $f : \Omega \to \mathbb{K}$ is $\mathbf{m}$-measurable and $(x^* \circ \mathbf{m})$-integrable for each $x^* \in H$. If, for each $\epsilon > 0$, there exists $g_\epsilon \in \mathcal{L}_1(\mathbf{m})$ such that $\sup_{x^* \in H} \int_T |f - g_\epsilon| dv(x^* \circ \mathbf{m}) < \epsilon$, then $f \in \mathcal{L}_1(\mathbf{m})$.*

Proof. (i) This is proved by an argument similar to that in the proof of Proposition 1.2.15(iii) of Chapter 1.

(ii) Let $\nu_{x^*}(\cdot) = \int_{(\cdot)} f d(x^* \circ \mathbf{m})$, $x^* \in H$. Then by hypothesis and by Proposition 5, §8 of [Din1], ν_{x^*} is σ-additive on $\sigma(\mathcal{P})$ and by Proposition 1.1.19 of Chapter 1, $v(\nu_{x^*})(A) = \int_A |f| dv(x^* \circ \mathbf{m})$ for $A \in \sigma(\mathcal{P})$. Let

$$\eta(f) = \sup_{x^* \in H} \int_T |f| dv(x^* \circ \mathbf{m}).$$

If $f \in \mathcal{L}_1(\mathbf{m})$, then by (ii) and (iii) of Theorem 2.1.5 and by Remark 2.2.3 of Chapter 2, $\gamma(\cdot) = \int_{(\cdot)} f d\mathbf{m}$ is σ-additive on $\sigma(\mathcal{P})$ and $||\gamma||(T) = ||\gamma||(N(f)) = \sup_{|x^*| \leq 1} \int_T |f| dv(x^* \circ \mathbf{m})$. Consequently, by (i) above, by Theorem 3.1.3 of Chapter 3 and by Proposition 1.1.19 of Chapter 1 we have

$$\mathbf{m}_1^\bullet(f, T) = \sup_{|x^*| \leq 1} \int_T |f| dv(x^* \circ \mathbf{m}) = ||\gamma||(T) = ||\gamma||(N(f))$$
$$= \sup_{x^* \in H} v(x^* \circ \gamma)(N(f))$$
$$= \sup_{x^* \in H} v(x^* \circ \gamma)(T) = \sup_{x^* \in H} \int_T |f| dv(x^* \circ \mathbf{m}) = \eta(f). \quad (6.3.1.1)$$

Let $\Sigma = \{f : T \to \mathbb{K}, \ f \ \mathbf{m}\text{-measurable and } (x^* \circ \mathbf{m})\text{-integrable for each } x^* \in H \text{ with } \eta(f) < \infty\}$. For $f \in \mathcal{L}_1\mathcal{M}(\mathbf{m})$ (see Definition 3.1.9 of Chapter 3), we have

$\eta(f) = \sup_{x^* \in H} \int_T |f| dv(x^* \circ \mathbf{m}) \leq \mathbf{m}_1^\bullet(f, T) < \infty$ and hence $\mathcal{L}_1 \mathcal{M}(\mathbf{m}) \subset \Sigma$. Clearly, η is a seminorm on Σ.

Claim 1. $\mathcal{L}_1(\mathbf{m})$ is closed in (Σ, η).

In fact, let $(f_n)_1^\infty \subset \mathcal{L}_1(\mathbf{m})$ and let $f \in \Sigma$ such that $\eta(f_n - f) \to 0$. Then by (6.3.1.1), $(f_n)_1^\infty$ is Cauchy in $\mathcal{L}_1(\mathbf{m})$. Hence by Theorem 3.2.8 of Chapter 3, there exists $g \in \mathcal{L}_1(\mathbf{m})$ such that $\lim_n \mathbf{m}_1^\bullet(f_n - g, T) = 0$. Since $H \subset \{x^* \in X^* : |x^*| \leq 1\}$ by Lemma 5.2.13 of Chapter 5, $\eta(f_n - g) \leq \mathbf{m}_1^\bullet(f_n - g, T) \to 0$ as $n \to \infty$. Then $\eta(f - g) \leq \eta(f - f_n) + \eta(f_n - g) \to 0$ and hence $\eta(f - g) = 0$. Clearly, $f - g$ is $\mathbf{m}$-measurable and hence $N(f - g) = B \cup N$, where $B \in \sigma(\mathcal{P})$ and $N \subset M \in \sigma(\mathcal{P})$ with $||\mathbf{m}||(M) = 0$. Then by (i) or by the fact that $H \subset \{x^* : |x^*| \leq 1\}$, $v(x^* \circ \mathbf{m})(M) = 0$ for $x^* \in H$. Now $\sup_{x^* \in H} \int_{B \cup M} |f - g| dv(x^* \circ \mathbf{m}) = \sup_{x^* \in H} \int_B |f - g| dv(x^* \circ \mathbf{m}) \leq \eta(f - g) = 0$ and hence $v(x^* \circ \mathbf{m})(B) = 0$ for $x^* \in H$. Then by (i), $||\mathbf{m}||(B) = \sup_{x^* \in H} v(x^* \circ \mathbf{m})(B) = 0$ so that $||\mathbf{m}||(N(f - g)) = 0$. Therefore, $f = g$ $\mathbf{m}$-a.e. in T and hence $f \in \mathcal{L}_1(\mathbf{m})$. Thus the claim holds.

By hypothesis and by (6.3.1.1) we have

$$\sup_{x^* \in H} \int_T |f| dv(x^* \circ \mathbf{m}) \leq \sup_{x^* \in H} \int_T |g_\epsilon| dv(x^* \circ \mathbf{m}) + \sup_{x^* \in H} \int_T |f - g_\epsilon| dv(x^* \circ \mathbf{m})$$
$$< \mathbf{m}_1^\bullet(g_\epsilon, T) + \epsilon < \infty$$

and hence $f \in \Sigma$. Moreover, the hypothesis in (ii) implies that f belongs to the η-closure of $\mathcal{L}_1(\mathbf{m})$ in Σ. Then by Claim 1, $f \in \mathcal{L}_1(\mathbf{m})$. □

Lemma 6.3.2. *Let X and H be as in Lemma 6.3.1. Let $\mathbf{m} : \delta(\mathcal{C}) \to X$ (resp. $\mathbf{m} : \mathcal{B}(T) \to X$) be σ-additive. Let $V \in \mathcal{V}$. Then there exist a sequence (x_n^*) in H and a sequence (c_n) of positive numbers such that*

$$\lim_{\lambda(A) \to 0} ||\mathbf{m}||(A) = 0$$

for $A \in \mathcal{B}(V)$ (resp. for $A \in \mathcal{B}(T)$), where

$$\lambda = \sum_1^\infty c_n v(x_n^* \circ \mathbf{m})$$

is σ-additive and finite on $\mathcal{B}(V)$ (resp. on $\mathcal{B}(T)$). (Note that in the case of $\mathbf{m}$-defined on $\delta(\mathcal{C})$, (x_n^*) and λ depend on V). Consequently, $A \in \mathcal{B}(V)$ (resp. $A \in \mathcal{B}(T)$) is $\mathbf{m}$-null if and only if A is $(x^* \circ \mathbf{m})$-null for each $x^* \in H$. If $\mathbf{m}$ is further $\delta(\mathcal{C})$-regular (resp. $\mathcal{B}(T)$-regular) and if $f : T \to \mathbb{K}$ is $(x^* \circ \mathbf{m})$-measurable for each $x^* \in H$, then f is Lusin $\mathbf{m}$-measurable as well as $\mathbf{m}$-measurable.

Proof. Let $V \in \mathcal{V}$. As $\mathbf{m}$ is σ-additive on $\delta(\mathcal{C})$ (resp. on $\mathcal{B}(T)$) and as H is norm bounded by Lemma 5.2.13 of Chapter 5, $\{x^* \circ \mathbf{m} : x^* \in H\}$ is bounded and uniformly σ-additive on $\mathcal{B}(V)$ (resp. on $\mathcal{B}(T)$) and hence by the proof of Theorem IV.9.2 and by Theorem IV.9.1 of [DS1], there exist $(x_n^*)_1^\infty \subset H$ and $c_n > 0$, $n \in \mathbb{N}$, such that $\lambda = \sum_1^\infty c_n v(x_n^* \circ \mathbf{m})$ is σ-additive and finite on $\mathcal{B}(V)$ (resp. on $\mathcal{B}(T)$) and satisfies

$$\lim_{\lambda(A) \to 0} \sup_{x^* \in H} v(x^* \circ \mathbf{m})(A) = 0$$

for $A \in \mathcal{B}(V)$ (resp. $A \in \mathcal{B}(T)$). Then by Lemma 6.3.1(i),

$$\lim_{\lambda(A) \to 0} ||\mathbf{m}||(A) = 0 \text{ for } A \in \mathcal{B}(V) \quad (\text{resp. } A \in \mathcal{B}(T)). \tag{6.3.2.1}$$

If $A \in \mathcal{B}(V)$ (resp. $A \in \mathcal{B}(T)$) is $x^* \circ \mathbf{m}$-null for each $x^* \in H$, then $\lambda(A) = 0$ and hence $||\mathbf{m}||(A) = 0$ so that A is $\mathbf{m}$-null. The converse is trivial.

Now let us assume that $\mathbf{m}$ is further $\delta(\mathcal{C})$-regular (resp. $\mathcal{B}(T)$-regular). Let $K \in \mathcal{C}$ and $\epsilon > 0$. Choose $V \in \mathcal{V}$ such that $K \subset V$. Choose $(x_n^*)_1^\infty \subset H$ and $c_n > 0$, $n \in \mathbb{N}$, and λ as above. By (6.3.2.1), there exists $\delta > 0$ such that $||\mathbf{m}||(A) < \epsilon$ whenever $\lambda(A) < \delta$ for $A \in \mathcal{B}(V)$ (resp. for $A \in \mathcal{B}(T)$). By hypothesis, f is $(x_n^* \circ \mathbf{m})$-measurable so that $N(f) \in \mathcal{B}_c(T)$ and hence $N(f)$ is σ-bounded. Then by Theorem 6.2.6, f is Lusin $(x_n^* \circ \mathbf{m})$-measurable when $\mathbf{m}$ is defined on $\delta(\mathcal{C})$. In the other case, f is Lusin $(x_n^* \circ \mathbf{m})$-measurable by Theorem 6.2.5. Therefore, for each $n \in \mathbb{N}$, there exists $K_n \in \mathcal{C}$ such that $K_n \subset K$, $f|_{K_n}$ is continuous and $v(x_n^* \circ \mathbf{m})(K \backslash K_n) < \frac{\delta}{2^n c_n}$. Then $K_0 = \bigcap_1^\infty K_n \in \mathcal{C}$, $f|_{K_0}$ is continuous and $\lambda(K \backslash K_0) \leq \sum_1^\infty c_n v(x_n^* \circ \mathbf{m})(K \backslash K_n) < \delta$. Hence $||\mathbf{m}||(K \backslash K_0) < \epsilon$. Therefore, f is Lusin $\mathbf{m}$-measurable in both cases. When $\mathbf{m}$ is defined on $\mathcal{B}(T)$, then by Theorem 6.2.5, f is $\mathbf{m}$-measurable. If $\mathbf{m}$ is defined on $\delta(\mathcal{C})$, then by hypothesis, f is $(x^* \circ \mathbf{m})$-measurable for $x^* \in H$ and hence, as noted above, $N(f) \in \mathcal{B}_c(T)$ and hence $N(f)$ is σ-bounded. Consequently, f is $\mathbf{m}$-measurable by Theorem 6.2.6. $\qquad \square$

In the sequel, $\mathcal{K}(T)$ is as in Notation 5.3.1 of Chapter 5.

Theorem 6.3.3. *Let $\mu_i : \delta(\mathcal{C}) \to \mathbb{K}$ be σ-additive and $\delta(\mathcal{C})$-regular for $i \in I$. Suppose $\sum_{i \in I} |\int_T \varphi d\mu_i|^p < \infty$ for each $\varphi \in \mathcal{K}(T)$ and for $1 \leq p < \infty$. Let $u : \mathcal{K}(T) \to l_p(I)$ be defined by $u(\varphi) = (\int_T \varphi d\mu_i)_{i \in I}$. Then u is a prolongable Radon operator on $\mathcal{K}(T)$. Let $\mathbf{m}_u$ be the representing measure of u. (See Definition 5.3.5 of Chapter 5.) Let $f : T \to \mathbb{K}$ belong to $\mathcal{L}_1(\mu_i)$ for $i \in I$. Then f is $\mathbf{m}_u$-integrable in T if and only if*

$$\sum_{i \in I} \left| \int_U f d\mu_i \right|^p < \infty \tag{6.3.3.1}$$

for each open Baire set U in T. In that case, $\int_T f d\mathbf{m}_u = (\int_T f d\mu_i)_{i \in I}$.

Let $p = 1$ and let $f \in \mathcal{L}_1(\mathbf{m}_u)$. If $\theta(\varphi) = \sum_{i \in I} \int_T \varphi d\mu_i$ for $\varphi \in \mathcal{K}(T)$, then $\theta \in \mathcal{K}(T)^*$, f is μ_θ-integrable and

$$\int_A f d\mu_\theta = \sum_{i \in I} \int_A f d\mu_i$$

for $A \in \mathcal{B}_c(T)$, where μ_θ is the complex Radon measure induced by θ in the sense of Definition 4.3 of [P3].

Proof. Let us recall from Notation 5.3.1 of Chapter 5 that the topology of $\mathcal{K}(T)$ is the inductive limit locally convex topology on $C_c(T)$ induced by the family $(C_c(T,C), I_C)$ where $C_c(T,C)$ are provided with the topology τ_u of uniform convergence. Clearly, $C_c(T,C)$ are Banach spaces. Let $u : \mathcal{K}(T) \to \ell_p(I)$ be given by $u(\varphi) = (\int_T \varphi \, d\mu_i)_{i \in I}$. Clearly, u is linear. We claim that u has a closed graph. In fact, let $\varphi_\alpha \to \varphi$ in $\mathcal{K}(T)$. As $\mu_i \in \mathcal{K}(T)^*$, $\int_T \varphi_\alpha \, d\mu_i \to \int_T \varphi \, d\mu_i$ for each $i \in I$. Suppose $u(\varphi_\alpha) \to (f_i)_{i \in I} \in \ell_p(I)$. Then given $\epsilon > 0$, there exist $J \subset I$, J finite, and an α_0, such that $\sum_{i \in J} | \int_T \varphi_\alpha \, d\mu_i - f_i|^p < (\frac{\epsilon}{2})^p$ for $\alpha \geq \alpha_0$. Moreover, there exists $\alpha_1 \geq \alpha_0$ such that $\sum_{i \in J} | \int_T \varphi_\alpha \, d\mu_i - \int_T \varphi \, d\mu_i|^p < (\frac{\epsilon}{2})^p$ since $\int_T \varphi_\alpha \, d\mu_i \to \int_T \varphi \, d\mu_i$ for each i. Then

$$\left(\sum_{i \in J} \left| \int_T \varphi \, d\mu_i - f_i \right|^p \right)^{\frac{1}{p}} \leq \left(\sum_{i \in J} \left| \int_T \varphi \, d\mu_i - \int_T \varphi_\alpha \, d\mu_i \right|^p \right)^{\frac{1}{p}}$$

$$+ \left(\sum_{i \in J} \left| \int_T \varphi_\alpha \, d\mu_i - f_i \right|^p \right)^{\frac{1}{p}} < \epsilon$$

for $\alpha \geq \alpha_1$. Thus, for $i \in J$,

$$\left| \int_T \varphi \, d\mu_i - f_i \right| \leq \left(\sum_{j \in J} \left| \int_T \varphi \, d\mu_j - f_j \right|^p \right)^{\frac{1}{p}} < \epsilon.$$

Since ϵ is arbitrary, $\int_T \varphi \, d\mu_i = f_i$. If $i \notin J$, by the same argument with $J \cup \{i\}$ in place of J, we have $\int \varphi \, d\mu_i = f_i$ for each $i \in I$. Thus $(f_i)_i = u(\varphi)$ and hence the graph of u is closed.

Since $C_c(T,C)$ is a Banach space, any linear mapping from $C_c(T,C)$ into $\ell_p(I)$ with a closed graph is continuous by the closed graph theorem (see Theorem 2.15 of [Ru2]) and hence by Problem C(i), Sec. 16, Chapter 5 of [KN], u is a continuous linear mapping.

Let $V \in \mathcal{V}$ and let $u_V(\varphi) = u(\varphi)$ for $\varphi \in C_c(V)$. Since $u : \mathcal{K}(T) \to \ell_p(I)$ is continuous, clearly u_V is continuous on $C_c(V)$ and hence its continuous extension $\widetilde{u_V} : C_0(V) \to \ell_p(I)$ is weakly compact by Theorem 13 of [P9] or by Corollary 2 of [P12] since $\ell_p(I)$ is weakly sequentially complete for $1 \leq p < \infty$ so that $c_0 \not\subset \ell_p(I)$ for $1 \leq p < \infty$. Hence u is a prolongable Radon operator on $\mathcal{K}(T)$ and hence by Theorem 5.3.9 of Chapter 5 its representing measure $\mathbf{m}_u : \delta(\mathcal{C}) \to \ell_p(I)$, $1 \leq p < \infty$, is σ-additive and $\delta(\mathcal{C})$-regular and

$$u(\varphi) = \int_T \varphi \, d\mathbf{m}_u, \ \varphi \in C_c(T) \tag{6.3.3.2}$$

where the integral is a (BDS)-integral.

Now suppose (6.3.3.1) is satisfied. For $1 < p < \infty$, let $H_I^{(p)} = \{ (\alpha_i)_{i \in I} \in \ell_q(I) : \sum_{i \in I} |\alpha_i|^q \leq 1, \alpha_i = 0 \text{ for } i \in I \backslash J, \text{ for some } J \subset I, J \text{ finite} \}$ where $\frac{1}{p} +$

$\frac{1}{q} = 1$. For $p = 1$, let $H_I^{(1)} = \{(\alpha_i)_{i \in I} \in \ell_\infty(I) : \sup_{i \in I} |\alpha_i| \leq 1, \alpha_i = 0$ for $i \in I \backslash J$, for some $J \subset I, J$ finite$\}$. Clearly, $H_I^{(p)}$ is a norm determining set for $\ell_p(I)$, $1 \leq p < \infty$.

Claim 1. Let $x^* = (\alpha_i)_{i \in I} \in H_I^{(p)}$, $1 \leq p < \infty$, where $\alpha_i = 0$ for $i \in I \backslash J_{x^*}$, for some $J_{x^*} \subset I$ and finite. Then $x^* \circ \mathbf{m}_u = \sum_{i \in I} \alpha_i \mu_i = \sum_{i \in J_{x^*}} \alpha_i \mu_i$.

In fact, by Theorem 4.1.8(v) and Remark 4.1.5 of Chapter 4 and by (6.3.3.2) we have

$$\int_T \varphi d(x^* \circ \mathbf{m}_u) = x^* u(\varphi) = \sum_{i \in I} \alpha_i \int_T \varphi d\mu_i = \int_T \varphi d \left(\sum_{i \in J_{x^*}} \alpha_i \mu_i \right) \qquad (6.3.3.3)$$

for $\varphi \in \mathcal{K}(T)$. Then for $\varphi \in C_c(V)$, by (6.3.3.3) we have $\int_T \varphi d(x^* \circ (\mathbf{m}_u)_V) = x^* u_V(\varphi) = x^* u(\varphi) = \int_T \varphi d(\sum_{i \in J_{x^*}} \alpha_i \mu_i)$, where $(\mathbf{m}_u)_V = \mathbf{m}_u|_{\mathcal{B}(V)}$. As $x^* \circ (\mathbf{m}_u)_V$ and $\mu_i|_{\mathcal{B}(V)}$ are σ-additive and $\mathcal{B}(V)$-regular, by the uniqueness part of the Riesz representation theorem we conclude that $x^* \circ (\mathbf{m}_u)_V = \sum_{i \in I} \alpha_i \mu_i|_{\mathcal{B}(V)}$. Since V is arbitrary in $\mathcal{V}$ and since $\delta(\mathcal{C}) = \bigcup_{V \in \mathcal{V}} \mathcal{B}(V)$, it follows that $x^* \circ \mathbf{m}_u = \sum_{i \in J_{x^*}} \alpha_i \mu_i = \sum_{i \in I} \alpha_i \mu_i$. Hence the claim holds.

Let $\varphi \in C_0(T)$. By hypothesis, $f \in \mathcal{L}_1(\mu_i)$ for $i \in I$ and φ is $\mathcal{B}_c(T)$-measurable by Theorem 51.B of [H] and is bounded. Hence $f\varphi \in \mathcal{L}_1(\mu_i)$ for $i \in \mathcal{I}$. Let

$$\theta_i(\varphi) = \int_T f\varphi d\mu_i, \quad \varphi \in C_0(T)$$

for $i \in I$. Then θ_i is a bounded linear functional on $C_0(T)$ and hence the complex Radon measure μ_{θ_i} induced by θ_i in the sense of Definition 4.3 of [P3] is a σ-additive $\mathcal{B}(T)$-regular scalar measure on $\mathcal{B}(T)$ by Theorems 3.3 and 4.6 of [P4]. Then it is easy to check that $\mu_{\theta_i}|_{\mathcal{B}(V)}$ is $\mathcal{B}(V)$-regular for each $V \in \mathcal{V}$.

Claim 2. $\eta_i(\cdot) = \int_{(\cdot)} f d\mu_i$ is σ-additive on $\mathcal{B}_c(T)$ and η_i is $v(\mu_i)$-continuous, $i \in I$.

In fact, $v(\mu_i) : \mathcal{B}_c(T) \to [0, \infty]$ is σ-additive by Property 9, § 3, Chapter I of [Din1] and $v(\mu_i)(E) = 0$, $E \in \mathcal{B}_c(T)$ implies $v(\eta_i)(E) = 0$. Then by Theorem 6.11 of [Ru1] (whose proof is valid for σ-rings too) we conclude that $v(\eta_i)$ and hence η_i is $v(\mu_i)$-continuous on $\mathcal{B}_c(T)$.

Since $v(\eta_i)$ is $v(\mu_i)$-continuous on $\delta(\mathcal{C})$ and since μ_i is $\delta(\mathcal{C})$-regular by hypothesis, η_i, $i \in I$, are $\delta(\mathcal{C})$-regular. Moreover, for $\varphi \in C_0(T)$, we have

$$\int_T \varphi d\eta_i = \int_T \varphi f d\mu_i = \theta_i(\varphi) = \int_T \varphi d\mu_{\theta_i}, \quad i \in I. \qquad (6.3.3.4)$$

Thus, for $V \in \mathcal{V}$ and $\varphi \in C_c(V)$, we have $\int_T \varphi d\eta_i|_{\mathcal{B}(V)} = \int_T \varphi d\mu_{\theta_i}|_{\mathcal{B}(V)}$ and hence by the uniqueness part of the Riesz representation theorem, we have $\eta_i|_{\mathcal{B}(V)} = \mu_{\theta_i}|_{\mathcal{B}(V)}$. As V is arbitrary in $\mathcal{V}$, we conclude that

$$\eta_i|_{\delta(\mathcal{C})} = \mu_{\theta_i}|_{\delta(\mathcal{C})}. \qquad (6.3.3.5)$$

Since $v(\eta_i)(T) = \int_T |f| dv(\mu_i) < \infty$ by Proposition 1.1.19 of Chapter 1 and by the hypothesis that $f \in \mathcal{L}_1(\mu_i)$ and since $\sigma(\delta(\mathcal{C})) = \mathcal{B}_c(T)$, we conclude that

$$\eta_i = \mu_{\theta_i}|_{\mathcal{B}_c(T)} \tag{6.3.3.6}$$

for $i \in I$. Then by Theorem 2.4 of [P4], η_i is $\mathcal{B}_c(T)$-regular for $i \in I$.

Let $x^* = (\alpha_i) \in H_I^{(p)}$. Then there exists a finite set $J_{x^*} \subset I$ such that $\alpha_i = 0$ for $i \in I \backslash J_{x^*}$. Let $\Psi_{x^*} = \sum_{i \in J_{x^*}} \alpha_i \theta_i$. Then Ψ_{x^*} is a bounded linear functional on $C_0(T)$ and for $\varphi \in C_0(T)$, we have $\Psi_{x^*}(\varphi) = \sum_{i \in J_{x^*}} \alpha_i \theta_i(\varphi) = \int_T \varphi d(\sum_{i \in J_{x^*}} \alpha_i \mu_{\theta_i})$ by (6.3.3.4). Then by (6.3.3.6), we have

$$\mu_{\Psi_{x^*}} = \sum_{i \in J_{x^*}} \alpha_i \mu_{\theta_i} \text{ on } \mathcal{B}(T) \text{ and } \mu_{\psi_{x^*}}|_{\mathcal{B}_c(T)} = \sum_{i \in J_{x^*}} \alpha_i \eta_i \tag{6.3.3.7}$$

where $\mu_{\Psi_{x^*}}$ is the bounded complex Borel measure induced by Ψ_{x^*} in the sense of Definition 4.3 of [P3].

Claim 3. $\sup_{x^* \in H_I^{(p)}} v(\mu_{\Psi_{x^*}}, \mathcal{B}(T))(T) = M$ (say) $< \infty$ for $1 \le p < \infty$.

In fact, let U be an open Baire set in T and let $1 < p < \infty$. Then by hypothesis (6.3.3.1), by (6.3.3.7) and by Hölder's inequality, we have

$$\sup_{x^* \in H_I^{(p)}} |\mu_{\Psi_{x^*}}(U)| = \sup_{x^* = (\alpha_i)_i \in H_I^{(p)}} \left| \sum_{i \in J_{x^*}} \alpha_i \eta_i(U) \right| = \sup_{x^* \in H_I^{(p)}} \left| x^* \left(\int_U f d\mu_i \right)_{i \in I} \right|$$

$$\le \sup_{x^* \in H_I^{(p)}} |x^*|_q \left(\sum_{i \in I} \left| \int_U f d\mu_i \right|^p \right)^{\frac{1}{p}} \le \left(\sum_{i \in I} \left| \int_U f d\mu_i \right|^p \right)^{\frac{1}{p}} < \infty \tag{6.3.3.8}$$

where $\frac{1}{p} + \frac{1}{q} = 1$. If $p = 1$, by (6.3.3.1) and by (6.3.3.7) we have

$$\sup_{x^* \in H_I^{(1)}} |\mu_{\Psi_{x^*}}(U)| = \sup_{x^* = (\alpha_i)_i \in H_I^{(1)}} \left| \sum_{i \in J_{x^*}} \alpha_i \eta_i(U) \right| \le \sup_{x^* \in H_I^{(1)}} \left| x^* \left(\int_U f d\mu_i \right)_{i \in I} \right|$$

$$= \sup_{x^* \in H_I^{(1)}, x^* = (\alpha_i)_{i \in I}} \sum_{i \in I} \left| \alpha_i \int_U f d\mu_i \right| \le \sum_{i \in I} \left| \int_U f d\mu_i \right| < \infty. \tag{6.3.3.8$'$}$$

As $\{\mu_{\Psi_{x^*}} : x^* \in H_I^{(p)}\} \subset M(T)$ for $1 \le p < \infty$, the claim holds by (6.3.3.8) and (6.3.3.8$'$) and by Corollary 5.2.5 of Chapter 5.

Claim 4. Given $\varphi \in C_0(T)$ and $\epsilon > 0$, there exists a simple function s as a complex linear combination of the characteristic functions of relatively compact open Baire sets in T such that

$$\|s - \varphi\|_T < \frac{\epsilon}{2M} \tag{6.3.3.9}$$

where M is as in Claim 3.

In fact, in the proof of Lemma 5.2.20(i) of Chapter 5, each of the sets $E_{i,n}$ is a difference of two F_σ open sets $U_{i,n}$, $V_{i,n}$ in T. Clearly, $U_{i,n}$ and $V_{i,n}$ are σ-compact and relatively compact as supp φ is compact. Then by Lemma 5.2.3 of Chapter 5, $U_{i,n}$ and $V_{i,n}$ are relatively compact open Baire sets in T. Then, the functions (s'_n) in the proof of Lemma 5.2.20(i) of Chapter 5 are complex linear combinations of the characteristic functions of relatively compact open Baire sets in T. As $s'_n \to \varphi$ uniformly in T, the claim holds.

Claim 5. Let $\varphi \in C_0(T)$. Then

$$\sum_{i \in I} \left| \int_T f\varphi d\mu_i \right|^p < \infty. \tag{6.3.3.10}$$

In fact, given $\epsilon > 0$, choose s as in Claim 4. By hypothesis (6.3.3.1),

$$\sum_{i \in I} \left| \int_T sf d\mu_i \right|^p < \infty.$$

Then there exists a finite set $J_0 \subset I$ such that

$$\sum_{I \setminus J_0} \left| \int_T sf d\mu_i \right|^p < \left(\frac{\epsilon}{2} \right)^p. \tag{6.3.3.11}$$

Let $I_1 = I \setminus J_0$. Let $H_{I_1}^{(p)} = \{(\alpha_i)_{i \in I_1}: \text{there exists a finite set } J \subset I_1 \text{ such that } \alpha_i = 0 \text{ for } i \in I_1 \setminus J \text{ and } \|(\alpha_i)_{i \in I_1}\|_q \leq 1\}$ where $\frac{1}{p} + \frac{1}{q} = 1$ when $1 < p < \infty$; and $q = \infty$ when $p = 1$. Let $x^* = (\alpha_i)_{i \in I_1} \in H_{I_1}^{(p)}$ be fixed. Then there exists a finite set $J_{x^*} \subset I_1$ such that $\alpha_i = 0$ for $i \in I_1 \setminus J_{x^*}$. Let

$$\Phi_{x^*}(\varphi) = \sum_{i \in J_{x^*}} \alpha_i \theta_i(\varphi).$$

Then by (6.3.3.4) we have

$$\Phi_{x^*}(\varphi) = \sum_{i \in J_{x^*}} \alpha_i \int_T \varphi d\mu_{\theta_i} = \sum_{i \in J_{x^*}} \int_T \alpha_i f\varphi d\mu_i \tag{6.3.3.12}$$

for $\varphi \in C_0(T)$. Then Φ_{x^*} is a bounded linear functional on $C_0(T)$ and the complex Radon measure $\mu_{\Phi_{x^*}}$ induced by Φ_{x^*} in the sense of Definition 4.3 of [P3] is given by $\mu_{\Phi_{x^*}} = \sum_{i \in J_{x^*}} \alpha_i \mu_{\theta_i}$. Then $|\Phi_{x^*}(\varphi)| \leq |\int_T (\varphi - s) d\mu_{\Phi_{x^*}}| + |\int_T s d\mu_{\Phi_{x^*}}| \leq \|\varphi - s\|_T v(\mu_{\Phi_{x^*}}, \mathcal{B}(T))(T) + |\sum_{i \in J_{x^*}} \alpha_i \int_T sf d\mu_i|$ since $\mu_{\Phi_{x^*}} = \sum_{i \in J_{x^*}} \alpha_i \mu_{\theta_i} = \sum_{i \in J_{x^*}} \alpha_i \eta_i$ on $\mathcal{B}_c(T)$ by (6.3.3.6) and since s is a $\mathcal{B}_c(T)$-simple function. Taking $y^* = (\alpha_i)_{i \in I}$ with $\alpha_i = 0$ for $i \in I \setminus J_{x^*}$, we observe that Φ_{x^*} is the same as Ψ_{y^*} defined before Claim 3 and hence by Claim 3 we have $\sup_{x^* \in H_{I_1}^{(p)}} v(\mu_{\Phi_{x^*}}, \mathcal{B}(T))(T) \leq$

M for $1 \leq p < \infty$ where M is as in Claim 3. Hence by (6.3.3.9) and (6.3.3.11) we have

$$|\Phi_{x^*}(\varphi)| \leq ||\varphi - s||_T \cdot v(\mu_{\Phi_{x^*}}, \mathcal{B}(T))(T) + \left| \left(\sum_{i \in J_{x^*}} \alpha_i \int_T sfd\mu_i \right) \right|$$

$$\leq \frac{\epsilon}{2M} \cdot M + \left| x^* \left(\int_T sfd\mu_i \right)_{i \in I_1} \right|$$

$$\leq \frac{\epsilon}{2} + ||x^*||_q \left\| \left(\int_T sfd\mu_i \right)_{i \in I_1} \right\|_p$$

$$< \epsilon$$

since $(\int_T sfd\mu_i)_{i \in I_1} \in \ell_p(I_1)$ by (6.3.3.11).

Varying $x^* \in H_{I_1}^{(p)}$, we have $\sup_{x^* \in H_1^{(p)}} |\Phi_{x^*}(\varphi)| \leq \epsilon$. As $H_{I_1}^{(p)}$ is a norm determining set for $\ell_p(I_1)$ for $1 \leq p < \infty$ and as $\Phi_{x^*}(\varphi) = x^*(\int_T f\varphi d\mu_i)_{i \in I_1}$ by (6.3.3.12), we have

$$\left(\sum_{i \in I_1} \left| \int_T f\varphi d\mu_i \right|^p \right)^{\frac{1}{p}} = \left\| \left(\int_T f\varphi d\mu_i \right)_{i \in I_1} \right\|_p = \sup_{x^* \in H_{I_1}^{(p)}} |\Phi_{x^*}(\varphi)| \leq \epsilon$$

for $1 \leq p < \infty$. Hence Claim 5 holds.

Then by Claim 5, the mapping $\Phi : C_0(T) \to \ell_p(I)$ given by

$$\Phi(\varphi) = \left(\int_T f\varphi d\mu_i \right)_{i \in I} = \left(\int_T \varphi d\eta_i \right)_{i \in I}$$

is well defined and linear. It is easy to check that Φ has a closed graph. (See the proof of the claim that u has a closed graph.) Then by the closed graph theorem Φ is continuous. Since $c_0 \not\subset \ell_p(I)$ for $1 \leq p < \infty$, Φ is weakly compact by Theorem 13 of [P9] for $1 \leq p < \infty$. Then by Theorems 2 and 6 of [P9] its representing measure $\mathbf{m}_\Phi : \mathcal{B}(T) \to \ell_p(I)^{**}$ has range in $\ell_p(I)$, and is σ-additive and $\mathcal{B}(T)$-regular.

Claim 6. Let $x^* = (\alpha_i)_{i \in I} \in H_I^{(p)}$, $1 \leq p < \infty$, so that there exists a finite set $J_{x^*} \subset I$ such that $\alpha_i = 0$ for $i \in I \backslash J_{x^*}$. Then

$$(x^* \circ \mathbf{m}_\Phi)|_{\mathcal{B}_c(T)} = \sum_{i \in J_{x^*}} \alpha_i \eta_i = \sum_{i \in I} \alpha_i \eta_i.$$

In fact, for $\varphi \in C_0(T)$, by Theorem 1 of [P9] we have $\int_T \varphi d(x^* \circ \mathbf{m}_\Phi) = x^* \Phi(\varphi) = \sum_{i \in J_{x^*}} \alpha_i \int_T \varphi d\eta_i = \int_T \varphi d(\sum_{i \in J_{x^*}} \alpha_i \eta_i)$; $x^* \circ \mathbf{m}_\Phi$ is $\mathcal{B}_c(T)$-regular since $\mathbf{m}_\Phi|_{\mathcal{B}_c(T)}$ is $\mathcal{B}_c(T)$-regular by Theorem 7(xxiii) of [P9] and $\sum_{i \in J_{x^*}} \alpha_i \eta_i$ is $\mathcal{B}_c(T)$-regular as observed after (6.3.3.6). Consequently, by the uniqueness part of the σ-Borel version of the Riesz representation theorem the claim holds.

By hypothesis, f is μ_i-measurable for $i \in I$ and hence f is $\sum_{i \in J_{x^*}} \alpha_i \mu_i$-measurable for $x^* = (\alpha_i)_{i \in I} \in H_I^{(p)}$ for $1 \leq p < \infty$, where $\alpha_i = 0$ for $i \in I \backslash J_{x^*}$, $J_{x^*} \subset I$ and J_{x^*} is finite. Since η_i is $v(\mu_i)$-continuous for $i \in I$ by Claim 2, f is also η_i-measurable for $i \in I$. Hence by Claim 6, f is $(x^* \circ \mathbf{m}_\Phi)|_{\mathcal{B}_c(T)}$-measurable for $x^* \in H_I^{(p)}$. $\mathbf{m}_\Phi$ is $\ell_p(I)$-valued, σ-additive and Borel regular for $1 \leq p < \infty$, and hence by Theorem 6.1.12, $\mathbf{m}_\Phi|_{\delta(\mathcal{C})}$ is $\ell_p(I)$-valued, σ-additive and $\delta(\mathcal{C})$-regular for $1 \leq p < \infty$. As $H_I^{(p)}$ is a norm determining set for $\ell_p(I)$, then by Lemma 6.3.2, f is $\mathbf{m}_\Phi|_{\delta(\mathcal{C})}$-measurable as well as Lusin $\mathbf{m}_\Phi|_{\delta(\mathcal{C})}$-measurable.

Let $\epsilon > 0$. As $\mathbf{m}_\Phi$ is $\mathcal{B}(T)$-regular, there exists $K \in \mathcal{C}$ such that $||\mathbf{m}_\Phi||(T \backslash K) < \frac{\epsilon}{2}$. As f is Lusin $\mathbf{m}_\Phi|_{\delta(\mathcal{C})}$-measurable, there exists $K_0 \in \mathcal{C}$ with $K_0 \subset K$ such that $f|_{K_0}$ is continuous and $||\mathbf{m}_\Phi||(K \backslash K_0) < \frac{\epsilon}{2}$. Then

$$||\mathbf{m}_\Phi||(T \backslash K_0) < \epsilon \qquad (6.3.3.13)$$

and $f \chi_{K_0}$ is bounded and $\mathcal{B}(K_0)$-measurable, as it is continuous on the compact K_0. Consequently, $f \chi_{K_0}$ is a bounded $\mathcal{B}_c(T)$-measurable function with compact support. As u is prolongable with the representing measure $\mathbf{m}_u$, by (18) of Theorem 5.3.12 of Chapter 5,

$$f \chi_{K_0} \in \mathcal{L}_1(\mathbf{m}_u). \qquad (6.3.3.14)$$

By Claim 1 and by the hypothesis that $f \in \mathcal{L}_1(\mu_i)$ for $i \in I$, f is $(x^* \circ \mathbf{m}_u)$-measurable for each $x^* \in H_I^{(p)}$, $1 \leq p < \infty$ and as observed in the beginning of the proof, $\mathbf{m}_u$ is $\ell_p(I)$-valued, σ-additive and $\delta(\mathcal{C})$-regular. As $H_I^{(p)}$ is a norm determining set for $\ell_p(I)$, $1 \leq p < \infty$, by Lemma 6.3.2,

$$f \text{ is } \mathbf{m}_u\text{-measurable.} \qquad (6.3.3.15)$$

Consequently, $f - f \chi_{K_0}$ is also $\mathbf{m}_u$-measurable. Now by Claims 1 and 6, by Lemma 6.3.2 and by the fact that

$$\int_{N(f) \backslash K_0} |f| dv \left(\sum_{i \in J_{x^*}} \alpha_i \mu_i \right) = v \left(\int_{N(f) \backslash K_0} f d \left(\sum_{i \in J_{x^*}} \alpha_i \mu_i \right) \right)$$

$$= v \left(\sum_{i \in J_{x^*}} \alpha_i \eta_i \right) (N(f) \backslash K_0)$$

(by Proposition 1.1.19 of Chapter 1), we have

$$\sup_{x^* \in H_I^{(p)}} \int_T |f - f \chi_{K_0}| dv(x^* \circ \mathbf{m}_u) = \sup_{x^* \in H_I^{(p)}} \int_{N(f) \backslash K_0} |f| dv(x^* \circ \mathbf{m}_u)$$

$$= \sup_{x^* = (\alpha_i)_i \in H_I^{(p)}} \int_{N(f) \backslash K_0} |f| dv \left(\sum_{i \in J_{x^*}} \alpha_i \mu_i \right)$$

$$= \sup_{x^*=(\alpha_i)_{i\in I}\in H_I^{(p)}} v\left(\sum_{i\in J_{x^*}} \alpha_i\eta_i\right)(N(f)\backslash K_0)$$

$$= \sup_{x^*\in H_I^{(p)}} v(x^*\circ \mathbf{m}_\Phi)(N(f)\backslash K_0) \le \sup_{x^*\in H_I^{(p)}} v(x^*\circ \mathbf{m}_\Phi)(T\backslash K_0)$$

$$\le ||\mathbf{m}_\Phi||(T\backslash K_0) < \epsilon \qquad (6.3.3.16)$$

by (6.3.3.13) for $1 \le p < \infty$. Since f is $\mathbf{m}_u$-measurable by (6.3.3.15), since $f\chi_{K_0} \in \mathcal{L}_1(\mathbf{m}_u)$ by (6.3.3.14), since $H_I^{(p)}$ is a norm determining set for $\ell_p(I)$ for $1 \le p < \infty$ and since $\epsilon > 0$ is arbitrary, by (6.3.3.16) and by Lemma 6.3.1(ii) we conclude that $f \in \mathcal{L}_1(\mathbf{m}_u)$. Hence the condition (6.3.3.1) is sufficient.

Moreover, for $f \in \mathcal{L}_1(\mathbf{m}_u)$, let $x_i^* = (\alpha_j)_{j\in I} \in H_I^{(p)}$, where $\alpha_i = 1$ and $\alpha_j = 0$, $j \ne i$. Then by Claim 1, $x_i^* \circ \mathbf{m}_u = \mu_i$ and hence by Theorem 5.3.11(iii) of Chapter 5, $f \in \mathcal{L}_1(\mu_i)$ and $x_i^*(\int_T f d\mathbf{m}_u) = \int_T f d(x_i^*\circ \mathbf{m}_u) = \int_T f d\mu_i$. Hence

$$\int_T f d\mathbf{m}_u = \left(\int_T f d\mu_i\right)_{i\in I} \in \ell_p(I) \qquad (6.3.3.17)$$

for $1 \le p < \infty$.

Conversely, let $f \in \mathcal{L}_1(\mathbf{m}_u)$ and let U be an open Baire set in T. Then by Theorem 2.1.5(vi) and by Remark 2.2.3 of Chapter 2, $f\chi_U \in \mathcal{L}_1(\mathbf{m}_u)$ and hence by (6.3.3.17), $\int_T f\chi_U d\mathbf{m}_u = (\int_U f d\mu_i)_{i\in I} \in \ell_p(I)$. Therefore $\sum_{i\in I} |\int_U f d\mu_i|^p < \infty$ for $1 \le p < \infty$. Thus the condition (6.3.3.1) is also necessary.

Let $p = 1$, $f \in \mathcal{L}_1(\mathbf{m}_u)$ and $x^* = (\alpha_i)_{i\in I} \in \ell_\infty(I)$, where $\alpha_i = 1$ for each i. Then θ given in the last part of the theorem is the same as x^*u and hence $\theta \in \mathcal{K}(T)^*$. Then by Corollary 5.3.10 of Chapter 5, $\mu_\theta = \mathbf{m}_{x^*u}$ considering x^*u as a scalar-valued prolongable Radon operator. Then by Theorem 5.3.11 of Chapter 5 and by hypothesis, we have $f \in \mathcal{L}_1(x^*\mathbf{m}_u) = \mathcal{L}_1(\mathbf{m}_{x^*u}) = \mathcal{L}_1(\mu_\theta)$ and

$$\int_A f d\mu_\theta = \int_A f d\mathbf{m}_{x^*u} = x^*\left(\int_A f d\mathbf{m}_u\right) = x^*\left(\left(\int_A f d\mu_i\right)_{i\in I}\right) = \sum_{i\in I}\int_A f d\mu_i$$

for each $A \in \mathcal{B}_c(T)$.

This completes the proof of the theorem. $\qquad\square$

The above theorem for the case $I = \mathbb{N}$ and $p = 1$ is used in the proof of the following result, the last part of which strengthens Theorem 2.2.2 of Chapter 2 when $\mathcal{P} = \delta(\mathcal{C})$ and $\mathbf{m}$ is a Banach space-valued σ-additive $\mathcal{P}$-regular measure on $\mathcal{P}$.

Theorem 6.3.4. *Let X be a Banach space and let $\mathbf{m} : \delta(\mathcal{C}) \to X$ be σ-additive and $\delta(\mathcal{C})$-regular. Let H be a norm determining set for X with the Orlicz property. Then a function $f : T \to \mathbb{K}$ is $\mathbf{m}$-integrable in T if and only if $f \in \mathcal{L}_1(x^*\circ \mathbf{m})$ for*

*each $x^* \in H$ and, for each open Baire set U in T, there exists a vector $x_U \in X$
such that*

$$x^*(x_U) = \int_U f d(x^* \circ \mathbf{m}) \tag{6.3.4.1}$$

*for $x^* \in H$. In that case, $f\varphi \in \mathcal{L}_1(\mathbf{m})$ for each $\varphi \in C_0(T)$ and the mapping
$\Psi : C_0(T) \to X$ given by $\Psi(\varphi) = \int_T f\varphi d\mathbf{m}$ is a weakly compact operator. Finally,
f is $\mathbf{m}$-integrable in T if and only if $f \in \mathcal{L}_1(x^* \circ \mathbf{m})$ for $x^* \in X^*$ and (6.3.4.1)
holds for each $x^* \in X^*$ and for each open Baire set U in T.*

Proof. If $f \in \mathcal{L}_1(\mathbf{m})$, then f is (KL) $\mathbf{m}$-integrable in T by Theorem 2.2.2 of Chapter
2 and hence the conditions hold.

Conversely, let the conditions hold. Let $<H>$ be the vector space spanned
by H and let F be the norm closure of $<H>$ in X^*. By hypothesis, for each open
Baire set U in T there exists $x_U \in X$ such that (6.3.4.1) holds for $x^* \in H$ and
consequently,

$$x^*(x_U) = \int_U f d(x^* \circ \mathbf{m}) \tag{6.3.4.2}$$

for $x^* \in <H>$.

Claim 1. (6.3.4.2) holds for each $x^* \in F$ and for each open Baire set U in T.

In fact, given $x^* \in F$, there exists a sequence $(x_n^*) \subset <H>$ such that $x^* = \sum_1^\infty x_n^*$ with $\sum_1^\infty |x_n^*| < \infty$. Given $\varphi \in \mathcal{K}(T)$, φ is $\mathcal{B}_c(T)$-measurable by Theorem
51.B of [H] and is bounded and hence $\varphi \in \mathcal{L}_1(\mathbf{m})$ by Theorem 2.1.5(v) and Remark
2.2.3 of Chapter 2. Then by Theorem 2.1.5(viii) followed by Remark 2.2.3 of
Chapter 2 we have

$$\sum_1^\infty \left| \int_T \varphi d(x_n^* \circ \mathbf{m}) \right| = \sum_1^\infty \left| x_n^* \left(\int_T \varphi d\mathbf{m} \right) \right| \le \left| \int_T \varphi d\mathbf{m} \right| \left(\sum_1^\infty |x_n^*| \right) < \infty.$$
$$\tag{6.3.4.3}$$

Moreover, by (6.3.4.2) we have

$$\sum_1^\infty \left| \int_U f d(x_n^* \circ \mathbf{m}) \right| = \sum_1^\infty |x_n^*(x_U)| \le \left(\sum_1^\infty |x_n^*| \right) |x_U| < \infty. \tag{6.3.4.4}$$

Clearly, $x^* \circ \mathbf{m} = \lim_n(\sum_{k=1}^n x_k^*) \circ \mathbf{m} = \lim_n \sum_{k=1}^n (x_k^* \circ \mathbf{m}) = \sum_1^\infty (x_n^* \circ \mathbf{m})$ as
$\mathbf{m}$ has range in X. As $\mathbf{m}$ is σ-additive and $\delta(\mathcal{C})$-regular on $\delta(\mathcal{C})$, $x^* \circ \mathbf{m}$ is σ-additive
and $\delta(\mathcal{C})$-regular on $\delta(\mathcal{C})$.

The mapping $u : \mathcal{K}(T) \to \ell_1(\mathbb{N})$ given by

$$u(\varphi) = \left(\int_T \varphi d(x_n^* \circ \mathbf{m}) \right)_{n=1}^\infty$$

is well defined by (6.3.4.3) and is clearly linear. By (6.3.4.3) and (6.3.4.4), the
complex measures $(x_n^* \circ \mathbf{m})_{n=1}^\infty$ and the function f satisfy the sufficiency part

of the hypotheses of Theorem 6.3.3 for $p = 1$ and $I = \mathbb{N}$ and consequently, u is a prolongable Radon operator and $f \in \mathcal{L}_1(\mathbf{m}_u)$. By the last part of the said theorem, $\theta : \mathcal{K}(T) \to \mathbb{K}$, given by $\theta(\varphi) = \sum_{n=1}^{\infty} \int_T \varphi d(x_n^* \circ \mathbf{m})$, belongs to $\mathcal{K}(T)^*$, f is μ_θ-integrable and

$$\int_U f d\mu_\theta = \sum_{n=1}^{\infty} \int_U f d(x_n^* \circ \mathbf{m}) \qquad (6.3.4.5)$$

for each open Baire set U in T and for the set $U = N(f)$ since $f \in \mathcal{L}_1(\mathbf{m}_u)$ so that f is $\mathbf{m}_u$-measurable and hence $N(f) \in \widetilde{\mathcal{B}_c(T)}$ with respect to $\|\mathbf{m}_u\|$.

Now $\int_T \varphi d\mu_\theta = \theta(\varphi) = \sum_{n=1}^{\infty} \int_T \varphi d(x_n^* \circ \mathbf{m}) = \int_T \varphi d(x^* \circ \mathbf{m})$ for $\varphi \in \mathcal{K}(T)$, since $|\int_T \varphi d(x^* \circ \mathbf{m}) - \int_T \varphi d(\sum_1^k x_n^* \circ \mathbf{m})| = |(x^* - \sum_1^k x_n^*) \int_T \varphi d\mathbf{m}| \leq |x^* - \sum_1^k x_n^*| \|\int_T \varphi d\mathbf{m}| \to 0$ as $k \to \infty$ by (v) and (viii) of Theorem 2.1.5 and Remark 2.2.3 of Chapter 2 and since $x^* = \sum_1^\infty x_n^*$. Since $x^* \circ \mathbf{m}$ is σ-additive and $\delta(\mathcal{C})$-regular, and since $\mu_\theta|_{\delta(\mathcal{C})}$ is σ-additive and $\delta(\mathcal{C})$-regular by Theorem 4.4(i) of [P4], by an argument based on the uniqueness part of the Riesz representation theorem which is similar to that in the proof of Claim 1 in the proof of Theorem 6.3.3 we have $(x^* \circ \mathbf{m})|_{\delta(\mathcal{C})} = \mu_\theta|_{\delta(\mathcal{C})}$. Then by (6.3.4.5) and (6.3.4.2) we have

$$\int_U f d(x^* \circ \mathbf{m}) = \int_U f d\mu_\theta = \sum_{n=1}^{\infty} \int_U f d(x_n^* \circ \mathbf{m}) = \sum_{n=1}^{\infty} x_n^*(x_U) = x^*(x_U)$$

for any open Baire set U in T and hence Claim 1 holds. Moreover, as f is μ_θ-integrable and as $\mu_\theta|_{\delta(\mathcal{C})} = x^* \circ \mathbf{m}$, f is $(x^* \circ \mathbf{m})$-integrable and hence $\int_T |f| dv(x^* \circ \mathbf{m}) < \infty$. Thus, for $x^* \in F$, $f \in \mathcal{L}_1(x^* \circ \mathbf{m})$ and

$$\int_T |f| dv(x^* \circ \mathbf{m}) < \infty. \qquad (6.3.4.6)$$

Let $\mathcal{F}$ be the vector space spanned by the characteristic functions of open Baire sets in T. Then for each $g \in \mathcal{F}$, by Claim 1 there exists $x_g \in X$ such that

$$x^*(x_g) = \int_T f g d(x^* \circ \mathbf{m}) \qquad (6.3.4.7)$$

for each $x^* \in F$. Let $\mathcal{G} = \{x_g : g \in \mathcal{F}, \|g\|_T \leq 1\}$.

Claim 2. $\sup_{x_g \in \mathcal{G}} |x_g| = M$ (say) $< \infty$.

In fact, for $x^* \in F$, by (6.3.4.6) and (6.3.4.7) we have

$$\sup_{x_g \in \mathcal{G}} |x^*(x_g)| = \sup_{x_g \in \mathcal{G}} |\int_T f g d(x^* \circ \mathbf{m})| \leq \int_T |f| dv(x^* \circ \mathbf{m}) < \infty$$

as $\|g\|_T \leq 1$ for $g \in \mathcal{G}$. Hence $\mathcal{G}$ is $\sigma(X, F)$-bounded. Since H is a norm determining set for X, by Lemma 5.2.13 of Chapter 5 we have $|x^*| \leq 1$ for $x^* \in H$ and hence

$|x| = \sup_{|x^*| \le 1} |x^*(x)| \ge \sup_{x^* \in F, |x^*| \le 1} |x^*(x)| \ge \sup_{x^* \in H} |x^*(x)| = |x|$ for $x \in X$ (since $H \subset F$). Hence X can be considered as a subspace of F^* with the restriction of the norm of F^*. Then by the Banach-Steinhaus theorem (Theorem II.3.21 of [DS1]) applied to the Banach space F, the set $\mathcal{G}$ is norm bounded and hence the claim holds.

For $g \in \mathcal{F}$, let $\Phi(g) = x_g$. By (6.3.4.7) and by the hypothesis that H is norm determining, $\Phi : \mathcal{F} \to X$ is well defined and clearly, it is linear. By Claim 2, Φ is continuous. Hence Φ has a unique continuous linear extension $\hat{\Phi}$ on the closure $\bar{\mathcal{F}}$ of $\mathcal{F}$ in the Banach space of all bounded scalar functions on T, with the supremum norm. $C_0(T) \subset \bar{\mathcal{F}}$ by Claim 4 in the proof of Theorem 6.3.3 and hence let $\Phi_0 = \hat{\Phi}|_{C_0(T)}$. Then $\Phi_0 : C_o(T) \to X$ is linear and continuous and hence by Theorem 1 of [P9] its representing measure $\boldsymbol{\eta}$ is given by $\Phi_0^{**}|_{\mathcal{B}(T)}$. Moreover, by the same theorem, $x^* \circ \boldsymbol{\eta} \in M(T)$ for each $x^* \in X^*$ and

$$x^*\Phi_0(\varphi) = \int_T \varphi d(x^* \circ \boldsymbol{\eta}) \text{ for } \varphi \in C_0(T). \tag{6.3.4.8}$$

By hypothesis, $f \in \mathcal{L}_1(x^* \circ \mathbf{m})$ for $x^* \in H$ and hence $\nu_{x^*}(\cdot) = \int_{(\cdot)} f d(x^* \circ \mathbf{m})$ is σ-additive on $\mathcal{B}_c(T)$ for $x^* \in H$. Clearly, ν_{x^*} is $v(x^* \circ \mathbf{m})$-continuous by Claim 2 in the proof of Theorem 6.3.3. Then f is ν_{x^*}-measurable since f is $(x^* \circ \mathbf{m})$-measurable and since ν_{x^*} is $v(x^* \circ \mathbf{m})$-continuous.

Let $\varphi \in C_0(T)$. Since $C_0(T) \subset \bar{\mathcal{F}}$, there exists $(g_n)_1^\infty \subset \mathcal{F}$ such that $g_n \to \varphi$ uniformly in T so that $\Phi_0(\varphi) = \lim_n \Phi(g_n)$. Then by (6.3.4.7) and by the definition of Φ we have

$$x^*\Phi_0(\varphi) = \lim_n x^*\Phi(g_n) = \lim_n x^*(x_{g_n}) = \lim_n \int_T f g_n d(x^* \circ \mathbf{m}) \tag{6.3.4.9}$$

for $x^* \in F$. On the other hand, as φ is $\mathcal{B}_c(T)$-measurable by Theorem 51.B of [H] and bounded, $f\varphi \in \mathcal{L}_1(x^* \circ \mathbf{m})$ for $x^* \in H$ and by (6.3.4.6),

$$\left| \int_T f\varphi d(x^* \circ \mathbf{m}) - \int_T f g_n d(x^* \circ \mathbf{m}) \right| \le ||\varphi - g_n||_T \left(\int_T |f| dv(x^* \circ \mathbf{m}) \right) \to 0$$

as $n \to \infty$ for each $x^* \in H$ and therefore

$$\int_T f\varphi d(x^* \circ \mathbf{m}) = \lim_n \int_T f g_n d(x^* \circ \mathbf{m})$$

for $x^* \in H$. Then by (6.3.4.9) and (6.3.4.8) and by the definition of $\nu_{x^*}(\cdot)$ we have

$$\int_T \varphi d\nu_{x^*} = \int_T f\varphi d(x^* \circ \mathbf{m}) = \lim_n \int_T f g_n d(x^* \circ \mathbf{m})$$

$$= x^*\Phi_0(\varphi) = \int_T \varphi d(x^* \circ \boldsymbol{\eta}) \tag{6.3.4.10}$$

for $\varphi \in C_0(T)$ and $x^* \in H$. As $x^* \circ \boldsymbol{\eta} \in M(T)$ and as ν_{x^*} is σ-additive on $\mathcal{B}_c(T)$ and $\delta(\mathcal{C})$-regular for $x^* \in H$, by an argument similar to that in the proof of Claim 1 in the proof of Theorem 6.3.3 and by (6.3.4.10), we have

$$(x^* \circ \boldsymbol{\eta})|_{\delta(\mathcal{C})} = \nu_{x^*}|_{\delta(\mathcal{C})} \qquad (6.3.4.11)$$

for $x^* \in H$. As $v(x^* \circ \boldsymbol{\eta}, \mathcal{B}(T))(T) < \infty$, by Theorem 4.5 of [P13] or by Theorem 2.5 of [P4] we have $\nu_{x^*} = (x^* \circ \boldsymbol{\eta})|_{\mathcal{B}_c(T)}$ and

$$v(\nu_{x^*}, \mathcal{B}_c(T))(\cdot) = v(x^* \circ \boldsymbol{\eta}, \mathcal{B}(T))|_{\mathcal{B}_c(T)}(\cdot) \qquad (6.3.4.12)$$

for $x^* \in H$. Consequently, as f is ν_{x^*}-measurable, it is $(x^* \circ \boldsymbol{\eta})|_{\mathcal{B}_c(T)}$-measurable for $x^* \in H$ and consequently, f is $(x^* \circ \boldsymbol{\eta})$-measurable for $x^* \in H$.

Let U be an open Baire set in T. Then by §14 of [Din1] there exists an increasing sequence (φ_n) of functions in $C_c(T)$ such that $\varphi_n \nearrow \chi_U$. Then by LDCT, by (6.3.4.1), (6.3.4.10) and (6.3.4.12) we have

$$(x^* \circ \boldsymbol{\eta})(U) = \int_T \chi_U d(x^* \circ \boldsymbol{\eta}) = \lim_n \int_T \varphi_n d(x^* \circ \boldsymbol{\eta}) = \lim_n \int_T \varphi_n d\nu_{x^*}$$

$$= \lim_n \int_T \varphi_n f d(x^* \circ \mathbf{m}) = \int_T \chi_U f d(x^* \circ \mathbf{m}) = x^*(x_U) \qquad (6.3.4.13)$$

for $x^* \in H$. Then by Theorem 5.2.14 of Chapter 5, Φ_0 is weakly compact and hence by Theorem 2 of [P9], $\boldsymbol{\eta}$ is σ-additive and has range in X and by Theorem 6(xix) of [P9], $\boldsymbol{\eta}$ is $\mathcal{B}(T)$-regular. (Note that only here do we use the hypothesis that H has the Orlicz property to assert that Φ_0 is weakly compact.) Then, as f is $(x^* \circ \boldsymbol{\eta})$-measurable for $x^* \in H$, by Lemma 6.3.2 f is $\boldsymbol{\eta}$-measurable as well as Lusin $\boldsymbol{\eta}$-measurable. Given $\epsilon > 0$, by the $\mathcal{B}(T)$-regularity of $\boldsymbol{\eta}$ there exists $K \in \mathcal{C}$ such that $||\boldsymbol{\eta}||(T \backslash K) < \frac{\epsilon}{2}$. As f is Lusin $\boldsymbol{\eta}$-measurable, there exists $K_0 \in \mathcal{C}$ such that $K_0 \subset K$, $f|_{K_0}$ is continuous, and $||\boldsymbol{\eta}||(K \backslash K_0) < \frac{\epsilon}{2}$. Then

$$||\boldsymbol{\eta}||(T \backslash K_0) < \epsilon. \qquad (6.3.4.14)$$

Moreover, $f \chi_{K_0}$ is bounded as $f \chi_{K_0}$ is continuous on the compact K_0 and is $\mathcal{B}_c(T)$-measurable with compact support so that $f \chi_{K_0} \in \mathcal{L}_1(\mathbf{m})$ by Theorem 2.1.5 and Remark 2.2.3 of Chapter 2. As f is $(x^* \circ \mathbf{m})$-measurable for $x^* \in H$, $f - f \chi_{K_0}$ is also $(x^* \circ \mathbf{m})$-measurable for $x^* \in H$. Moreover, as H is a norm determining set, by Lemma 6.3.2 $f - f \chi_{K_0}$ is $\mathbf{m}$-measurable. Then by (6.3.4.12), (6.3.4.14), by the fact that $v(\nu_{x^*})(\cdot) = \int_{(\cdot)} |f| dv(x^* \circ \mathbf{m})$ by Proposition 1.1.19 of Chapter 1 and by Lemma 6.3.1(i) we have

$$\sup_{x^* \in H} \int_T |f - f \chi_{K_0}| dv(x^* \circ \mathbf{m})$$

$$= \sup_{x^* \in H} \int_{N(f) \backslash K_0} |f| dv(x^* \circ \mathbf{m}) = \sup_{x^* \in H} v(\nu_{x^*})(N(f) \backslash K_0)$$

$$= \sup_{x^* \in H} v(x^* \circ \boldsymbol{\eta})(N(f) \backslash K_0) = ||\boldsymbol{\eta}||(N(f) \backslash K_0) \leq ||\boldsymbol{\eta}||(T \backslash K_0) < \epsilon.$$

Consequently, by Lemma 6.3.1(ii), $f \in \mathcal{L}_1(\mathbf{m})$. Hence the conditions are also sufficient.

If $f \in \mathcal{L}_1(\mathbf{m})$, then by Theorem 2.1.5(vii) and Remark 2.3.3 of Chapter 2 and by Theorem 51.B of [H], $f\varphi \in \mathcal{L}_1(\mathbf{m})$ for each $\varphi \in C_0(T)$ and hence $\Psi : C_0(T) \to X$ given by $\Psi(\varphi) = \int_T \varphi f d\mathbf{m}$ is well defined for $\varphi \in C_0(T)$. Consequently, by (6.3.4.10) and by (viii) of the said theorem in Chapter 2 we have

$$x^* \Psi(\varphi) = \int_T f\varphi d(x^* \circ \mathbf{m}) = x^* \Phi_0(\varphi)$$

for $x^* \in H$ and for $\varphi \in C_0(T)$. As H is norm determining, it follows that $\Psi = \Phi_0$ and hence Ψ is weakly compact.

If $H = \{x^* \in X^* : |x^*| \leq 1\}$, then H is a norm determining set for X and has the Orlicz property by the Orlicz-Pettis theorem. Hence the last part holds by the first part.

This completes the proof of the theorem.   □

The following result which is deduced from the last part of Theorem 6.3.4 improves Theorem 4.2.2(i) of Chapter 4 for $\delta(\mathcal{C})$-regular σ-additive vector measures on $\delta(\mathcal{C})$.

Theorem 6.3.5. *Let X be a quasicomplete lcHs and let $\mathbf{m} : \delta(\mathcal{C}) \to X$ be σ-additive and $\delta(\mathcal{C})$-regular. Let $f : T \to \mathbb{K}$. Then f is $\mathbf{m}$-integrable in T if and only if $f \in \mathcal{L}_1(x^* \circ \mathbf{m})$ for each $x^* \in X^*$ and, for each open Baire set U in T, there exists a vector $x_U \in X$ such that*

$$x^*(x_U) = \int_U f d(x^* \circ \mathbf{m}) \tag{6.3.5.1}$$

for $x^* \in X^*$.

Proof. Clearly, the conditions are necessary. Conversely, let the conditions hold. For each $q \in \Gamma$, $\Pi_q : X \to X_q \subset \widetilde{X}_q$ is continuous. Hence $(y^* \circ \Pi_q) \in X^*$ for $y^* \in X_q^* = (\widetilde{X}_q)^*$ and hence by hypothesis $f \in \mathcal{L}_1(y^* \circ \mathbf{m}_q)$ for each $y^* \in X_q^*$ and by (6.3.5.1) we have

$$(y^* \circ \Pi_q)(x_U) = \int_U f d(y^* \circ \Pi_q \mathbf{m}) = \int_U f d(y^* \circ \mathbf{m}_q)$$

for each open Baire set U in T. Then by the last part of Theorem 6.3.4, $f \in \mathcal{L}_1(\mathbf{m}_q)$. As q is arbitrary in Γ, it follows that f is $\mathbf{m}$-integrable in T and moreover, by Definition 4.2.1 of Chapter 4,

$$\int_A f d\mathbf{m} = \underleftarrow{\lim} \int_A f d\mathbf{m}_q, \ A \in \mathcal{B}_c(T)$$

and

$$\int_T f d\mathbf{m} = \lim_{\longleftarrow} \int_{N(f)\setminus N_q} f d\mathbf{m}_q$$

(see Definition 4.2.1 of Chapter 4). □

In order to generalize Theorem 6.3.4 to complete lcHs-valued vector measures, we first generalize Lemmas 6.3.1 and 6.3.2 as follows.

Lemma 6.3.6. *Let X be an lcHs with topology τ and let H be a subset of X^* such that τ is identical with the topology of uniform convergence in equicontinuous subsets of H. Let $\mathcal{P}$ be a δ-ring of subsets of a set $\Omega(\emptyset)$ and let $\mathbf{m} : \mathcal{P} \to X$ be additive. Let $\mathcal{E}_H = \{E \subset H : E\ equicontinuous\}$. Then:*

(i) *$\|\mathbf{m}\|_{q_E}(A) = \sup_{x^* \in E} v(x^* \circ \mathbf{m})(A),\ A \in \sigma(\mathcal{P})$.*

(ii) *Suppose X is further quasicomplete, $\mathbf{m}$ is σ-additive and $f : \Omega \to \mathbb{K}$ is $\mathbf{m}$-measurable and $(x^* \circ \mathbf{m})$-integrable for each $x^* \in H$. Then $f \in \mathcal{L}_1(\mathbf{m})$ if, for each $E \in \mathcal{E}_H$ and $\epsilon > 0$, there exists $g_E^{(\epsilon)} \in \mathcal{L}_1(\mathbf{m}_{q_E})$ such that*

$$\sup_{x^* \in E} \int_T |f - g_E^{(\epsilon)}| dv(x^* \circ \mathbf{m}) < \epsilon.$$

Proof. (i) is due to Proposition 1.2.13(iii) of Chapter 1.

(ii) For $x^* \in E$, let Ψ_{x^*} be as in Proposition 1.2.13 of Chapter 1. Then by hypothesis and by the latter proposition, we have

$$\sup_{x^* \in E} \int_T |f - g_E^{(\epsilon)}| dv(\Psi_{x^*} \circ \mathbf{m}_{q_E}) = \sup_{x^* \in E} \int_T |f - g_E^{(\epsilon)}| dv(x^* \circ \mathbf{m}) < \epsilon. \quad (6.3.6.1)$$

Then by (6.3.6.1) and by Lemma 6.3.1(ii) applied to $\mathbf{m}_{q_E} : \mathcal{P} \to X_{q_E} \subset \widetilde{X_{q_E}}$, $f \in \mathcal{L}_1(\mathbf{m}_{q_E})$ since $\{\Psi_{x^*} : x^* \in E\}$ is a norm determining set for $\widetilde{X_{q_E}}$ by Proposition 1.2.13 of Chapter 1 and since f is $\mathbf{m}_{q_E}$-measurable by hypothesis. Since E is arbitrary in $\mathcal{E}_H$ and since $\{q_E : E \in \mathcal{E}_H\}$ generates the topology τ, by Definition 4.2.1 and by Remark 4.1.5 of Chapter 4 we conclude that $f \in \mathcal{L}_1(\mathbf{m})$. □

Lemma 6.3.7. *Let X be an lcHs and let H satisfy the hypothesis of Lemma 6.3.6. Let $\mathcal{E}_H$ be as in Lemma 6.3.6. Suppose $\mathbf{m} : \delta(\mathcal{C}) \to X$ (resp. $\mathbf{m} : \mathcal{B}(T) :\to X$) is σ-additive and let $V \in \mathcal{V}$. Then a set $A \in \mathcal{B}(V)$ (resp. $A \in \mathcal{B}(T)$) is $\mathbf{m}$-null if and only if A is $(x^* \circ \mathbf{m})$-null for each $x^* \in H$. If $\mathbf{m}$ is further $\delta(\mathcal{C})$-regular (resp. $\mathcal{B}(T)$-regular) and if $f : T \to \mathbb{K}$ is $(x^* \circ \mathbf{m})$-measurable for $x^* \in H$, then f is Lusin $\mathbf{m}$-measurable as well as $\mathbf{m}$-measurable.*

Proof. Let A be $(x^* \circ \mathbf{m})$-null for each $x^* \in H$ and let $E \in \mathcal{E}_H$. For $x^* \in E$, let Ψ_{x^*} be as in Proposition 1.2.13 of Chapter 1. Then by the said proposition $H_E = \{\Psi_{x^*} : x^* \in E\}$ is a norm determining set for X_{q_E} and hence for $\widetilde{X_{q_E}}$ and therefore, by hypothesis and by Lemma 6.3.2, A is $\mathbf{m}_{q_E}$-null. As E is arbitrary in $\mathcal{E}_H$ and as $\{q_E : E \in \mathcal{E}_H\}$ generates the topology τ, A is $\mathbf{m}$-null by Remark 1.2.5 of Chapter 1. Conversely, if A is $\mathbf{m}$-null, clearly A is $(x^* \circ \mathbf{m})$-null for each $x^* \in H$.

Let f be $(x^* \circ \mathbf{m})$-measurable for $x^* \in H$. Let $E \in \mathcal{E}_H$. Then, f is $(\Psi_{x^*} \circ \mathbf{m})$-measurable for $\Psi_{x^*} \in H_E$. Hence, by Lemma 6.3.2 applied to $\mathbf{m}_{q_E} : \mathcal{P} \to X_{q_E} \subset \widetilde{X_{q_E}}$, f is $\mathbf{m}_{q_E}$-measurable, where $\mathcal{P} = \delta(\mathcal{C})$ (resp. $\mathcal{B}(T)$). As f is $x^* \circ \mathbf{m}$-measurable (for $x^* \in H$), $N(f)$ is σ-bounded. Then by Definition 1.2.6 and by Remark 1.2.5 of Chapter 1, f is $\mathbf{m}$-measurable and consequently, f is also Lusin $\mathbf{m}$-measurable by Theorem 6.2.6 (resp. by Theorem 6.2.5). □

Now we shall generalize Theorem 6.3.4 to complete lcHs-valued σ-additive $\delta(\mathcal{C})$-regular measures.

Theorem 6.3.8. (Generalization of Theorem 6.3.4 to complete lcHs-valued measures). *Let X be a complete lcHs with topology τ and let H be a subset of X^* with the Orlicz property such that τ is identical with the topology of uniform convergence in equicontinuous subsets of H. Let $\mathbf{m} : \delta(\mathcal{C}) \to X$ be σ-additive and $\delta(\mathcal{C})$-regular and let $f : T \to \mathbb{K}$. Then $f \in \mathcal{L}_1(\mathbf{m})$ if and only if $f \in \mathcal{L}_1(x^* \circ \mathbf{m})$ for each $x^* \in H$ and, for each open Baire set U in T, there exists $x_U \in X$ such that*

$$x^*(x_U) = \int_U f d(x^* \circ \mathbf{m}) \tag{6.3.8.1}$$

for $x^* \in H$.

Proof. Let $\mathcal{E}_H = \{E \subset H : E \text{ equicontinuous}\}$. By hypothesis, the seminorms q_E, $E \in \mathcal{E}_H$, generate the topology τ. Let $E \in \mathcal{E}_H$ be fixed. By Proposition 1.2.13 of Chapter 1, $(\Psi_{x^*} \circ \Pi_{q_E})(x) = x^*(x)$ for $x \in X$ and for $x^* \in E$ and hence we identify $x^* \in E$ with Ψ_{x^*}. Let $H_E = U^0_{q_E} \cap H$ for $E \in \mathcal{E}_H$. Since $E \subset H_E \subset U^o_{q_E}$ and since $\{\Psi_{x^*} : x^* \in E\}$ is a norm determining set for X_{q_E} by Proposition 1.2.13 of Chapter 1, it follows that H_E is also a norm determining set for X_{q_E} and hence for $\widetilde{X_{q_E}}$. Let $\mathcal{F}$ be the vector space spanned by the characteristic functions of open Baire sets in T. Then by (6.3.8.1), for each $g \in \mathcal{F}$, there exists $x_g \in X$ such that

$$x^*(x_g) = \int_T f g d(x^* \circ \mathbf{m}) \tag{6.3.8.2}$$

for $x^* \in H$ and x_g is unique as the Hausdorff topology τ is generated by $\{q_E : E \in \mathcal{E}_H\}$. Then the mapping $\Phi : \mathcal{F} \to X$ given by $\Phi(g) = x_g$ for $g \in \mathcal{F}$ is well defined and linear. For $E \in H_\mathcal{E}$, arguing as in the proof of Theorem 6.3.4 with $\widetilde{X_{q_E}}, \mathbf{m}_{q_E}$ and H_E in place of X, H and $\mathbf{m}$, respectively, and using (6.3.8.1) in place of (6.3.4.1) we can show that $\Pi_{q_E} \circ \Phi : \mathcal{F} \to \widetilde{X_{q_E}}$ is continuous for $E \in \mathcal{E}_H$. Therefore, there exists a unique continuous linear extension $\widehat{\Phi^{(E)}}$ of $\Pi_{q_E} \circ \Phi$ to the whole of $\bar{\mathcal{F}}$ with values in $\widetilde{X_{q_E}}$ where $\bar{\mathcal{F}}$ is the closure of $\mathcal{F}$ in the Banach space of all bounded scalar functions on T with supremum norm. Then there exists a constant M_E such that

$$|\widehat{\Phi^{(E)}}(\varphi)|_{q_E} \le M_E \|\varphi\|_T \tag{6.3.8.3}$$

for $\varphi \in \bar{\mathcal{F}}$. Then by Claim 4 in the proof of Theorem 6.3.3, $C_0(T) \subset \bar{\mathcal{F}}$ and hence $\widehat{\Phi_0^{(E)}} = \widehat{\Phi^{(E)}}|_{C_0(T)}$ is continuous and linear and has range in $\widehat{X_{q_E}}$. Then (6.3.8.3) also holds for $\widehat{\Phi_0^{(E)}}$ with $\varphi \in C_0(T)$. Moreover, (6.3.8.3) holds for all $E \in \mathcal{E}_H$.

By hypothesis, X is a complete lcHs and $\{q_E : E \in \mathcal{E}_H\}$ generates the topology τ and hence by Theorem 5.4, §5, Chapter II of [Scha], $X = \varprojlim \widehat{X_{q_E}}$. Let us define $\hat{\Phi} : \bar{\mathcal{F}} \to X$ by $\hat{\Phi}(\varphi) = \varprojlim \widehat{\Phi^{(E)}}(\varphi)$ for $\varphi \in \mathcal{F}$. Let $\Phi_0 : C_0(T) \to X$ be given by $\Phi_0 = \hat{\Phi}|_{C_0(T)}$. Then by (6.3.8.3) we have $|\hat{\Phi}(\varphi)|_{q_E} = |(\Pi_{q_E} \circ \hat{\Phi})(\varphi)|_{q_E} = |\widehat{\Phi^{(E)}}(\varphi)|_{q_E} \le M_E \|\varphi\|_T$ for each $E \in \mathcal{E}_H$ and for each $\varphi \in \bar{\mathcal{F}}$. Hence $\hat{\Phi}$ and Φ_0 are X-valued continuous linear mappings.

Let $\boldsymbol{\eta}$ be the representing measure of Φ_0 in the sense of Definition 4 of [P9]. Then by Theorem 1 of [P9],

$$x^* \Phi_0(\varphi) = \int_T \varphi d(x^* \circ \boldsymbol{\eta}) \tag{6.3.8.4}$$

for $x^* \in X^*$ and for $\varphi \in C_0(T)$.

Claim 1. $\int_T f\varphi d(x^* \circ \mathbf{m}) = x^* \Phi_0(\varphi) = \int_T \varphi d(x^* \circ \boldsymbol{\eta})$ for $x^* \in H$ and for $\varphi \in C_0(T)$.

In fact, let $\varphi \in C_0(T)$. As $f \in \mathcal{L}_1(x^* \circ \mathbf{m})$ for $x^* \in H$ and as φ is bounded and $\mathcal{B}_c(T)$-measurable by Theorem 51.B of [H], $f\varphi \in \mathcal{L}_1(x^* \circ \mathbf{m})$ for $x^* \in H$. By Claim 4 in the proof of Theorem 6.3.3, there exists $(g_n)_1^\infty \subset \mathcal{F}$ such that $\|\varphi - g_n\|_T \to 0$. Then, for $x^* \in H$,

$$\left| \int_T f\varphi d(x^* \circ \mathbf{m}) - \int_T f g_n d(x^* \circ \mathbf{m}) \right| \le \|\varphi - g_n\|_T \int_T |f| dv(x^* \circ \mathbf{m}) \to 0 \tag{6.3.8.5}$$

as $n \to \infty$, since $f \in \mathcal{L}_1(x^* \circ \mathbf{m})$. Observing that $\hat{\Phi}(g) = \Phi(g) = x_g$ for $g \in \mathcal{F}$, by (6.3.8.2), (6.3.8.4) and (6.3.8.5) and by the fact that $\|\varphi - g_n\|_T \to 0$ as $n \to \infty$ we have $\int_T f\varphi d(x^* \circ \mathbf{m}) = \lim_n \int_T f g_n d(x^* \circ \mathbf{m}) = \lim_n x^*(x_{g_n}) = \lim_n x^* \Phi(g_n) = \lim_n x^* \hat{\Phi}(g_n) = x^* \hat{\Phi}(\varphi) = x^* \Phi_0(\varphi) = \int_T \varphi d(x^* \circ \boldsymbol{\eta})$ for $x^* \in H$. Hence the claim holds.

Let U be an open Baire set in T and choose $(\varphi_n)_1^\infty \subset C_0(T)$ such that $\varphi_n \nearrow \chi_U$ (see the proof (6.3.4.13)). Then by Claim 1, by LDCT and by (6.3.8.1) we have

$$(x^* \circ \boldsymbol{\eta})(U) = \lim_n \int_T \varphi_n d(x^* \circ \boldsymbol{\eta}) = \lim_n x^* \Phi_0(\varphi_n) = \lim_n \int_T f\varphi_n d(x^* \circ \mathbf{m})$$

$$= \int_T \chi_U f d(x^* \circ \mathbf{m}) = x^*(x_U) \tag{6.3.8.6}$$

for $x^* \in H$. Since H has the Orlicz property, by (6.3.8.6) and by Theorem 5.2.16 of Chapter 5, Φ_0 is weakly compact.

Then by Theorems 2 and 6 of [P9], η is X-valued, σ-additive and $\mathcal{B}(T)$-regular. Let $\nu_{x^*}(\cdot) = \int_{(\cdot)} f d(x^* \circ \mathbf{m})$ for $x^* \in H$. Then for $\varphi \in C_0(T)$ and for $x^* \in H$, by Claim 1 we have

$$\int_T \varphi d\nu_{x^*} = \int_T f\varphi d(x^* \circ \mathbf{m}) = x^* \Phi_0(\varphi) = \int_T \varphi d(x^* \circ \boldsymbol{\eta})$$

and hence an argument similar to that in the paragraph following (6.3.4.10) in the proof of Theorem 6.3.4 shows that f is $(x^* \circ \boldsymbol{\eta})$-measurable for $x^* \in H$. Then by Lemma 6.3.7, f is $\boldsymbol{\eta}$-measurable as well as Lusin $\boldsymbol{\eta}$-measurable. Thus, given $E \in \mathcal{E}_H$ and $\epsilon > 0$, arguing as in the proof of Theorem 6.3.4, there exists a compact K_0 in T such that $\|\boldsymbol{\eta}\|_{q_E}(T \backslash K_0) < \epsilon$ and $f|_{K_0}$ is continuous. Then $f\chi_{K_0}$ is bounded and $\mathcal{B}_c(T)$-measurable with compact support. Consequently, by Theorems 4.1.9(i)(b) and 4.2.3 of Chapter 4, $f\chi_{K_0} \in \mathcal{L}_1(\mathbf{m})$ and hence $f\chi_{K_0} \in \mathcal{L}_1(\mathbf{m}_{q_E})$. Moreover, by hypothesis f is $(x^* \circ \mathbf{m})$-measurable for each $x^* \in H$ and as H satisfies the hypothesis of Lemma 6.3.7, it follows that f is $\mathbf{m}$-measurable. Then an argument similar to that in the last part of the proof of Theorem 6.3.4, invoking Lemma 6.3.6(ii) in place of Lemma 6.3.1(ii), shows that $f \in \mathcal{L}_1(\mathbf{m}_{q_E})$. Since E is arbitrary in $\mathcal{E}_H$ and since $(q_E)_{E \in \mathcal{E}_H}$ generates the topology τ, by Definition 4.2.1 and by Theorem 4.2.3 of Chapter 4, $f \in \mathcal{L}_1(\mathbf{m})$, i.e., f is $\mathbf{m}$-integrable in T with values in X.

This completes the proof of the theorem. □

Now we give an analogue of Theorem 6.3.3 when $\mu_i : \mathcal{B}(T) \to \mathbb{K}$, $i \in I$, are σ-additive and Borel regular.

Theorem 6.3.9. *Let $\mu_i : \mathcal{B}(T) \to \mathbb{K}$ be σ-additive and $\mathcal{B}(T)$-regular for $i \in I$. Suppose $\sum_{i \in I} |\int_T \varphi d\mu_i|^p < \infty$ for each $\varphi \in C_0(T)$ and for $1 \leq p < \infty$. Let $u : C_0(T) \to \ell_p(I)$ be defined by $u(\varphi) = (\int_T \varphi d\mu_i)_{i \in I}$. Then u is a weakly compact operator on $C_0(T)$. Let $\mathbf{m}_u$ be the representing measure of u in the sense of Definition 4 of [P9] and let $f : T \to \mathbb{K}$ belong to $\mathcal{L}_1(\mu_i)$ for $i \in I$. Then f is $\mathbf{m}_u$-integrable in T if and only if*

$$\sum_{i \in I} \left| \int_U f d\mu_i \right|^p < \infty \tag{6.3.9.1}$$

for each open Baire set U in T. In that case, $\int_T f d\mathbf{m}_u = (\int_T f d\mu_i)_{i \in I}$.

Let $p = 1$ and let $f \in \mathcal{L}_1(\mathbf{m}_u)$. If $\theta(\varphi) = \sum_{i \in I} \int_T \varphi d\mu_i$ for $\varphi \in C_0(T)$, then $\theta \in \mathcal{K}(T)_b^*$ (see pp. 65 and 69 of [P4]), f is μ_θ-integrable and

$$\int_A f d\mu_\theta = \sum_{i \in I} \int_A f d\mu_i$$

for $A \in \mathcal{B}(T)$, where μ_θ is the bounded complex Radon measure induced by θ in the sense of Definition 4.3 of [P3].

Proof. By an argument similar to that in the proof of Theorem 6.3.3 we can show that the linear mapping u has a closed graph and hence by Theorem 2.15 of [Ru2], u is continuous. Since $c_0 \not\subset \ell_p(I)$ for $1 \leq p < \infty$, by Theorem 13 of [P9] or by Corollary 2 of [P12], u is weakly compact. Then by Theorems 2 and 6 of [P9], the representing measure $\mathbf{m}_u$ of u is $\ell_p(I)$-valued, σ-additive and Borel regular on $\mathcal{B}(T)$.

Let $H_I^{(p)}$ be as in the proof of Theorem 6.3.3 for $1 \leq p < \infty$. If $x^* = (\alpha_i) \in H_I^{(p)}$, then there exists $J_{x^*} \subset I$, J_{x^*} finite, such that $\alpha_i = 0$ for $i \in I \backslash J_{x^*}$. Then by Theorem 1 of [P9] we have $x^* u(\varphi) = \int_T \varphi d(x^* \circ \mathbf{m}_u) = \int_T \varphi d(\sum_{i \in J_{x^*}} \alpha_i \mu_i)$ for $\varphi \in C_0(T)$. Since $x^* \circ \mathbf{m}_u$ and μ_i, $i \in I$, are σ-additive and Borel regular, by the uniqueness part of the Riesz representation theorem we conclude that

$$x^* \circ \mathbf{m}_u = \sum_{i \in J_{x^*}} \alpha_i \mu_i = \sum_{i \in I} \alpha_i \mu_i. \tag{6.3.9.2}$$

For $i \in I$, let θ_i and η_i be defined as in the proof of Theorem 6.3.3. Note that η_i is σ-additive on $\mathcal{B}(T)$ and η_i is $v(\mu_i)$-continuous for $i \in I$ so that η_i is $\mathcal{B}(T)$-regular for $i \in I$. Since

$$\int_T \varphi d\eta_i = \int_T \varphi f d\mu_i = \theta_i(\varphi) = \int_T \varphi d\mu_{\theta_i}$$

for $\varphi \in C_0(T)$ and for $i \in I$, by the uniqueness part of the Riesz representation theorem

$$\eta_i = \mu_{\theta_i} \quad \text{for } i \in I. \tag{6.3.9.3}$$

Defining Ψ_{x^*} as in the proof of Theorem 6.3.3, by (6.3.9.3) we have

$$\mu_{\Psi_{x^*}} = \sum_{i \in J_{x^*}} \alpha_i \mu_{\theta_i} = \sum_{i \in J_{x^*}} \alpha_i \eta_i \text{ on } \mathcal{B}(T). \tag{6.3.9.4}$$

Using (6.3.9.1), (6.3.9.4) and Hölder's inequality and arguing as in the proof of Claim 3 in the proof of Theorem 6.3.3 one can show that

$$\sup_{x^* \in H_I^{(p)}} |\mu_{\Psi_{x^*}}(U)| \leq \left(\sum_{i \in I} \left| \int_U f d\mu_i \right|^p \right)^{\frac{1}{p}} < \infty \tag{6.3.9.5}$$

for $1 \leq p < \infty$ and for an open Baire set U in T. As $\{\mu_{\Psi_{x^*}} : x^* \in H_I^{(p)}\} \subset M(T)$ for $1 \leq p < \infty$, by (6.3.9.5) and by Corollary 5.2.5 of Chapter 5 we have

$$\sup_{x^* \in H_I^{(p)}} v(\mu_{\Psi_{x^*}}, \mathcal{B}(T))(T) = M_p \text{ (say) } < \infty \tag{6.3.9.6}$$

for $1 \leq p < \infty$.

By an argument quite similar to the proof of Claim 5 in the proof of Theorem 6.3.3 and by the use of (5.3.9.6) in place of Claim 3 in the proof of the said theorem, we have

$$\sum_{i\in I}\left|\int_T f\varphi d\mu_i\right|^p < \infty \qquad (6.3.9.7)$$

for $\varphi \in C_0(T)$ and for $1 \le p < \infty$.

Then the mapping $\Phi : C_0(T) \to \ell_p(I)$ given by

$$\Phi(\varphi) = \left(\int_T f\varphi d\mu_i\right)_{i\in I} = \left(\int_T \varphi d\eta_i\right)_{i\in I}$$

is well defined and linear and has a closed graph. Then by the closed graph theorem, Φ is continuous. Since $c_0 \not\subset \ell_p(I)$ for $1 \le p < \infty$, by Theorem 13 of [P9], Φ is weakly compact. Then by Theorems 2 and 6 of [P9], the representing measure $\mathbf{m}_\Phi$ of Φ is $\ell_p(I)$-valued, σ-additive and Borel regular on $\mathcal{B}(T)$.

Then arguing as in the proof of Claim 6 in the proof of Theorem 6.3.3 and using the uniqueness part of the Riesz representation theorem (Borel version), we conclude that

$$(x^* \circ \mathbf{m}_\Phi) = \sum_{i\in J_{x^*}} \alpha_i \eta_i = \sum_{i\in I} \alpha_i \eta_i \qquad (6.3.9.8)$$

for each $x^* \in H_I^{(p)}$, $1 \le p < \infty$. Invoking the Borel case of Lemma 6.3.2 and using (6.3.9.8) we conclude that f is $\mathbf{m}$-measurable as well as Lusin $\mathbf{m}$-measurable. The rest of the argument in the proof of Theorem 6.3.3 holds here verbatim with $\mathcal{B}(T)$ in place of $\mathcal{B}_c(T)$ and $\mathcal{B}(T)$-regular in place of $\delta(\mathcal{C})$-regular.

This completes the proof of the theorem. $\qquad\qquad \square$

Using the above theorem for $I = \mathbb{N}$ and for $p = 1$, we obtain below the analogue of Theorem 6.3.4 for a Banach space-valued σ-additive Borel regular measure.

Theorem 6.3.10. *Let X be a Banach space and let $\mathbf{m} : \mathcal{B}(T) \to X$ be σ-additive and Borel regular. Let H be a norm determining set for X with the orlicz property and let $f : T \to \mathbb{K}$. Then $f \in \mathcal{L}_1(\mathbf{m})$ if and only if $f \in \mathcal{L}_1(x^* \circ \mathbf{m})$ for each $x^* \in H$ and, for each open Baire set U in T there exists a vector $x_U \in X$ such that*

$$x^*(x_U) = \int_U f d(x^* \circ \mathbf{m})$$

for $x^* \in H$. Moreover, other conclusions of Theorem 6.3.4 hold. (By Theorem 6.1.12, $\mathbf{m}|_{\delta(\mathcal{C})}$ satisfies the hypothesis of Theorem 6.3.4 and hence the conclusions of Theorem 6.3.4 hold for $\mathbf{m}|_{\delta(\mathcal{C})}$ but it requires a proof to show that they hold for $\mathbf{m}$ itself.)

Proof. By Theorem 2.2.2 of Chapter 2, the conditions are necessary. Conversely, let $<H>$ and F be as in the proof of Theorem 6.3.4. If $x^* \in F$ and $x^* = \sum_1^\infty x_n^*$ with $(x_n^*) \subset H$ and with $\sum_1^\infty |x_n^*| < \infty$, then arguing as in the proof of the said theorem, we have $x^* \circ \mathbf{m} = \sum_1^\infty x_n^* \circ \mathbf{m}$ and moreover, $x^* \circ \mathbf{m}$ is σ-additive and $\mathcal{B}(T)$-regular. Further, (6.3.4.3) holds for $\varphi \in C_0(T)$ and (6.3.4.4) also holds. Consequently, the mapping $u : C_0(T) \to \ell_1(\mathbb{N})$ given by

$$u(\varphi) = \left(\int_T \varphi d(x_n^* \circ \mathbf{m}) \right)_{n=1}^\infty$$

is well defined and linear. Then the complex measures $(x_n^* \circ \mathbf{m})_{n=1}^\infty$ and the function f satisfy the sufficiency part of the hypotheses of Theorem 6.3.9 for $p = 1$ and for $I = \mathbb{N}$ and hence u is a weakly compact operator on $C_0(T)$ and $f \in \mathcal{L}_1(\mathbf{m}_u)$, where $\mathbf{m}_u$ is the representing measure of u. Then by the last part of the said theorem, $\theta : C_0(T) \to \mathbb{K}$, given by $\theta(\varphi) = \sum_{n=1}^\infty \int_T \varphi d(x_n^* \circ \mathbf{m})$, belongs to $\mathcal{K}(T)_b^*$, f is μ_θ-integrable and

$$\int_U f d\mu_\theta = \sum_{n=1}^\infty \int_U f d(x_n^* \circ \mathbf{m})$$

for each open Baire set U in T and for the set $U = N(f)$ since $f \in \mathcal{L}_1(\mathbf{m}_u)$ so that $N(f) \in \widetilde{\mathcal{B}(T)}$ with respect to $||\mathbf{m}_u||$. Let $x^* \in F$, where F is the norm closure of $\langle H \rangle$, the vector space spanned by H. Observing that $x^* \circ \mathbf{m}$ is σ-additive and $\mathcal{B}(T)$-regular by hypothesis and μ_θ is σ-additive and $\mathcal{B}(T)$-regular by Theorem 5.3 of [P4], and proving that $\int_T \varphi d\mu_\theta = \int_T \varphi d(x^* \circ \mathbf{m})$ for $\varphi \in C_0(T)$ as in the proof of Claim 1 in the proof of Theorem 5.3.4, by the uniqueness part of the Riesz representation theorem we conclude that $\mu_\theta = x^* \circ \mathbf{m}$. Moreover, by (6.3.4.2) and (6.3.4.5) we have $x^*(x_U) = \int_U f d(x^* \circ \mathbf{m})$, where by hypothesis $y^*(x_U) = \int_T f d(y^* \circ \mathbf{m})$ for each open Baire set U in T and for each $y^* \in H$.

Arguing as in the proof of Theorem 6.3.4, one can define the continuous linear mapping $\Phi_0 : C_0(T) \to X$ with the representing measure η. Let $\nu_{x^*}(\cdot) = \int_{(\cdot)} f d(x^* \circ \mathbf{m})$ for $x^* \in H$. Then ν_{x^*} is $v(x^* \circ \mathbf{m})$-continuous. Then f is ν_{x^*}-measurable since f is $(x^* \circ \mathbf{m})$-measurable and ν_{x^*} is $\mathcal{B}(T)$-regular for $x^* \in H$. Arguing as in the proof of Theorem 6.3.4, by the uniqueness part of the Riesz representation theorem we have $\nu_{x^*} = x^* \circ \eta$, $x^* \in H$ and hence f is $(x^* \circ \eta)$-measurable for $x^* \in H$. The remaining arguments in the proof of Theorem 6.3.4 hold here excepting that the Borel case of Lemma 6.3.2 has to be invoked here.

This completes the proof of the theorem.   □

The following theorem, which improves Theorem 4.2.2(i) of Chapter 4 for $\mathcal{B}(T)$-regular σ-additive vector measures, is immediate from the last part of Theorem 6.3.10 by an argument similar to that in the proof of Theorem 6.3.5.

Theorem 6.3.11. *Let X be a quasicomplete lcHs and let $\mathbf{m} : \mathcal{B}(T) \to X$ be σ-additive and $\mathcal{B}(T)$-regular. Then a function $f : T \to \mathbb{K}$ is $\mathbf{m}$-integrable in T if and* *only if* $f \in \mathcal{L}_1(x^* \circ \mathbf{m})$ *for each* $x^* \in X^*$ *and, for each open Baire set* U *in* T, *there exists a vector* $x_U \in X$ *such that*

$$x^*(x_U) = \int_U f d(x^* \circ \mathbf{m})$$

for $x^* \in X^*$.

The following result is an analogue of Theorem 6.3.8 for the Borel-regular σ-additive X-valued vector measure $\mathbf{m}$ where X is a complete lcHs.

Theorem 6.3.12. *Let* X, τ *and* H *be as in Theorem 6.3.8 and let* $\mathbf{m} : \mathcal{B}(T) \to X$ *be* σ-*additive and Borel regular. Then a function* $f : T \to \mathbb{K}$ *belongs to* $\mathcal{L}_1(\mathbf{m})$ *if and only if* $f \in \mathcal{L}_1(x^* \circ \mathbf{m})$ *for each* $x^* \in H$ *and, for each open Baire set* U *in* T, *there exists a vector* $x_U \in X$ *such that*

$$x^*(x_U) = \int_U f d(x^* \circ \mathbf{m})$$

for $x^* \in H$.

Proof. The proof of Theorem 6.3.8 holds here verbatim excepting that we have to invoke the uniqueness part of the Riesz representation theorem (Borel version) to show that $\nu_{x^*} = x^* \circ \boldsymbol{\eta}$ for $x^* \in H$ so that f is $(x^* \circ \boldsymbol{\eta})$-measurable for $x^* \in H$ and then invoke the Borel case of Lemma 6.3.7. The details are left to the reader. $\square$

6.4   Additional convergence theorems

First we obtain a generalization of the Bourbaki version of the Egoroff theorem for an lcHs-valued σ-additive $\delta(\mathcal{C})$-regular measure. Then Proposition 4.3 of [T] is suitably generalized in Theorem 6.4.4. Corollary T_2 on p. 176 of [T] is improved in Theorem 6.4.6 which is generalized to vector measures in Theorems 6.4.8 and 6.4.12. The latter generalizations are based on Theorems 6.3.4 and 6.3.5.

Theorem 6.4.1. (Generalization of the Bourbaki version of the Egoroff theorem). *Let* X *be an lcHs,* $\mathbf{n} : \delta(\mathcal{C}) \to X$ *be* σ-*additive and* $\delta(\mathcal{C})$-*regular and* $f_0 : T \to \mathbb{K}$. *For each* $q \in \Gamma$, *let* $f_n^{(q)} : T \to \mathbb{K}$ *be* $\mathbf{n}_q$-*measurable for* $n \in \mathbb{N}$ *and let* $f_n^{(q)} \to f_0$ $\mathbf{n}_q$-*a.e. in* T. *Then:*

(i) f_0 *is* $\mathbf{n}$-*measurable.*

(ii) *Given* $K \in \mathcal{C}$, $q \in \Gamma$ *and* $\epsilon > 0$, *there exists* $K_0^{(q)} \in \mathcal{C}$ *such that* $K_0^{(q)} \subset K$, $\|\mathbf{n}\|_q(K \backslash K_0^{(q)}) < \epsilon$, $f_n^{(q)}|_{K_0^{(q)}}$, $n \in \mathbb{N} \cup \{0\}$, *are continuous and* $f_n^{(q)} \to f_0$ *uniformly in* $K_0^{(q)}$.

Proof. (i) By hypothesis, f_0 is $\mathbf{n}_q$-measurable for each $q \in \Gamma$ and hence f_0 is $\mathbf{n}$-measurable.

(ii) Without loss of generality we shall assume X to be a normed space. By Lemma 5.2.2 of Chapter 5, $\delta(\mathcal{C}) \cap K = \mathcal{B}(K)$ is a σ-algebra and hence by Theorem 3.1.18(viii) of Chapter 3, given $\epsilon > 0$, there exists $A_\epsilon \in \mathcal{B}(K)$ such that $||\mathbf{n}||(A_\epsilon) < \frac{\epsilon}{3}$ and $f_n \to f$ uniformly in $K \backslash A_\epsilon$. As $K \backslash A_\epsilon \in \mathcal{B}(K) \subset \delta(\mathcal{C})$, by the $\delta(\mathcal{C})$-regularity of $\mathbf{n}$ there exists a compact set $K_1 \subset K \backslash A_\epsilon$ such that $||\mathbf{n}||(K \backslash A_\epsilon \backslash K_1) < \frac{\epsilon}{3}$. Then particularly, $f_n \to f_0$ uniformly in K_1. Moreover, by hypothesis, by Theorem 6.2.6 and by Definition 6.2.3, for each n there exists $C_n \in \mathcal{C}$ with $C_n \subset K_1$ such that $f_n|_{C_n}$ is continuous and $||\mathbf{n}||(K_1 \backslash C_n) < \frac{\epsilon}{3} \cdot \frac{1}{2^n}$. Then $K_0 = \bigcap_1^\infty C_n \in \mathcal{C}$, $K_0 \subset K_1 \subset K$ and $f_n|_{K_0}$ is continuous for each $n \in \mathbb{N}$ so that their uniform limit f_0 is also continuous in K_0. Moreover, $||\mathbf{n}||(K_1 \backslash K_0) \leq \sum_{n=1}^\infty ||\mathbf{n}||(K_1 \backslash C_n) < \frac{\epsilon}{3}$ so that $||\mathbf{n}||(K \backslash K_0) < \epsilon$. $\qquad\square$

The following definition is motivated by that on p. 122 of [T].

Definition 6.4.2. Let X be an lcHs, $\mathbf{n} : \delta(\mathcal{C}) \to X$ be σ-additive, $q \in \Gamma$ and $f : T \to \mathbb{K}$ be $\mathbf{n}$-measurable. A sequence $(f_n^{(q)})$ of $\mathbf{n}_q$-measurable scalar functions is said to converge to f in measure $\mathbf{n}_q$ over compacts, if, given $K \in \mathcal{C}$ and $\eta > 0$, the sequence $||\mathbf{n}||_q(\{t \in K : |f_n^{(q)}(t) - f(t)| \geq \eta\}) \to 0$ as $n \to \infty$.

Theorem 6.4.3. *Let X be an lcHs, $\mathbf{n} : \delta(\mathcal{C}) \to X$ be σ-additive, $q \in \Gamma$ and $f_0, g_0, g, f_n^{(q)}, g_n^{(q)} : T \to \mathbb{K}$, $n \in \mathbb{N}$, be $\mathbf{n}_q$-measurable for $n \in \mathbb{N}$. Let $f : T \to \mathbb{K}$. If $f_n^{(q)} \to f_0$ and $g_n^{(q)} \to g_0$ in measure $\mathbf{n}_q$ over compacts, then the following hold:*

(i) *$\lambda f_n^{(q)} \to \lambda f_0$ in measure $\mathbf{n}_q$ over compacts for $\lambda \in \mathbb{K}$.*

(ii) *$f_n^{(q)} + g_n^{(q)} \to f_0 + g_0$ in measure $\mathbf{n}_q$ over compacts.*

(iii) *If $f_n^{(q)} \to g$ in measure $\mathbf{n}_q$ over compacts, then $f_0 = g$ $\mathbf{n}_q$-a.e. in T.*

(iv) *If $f_n^{(q)} \to f$ $\mathbf{n}_q$-a.e. in T and if $\mathbf{n}$ is $\delta(\mathcal{C})$-regular, then $f_n^{(q)} \to f$ in measure $\mathbf{n}_q$ over compacts.*

Proof. (i) is obvious. Let $K \in \mathcal{C}$ and $\eta > 0$. For any three $\mathbf{n}_q$-measurable scalar functions φ, ψ, and h on T, it is easy to verify that $||\mathbf{n}||_q(\{t \in K : |\varphi(t) - \psi(t)| \geq \eta\}) \leq ||\mathbf{n}||_q(\{t \in K : |\varphi(t) - h(t)| \geq \frac{\eta}{2}\}) + ||\mathbf{n}||_q(\{t \in K : |h(t) - \psi(t)| \geq \frac{\eta}{2}\})$. Using this inequality, one can prove (iii) and that $||\mathbf{n}||_q(N(f_0 - g) \cap K) = 0$ in (iii) for each $K \in \mathcal{C}$, since $N(f_0 - g) \cap K = \bigcup_{n=1}^\infty \{t \in K : |f_0(t) - g(t)| \geq \frac{1}{n}\}$. If $A = N(f_0 - g)$, then $A \in \widetilde{\mathcal{B}_c(T)_q}$ so that A is of the form $A = B \cup N$, $B \in \mathcal{B}_c(T)$, $N \subset M \in \mathcal{B}_c(T)$ and $||\mathbf{n}||_q(M) = 0$. Then there exists an increasing sequence $(B_n)_1^\infty \subset \delta(\mathcal{C})$ such that $B_n \nearrow B \cup M$. Then $K_n = \bar{B}_n \in \mathcal{C}$ where $\bar{B}_n$ is the closure of B_n in T and $B \cup M \subset \bigcup_1^\infty K_n$. Hence $||\mathbf{n}||_q(B \cup M) \leq \sum_1^\infty ||\mathbf{n}||_q(N(f_0 - g) \cap K_n) = 0$. Hence $||\mathbf{n}||_q(A) = 0$ so that $f_0 = g$ $\mathbf{n}_q$-a.e. in T.

(iv) By hypothesis, f is $\mathbf{n}_q$-measurable. Let $K \in \mathcal{C}$ and $\eta > 0$. By hypothesis and by Theorem 6.4.1 there exists $K_0^{(q)} \in \mathcal{C}$ with $K_0^{(q)} \subset K$ such that

$||\mathbf{n}||_q(K\backslash K_0^{(q)}) < \eta$, $f_n^{(q)}|_{K_0^{(q)}}$ and $f|_{K_0^{(q)}}$, $n \in \mathbb{N}$, are continuous and $f_n^{(q)} \to f$ uniformly in $K_0^{(q)}$. Hence, given $\epsilon > 0$, there exists n_0 such that $\sup_{t \in K_0^{(q)}} |f_n^{(q)}(t) - f(t)| < \epsilon$ for $n \geq n_0$ so that $||\mathbf{n}||_q(\{t \in K : |f_n^{(q)}(t) - f(t)| \geq \epsilon\}) \leq ||\mathbf{n}||_q(K\backslash K_0^{(q)}) < \eta$ for $n \geq n_0$. Hence (iv) holds.   □

Theorem 6.4.4. (A variant of Theorem 4.5.12(ii) of Chapter 4). *Let X be a quasicomplete (resp. sequentially complete) lcHs, let $1 \leq p < \infty$ and let $\mathbf{n} : \delta(\mathcal{C}) \to X$ be σ-additive. Let $(f_n^{(q)})_1^\infty \subset \mathcal{L}_p(\mathbf{n}_q)$ (resp. $\subset \mathcal{L}_p(\sigma(\delta(\mathcal{C})), \mathbf{n}_q)$) for each $q \in \Gamma$ and let $f : T \to \mathbb{K}$ (resp. and be $\mathcal{B}_c(T)$-measurable). Suppose $f_n^{(q)} \to f$ $\mathbf{n}_q$-a.e. in T for each $q \in \Gamma$. Then $f \in \mathcal{L}_p(\mathbf{n})$ (resp. $f \in \mathcal{L}_p(\sigma(\delta(\mathcal{C})), \mathbf{n})$) and $\lim_n (\mathbf{n}_q)_p^\bullet (f_n^{(q)} - f, T) = 0$ for each $q \in \Gamma$ if and only if the following conditions hold:*

(i) *$(\mathbf{n}_q)_p^\bullet(f_n^{(q)}, \cdot)$, $n \in \mathbb{N}$, are uniformly $\mathbf{n}_q$-continuous on $\mathcal{B}_c(T)$ for each $q \in \Gamma$. (See Definition 3.4.3 of Chapter 3.)*

(ii) *For each $\epsilon > 0$ and $q \in \Gamma$, there exists $K^{(q)} \in \mathcal{C}$ such that*

$$\sup_n (\mathbf{n}_q)_p^\bullet (f_n^{(q)}, T\backslash K^{(q)}) < \epsilon.$$

Proof. If $\mathcal{P} = \delta(\mathcal{C})$, note that $\sigma(\mathcal{P}) = \mathcal{B}_c(T)$. Then (i) is the same as condition (a) of Theorem 4.5.12(ii) of Chapter 4. (ii) is equivalent to condition (b) of Theorem 4.5.12(ii) of Chapter 4, since for $A_\epsilon^{(q)} \in \mathcal{P}$, $\overline{A_\epsilon^{(q)}} = K^{(q)} \in \mathcal{C}$ and $T\backslash A_\epsilon^{(q)} \supset T\backslash K^{(q)}$.   □

Remark 6.4.5. In the light of Lemma 6.1.5 and Theorem 6.1.12, Theorems 6.4.1 and 6.4.3(iv) hold for $\mathbf{n} = \mathbf{m}|_{\delta(\mathcal{C})}$ (resp. $\mathbf{n} = \boldsymbol{\omega}|_{\delta(\mathcal{C})}$) where $\mathbf{m} : \mathcal{B}(T) \to X$ (resp. $\boldsymbol{\omega} : \mathcal{B}_c(T) \to X$) is σ-additive and Borel regular (resp. and σ-Borel regular).

The following theorem is an improved version of Corollary T_2 on p. 176 of [T].

Theorem 6.4.6. *Let $\theta \in \mathcal{K}(T)^*$, let $\lambda = \mu_{|\theta|}^*|_{\mathcal{B}(T)}$ where $|\theta|$ is the absolute value of θ as on p. 55 of [B] and $\mu_{|\theta|}$ is the complex Radon measure induced by $|\theta|$ in the sense of Definition 4.3 of [P3]. Similarly, let μ_θ be the complex Radon measure induced by θ. Then $\lambda : \mathcal{B}(T) \to [0, \infty]$ is σ-additive and Radon-regular in the sense of Definition 3.3 of [P3]. Suppose $(f_n)_1^\infty \subset \mathcal{L}_1(\lambda)$ such that the sequence $(\int_U f_n d\mu_\theta)_1^\infty$ is convergent for each open Baire set U in T. Then there exists $f \in \mathcal{L}_1(\lambda)$ such that $\lim_n \int_A f_n d\mu_\theta = \int_A f d\mu_\theta$ for each Borel set A in T. Consequently,*

$$\lim_n \int_T g f_n d\mu_\theta = \int_T g f d\mu_\theta \qquad (6.4.6.1)$$

for each λ-measurable bounded scalar function g on T and consequently, f is unique in $L_1(\lambda)$. If $f_n \to h$ λ-a.e. in T, then $f = h$ λ-a.e. in T and $\lim_n \int_A f_n d\mu_\theta = \int_A h d\mu_\theta$ for each $A \in \mathcal{B}(T)$, the convergence being uniform with respect to $A \in \mathcal{B}(T)$.

Proof. λ is σ-additive and Radon-regular by Theorem 2.2 of [P3]. For $\varphi \in L_1(\lambda)$, let $\mu(\cdot) = \int_{(\cdot)} \varphi d\mu_\theta$. Since $\varphi \in L_1(\lambda)$, by Lemma 1, no. 6, § 5, Chapter IV of [B], there exist a sequence $(K_n)_1^\infty \subset \mathcal{C}$ and a λ-null set N such that $N(\varphi) \subset N \cup \bigcup_1^\infty K_n$. Hence, for $A \in \mathcal{B}(T)$, $\int_A \varphi d\mu_\theta = \int_{\bigcup_1^\infty (A \cap K_n)} \varphi d\mu_\theta$ and with this observation we note that $\mu : \mathcal{B}(T) \to \mathbb{K}$ given by $\mu(A) = \int_A \varphi d\mu_\theta$ is well de-fined, σ-additive and Borel regular, since μ is $v(\mu_\theta)$-continuous on $\mathcal{B}_c(T)$ and hence on $\mathcal{B}(T)$. (See Claim 2 in the proof of Theorem 6.3.3.) Therefore, $\mu \in M(T)$. Moreover, by Notation 4.2.1 of Chapter 4, by Proposition 1.1.19 of Chap-ter 1 and by Theorems 4.7 and 4.11 of [P3] we have $||\mu|| = v(\mu, \mathcal{B}(T))(T) = v(\mu, \mathcal{B}_c(T))(\bigcup_1^\infty K_n) = \int_{\bigcup_1^\infty K_n} |\varphi| dv(\mu_\theta) = \int_{\bigcup_1^\infty K_n} |\varphi| d\mu_{|\theta|} = \int_T |\varphi| d\lambda = ||\varphi||_1$. Then, the mapping $\Phi : L_1(\lambda) \to M(T)$ given by $\Phi(\varphi)(\cdot) = \int_{(\cdot)} \varphi d\mu_\theta$ is lin-ear and isometric so that $M_\lambda = \Phi(L_1(\lambda)) = \{\mu \in M(T) : \text{there exists } \varphi \in L_1(\lambda) \text{ such that } \mu(\cdot) = \int_{(\cdot)} \varphi d\mu_\theta\}$ is complete with respect to the norm on $M(T)$. Therefore, M_λ is a closed subspace of $M(T)$. Then by the Hahn-Banach theorem, M_λ is a weakly closed subspace of $M(T)$.

Let $\mu_n(\cdot) = \int_{(\cdot)} f_n d\mu_\theta$, $n \in \mathbb{N}$. Then by the foregoing argument μ_n, $n \in \mathbb{N}$, belong to $M(T)$. By hypothesis, $\lim_n \mu_n(U) \in \mathbb{K}$ for each open Baire set U in T and hence by Theorem 5.2.6 of Chapter 5 there exists $\mu_0 \in M(T)$ such that $\mu_n \to \mu_0$ weakly in $M(T)$. As $(\mu_n)_1^\infty \subset M_\lambda$ and as M_λ is weakly closed, $\mu_0 \in M_\lambda$ and hence there exists $f \in L_1(\lambda)$ such that $\mu_0(\cdot) = \int_{(\cdot)} f d\mu_\theta$ on $\mathcal{B}(T)$. Moreover, as $\mu_n \to \mu_0$ weakly in $M(T)$, $(\mu_n)_{n=0}^\infty$ is weakly bounded and hence by Theorem 3.18 of [Ru2], $\sup_{n \in \mathbb{N} \cup \{0\}} ||\mu_n|| = M$(say) $< \infty$ and further, $\mu_n(A) \to \mu_0(A)$ for each $A \in \mathcal{B}(T)$ by (5.2.6.1).

Let g be a bounded Borel measurable function on T and let $\epsilon > 0$. Then there exists a $\mathcal{B}(T)$-simple function s such that $||g - s||_T < \frac{\epsilon}{3M}$. Let $s = \sum_{i=1}^r \alpha_i \chi_{A_i}$ with $(A_i)_1^r \subset \mathcal{B}(T)$. Choose n_0 such that

$$|\alpha_i||\mu_n(A_i) - \mu_0(A_i)| < \frac{\epsilon}{3r} \tag{6.4.6.2}$$

for $i = 1, 2, \ldots, r$ and for $n \geq n_0$. Then,

$$\left| \int_T g f_n d\mu_\theta - \int_T s f_n d\mu_\theta \right| \leq ||g - s||_T \left(\int_T |f_n| d\lambda \right) = ||g - s||_T ||\mu_n|| < \frac{\epsilon}{3} \tag{6.4.6.3}$$

for all $n \in \mathbb{N}$ and

$$\left| \int_T g f d\mu_\theta - \int_T s f d\mu_\theta \right| \leq ||g - s||_T \left(\int_T |f| d\lambda \right) = ||g - s||_T ||\mu_0|| < \frac{\epsilon}{3}. \tag{6.4.6.4}$$

Then by (6.4.6.2), (6.4.6.3) and (6.4.6.4) we have

$$\left| \int_T g f_n d\mu_\theta - \int_T g f d\mu_\theta \right| < \frac{2\epsilon}{3} + \left| \int_T s(f_n - f) d\mu_\theta \right| < \epsilon$$

for $n \geq n_0$ and hence (6.4.6.1) holds for bounded Borel measurable functions g.

If g is bounded and λ-measurable, then there exists a λ-null set $N \in \mathcal{B}(T)$ such that $g\chi_{T\setminus N}$ is Borel measurable and hence (6.4.6.1) holds for bounded λ-measurable functions g. Moreover, since (6.4.6.1) implies that $f_n \to f$ weakly in $L_1(\lambda)$, f is unique in $L_1(\lambda)$.

If $f_n \to h$ λ-a.e. in T, then by (6.4.6.1) (with $g = \chi_A$, $A \in \mathcal{B}_c(T)$) and by Proposition 1.1.21 of Chapter 1, h is μ_θ-integrable and $\lim_n \int_A f_n d\mu_\theta = \int_A h d\mu_\theta$ for each $A \in \mathcal{B}_c(T)$ and consequently, again by (6.4.6.1), $\int_A h d\mu_\theta = \int_A f d\mu_\theta$ for $A \in \mathcal{B}_c(T)$. Let $\nu(A) = \int_A (f - h) d\mu_\theta$ for $A \in \mathcal{B}_c(T)$. Then ν is a null measure on $\mathcal{B}_c(T)$ and hence $v(\nu)(N(f - h)) = \int_{N(f-h)} |f - h| dv(\mu_\theta, \mathcal{B}_c(T)) = \int_{N(f-h)} |f - h| d\mu_{|\theta|} = \int_{N(f-h)} |f - h| d\lambda = 0$ by Proposition 1.1.19 of Chapter 1 and by Theorems 4.7 and 4.11 of [P3]. Therefore, $f = h$ λ-a.e. in T. Since $\int_A f d\mu_\theta = \int_A h d\mu_\theta = \int_{\bigcup_1^\infty (A \cap K_n)} h d\mu_\theta$ for $A \in \mathcal{B}(T)$, where $N(f) \subset N \cup \bigcup_1^\infty C_n$, $(C_n)_1^\infty \subset \mathcal{C}$ and N is λ-null by Lemma 1, no. 6, §5, Chapter IV of [B], the last part holds by the equivalence of (i) and (iii) of Proposition 1.1.21 of Chapter 1. $\square$

The rest of the section is devoted to generalizing Theorem 6.4.6 to vector measures. We begin with the following lemma.

Lemma 6.4.7. *Let X be a quasicomplete lcHs and let $\mathbf{m} : \delta(\mathcal{C}) \to X$ be σ-additive and $\delta(\mathcal{C})$-regular. Let $f : T \to \mathbb{K}$ belong to $\mathcal{L}_1(\mathbf{m})$. Then:*

(i) *There exists $B \in \mathcal{B}_c(T)$ such that $N(f) \subset B$.*

(ii) *Let $\gamma : \mathcal{B}(T) \to X$ be defined by $\gamma(A) = \int_{A \cap B} f d\mathbf{m} = \int_A f d\mathbf{m}$ for $A \in \mathcal{B}(T)$. Then γ is σ-additive and Borel regular.*

Proof. As f is $\mathbf{m}$-measurable, by Proposition 1.1.18 of Chapter 1 there exists $M \in \mathcal{B}_c(T)$ such that $\|\mathbf{m}\|(M) = 0$ and such that $f\chi_{T\setminus M}$ is $\mathcal{B}_c(T)$-measurable. Consequently, $N(f\chi_{T\setminus M}) \in \mathcal{B}_c(T)$ so that $N(f) \subset N(f\chi_{T\setminus M}) \cup M = B$ (say). Thus (i) holds. (For the validity of (i), it suffices that f be just $\mathbf{m}$-measurable.)

(ii) For $A \in \mathcal{B}(T)$, $A \cap B$ is σ-bounded and Borel measurable and hence $A \cap B \in \mathcal{B}_c(T)$. Since $\int_{(\cdot)} f d\mathbf{m}$ is σ-additive on $\mathcal{B}_c(T)$ by Theorem 4.1.8(ii) and by Remark 4.1.5 of Chapter 4, it follows that $\gamma(\cdot) = \gamma(\cdot \cap B)$ is σ-additive on $\mathcal{B}(T)$.

Claim 1. γ is Borel inner regular.

In fact, let $A \in \mathcal{B}(T)$ and let B be as in (i). Then there exists $(E_k)_1^\infty \subset \delta(\mathcal{C})$ such that $E_k \nearrow B \cap A$. Let $\epsilon > 0$ and $q \in \Gamma$. Since γ is σ-additive on $\mathcal{B}(T)$, by Proposition 1.1.5 of Chapter 1, $\|\gamma\|_q$ is continuous on $\mathcal{B}(T)$ and hence there exists k_0 such that $\|\gamma\|_q((A \cap B)\setminus E_k) < \frac{\epsilon}{2}$ for $k \geq k_0$. On the other hand, by Theorem 4.1.8(iii)(c) and by Remark 4.1.5 of Chapter 4, there exists $\delta > 0$ such that $\|\gamma\|_q(F) < \frac{\epsilon}{2}$ whenever $F \in \mathcal{B}_c(T)$ with $\|\mathbf{m}\|_q(F) < \delta$. Since $\mathbf{m}$ is $\delta(\mathcal{C})$-regular by hypothesis, there exists $C \in \mathcal{C}$ such that $C \subset E_{k_0}$ and $\|\mathbf{m}\|_q(E_{k_0}\setminus C) < \delta$. Then $\|\gamma\|_q(E_{k_0}\setminus C) < \frac{\epsilon}{2}$. Then $\|\gamma\|_q(A\setminus C) < \epsilon$ since $\|\gamma\|_q(A\setminus B) = 0$. Hence the claim holds.

$\Pi_q \circ \gamma : \mathcal{B}(T) \to \widetilde{X_q}$ is σ-additive and hence has bounded range. Then by (a) and (c) of Proposition 1.2.15 of Chapter 1, $\{x^* \circ \gamma : x^* \in U_q^0\} = \{\Psi_{x^*}(\Pi_q \circ \gamma) :$

$x^* \in U_q^0\}$ is bounded in $M(T)$. By Claim 1, $\boldsymbol{\gamma}$ is Borel inner regular on $\mathcal{B}(T)$ and hence, given $A \in \mathcal{B}(T)$, $q \in \Gamma$ and $\epsilon > 0$, there exists a compact $K \subset A$ such that $||\boldsymbol{\gamma}||_q(A\backslash K) < \epsilon$. Then by the said proposition of Chapter 1 we have

$$\sup_{x^* \in U_q^0} v(x^* \circ \boldsymbol{\gamma})(A\backslash K) = ||\Pi_q \circ \boldsymbol{\gamma}||(A\backslash K) = ||\boldsymbol{\gamma}||_q(A\backslash K) < \epsilon$$

and hence $\{(x^* \circ \boldsymbol{\gamma}) : x^* \in U_q^0\}$ is uniformly Borel inner regular on $\mathcal{B}(T)$. Consequently, by Theorem 2 of [P8], $\{x^* \circ \boldsymbol{\gamma} : x^* \in U_q^0\}$ is uniformly Borel regular on $\mathcal{B}(T)$ and arguing as in the above invoking Proposition 1.2.15 of Chapter 1, we conclude that $\Pi_q \circ \boldsymbol{\gamma}$ is Borel regular on $\mathcal{B}(T)$ for each $q \in \Gamma$. Then by Definition 4.6.7 of Chapter 4, $\boldsymbol{\gamma}$ is Borel regular on $\mathcal{B}(T)$. $\qquad\square$

We use the above lemma to give in the following theorem (Theorem 6.4.8) two generalizations of the a.e. convergence part of Theorem 6.4.6 to σ-additive $\delta(\mathcal{C})$-regular vector measures on $\delta(\mathcal{C})$. Theorem 6.4.8 is a strengthened vector measure analogue of Proposition 4.8 of [T].

Theorem 6.4.8. *Let X be a quasicomplete lcHs with topology τ and let $\mathbf{m} : \delta(\mathcal{C}) \to X$ be σ-additive and $\delta(\mathcal{C})$-regular. Let $(f_n)_1^\infty \subset \mathcal{L}_1(\mathbf{m})$ and let $f : T \to \mathbb{K}$ or $[-\infty, \infty]$ be such that $f_n \to f$ $\mathbf{m}$-a.e. in T (see Definition 1.2.4 of Chapter 1).*

(a) *Suppose $(\int_U f_n d\mathbf{m})_1^\infty$ converges in X in τ for each open Baire set U in T. Then the following hold:*

  (i) *$f \in \mathcal{L}_1(\mathbf{m})$.*

  (ii) *For $A \in \mathcal{B}(T)$,*

$$\int_A f_n d\mathbf{m} \to \int_A f d\mathbf{m} \quad in \ \tau$$

  and for each $q \in \Gamma$, $\int_A f_n d\mathbf{m}_q \to \int_A f d\mathbf{m}_q$ and for a fixed $q \in \Gamma$, the convergence is uniform with respect to $A \in \mathcal{B}(T)$.

  (iii) *For bounded $\mathbf{m}$-measurable scalar functions g on T,*

$$\int_T f_n g d\mathbf{m} \to \int_T f g d\mathbf{m} \quad in \ \tau.$$

(b) *Suppose $(\int_U f_n d\mathbf{m})_1^\infty$ converges weakly in X for each open Baire set U in T. Then the following hold:*

  (i) *$f \in \mathcal{L}_1(\mathbf{m})$.*

  (ii) *For $A \in \mathcal{B}(T)$,*

$$\int_A f_n d\mathbf{m} \to \int_A f d\mathbf{m} \quad weakly.$$

  (iii) *For bounded $\mathbf{m}$-measurable scalar functions g on T,*

$$\int_T f_n g d\mathbf{m} \to \int_T f g d\mathbf{m} \quad weakly.$$

Proof. (a)(i) and (b)(i). Let U be an open Baire set in T. By hypothesis (a) (resp. (b)) there exists a vector $x_U \in X$ such that

$$\lim_n \int_U f_n d\mathbf{m} = x_U \text{ in } \tau(\text{resp. weakly}).$$

Then in both cases, by Theorem 4.1.8(v) and Remark 4.1.5 of Chapter 4,

$$\lim_n \int_U f_n d(x^* \circ \mathbf{m}) = x^*(x_U) \qquad (6.4.8.1)$$

for $x^* \in X^*$. On the other hand, by hypothesis and by Theorem 4.4(i) of [P4], $x^* \circ \mathbf{m} = \mu_\theta$ for some $\theta \in \mathcal{K}(T)^*$ and hence by Theorem 6.4.6, $f \in \mathcal{L}_1(x^* \circ \mathbf{m})$ for $x^* \in X^*$ and we have

$$\lim_n \int_U f_n d(x^* \circ \mathbf{m}) = \int_U f d(x^* \circ \mathbf{m}) \qquad (6.4.8.2)$$

for $x^* \in X^*$. Then by (6.4..8.1) and (6.4.8.2) we have

$$x^*(x_U) = \int_U f d(x^* \circ \mathbf{m})$$

for each open Baire set U in T and for $x^* \in X^*$. Consequently, by Theorem 6.3.5, (a)(i) (resp. (b)(i)) holds.

(a)(ii) Let $f_0 = f$. By (a)(i) and by Lemma 6.4.7, there exists $B_n \in \mathcal{B}_c(T)$ such that $N(f_n) \subset B_n$ and $\gamma_n : \mathcal{B}(T) \to X$ given by $\gamma_n(A) = \int_{A \cap B_n} f_n d\mathbf{m}$ for $A \in \mathcal{B}(T)$, is σ-additive and Borel regular for $n \in \mathbb{N} \cup \{0\}$. By hypothesis, $\lim_n \gamma_n(U) \in X$ in τ for each open Baire set U in T. Since X is also sequentially complete, by Theorem 5.2.23 of Chapter 5 there exists a unique X-valued σ-additive Borel regular measure γ on $\mathcal{B}(T)$ such that

$$\lim_n \int_T g d\gamma_n = \int_T g d\gamma \in X \qquad (6.4.8.3)$$

in τ for each bounded Borel measurable scalar function g on T, the integrals being defined as in Definition 1 of [P7].

Claim 1. For a bounded Borel measurable scalar function g on T,

$$\int_T g d\gamma_n = \int_T f_n g d\mathbf{m} \qquad (6.4.8.4)$$

for $n \in \mathbb{N}$.

In fact, (6.4.8.4) clearly holds for $\mathcal{B}(T)$-simple functions. Choose a sequence $(s_n)_1^\infty$ of $\mathcal{B}(T)$- simple functions such that $s_n \to g$ and $|s_n| \nearrow |g|$ pointwise in T.

Then by LDCT (Theorem 4.5.3(i) of Chapter 4) and by the validity of (6.4.8.4) for $\mathcal{B}(T)$-simple functions we have

$$\int_T g d\boldsymbol{\gamma}_n = \lim_k \int_T s_k d\boldsymbol{\gamma}_n = \lim_k \int_T f_n s_k d\mathbf{m} = \int_T f_n g d\mathbf{m}$$

since g and $f_n g$ are $\mathbf{m}$-integrable in T by (i)(b) and (ii) of Theorem 4.1.9 and by Remark 4.1.5 of Chapter 4. Hence the claim holds.

Let $B = \bigcup_{n=0}^\infty B_n$. Then $B \in \mathcal{B}_c(T)$ and $f_n = 0$ in $T\backslash B$ for $n \in \mathbb{N} \cup \{0\}$. For $x^* \in X^*$, let θ be as in the proof of (a)(i). Then by hypothesis (a) and by Theorem 6.4.6,

$$\lim_n \int_T f_n g d(x^* \circ \mathbf{m}) = \int_T f g d(x^* \circ \mathbf{m}) \tag{6.4.8.5}$$

for each bounded Borel measurable function g on T. Then for $A \in \mathcal{B}(T)$, by (6.4.8.4) and (6.4.8.5) and by Theorem 4.1.8(v) and Remark 4.1.5 of Chapter 4 (since $f \in \mathcal{L}_1(\mathbf{m})$ by (a)(i)), we have

$$\lim_n (x^* \circ \boldsymbol{\gamma}_n)(A) = \lim_n \int_A f_n d(x^* \circ \mathbf{m}) = \lim_n \int_{A \cap B} f_n d(x^* \circ \mathbf{m})$$
$$= \int_{A \cap B} f d(x^* \circ \mathbf{m}) = \int_A f d(x^* \circ \mathbf{m})$$
$$= (x^* \circ \boldsymbol{\gamma}_0)(A). \tag{6.4.8.6}$$

Moreover, by (6.4.8.3) we have

$$\lim_n (x^* \circ \boldsymbol{\gamma}_n)(A \cap B) = (x^* \circ \boldsymbol{\gamma})(A \cap B) = (x^* \circ \boldsymbol{\gamma})(A) \tag{6.4.8.7}$$

for $x^* \in X^*$, since $\boldsymbol{\gamma}_n(A\backslash B) = 0$ for $n \in \mathbb{N} \cup \{0\}$. Consequently, by (6.4.8.6) and (6.4.8.7) we have

$$(x^* \circ \boldsymbol{\gamma}_0)(A) = (x^* \circ \boldsymbol{\gamma})(A)$$

for $A \in \mathcal{B}(T)$ and for $x^* \in X^*$. Then, by the Hahn-Banach theorem, $\boldsymbol{\gamma}(A) = \boldsymbol{\gamma}_0(A)$ and hence, for $A \in \mathcal{B}(T)$, by (6.4.8.3) we have

$$\lim_n \int_A f_n d\mathbf{m} = \lim_n \boldsymbol{\gamma}_n(A) = \boldsymbol{\gamma}(A) = \boldsymbol{\gamma}_0(A) = \int_A f d\mathbf{m} \tag{6.4.8.8}$$

in τ.

In Theorem 4.2.9 of Chapter 4, take $f_n^{(q)} = f_n$ for all $q \in \Gamma$. Note that $\sigma(\delta(\mathcal{C})) = \mathcal{B}_c(T)$. Then in the notation of the said theorem, by Theorems 4.1.8(v) and 4.2.9 and by Remark 4.1.5 of Chapter 4 and by (6.4.8.8) we have

$$\boldsymbol{\gamma}_n^{(q)}(A) = \boldsymbol{\gamma}_n^{(q)}(A \cap B) = \int_{A \cap B} f_n d\mathbf{m}_q = \Pi_q \left(\int_{A \cap B} f_n d\mathbf{m} \right)$$
$$\to (\Pi_q \circ \boldsymbol{\gamma}_0)(A \cap B) = (\Pi_q \circ \boldsymbol{\gamma}_0)(A) = \int_A f d\mathbf{m}_q$$

in $\widetilde{X_q}$ and for a fixed $q \in \Gamma$, the limit is uniform with respect to $A \in \mathcal{B}(T)$ since it is so with respect to $E \in \mathcal{B}_c(T)$. Hence (a)(ii) holds.

(a)(iii) By hypothesis, by Lemma 6.4.7 and by Theorem 5.2.8 of Chapter 5, for each $q \in \Gamma$, there exists a finite constant M_q such that

$$\sup_{n \in \mathbb{N} \cup \{0\}} ||\gamma_n||_q(T) = M_q. \tag{6.4.8.9}$$

Let g be a bounded **m**-measurable scalar function on T and let $q \in \Gamma$. Then by Proposition 1.1.18 of Chapter 1 there exists $N_q \in \mathcal{B}_c(T)$ with $||\mathbf{m}||_q(N_q) = 0$ such that $h_q = g\chi_{T \setminus N_q}$ is $\mathcal{B}_c(T)$-measurable and bounded. Hence, given $\epsilon > 0$, there exists a $\mathcal{B}_c(T)$-simple function $s^{(q)}$ such that $|s^{(q)}| \le |h_q|$ and

$$||h_q - s^{(q)}||_T < \frac{\epsilon}{3M_q}. \tag{6.4.8.10}$$

Let $s^{(q)} = \sum_{i=1}^r \alpha_i \chi_{A_i}$ with $(A_i)_1^r \subset \mathcal{B}_c(T)$. By (a)(ii) there exists n_1 such that

$$|\alpha_i||\gamma_n(A_i) - \gamma_0(A_i)|_q < \frac{\epsilon}{3r} \tag{6.4.8.11}$$

for $i = 1, 2, \ldots, r$ and for $n \ge n_1$. Then by (6.4.8.11) we have

$$\left| \int_T s^{(q)} d\gamma_n - \int_T s^{(q)} d\gamma_0 \right|_q \le \sum_{i=1}^r |\alpha_i||\gamma_n(A_i) - \gamma_0(A_i)|_q < \frac{\epsilon}{3} \tag{6.4.8.12}$$

for $n \ge n_1$. Moreover, by (6.4.8.9) and (6.4.8.10) we have

$$\left| \int_T s^{(q)} d\gamma_n - \int_T h_q d\gamma_n \right|_q \le ||s^{(q)} - h_q||_T ||\gamma_n||_q(T) < \frac{\epsilon}{3} \tag{6.4.8.13}$$

for $n \in \mathbb{N} \cup \{0\}$.

Consequently, by (6.4.8.9), (6.4.8.12) and by (6.4.8.13), by the definition of h_q and by Claim 1 we have

$$\left| \int f_n g d\mathbf{m} - \int_T f g d\mathbf{m} \right|_q = \left| \int_T f_n h_q d\mathbf{m} - \int_T f h_q d\mathbf{m} \right|_q$$

$$\le \left| \int_T (h_q - s^{(q)}) d\gamma_n \right|_q + \left| \int_T s^{(q)} d\gamma_n - \int_T s^{(q)} d\gamma_0 \right|_q$$

$$+ \left| \int_T (s^{(q)} - h_q) d\gamma_0 \right|_q$$

$$< \epsilon$$

for $n \ge n_1$. Hence (a)(iii) holds.

(b)(ii) Let $x^* \in X^*$. Then by hypothesis and by Theorem 4.4(i) of [P4] there exists $\theta \in \mathcal{K}(T)^*$ such that $\mu_\theta = x^* \circ \mathbf{m}$. Then by hypothesis, by(b)(i) and by Theorem 6.4.6 and by Theorem 4.1.8(v) and Remark 4.1.5 of Chapter 4, (b)(ii) holds.

(b)(iii) Let U be an open Baire set in T. By hypothesis, $\lim_n (x^* \circ \boldsymbol{\gamma}_n)(U) \in \mathbb{K}$ for each $x^* \in X^*$. Hence $((\boldsymbol{\gamma}_n)(U))_1^\infty$ is weakly bounded and hence, by Theorem 3.18 of [Ru2], is bounded in τ. By (b)(i) $f \in \mathcal{L}_1(\mathbf{m})$. Hence for $q \in \Gamma$, by Lemma 6.4.7(ii) and by Theorem 5.2.8 of Chapter 5, (6.4.8.9) holds.

Let $x^* \in X^*$ be fixed and let $q_{x^*}(x) = |x^*(x)|$ for $x \in X$. By hypothesis and by Proposition 1.1.18 of Chapter 1 there exists a $\mathcal{B}_c(T)$-measurable bounded function h_{x^*} such that $h_{x^*} = g\chi_{T \setminus N_{x^*}}$ where $N_{x^*} \in \mathcal{B}_c(T)$ with $v(x^* \circ \mathbf{m})(N_{x^*}) = 0$. Choose a $\mathcal{B}_c(T)$-simple function s such that

$$||s - h_{x^*}||_T < \frac{\epsilon}{3M_{q_{x^*}}}. \tag{6.4.8.14}$$

Let $s = \sum_1^r \alpha_i \chi_{A_i}$, $(A_i)_1^r \subset \mathcal{B}_c(T)$. Then by (b)(i) and (b)(ii) and by Theorem 4.1.8(v) and Remark 4.1.5 of Chapter 4 there exists n_2 such that

$$\left| \alpha_i \int_{A_i} f_n d(x^* \circ \mathbf{m}) - \int_{A_i} f d(x^* \circ \mathbf{m}) \right| < \frac{\epsilon}{3r} \tag{6.4.8.15}$$

for $i = 1, 2, \ldots, r$ and for $n \geq n_2$. Then by (6.4.8.15) we have

$$\left| \int_T s f_n d(x^* \circ \mathbf{m}) - \int_T s f d(x^* \circ \mathbf{m}) \right| \leq \sum_{i=1}^r |\alpha_i| \left| \int_{A_i} (f_n - f) d(x^* \circ \mathbf{m}) \right| < \frac{\epsilon}{3} \tag{6.4.8.16}$$

for $n \geq n_2$. Then by Claim 1, by (6.4.8.14), by Proposition 3.2.13(iv), by Theorem 4.1.8(v) and Remark 4.1.5 of Chapter 4 and by (6.4.8.9) we have

$$\left| \int_T s f_n d(x^* \circ \mathbf{m}) - \int_T h_{x^*} f_n d(x^* \circ \mathbf{m}) \right|$$
$$= \left| \int_T (s - h_{x^*}) d(x^* \circ \boldsymbol{\gamma}_n) \right|$$
$$\leq ||s - h_{x^*}||_T v(x^* \circ \boldsymbol{\gamma}_n)(T)$$
$$= ||s - h_{x^*}||_T ||\boldsymbol{\gamma}_n||_{q_{x^*}}(T) < \frac{\epsilon}{3} \tag{6.4.8.17}$$

for $n \in \mathbb{N} \cup \{0\}$. Then by (b)(i) and by Theorems 4.1.8(v) and 4.1.9(ii) and by Remark 4.1.5 of Chapter 4, and by (6.4.8.16) and (6.4.8.17),

$$x^* \left(\int_T f_n g d\mathbf{m} - \int_T f g d\mathbf{m} \right) \to 0$$

as $n \to \infty$ for each $x^* \in X^*$ and hence (b)(iii) holds. $\qquad\square$

Thus Theorem 6.4.8 generalizes Theorem 6.4.6 to $\delta(\mathcal{C})$-regular σ-additive quasicomplete lcHs-valued vector measures $\mathbf{m}$ when $f_n \to f$ $\mathbf{m}$-a.e. in T. In order to generalize the said theorem when f_n do not satisfy the $\mathbf{m}$-a.e. convergence hypothesis, we restrict $\mathbf{m}$ to be Banach space-valued. To this end, we adapt the proofs of Lemmas 1, 2 and 3 on pp. 126–129 of [T] and then we give a stronger vector measure version of Theorem 4.9 of [T] in Theorem 6.4.12 below.

Lemma 6.4.9. *Let X be a Banach space and let $\mathbf{m} : \delta(\mathcal{C}) \to X$ be σ-additive and $\delta(\mathcal{C})$-regular. Let $(f_n)_1^\infty \subset \mathcal{L}_1(\mathbf{m})$. Then there exists a sequence $(K_n)_1^\infty \subset \mathcal{C}$ such that each f_n vanishes in $T \backslash \bigcup_1^\infty K_k$.*

Proof. By Lemma 6.4.7(i), for each $n \in \mathbb{N}$, there exists $B_n \in \mathcal{B}_c(T)$ such that $N(f_n) \subset B_n$ so that $\bigcup_1^\infty N(f_n) \subset \bigcup_1^\infty B_n \in \mathcal{B}_c(T)$. Since $\bigcup_1^\infty B_n$ is σ-bounded, there exists a sequence $(K_n)_1^\infty \subset \mathcal{C}$ such that $\bigcup_1^\infty B_n \subset \bigcup_1^\infty K_n$. Then $f_n = 0$ in $T \backslash \bigcup_1^\infty K_k$ for each n. $\qquad \square$

Lemma 6.4.10. *Let X and $\mathbf{m}$ be as in Lemma 6.4.9 and let H be a norm determining set in X^*. Given a sequence $(K_n)_1^\infty \subset \mathcal{C}$, there exists a sequence $(x_n^*)_1^\infty \subset H$ such that every σ-Borel set $A \subset \bigcup_1^\infty K_n$ is $\mathbf{m}$-null whenever A is $(x_n^* \circ \mathbf{m})$-null for each $n \in \mathbb{N}$.*

Proof. By Lemma 5.2.2 of Chapter 5, $A \cap K_n \in \delta(\mathcal{C})$ for $n \in \mathbb{N}$. Choose a relatively compact open set V_n such that $K_n \subset V_n$. Arguing as in the proof of Lemma 6.3.2, we can find a sequence $(x_{n,r}^*)_{r=1}^\infty \subset H$ such that $E \in \mathcal{B}(V_n)$ is $\mathbf{m}$-null whenever E is $(x_{n,r}^* \circ \mathbf{m})$-null for all $r \in \mathbb{N}$. Let $(x_n^*)_1^\infty = \{x_{n,r}^* : n, r \in \mathbb{N}\}$. Then A is $\mathbf{m}$-null whenever A is $(x_n^* \circ \mathbf{m})$-null for each $n \in \mathbb{N}$ since $A \cap K_n \in \mathcal{B}(V_n)$ for all n and $A = \bigcup_1^\infty (A \cap K_n)$. $\qquad \square$

Lemma 6.4.11. *Let $\mu_k : \delta(\mathcal{C}) \to \mathbb{K}$, $k \in \mathbb{N}$, be σ-additive and $\delta(\mathcal{C})$-regular and let $(f_n)_1^\infty \subset \bigcap_{k=1}^\infty \mathcal{L}_1(\mu_k)$. Suppose $\lim_n \int_U f_n d\mu_k \in \mathbb{K}$ for each $k \in \mathbb{N}$ and for each open Baire set U in T. Then there exists a sequence $(g_n)_1^\infty$ such that each g_n is a convex combination of $(f_k)_{k \geq n}$ and such that $(g_n)_1^\infty$ converges in mean in $\mathcal{L}_1(\mu_k)$ and also converges pointwise μ_k-a.e. in T for each $k \in \mathbb{N}$.*

Proof. By hypothesis and by Theorem 6.4.6, for each $k \in \mathbb{N}$, (f_n) converges weakly to some $h_k \in \mathcal{L}_1(\mu_k)$. Let us embed $\bigcap_{k=1}^\infty \mathcal{L}_1(\mu_k)$ in the diagonal of the product space $P = \Pi_{k=1}^\infty \mathcal{L}_1(\mu_k)$. For each f_n, let $\widetilde{f_n} = (f_n, f_n, \dots) \in P$. Clearly, P is a pseudo-metrizable locally convex space with the pseudo-norm ρ given by

$$\rho((\varphi_k)_1^\infty, (\psi_k)_1^\infty) = \sum_{k=1}^\infty \frac{1}{2^k} \frac{\int_T |\varphi_k - \psi_k| dv(\mu_k)}{1 + \int_T |\varphi_k - \psi_k| dv(\mu_k)}$$

for $(\varphi_k)_1^\infty, (\psi_k)_1^\infty \in P$. Then by Theorem 4.3, Chapter IV of [Scha], $(\widetilde{f_n})_{n=1}^\infty$ converges weakly to some $\hat{h} = (h_k)_{k=1}^\infty \in P$. Then $(\widetilde{f_n})_{n \geq k}$ converges to $\hat{h}$ weakly for each k and hence by the Hahn-Banach theorem (see the proof of Theorem 3.13 of [Ru2]), $\hat{h} = (h_i)_1^\infty$ belongs to the closed convex hull of $(\widetilde{f_n})_{n \geq k}$ for each k. Then there exists a sequence $(\widetilde{g_n})$ such that $\widetilde{g_n} \in co(\widetilde{f_k}, \widetilde{f_{k+1}}, \dots)$ with $\rho(\widetilde{g_n}, \hat{h}) \to 0$. Let $g_n = \sum_{i=n}^{N(n)} \alpha_i f_i$, $\alpha_i \geq 0$, $\sum_{i=n}^{N(n)} \alpha_i = 1$ so that $\widetilde{g_n} = (g_n, g_n, \dots)$. Then $g_n \to h_k$ in mean in $\mathcal{L}_1(\mu_k)$ for each k.

For $k = 1$, there exists a subsequence $(g_{1,r})_{r=1}^{\infty}$ of $(g_n)_1^{\infty}$ such that $g_{1,r} \to h_1 \ \mu_1$-a.e. in T. Proceeding successively, there exists a subsequence $(g_{n,r})_{r=1}^{\infty}$ of $(g_{n-1,r})_{r=1}^{\infty}$ such that $g_{n,r} \to h_n \ \mu_n$-a.e. in T. Then the diagonal sequence $(g_{n,n})_{n=1}^{\infty}$ is a subsequence of $(g_n)_{n=1}^{\infty}$ and converges to $h_k \ \mu_k$-a.e. in T for each k. Clearly, $(g_{n,n})_{n=1}^{\infty}$ also converges in mean to h_k in $\mathcal{L}_1(\mu_k)$ for each k. Hence the lemma holds. $\qquad\square$

Theorem 6.4.12. (Full generalization of Theorem 6.4.6 to Banach space-valued measures on $\delta(\mathcal{C})$). *Let X be a Banach space with topology τ and let H be a norm determining set in X^* with the Orlicz property. Let $\mathbf{m} : \delta(\mathcal{C}) \to X$ be σ-additive and $\delta(\mathcal{C})$-regular.*

(a) *Suppose $(f_n)_1^{\infty} \subset \mathcal{L}_1(\mathbf{m})$ is such that for each open Baire set U in T, there exists a vector $x_U \in X$ such that $\int_U f_n d\mathbf{m} \to x_U$ in τ. Then the following hold:*

   (i) *There exists a function $f \in \mathcal{L}_1(\mathbf{m})$ such that $\int_U f_n d\mathbf{m} \to \int_U f d\mathbf{m}$ in τ.*

   (ii) *f in (a)(i) is unique up to $\mathbf{m}$-a.e. in T.*

   (iii) *For $A \in \mathcal{B}(T)$, $\int_A f_n d\mathbf{m} \to \int_A f d\mathbf{m}$ in τ.*

   (iv) *For each bounded $\mathbf{m}$-measurable scalar function g on T, $\int_T f_n g d\mathbf{m} \to \int_T f g d\mathbf{m}$ in τ.*

(b) *Suppose $(f_n)_1^{\infty} \subset \mathcal{L}_1(\mathbf{m})$ is such that for each open Baire set U in T, there exists a vector $x_U \in X$ such that $\int_U f_n d\mathbf{m} \to x_U$ in $\sigma(X, H)$. Then the following hold:*

   (i) *There exists $f \in \mathcal{L}_1(\mathbf{m})$ such that $\int_U f_n d\mathbf{m} \to \int_U f d\mathbf{m}$ in $\sigma(X, H)$.*

   (ii) *f in (b)(i) is unique up to $\mathbf{m}$-a.e. in T.*

   (iii) *For $A \in \mathcal{B}(T)$, $\int_A f_n d\mathbf{m} \to \int_A f d\mathbf{m}$ in $\sigma(X, H)$.*

   (iv) *For each bounded $\mathbf{m}$-measurable scalar function g on T, $\int_T f_n g d\mathbf{m} \to \int_T f g d\mathbf{m}$ in $\sigma(X, H)$.*

Proof. Let $H_0 = \{x^* \in X^* : |x^*| \leq 1\}$. By hypothesis (a) (resp. (b)), and by Theorem 2.1.5(viii) and Remark 2.2.3 of Chapter 2, $\lim_n \int_U f_n d(x^* \circ \mathbf{m}) = x^*(x_U) \in \mathbb{K}$ for each $x^* \in H_0$ (resp. for $x^* \in H$). By Lemma 6.4.9, there exists $(K_n)_1^{\infty} \subset \mathcal{C}$ such that each f_n vanishes in $T \backslash \bigcup_1^{\infty} K_k$. By Lemma 6.4.10, we associate the sequence $(K_n)_1^{\infty}$ with a sequence $(x_n^*)_1^{\infty} \subset H_0$ (resp. $\subset H$) satisfying the property mentioned in the said lemma. Then by Lemma 6.4.11 there exists a sequence $(g_n)_1^{\infty}$ such that each g_n is of the form

$$g_n = \sum_{i=n}^{N(n)} \alpha_i^{(n)} f_i, \ \alpha_i^{(n)} \geq 0 \text{ and } \sum_{i=n}^{N(n)} \alpha_i^{(n)} = 1 \qquad (6.4.12.1)$$

and such that (g_n) converges in mean in $\mathcal{L}_1(x_i^* \circ \mathbf{m})$ and converges $(x_i^* \circ \mathbf{m})$-a.e. in T for each $i \in \mathbb{N}$. Then by Lemma 6.4.10, $(g_n)_1^\infty$ converges $\mathbf{m}$-a.e. in T. Let f be the $\mathbf{m}$-a.e. pointwise limit of $(g_n)_1^\infty$.

(a)(i) Let U be an open Baire set in T and let $x_U \in X$ be as in the hypothesis (a). Then, given $\epsilon > 0$, there exists n_0 such that

$$\left| \int_U f_n d\mathbf{m} - x_U \right| < \epsilon \qquad (6.4.12.2)$$

for $n \geq n_0$. Let $(g_n)_1^\infty$ be as in (6.4.12.1). Then

$$\left| \int_U g_n d\mathbf{m} - x_U \right| = \left| \sum_{i=n}^{N(n)} (\alpha_i^{(n)} \int_U f_i d\mathbf{m} - \alpha_i^{(n)} x_U) \right| \leq \sum_{i=n}^{N(n)} \alpha_i^{(n)} \epsilon = \epsilon$$

for $n \geq n_0$ and hence

$$\lim_n \int_U g_n d\mathbf{m} = x_U \text{ in } \tau. \qquad (6.4.12.3)$$

Then by Theorem 6.4.8(a), $f \in \mathcal{L}_1(\mathbf{m})$ and

$$\int_U g_n d\mathbf{m} \to \int_U f d\mathbf{m} \text{ in } \tau. \qquad (6.4.12.4)$$

Then by (6.4.12.3) and (6.4.12.4) we have

$$\int_U f d\mathbf{m} = x_U$$

and hence by hypothesis,

$$\lim_n \int_U f_n d\mathbf{m} = \int_U f d\mathbf{m} \text{ in } \tau.$$

Thus (a)(i) holds.

(b)(i) Let $x^* \in H$. By hypothesis (b) and by Theorem 2.1.5(viii) and Remark 2.2.3 of Chapter 2,

$$\lim_n \int_U f_n d(x^* \circ \mathbf{m}) = \lim_n x^* \left(\int_U f_n d\mathbf{m} \right) = x^*(x_U) \qquad (6.4.12.5)$$

for each open Baire set U in T. Hence, given $\epsilon > 0$, there exists n_1 such that

$$\left| \int_U f_n d(x^* \circ \mathbf{m}) - x^*(x_U) \right| < \epsilon \qquad (6.4.12.6)$$

for $n \geq n_1$. Then by (6.4.12.1) and (6.4.12.6) we have

$$\left| \int_U g_n d(x^* \circ \mathbf{m}) - x^*(x_U) \right| = \left| \sum_{i=n}^{N(n)} (\alpha_i^{(n)} \int_U f_i d(x^* \circ \mathbf{m}) - \alpha_i^{(n)} x^*(x_U)) \right| < \epsilon$$

for $n \geq n_1$. Hence

$$\lim_n \int_U g_n d(x^* \circ \mathbf{m}) = x^*(x_U) \qquad (6.4.12.7)$$

for $x^* \in H$. By hypothesis and by Theorem 4.4(i) of [P4] there exists $\theta \in \mathcal{K}(T)^*$ such that $\mu_\theta = x^* \circ \mathbf{m}$ and hence by (6.4.12.7), by the fact that $g_n \to f\mathbf{m}$-a.e. in T so that $g_n \to f$ $(x^* \circ \mathbf{m})$-a.e. in T and hence by Theorem 6.4.6, $f \in \mathcal{L}_1(x^* \circ \mathbf{m})$ and

$$\lim_n \int_U g_n d(x^* \circ \mathbf{m}) = \int_U f d(x^* \circ \mathbf{m}) \qquad (6.4.12.8)$$

for $x^* \in H$. Then by (6.4.12.7) and (6.4.12.8),

$$\int_U f d(x^* \circ \mathbf{m}) = x^*(x_U)$$

for $x^* \in H$. Consequently, by Theorem 6.3.4, $f \in \mathcal{L}_1(\mathbf{m})$ and by Theorem 2.1.5(viii) and Remark 2.2.3 of Chapter 2,

$$x^* \left(\int_U f d\mathbf{m} \right) = \int_U f d(x^* \circ \mathbf{m}) = x^*(x_U)$$

for $x^* \in H$. As H is norm determining, we conclude that

$$\int_U f d\mathbf{m} = x_U \qquad (6.4.12.9)$$

and hence by (6.4.12.5) and (6.4.12.9), (b)(i) holds.

(a)(ii) (resp. (b)(ii))

Claim 1. Let μ_1 and μ_2 be in $M(T)$. If $\mu_1(U) = \mu_2(U)$ for each open Baire set U in T, then $\mu_1 = \mu_2$.

In fact, let $\nu_i = \mu_i|_{\mathcal{B}_0(T)}$ for $i = 1, 2$. By Proposition 1 of [DP1], ν_1 and ν_2 are Baire regular. For $E \in \mathcal{B}_0(T)$, let $\mathcal{U}_E = \{U \in \mathcal{B}_0(T) : U \text{ open}, U \supset E\}$ and let $U_1 \geq U_2$ for $U_1, U_2 \in \mathcal{U}_E$ if $U_1 \subset U_2$. Then $\mathcal{U}_E$ is a directed set and by the Baire regularity of ν_1 and ν_2 we have

$$\nu_1(E) = \lim_{U \in \mathcal{U}_E, U \to E} \nu_1(U) = \lim_{U \in \mathcal{U}_E, U \to E} \nu_2(U) = \nu_2(E)$$

and hence $\nu_1 = \nu_2$ on $\mathcal{B}_0(T)$. Then the claim is immediate from the uniqueness part of Proposition 1 of [DP1].

Suppose h is another function in $\mathcal{L}_1(\mathbf{m})$ such that $\lim_n \int_U f_n d\mathbf{m} = \int_U h d\mathbf{m}$ in τ (resp. in $\sigma(X, H)$) for open Baire sets U in T. Let $x^* \in H_0$ (resp. $x^* \in H$). By hypothesis and by Theorem 4.4(i) of [P4] there exists $\theta \in \mathcal{K}(T)^*$ such that $x^* \circ \mathbf{m} = \mu_\theta$ and as seen in the proof of Theorem 6.4.6, $\int_{(\cdot)} h d(x^* \circ \mathbf{m})$ and $\int_{(\cdot)} f d(x^* \circ \mathbf{m})$ belong to $M(T)$. Then by hypothesis (a) (resp. (b)), by Theorem

2.1.5(viii) and Remark 2.2.3 of Chapter 2 and by Claim 1 above, ν is a null measure where $\nu(A) = \int_A (h - f) d(x^* \circ \mathbf{m})$ for $A \in \mathcal{B}(T)$. Then by Proposition 1.1.19 of Chapter 1 and by Theorem 1.39 of [Ru1], $h = f$ $(x^* \circ \mathbf{m})$-a.e. in T. This holds for each $x^* \in H_0$ (resp. $x^* \in H$) and hence by Lemmas 6.4.9 and 6.4.10, $h = f$ $\mathbf{m}$-a.e. in T. Therefore, (a)(ii) (resp. (b)(ii)) holds.

(b)(iii) and (b)(iv) Let $f_0 = f$. For $x^* \in X^*$, by (b)(i) and by Theorem 2.1.5(viii) and Remark 2.2.3 of Chapter 2 we have

$$\lim_n \int_U f_n d(x^* \circ \mathbf{m}) = \int_U f d(x^* \circ \mathbf{m})$$

for each open Baire set U in T. Moreover, by hypothesis and by Theorem 4.4(i) of [P4] there exists $\theta \in \mathcal{K}(T)^*$ such that $x^* \circ \mathbf{m} = \mu_\theta$ and hence by Theorem 6.4.6 and by the uniqueness of f in $L_1(x^* \circ \mathbf{m})$, we have

$$\lim_n \int_T g f_n d(x^* \circ \mathbf{m}) = \int_T g f d(x^* \circ \mathbf{m}) \qquad (6.4.12.10)$$

for each bounded $\mathbf{m}$-measurable scalar function g on T. Then by (6.4.12.10), by Theorem 2.1.5 and Remark 2.2.3 of Chapter 2 and by the fact that $(f_n)_{n=0}^\infty \subset \mathcal{L}_1(\mathbf{m})$, we conclude that $\int_T f_n g d\mathbf{m} \to \int_T f g d\mathbf{m}$ in $\sigma(X, H)$. Hence (b)(iv) holds. Let $A \in \mathcal{B}(T)$ and let B_n, $n \in \mathbb{N} \cup \{0\}$, be as in the proof of Theorem 6.4.8(a). Let $B = \bigcup_{n=0}^\infty B_n$. Then, by (b)(iv) we have $x^*(\int_A f_n d\mathbf{m}) = x^*(\int_{A \cap B} f_n d\mathbf{m}) \to x^*(\int_{A \cap B} f d\mathbf{m}) = x^*(\int_A f d\mathbf{m})$ for $x^* \in H$, since $A \cap B \in \mathcal{B}_c(T)$ so that $\chi_{A \cap B}$ is $\mathcal{B}_c(T)$-measurable and hence $\mathbf{m}$-measurable. Hence (b)(iii) also holds.

(a)(iii) Let B_n, $n \in \mathbb{N} \cup \{0\}$, B and f_0 be as in the proof of (b)(iii) and (b)(iv). Let $\gamma_n, n \in \mathbb{N} \cup \{0\}$, be as in the proof of Theorem 6.4.8(a)(ii). Then by hypothesis, $\lim_n \gamma_n(U) \in X$ in τ for each open Baire set U in T and hence, by Theorem 5.2.21 of Chapter 5 there exists a unique X-valued σ-additive Borel regular measure γ on $\mathcal{B}(T)$ such that

$$\lim_n \int_T g d\gamma_n = \int_T g d\gamma \qquad (6.4.12.11)$$

in τ for each bounded Borel measurable scalar function g on T. Then by Claim 1 in the proof of Theorem 6.4.8, by Theorem 2.1.5(viii) and Remark 2.2.3 of Chapter 2 and by (6.4.12.11) we have

$$\lim_n \int_T g f_n d(x^* \circ \mathbf{m}) = \lim_n x^* \left(\int_T g f_n d\mathbf{m} \right) = \lim_n x^* \left(\int_T g d\gamma_n \right)$$

$$= x^* \left(\int_T g d\gamma \right) \qquad (6.4.12.12)$$

for $x^* \in X^*$. By (6.4.12.10) which holds for $x^* \in X^*$, and by Theorem 2.1.5 and Remark 2.2.3 of Chapter 2 we have

$$\lim_n \int_T g f_n d(x^* \circ \mathbf{m}) = \int_T g f d(x^* \circ \mathbf{m}) = x^* \left(\int_T g d\gamma_0 \right) \qquad (6.4.12.13)$$

and hence by (6.4.12.12) and (6.4.12.13) we have

$$x^* \left(\int_T g d\boldsymbol{\gamma} \right) = x^* \left(\int_T g d\boldsymbol{\gamma}_0 \right)$$

for $x^* \in X^*$. Then by the Hahn-Banach theorem and by Claim 1 in the proof of Theorem 6.4.8 we have

$$\int_T g d\boldsymbol{\gamma} = \int_T g d\boldsymbol{\gamma}_0 = \int_T g f d\mathbf{m} \qquad (6.4.12.14)$$

for each bounded Borel measurable function g on T and hence by Claim 1, by (6.4.12.11) and by (6.4.12.14) we have

$$\lim_n \int_T f_n g d\mathbf{m} = \int_T f g d\mathbf{m} \qquad \text{in } \tau. \qquad (6.4.12.15)$$

If g is a bounded $\mathbf{m}$-measurable scalar function on T, then there exists a bounded Borel measurable function h such that $g = h$ $\mathbf{m}$-a.e. in T and hence by (6.4.12.15), (a)(iv) holds. (a)(iii) is immediate from (a)(iv). □

Remark 6.4.13. Suppose X is a quasicomplete lcHs with topology τ and $\mathbf{m}$: $\mathcal{B}(T) \to X$ is σ-additive and Borel regular. Then, in the light of Theorem 6.3.11, the analogue of Theorem 6.4.8 for $\mathbf{m}$ holds here verbatim. Moreover, if X is a Banach space, then in view of Theorem 6.2.1(iv), Lemma 6.4.9 holds here with f_n vanishing in $\mathbf{m}$-a.e. in $T \backslash \bigcup_1^\infty K_k$ since Lemma 1, no. 6, §5, Chapter IV of [B] holds here for each $N(f_n)$. Then the analogue of Theorem 6.4.12 for $\mathbf{m}$ holds here verbatim if we use Theorem 6.3.10 in place of Theorem 6.3.4. The details are left to the reader.

6.5   Duals of $\mathcal{L}_1(\mathbf{m})$ and $\mathcal{L}_1(\mathbf{n})$

Let X be a Banach space. Suppose $\mathbf{m} : \mathcal{B}(T) \to X$ (resp. $\mathbf{n} : \delta(\mathcal{C}) \to X$) is σ-additive and $\mathcal{B}(T)$-regular (resp. and $\delta(\mathcal{C})$-regular). The present section is devoted to the study of the duals of $\mathcal{L}_1(\mathbf{m})$ and $\mathcal{L}_1(\mathbf{n})$. Also are given vector measure analogues of Theorem 4.1 and Proposition 5.9 of [T]. When $\mathbf{m}$ is defined on a σ-algebra of sets, the dual of $\mathcal{L}_1(\mathbf{m})$ was studied by Okada in [O].

Lemma 6.5.1. *Let X be a Banach space and let $\mathbf{m} : \mathcal{B}(T) \to X$ (resp. $\mathbf{n} : \delta(\mathcal{C}) \to X$) be σ-additive and $\mathcal{B}(T)$-regular (resp. and $\delta(\mathcal{C})$-regular). If $u \in \mathcal{L}_1(\mathbf{m})^*$ (resp. $v \in \mathcal{L}_1(\mathbf{n})^*$), then there exists a unique σ-additive and $\mathcal{B}(T)$-regular (resp. and $\delta(\mathcal{C})$-regular) scalar measure η_u on $\mathcal{B}(T)$ (resp. ζ_v on $\delta(\mathcal{C})$) such that*

$$u(f) = \int_T f d\eta_u \text{ and } \int_T |f| dv(\eta_u, \mathcal{B}(T)) \le ||u|| \mathbf{m}_1^\bullet(f, T) \qquad (6.5.1.1)$$

for $f \in \mathcal{L}_1(\mathbf{m})$ where

$$||u|| = \sup\{|u(f)| : f \in \mathcal{L}_1(\mathbf{m}), \mathbf{m}_1^\bullet(f, T) \le 1\}$$

$$(resp.\ v(f) = \int_T f d\zeta_v \text{ and } \int_T |f| dv(\zeta_v, \delta(\mathcal{C})) \le ||v|| \mathbf{n}_1^\bullet(f, T) \qquad (6.5.1.1')$$

for $f \in \mathcal{L}_1(\mathbf{n})$ where

$$||v|| = \sup\{|v(f)| : f \in \mathcal{L}_1(\mathbf{n}), \mathbf{n}_1^\bullet(f, T) \le 1\}).$$

Proof. Let $u \in \mathcal{L}_1(\mathbf{m})^*$ (resp. $v \in \mathcal{L}_1(\mathbf{n})^*$). Then

$$|u(f)| \le ||u|| \mathbf{m}_1^\bullet(f, T) \qquad (6.5.1.2)$$

for $f \in \mathcal{L}_1(\mathbf{m})$

$$(resp.\ |v(f)| \le ||v|| \mathbf{n}_1^\bullet(f, T) \qquad (6.5.1.2')$$

for $f \in \mathcal{L}_1(\mathbf{n})$).

Let $\varphi \in \mathcal{K}(T)$ (resp. $\varphi \in C_c(T, C)$ with $C \in \mathcal{C}$ – see Notation 5.2.1 of Chapter 5). Then by (6.5.1.2) (resp. by (6.5.1.2')), we have

$$|u(\varphi)| \le ||u|| \mathbf{m}_1^\bullet(\varphi, T) = ||u|| \left(\sup_{|x^*| \le 1} \int_T |\varphi| dv(x^* \circ \mathbf{m}) \right)$$

$$\le ||u|| ||\varphi||_T ||\mathbf{m}||(T)$$

$$(resp.\ |v(\varphi)| \le ||v|| \mathbf{n}_1^\bullet(\varphi, T) \le ||v|| ||\varphi||_T ||\mathbf{n}||(C)).$$

Hence $u|_{\mathcal{K}(T)} \in \mathcal{K}(T)_b^*$ (see pp. 65 and 69 of [P4]). Let $u|_{\mathcal{K}(T)} = \theta_u$ (by Theorem 4.4(i) of [P4], the complex Radon measure μ_{θ_u} induced by θ_u is determined uniquely by u) and let $\eta_u = \mu_{\theta_u}|_{\mathcal{B}(T)}$. Then η_u is σ-additive on $\mathcal{B}(T)$ and is Borel regular by Theorem 5.3 of [P4] and moreover,

$$u(\varphi) = \int_T \varphi d\eta_u = \int_T \varphi d\mu_{\theta_u} = \theta_u(\varphi) \qquad (6.5.1.3)$$

for $\varphi \in C_0(T)$. (Resp. $v|_{\mathcal{K}(T)} \in \mathcal{K}(T)^*$. Let $v|_{\mathcal{K}(T)} = \theta'_v$. By Theorem 4.4(i) of [P4], this is uniquely determined by v and $\zeta_v = \mu_{\theta'_v}|_{\delta(\mathcal{C})}$ is σ-additive on $\delta(\mathcal{C})$ and is $\delta(\mathcal{C})$-regular by Theorem 4.7 of [P3] and

$$v(\varphi) = \int_T \varphi d\zeta_v = \int_T \varphi d\mu_{\theta'_v} = \theta'_v(\varphi) \qquad (6.5.1.3')$$

for $\varphi \in \mathcal{K}(T)$). Thus η_u is unique (resp. ζ_v is unique).

Then by (12) on p. 55 of [B] and by (6.5.1.2) and (6.5.1.3) (resp. and by (6.5.1.2$'$) and (6.5.1.3$'$)) we have

$$
|\theta_u|(|\varphi|) = \sup_{\Psi \in \mathcal{K}(T), |\Psi| \le |\varphi|} |\theta_u(\Psi)| = \sup_{\Psi \in \mathcal{K}(T), |\Psi| \le |\varphi|} |u(\Psi)|
$$
$$
\le ||u|| \sup_{\Psi \in \mathcal{K}(T), |\Psi| \le |\varphi|} \mathbf{m}_1^{\bullet}(\Psi, T) = ||u|| \mathbf{m}_1^{\bullet}(\varphi, T) \qquad (6.5.1.4)
$$

(resp.

$$
|\theta'_v|(|\varphi|) = \sup_{\Psi \in \mathcal{K}(T), |\Psi| \le |\varphi|} |v(\Psi)| \le ||v|| \sup_{\Psi \in \mathcal{K}(T), |\Psi| \le |\varphi|} \mathbf{n}_1^{\bullet}(\Psi, T)
$$
$$
= ||v|| \mathbf{n}_1^{\bullet}(\varphi, T) \qquad (6.5.1.4'))
$$

for $\varphi \in \mathcal{K}(T)$. Then by Theorems 4.7 and 4.11 of [P3] and by the last part of Theorem 3.3 of [P4] (resp. by Theorems 4.7 and 4.11 of [P3]) and by (6.5.1.4) (resp. by (6.5.1.4$'$)) we have

$$
|\theta_u|(|\varphi|) = \int_T |\varphi| d\mu_{|\theta_u|} = \int_T |\varphi| dv(\mu_{\theta_u}, \mathcal{B}(T)) \le ||u|| \mathbf{m}_1^{\bullet}(\varphi, T) \qquad (6.5.1.5)
$$

$$
(\text{resp. } |\theta'_v|(|\varphi|) = \int_T |\varphi| d\mu_{|\theta'_v|} = \int_T |\varphi| dv(\mu_{\theta'_v}, \delta(\mathcal{C})) \le ||v|| \mathbf{n}_1^{\bullet}(\varphi, T) \quad (6.5.1.5'))
$$

for $\varphi \in \mathcal{K}(T)$.

Let $\Im^+$ be the set of all nonnegative lower semicontinuous functions on T.

Claim 1. For $f \in \Im^+ \cap \mathcal{L}_1(\mathbf{m})$ (resp. $f \in \Im^+ \cap \mathcal{L}_1(\mathbf{n})$),

$$
|\theta_u|^*(f) \le ||u|| \mathbf{m}_1^{\bullet}(f, T) \qquad (6.5.1.6)
$$
$$
(\text{resp. } |\theta'_v|^*(f) \le ||v|| \mathbf{n}_1^{\bullet}(f, T) \qquad (6.5.1.6'))
$$

where $|\theta_u|^*$ and $|\theta'_v|^*$ are as in Definition 1, § 1, Chapter IV of [B].

In fact, by the said definition of [B] and by (6.5.1.5) we have

$$
|\theta_u|^*(f) = \sup_{\varphi \in \mathcal{K}(T)^+, \varphi \le f} |\theta_u|(\varphi) \le ||u|| \sup_{\varphi \in \mathcal{K}(T)^+, \varphi \le f} \mathbf{m}_1^{\bullet}(\varphi, T) \le ||u|| \mathbf{m}_1^{\bullet}(f, T)
$$

for $f \in \Im^+ \cap \mathcal{L}_1(\mathbf{m})$. Similarly, by (6.5.1.5$'$), (6.5.1.6$'$) holds.

Claim 2. Let $f \in \mathcal{L}_1(\mathbf{m})$ (resp. $f \in \mathcal{L}_1(\mathbf{n})$). Then

$$
\int_T |f| dv(\eta_u, \mathcal{B}(T)) = \int_T |f| dv(\mu_{\theta_u}, \mathcal{B}(T)) \le ||u|| \mathbf{m}_1^{\bullet}(f, T) \qquad (6.5.1.7)
$$

and hence $f \in \mathcal{L}_1(\eta_u)$

$$
\left(\text{resp. } \int_T |f| dv(\zeta_v, \delta(\mathcal{C})) = \int_T |f| dv(\mu_{\theta'_v}, \delta(\mathcal{C})) \le ||v|| \mathbf{n}_1^{\bullet}(f, T) \right. \qquad (6.5.1.7'))
$$

and hence $f \in \mathcal{L}_1(\zeta_v))$.

In fact, as $\eta_u = \mu_{\theta_u}|_{\mathcal{B}(T)}$ (resp. $\zeta_v = \mu_{\theta'_v}|_{\delta(\mathcal{C})}$) it suffices to prove the claim for μ_{θ_u} (resp. $\mu_{\theta'_v}$). Note that $|f| \in \mathcal{L}_1(\mathbf{m})$ (resp. $\mathcal{L}_1(\mathbf{n})$) by Theorem 2.1.5(vii) and Remark 2.2.3 of Chapter 2 . Given $\epsilon > 0$, by Theorem 5.1.2 (resp. by Theorem 5.1.3) of Chapter 5, there exist functions g and h on T such that $0 \leq g \leq |f| \leq h$ $\mathbf{m}$-a.e. (resp. $\mathbf{n}$-a.e.) in T, g is upper semicontinuous, bounded and $\mathbf{m}$-integrable in T (resp. and $\mathbf{n}$-integrable in T), h is lower semicontinuous and $\mathbf{m}$-integrable in T (resp. and $\mathbf{n}$-integrable in T) and

$$\mathbf{m}_1^\bullet(h - g, T) < \frac{\epsilon}{||u||} \qquad (6.5.1.8)$$

(resp. g and h are $\mathcal{B}_c(T)$-measurable as mentioned in the beginning paragraph of the proof of Theorem 5.1.3 of Chapter 5 and

$$\mathbf{n}_1^\bullet(h - g, T) < \frac{\epsilon}{||v||} \qquad (6.5.1.8')).$$

As h and $-g$ are lower semicontinuous, $h - g \in \mathfrak{S}^+ \cap \mathcal{L}_1(\mathbf{m})$ (resp. $h - g \in \mathfrak{S}^+ \cap \mathcal{L}_1(\mathbf{n})$) by Theorems 3.3 and 3.4, § 3, Chapter III of [MB]. Then by Proposition 1, no.2, § 4, Chapter IV of [B] and by (6.5.1.6) and (6.5.1.8) (resp. and by (6.5.1.6') and (6.5.1.8')) we have

$$0 \leq \int_T (h - g) dv(\mu_{\theta_u}, \mathcal{B}(T)) = |\theta_u|^*(h - g) \leq ||u|| \mathbf{m}_1^\bullet(h - g, T) < \epsilon \qquad (6.5.1.9)$$

$$(\text{resp. } 0 \leq \int_T (h - g) dv(\mu_{\theta'_v}, \delta(\mathcal{C})) = |\theta'_v|^*(h - g) \leq ||v|| \mathbf{n}_1^\bullet(h - g, T) < \epsilon. \qquad (6.5.1.9'))$$

Then, as $h \in \mathfrak{S}^+ \cap \mathcal{L}_1(\mathbf{m})$ (resp. $h \in \mathfrak{S}^+ \cap \mathcal{L}_1(\mathbf{n})$), by Claim 1, by Proposition 1, no. 2, § 4, Chapter IV of [B] and by (6.5.1.9) (resp. and by (6.5.1.9')) we have

$$\int_T |f| dv(\mu_\rho, \mathcal{R}) \leq \int_T h dv(\mu_\rho, \mathcal{R}) = |\rho|^*(h) \leq ||w|| \boldsymbol{\omega}_1^\bullet(h, T)$$
$$\leq ||w|| \{ \boldsymbol{\omega}_1^\bullet(h - g, T) + \boldsymbol{\omega}_1^\bullet(g, T) \} < \epsilon + ||w|| \boldsymbol{\omega}_1^\bullet(g, T)$$
$$\leq \epsilon + ||w|| \boldsymbol{\omega}_1^\bullet(f, T)$$

since $0 \leq g \leq |f|$, where $\mathcal{R} = \mathcal{B}(T)$, $w = u$, $\rho = \theta_u$ and $\boldsymbol{\omega}_1^\bullet(h, T) = \mathbf{m}_1^\bullet(h, T)$ (resp. $\mathcal{R} = \delta(\mathcal{C})$, $w = v$, $\rho = \theta'_v$ and $\boldsymbol{\omega}_1^\bullet(h, T) = \mathbf{n}_1^\bullet(h, T)$). As $\epsilon > 0$ is arbitrary, the claim holds.

Claim 3. For $f \in \mathcal{L}_1(\mathbf{m})$, $u(f) = \int_T f d\eta_u = \int_T f d\mu_{\theta_u}$ and for $f \in \mathcal{L}_1(\mathbf{n})$, $v(f) = \int_T f d\zeta_v = \int_T f d\mu_{\theta'_v}$.

In fact, it suffices to prove the claim for $f \in \mathcal{L}_1(\mathbf{m})$ since the proof for $f \in \mathcal{L}_1(\mathbf{n})$ is similar.

Let $f \in \mathcal{L}_1(\mathbf{m})$. Since $(C_c(T), \mathbf{m}_1^\bullet(\cdot, T))$ is dense in $\mathcal{L}_1(\mathbf{m})$ by Theorem 6.1.10, there exists $(\varphi_n)_1^\infty \subset C_c(T)$ such that

$$\lim_n \mathbf{m}_1^\bullet(f - \varphi_n, T) = 0. \qquad (6.5.1.10)$$

Then by Claim 2 and by (6.5.1.10), we have

$$\lim_n \int_T |\varphi_n - f| dv(\mu_{\theta_u}, \mathcal{B}(T)) \le ||u||(\lim_n \mathbf{m}_1^\bullet(f - \varphi_n, T)) = 0$$

and hence by (6.5.1.3) we have

$$\int_T f d\mu_{\theta_u} = \lim_n \int_T \varphi_n d\mu_{\theta_u} = \lim_n \int_T \varphi_n d\eta_u = \lim_n u(\varphi_n) = u(f) \qquad (6.5.1.11)$$

since $u \in \mathcal{L}_1(\mathbf{m})^*$. By Claim 2, $f \in \mathcal{L}_1(\eta_u)$ and since $\eta_u = \mu_{\theta_u}|_{\mathcal{B}(T)}$, we have

$$\int_T f d\eta_u = \int_T f d\mu_{\theta_u} = u(f)$$

by (6.5.1.11). Hence the claim holds.

Now the lemma is immediate from Claims 2 and 3. $\qquad\qquad\square$

Theorem 6.5.2. *Let X be a Banach space and let $\mathbf{m} : \mathcal{B}(T) \to X$ be σ-additive and $\mathcal{B}(T)$-regular (resp. and let $\mathbf{n} : \delta(\mathcal{C}) \to X$ be σ-additive and $\delta(\mathcal{C})$-regular). Let $Y = \{\eta : \mathcal{B}(T) \to \mathbb{K} : \sigma\text{-additive and } \mathcal{B}(T)\text{-regular such that there exists a constant } M \text{ satisfying } \int_T |f| dv(\eta, \mathcal{B}(T)) \le M\mathbf{m}_1^\bullet(f, T) \text{ for } f \in \mathcal{L}_1(\mathbf{m})\}$ and let $Z = \{\zeta : \delta(\mathcal{C}) \to \mathbb{K} : \sigma\text{-additive and } \delta(\mathcal{C})\text{-regular such that there exists a constant } M \text{ satisfying } \int_T |f| dv(\zeta, \delta(\mathcal{C})) \le M\mathbf{n}_1^\bullet(f, T) \text{ for } f \in \mathcal{L}_1(\mathbf{n})\}$. Let $|||\eta||| = \sup\{|\int_T f d\eta| : f \in \mathcal{L}_1(\mathbf{m}), \mathbf{m}_1^\bullet(f, T) \le 1\}$ for $\eta \in Y$ and let $|||\zeta||| = \sup\{|\int_T f d\zeta| : f \in \mathcal{L}_1(\mathbf{n}), \mathbf{n}_1^\bullet(f, T) \le 1\}$ for $\zeta \in Z$. Then:*

(i) *$\mathcal{L}_1(\mathbf{m})^*$ (resp. $\mathcal{L}_1(\mathbf{n})^*$) is isometrically isomorphic with $(Y, |||\cdot|||)$ (resp. $(Z, |||\cdot|||)$) so that $\mathcal{L}_1(\mathbf{m})^* = Y$ (resp. $\mathcal{L}_1(\mathbf{n})^* = Z$). Consequently, $(Y, |||\cdot|||)$ (resp. $(Z, |||\cdot|||)$) is a Banach space.*

(ii) *The closed unit ball B_Y of Y (resp. B_Z of Z) is given by*

$$A = \{\eta \in Y : \int_T |f| dv(\eta, \mathcal{B}(T)) \le \mathbf{m}_1^\bullet(f, T) \quad \text{for } f \in \mathcal{L}_1(\mathbf{m})\}$$

$$\left(\text{resp.} \quad B = \{\zeta \in Z : \int_T |f| dv(\zeta, \delta(\mathcal{C})) \le \mathbf{n}_1^\bullet(f, T) \text{ for } f \in \mathcal{L}_1(\mathbf{n})\}\right).$$

(iii) *If $B_Y^+ = \{\eta \in B_Y : \eta \ge 0\}$ (resp. $B_Z^+ = \{\zeta \in B_Z : \zeta \ge 0\}$), then*

$$\mathbf{m}_1^\bullet(f, T) = \sup_{\eta \in B_Y^+} \int_T |f| d\eta \quad \text{for } f \in \mathcal{L}_1(\mathbf{m})$$

$$\left(\text{resp. } \mathbf{n}_1^\bullet(f, T) = \sup_{\zeta \in B_Z^+} \int_T |f| d\zeta \quad \text{for } f \in \mathcal{L}_1(\mathbf{n})\right).$$

Proof. Let $\eta \in Y$ and let $u_\eta : \mathcal{L}_1(\mathbf{m}) \to \mathbb{K}$ be given by

$$u_\eta(f) = \int_T f\, d\eta \qquad (6.5.2.1)$$

(resp. let $\zeta \in Z$ and let $v_\zeta : \mathcal{L}_1(\mathbf{n}) \to \mathbb{K}$ be given by

$$v_\zeta(f) = \int_T f\, d\zeta. \qquad (6.5.2.1'))$$

Then by hypothesis, there exists $M > 0$ such that

$$|u_\eta(f)| \le \int_T |f|\, dv(\eta, \mathcal{B}(T)) \le M\mathbf{m}_1^\bullet(f, T) \qquad (6.5.2.2)$$

for $f \in \mathcal{L}_1(\mathbf{m})$

$$(\text{resp. } |v_\zeta(f)| \le \int_T |f|\, dv(\zeta, \delta(\mathcal{C})) \le M\mathbf{n}_1^\bullet(f, T) \qquad (6.5.2.2')$$

for $f \in \mathcal{L}_1(\mathbf{n})$). Hence $u_\eta \in \mathcal{L}_1(\mathbf{m})^*$ (resp. $v_\zeta \in \mathcal{L}_1(\mathbf{n})^*$).

Conversely, let $u \in \mathcal{L}_1(\mathbf{m})^*$ (resp. $v \in \mathcal{L}_1(\mathbf{n})^*$). Then by Lemma 6.5.1, there exists a unique $\eta_u \in Y$ (resp. $\zeta_v \in Z$) such that

$$u(f) = \int_T f\, d\eta_u \text{ with } \int_T |f|\, dv(\eta_u, \mathcal{B}(T)) \le \|u\| \mathbf{m}_1^\bullet(f, T) \text{ for } f \in \mathcal{L}_1(\mathbf{m})$$

$$(\text{resp. } v(f) = \int_T f\, d\zeta_v \text{ with } \int_T |f|\, dv(\zeta_v, \delta(\mathcal{C})) \le \|v\| \mathbf{n}_1^\bullet(f, T) \text{ for } f \in \mathcal{L}_1(\mathbf{n})).$$

Let $\Phi : \mathcal{L}_1(\mathbf{m})^* \to Y$ (resp. $\Psi : \mathcal{L}_1(\mathbf{n})^* \to Z$) be given by $\Phi(u) = \eta_u$ (resp. $\Psi(v) = \zeta_v$) so that

$$u(f) = \int_T f\, d\eta_u, \ f \in \mathcal{L}_1(\mathbf{m}) \qquad \text{by (6.5.1.11)}$$

$$(\text{resp. } v(f) = \int_T f\, d\zeta_v, \ f \in \mathcal{L}_1(\mathbf{n}) \qquad \text{by (6.5.1.11')}).$$

Then by Lemma 6.5.1, Φ (resp. Ψ) is linear since $\alpha\eta_u + \beta\eta_v = \eta_{\alpha u + \beta v}$ on $C_c(T)$ by the Riesz representation theorem and since $C_c(T)$ is dense in $\mathcal{L}_1(\mathbf{m})$ (resp. in $\mathcal{L}_1(\mathbf{n})$) by Theorem 6.1.10. By the uniqueness part of Lemma 6.5.1, Φ and Ψ are injective. To show that Φ (resp. Ψ) is surjective, let $\eta \in Y$ (resp. $\zeta \in Z$). Then u_η (resp. v_ζ) given by $u_\eta(f) = \int_T f\, d\eta$ for $f \in \mathcal{L}_1(\mathbf{m})$ (resp. $v_\zeta(f) = \int_T f\, d\zeta$ for $f \in \mathcal{L}_1(\mathbf{n})$) belongs to $\mathcal{L}_1(\mathbf{m})^*$ (resp. $\mathcal{L}_1(\mathbf{n})^*$) and arguing as in the beginning of the proof of Lemma 6.5.1 one can show that $u_\eta|_{\mathcal{K}(T)} = \theta_{u_\eta} \in \mathcal{K}(T)^*$ (resp. $v_\zeta|_{\mathcal{K}(T)} = \theta'_{v_\zeta} \in \mathcal{K}(T)^*$) and hence $\int_T \varphi\, d\eta = u_\eta(\varphi) = \int_T \varphi\, d\mu_{\theta_{u_\eta}}$ for $\varphi \in C_0(T)$ (resp. $\int_T \varphi\, d\zeta = v_\zeta(\varphi) = \int_T \varphi\, d\mu_{\theta'_{v_\zeta}}$ for $\varphi \in \mathcal{K}(T)$). Then by the uniqueness part of the Riesz representation theorem for $C_0(T)$ (resp. for $B(V)$ for $V \in \mathcal{V}$ since $\delta(\mathcal{C}) = \bigcup_{V \in \mathcal{V}} \mathcal{B}(V)$) we conclude that $\eta = \mu_{\theta_{u_\eta}}$ (resp. $\zeta = \mu_{\theta'_{v_\zeta}}$) so that $u_\eta(\varphi) = \int_T \varphi d\eta = \int_T \varphi d\mu_{\theta_{u_\eta}}$ for $\varphi \in C_c(T)$ (resp. $v_\zeta(\varphi) = \int_T \varphi d\zeta = \int_T \varphi d\mu_{\theta'_{v_\zeta}}$ for $\varphi \in C_c(T)$). Since $(C_c(T), \mathbf{m}_1^{\bullet}(\cdot, T))$ (resp. $C_c(T), \mathbf{n}_1^{\bullet}(\cdot, T))$ is dense in $\mathcal{L}_1(\mathbf{m})$ (resp. $\mathcal{L}_1(\mathbf{n})$) by Theorem 6.1.10, we conclude that $\int_T f d\eta = \int_T f d\mu_{\theta_{u_\eta}}$ for $f \in \mathcal{L}_1(\mathbf{m})$(resp. $\int_T f d\zeta = \int_T f d\mu_{\theta'_{v_\zeta}}$ for $f \in \mathcal{L}_1(\mathbf{n})$) so that $\eta = \Phi(\theta_{u_\eta})$ (resp. $\zeta = \Psi(\theta'_{v_\zeta})$). Hence Φ (resp. Ψ) is bijective. The remaining parts of (i) are immediate.

(ii) If $\eta \in A$, then

$$|||\eta||| = \sup_{f \in \mathcal{L}_1(\mathbf{m}), \mathbf{m}_1^{\bullet}(f,T) \leq 1} \left| \int_T f d\eta \right| \leq \sup_{f \in \mathcal{L}_1(\mathbf{m}), \mathbf{m}_1^{\bullet}(f,T) \leq 1} \int_T |f| dv(\eta, \mathcal{B}(T)) \leq 1$$

by the definition of A and hence $A \subset B_Y$. Similarly, $|||\zeta||| \leq 1$ for $\zeta \in B$ and hence $B \subset B_Z$.

Conversely, let $\eta \in B_Y$. Then $u_\eta(f) = \int_T f d\eta$ for $f \in \mathcal{L}_1(\mathbf{m})$ and

$$||u_\eta|| = \sup_{f \in \mathcal{L}_1(\mathbf{m}), \mathbf{m}_1^{\bullet}(f,T) \leq 1} \left| \int_T f d\eta \right| \leq 1.$$

Then by (6.5.1.1), $\int_T |f| dv(\eta, \mathcal{B}(T)) \leq ||u_\eta|| \mathbf{m}_1^{\bullet}(f,T) \leq \mathbf{m}_1^{\bullet}(f,T)$ for $f \in \mathcal{L}_1(\mathbf{m})$ and hence $|||\eta||| \leq 1$ and thus $\eta \in A$. Therefore, $A = B_Y$. Similarly, $B = B_Z$.

(iii) By (ii) we have

$$\boldsymbol{\omega}_1^{\bullet}(f,T) = \sup_{\eta \in D} | \int_T f d\eta | \leq \sup_{\eta \in D} \int_T |f| dv(\eta, \mathcal{R})$$

$$\leq \sup_{\eta \in D^+} \int_T |f| d\eta \leq \boldsymbol{\omega}_1^{\bullet}(f,T)$$

where $\boldsymbol{\omega} = \mathbf{m}$, $f \in \mathcal{L}_1(\mathbf{m})$ and $D = B_Y$ (resp. $\boldsymbol{\omega} = \mathbf{n}$, $f \in \mathcal{L}_1(\mathbf{n})$ and $D = B_Z$). Hence (iii) holds.   □

Remark 6.5.3. In the proof of Proposition 4.2 of [T], it is claimed that $\alpha^{\bullet} \leq M\mu^{\bullet}$ whenever $|\alpha(\varphi)| \leq M\mu^{\bullet}(|\varphi|)$. The proof of this statement is based on the continuity of L on $\mathcal{K}(T)$.

Using Theorem 6.5.2, we now give the vector measure analogues of Theorem 4.1 of [T].

Theorem 6.5.4. *Let X be a Banach space (resp. a quasicomplete lcHs). Let $\mathbf{m}$: $\mathcal{B}(T) \to X$ (resp. $\mathbf{n} : \delta(\mathcal{C}) \to X$) be σ-additive and $\mathcal{B}(T)$-regular (resp. and $\delta(\mathcal{C})$-regular). Let $(f_\alpha)_{\alpha \in (D, \geq)}$ be an increasing net of nonnegative lower semicontinuous functions with $f = \sup_\alpha f_\alpha$ also being $\mathbf{m}$-integrable (resp. $\mathbf{n}$-integrable) in T. Then $f_\alpha \to f$ in $\mathcal{L}_1(\mathbf{m})$ (resp. $\mathcal{L}_1(\mathbf{n})$) and consequently, $\lim_\alpha \int_T f_\alpha d\mathbf{m} = \int_T f d\mathbf{m}$ (resp. $\lim_\alpha \int_T f_\alpha d\mathbf{n} = \int_T f d\mathbf{n}$) in X.*

Proof. Case 1. X is a Banach space.

Then by Alaoglu's theorem, B_Y and B_Y^+ (resp. B_Z and B_Z^+) are compact in $\sigma(B_Y, \mathcal{L}_1(\mathbf{m}))$ (resp. in $\sigma(B_Z, \mathcal{L}_1(\mathbf{n}))$), where Y, Z, B_Y, B_Y^+, B_Z and B_Z^+ are as in Theorem 6.5.2. Then, for $\eta \in B_Y^+$ (resp. $\zeta \in B_Z^+$), by Theorem 5.3 of [P4] there exists $\theta \in \mathcal{K}_b(T)^*$ with $\theta \geq 0$ such that $\mu_\theta|_{\mathcal{B}(T)} = \eta$ (resp. by Theorem 4.4 of [P4] there exists $\theta' \in \mathcal{K}(T)^*$ with $\theta' \geq 0$ such that $\mu_{\theta'}|_{\delta(\mathcal{C})} = \zeta$). Then by Theorem 1, no. 1, § 1, Chapter IV of [B],

$$\int_T f d\eta = \sup_\alpha \int_T f_\alpha d\eta = \lim_\alpha \int_T f_\alpha d\eta$$

$$\left(\text{resp. } \int_T f d\zeta = \sup_\alpha \int_T f_\alpha d\zeta = \lim_\alpha \int_T f_\alpha d\zeta\right).$$

As f and f_α, $\alpha \in (D, \geq)$, belong to $\mathcal{L}_1(\mathbf{m})$ (resp. $\mathcal{L}_1(\mathbf{n})$), the mappings $\eta \to \int_T f d\eta$ and $\eta \to \int_T f_\alpha d\eta$ (resp. $\zeta \to \int_T f d\zeta$ and $\zeta \to \int_T f_\alpha d\zeta$) are continuous in $\sigma(B_Y^+, \mathcal{L}_1(\mathbf{m}))$ (resp. in $\sigma(B_Z^+, \mathcal{L}_1(\mathbf{n}))$). Consequently, by Dini's lemma, the limit is uniform with respect to $\eta \in B_Y^+$ (resp. $\zeta \in B_Z^+$). Then by (iii) of Theorem 6.5.2 we have

$$\lim_\alpha \mathbf{m}_1^\bullet(f - f_\alpha, T) = \lim_\alpha \sup_{\eta \in B_Y^+} \int_T |f - f_\alpha| d\eta = 0$$

and hence by (3.1.3.1) of Chapter 3 we have $\lim_\alpha \int_T f_\alpha d\mathbf{m} = \int_T f d\mathbf{m}$. Similarly, the results for $\mathbf{n}$ are proved.

Case 2. X is a quasicomplete lcHs.

By Theorem 4.5.13(i) of Chapter 4, $\mathcal{L}_1(\mathbf{m}) = \bigcap_{q \in \Gamma} \mathcal{L}_1(\mathbf{m}_q)$. Then by case 1, $(\mathbf{m}_q)_1^\bullet(f_\alpha - f, T) \to 0$ for $q \in \Gamma$ and hence $f_\alpha \to f$ in $\mathcal{L}_1(\mathbf{m})$. Consequently, by (4.3.2.1) and Remark 4.1.5 of Chapter 4, $\int_T f_\alpha d\mathbf{m} \to \int_T f d\mathbf{m}$ as $\alpha \to \infty$. Similarly, the results for $\mathbf{n}$ are proved. $\square$

Lemma 6.5.5. *Let X be a Banach space. Let $\mathbf{m} : \mathcal{B}(T) \to X$ (resp. $\mathbf{n} : \delta(\mathcal{C}) \to X$) be σ-additive and Borel regular (resp. and $\delta(\mathcal{C})$-regular). Let $\mathcal{R} = \mathcal{B}(T)$ or $\delta(\mathcal{C})$ and let $\boldsymbol{\omega} = \mathbf{m}$ when $\mathcal{R} = \mathcal{B}(T)$ and $\boldsymbol{\omega} = \mathbf{n}$ when $\mathcal{R} = \delta(\mathcal{C})$. If $\eta \in \mathcal{L}_1(\boldsymbol{\omega})^*$ (so that η satisfies the conditions in Theorem 6.5.2), then, for each bounded Borel function g on T, $g \cdot \eta$ given by*

$$(g \cdot \eta)(f) = \int_T f g d\eta \qquad \text{for } f \in \mathcal{L}_1(\boldsymbol{\omega})$$

is well defined and belongs to $\mathcal{L}_1(\boldsymbol{\omega})^*$.

Proof. Let $f \in \mathcal{L}_1(\mathbf{m})$. Then there exist A, N, M such that $N(f) = A \cup N$, $A \in \mathcal{B}(T)$, $N \subset M \in \mathcal{B}(T)$ with $||\mathbf{m}||(M) = 0$. Let $B = A \cup M$. Then $N(f) \subset B \in \mathcal{B}(T)$. If $f \in \mathcal{L}_1(\mathbf{n})$, then by Lemma 6.4.7 there exists $B \in \mathcal{B}_c(T)$ such that $N(f) \subset B$. By hypothesis, g is a bounded Borel function (resp. $g\chi_B$ is a bounded σ-Borel function when $f \in \mathcal{L}_1(\mathbf{n})$) and hence by Theorem 2.1.5(vi) and Remark 2.2.3 of Chapter 2, $fg\chi_B \in \mathcal{L}_1(\boldsymbol{\omega})$ and hence

$$(g \cdot \eta)(f) = \int_T fg\,d\eta = \int_T fg\chi_B\,d\eta$$

is well defined and $g \cdot \eta$ is a linear functional on $\mathcal{L}_1(\boldsymbol{\omega})$. Let $u_\eta(f) = \int_T f\,d\eta$ for $f \in \mathcal{L}_1(\boldsymbol{\omega})$ and $\eta \in \mathcal{L}_1(\boldsymbol{\omega})^*$ with η satisfying the conditions of Theorem 6.5.2. Then by Lemma 6.5.1 we have

$$|(g \cdot \eta)(f)| \leq ||g\chi_B||_T \int_T |f|\,dv(\eta, \mathcal{R}) \leq ||g||_T ||u_\eta||\boldsymbol{\omega}_1^\bullet(f, T).$$

Hence $g \cdot \eta \in \mathcal{L}_1(\boldsymbol{\omega})^*$. $\qquad\square$

The following lemmas, which are the same as or similar to Lemmas 6.4.9, 6.4.10 and 6.4.11, are needed to prove the vector measure analogue of Proposition 5.9 of [T].

Lemma 6.5.6. *Let $\mathcal{R}$ and $\boldsymbol{\omega}$ be as in Lemma 6.5.5. If $(f_n)_1^\infty \subset \mathcal{L}_1(\boldsymbol{\omega})$, there exists a sequence $(K_n)_1^\infty \subset \mathcal{C}$ such that each f_n vanishes $\boldsymbol{\omega}$-a.e. in $T\backslash\bigcup_1^\infty K_k$.*

Proof. If $\mathcal{R} = \delta(\mathcal{C})$ and $\boldsymbol{\omega} = \mathbf{n}$, then the result holds by Lemma 6.4.9. If $\mathcal{R} = \mathcal{B}(T)$ and $\boldsymbol{\omega} = \mathbf{m}$, then by the proof of Lemma 6.5.5 there exist $(B_n)_1^\infty \subset \mathcal{B}(T)$ such that $N(f_n) \subset B_n$ for each n. Let $B = \bigcup_1^\infty B_n$. Then $\bigcup_1^\infty N(f_n) \subset B \in \mathcal{B}(T)$. Then by Theorem 6.2.1(iv), there exists a sequence $(K_n)_1^\infty \subset \mathcal{C}$ such that $\bigcup_1^\infty K_n \subset B$ and $||\mathbf{m}||(B\backslash\bigcup_1^\infty K_n) = 0$. Since $f_n|_{T\backslash B} = 0$ for each n, it follows that $f_n = 0$ $\mathbf{m}$-a.e. in $T\backslash\bigcup_1^\infty K_n$. $\qquad\square$

Lemma 6.5.7. *Let X, $\mathcal{R}$,   and $\boldsymbol{\omega}$ be as in Lemma 6.5.5 and let $H = \{x^* \in X^* : |x^*| \leq 1\}$. Given a sequence $(K_n)_1^\infty \subset \mathcal{C}$, there exists a sequence $(x_n^*)_1^\infty \subset H$ such that every set $A \in \sigma(\mathcal{R})$ with $A \subset \bigcup_1^\infty K_n$ is $\boldsymbol{\omega}$-null whenever A is $(x_n^* \circ \boldsymbol{\omega})$-null for each $n \in \mathbb{N}$.*

Proof. For each n, $A\cap K_n \in \delta(\mathcal{C})$ by Lemma 5.2.2 of Chapter 5 whenever $A \in \mathcal{B}(T)$ or $A \in \mathcal{B}_c(T)$. Hence the proof of Lemma 6.4.10 holds here verbatim in both cases of $\boldsymbol{\omega}$. Hence the lemma holds. $\qquad\square$

Lemma 6.5.8. *Let $\mathcal{R} = \mathcal{B}(T)$ or $\delta(\mathcal{C})$. Let $\mu_k : \mathcal{R} \to \mathbb{K}$, $k \in \mathbb{N}$, be σ-additive and $\mathcal{R}$-regular and let $(f_n)_1^\infty \subset \bigcap_{k=1}^\infty \mathcal{L}_1(\mu_k)$. Suppose $(f_n)_1^\infty$ converges weakly to some $h_k \in \mathcal{L}_1(\mu_k)$ for each k. Then there exists a sequence $(g_n)_1^\infty$ such that g_n is a convex combination of $(f_k)_{k\geq n}$ and such that $(g_n)_1^\infty$ converges in mean in $\mathcal{L}_1(\mu_k)$ and also converges pointwise μ_k-a.e. in T for each $k \in \mathbb{N}$.*

Proof. The proof of Lemma 6.4.11 holds here verbatim. $\qquad\square$

The following theorem gives the vector measure analogue of Proposition 5.9 of [T].

Theorem 6.5.9. *Let X be a Banach space with $c_0 \not\subset X$ and let $\mathbf{m} : \mathcal{B}(T) \to X$ be σ-additive and Borel regular (resp. $\mathbf{n} : \delta(\mathcal{C}) \to X$ be σ-additive and $\delta(\mathcal{C})$-regular). Then $L_1(\mathbf{m})$ (resp. $L_1(\mathbf{n})$) is a weakly sequentially complete Banach space.*

Proof. In the light of Theorem 3.2.8 and Notation 3.3.6 of Chapter 3, $L_1(\mathbf{m})$ (resp. $L_1(\mathbf{n})$) is a Banach space. To show that these spaces are weakly sequentially complete, let $(f_n)_1^\infty$ be weakly Cauchy in $\mathcal{L}_1(\boldsymbol{\omega})$, where $\boldsymbol{\omega}$ and its domain $\mathcal{R}$ are as in Lemma 6.5.5. By Lemma 6.5.6, there exists $(K_n)_1^\infty \subset \mathcal{C}$ such that each f_n vanishes $\boldsymbol{\omega}$-a.e. in $T \backslash \bigcup_1^\infty K_k$. Let $(x_n^*)_1^\infty \subset H = \{x^* \in X^* : |x^*| \le 1\}$ be chosen so as to satisfy the property mentioned in Lemma 6.5.7. Let $\mu_n = x_n^* \circ \boldsymbol{\omega}$, $n \in \mathbb{N}$. Then by Theorem 2.1.5(viii) and Remark 2.2.3 of Chapter 2, $f_n \in \bigcap_{k=1}^\infty L_1(\mu_k)$ for each n.

Let $\eta \in \mathcal{L}_1(\boldsymbol{\omega})^*$ (satisfying the condition in Theorem 6.5.2). Then by Lemma 6.5.5, $g \cdot \eta \in \mathcal{L}_1(\boldsymbol{\omega})^*$ for each bounded Borel function g. As $(f_n)_1^\infty$ is weakly Cauchy in $\mathcal{L}_1(\boldsymbol{\omega})$, and as $g \cdot \eta \in \mathcal{L}_1(\boldsymbol{\omega})^*$, $(\int_T f_n g d\eta)_1^\infty$ is Cauchy in $\mathbb{K}$ and hence $(f_n)_1^\infty$ is weakly Cauchy in $\mathcal{L}_1(\eta)$ for each $\eta \in \mathcal{L}_1(\boldsymbol{\omega})^*$. As $\mathcal{L}_1(\eta)$ is weakly sequentially complete, there exists $f_\eta \in \mathcal{L}_1(\eta)$ such that $f_n \to f_\eta$ weakly in $\mathcal{L}_1(\eta)$. Thus $(f_n)_1^\infty$ is weakly convergent to f_{η_k} in $\mathcal{L}_1(\mu_k)$ for each $k \in \mathbb{N}$. Then by Lemma 6.5.8 there exists a sequence $(g_n)_1^\infty$ such that each g_n is of the form

$$g_n = \sum_{i=n}^{N(n)} \alpha_i^{(n)} f_i, \ \alpha_i^{(n)} \ge 0, \quad \text{and} \quad \sum_{i=n}^{N(n)} \alpha_i^{(n)} = 1$$

and such that $(g_n)_1^\infty$ converges in mean in $\mathcal{L}_1(\mu_k)$ and also converges μ_k-a.e. in T for each $k \in \mathbb{N}$. Then by Lemma 6.5.7, $(g_n)_1^\infty$ converges $\boldsymbol{\omega}$-a.e. in T. Let f be the $\boldsymbol{\omega}$-a.e. pointwise limit of $(g_n)_1^\infty$.

As $f_n \to f_\eta$ weakly in $\mathcal{L}_1(\eta)$, $g_n \to f_\eta$ weakly in $\mathcal{L}_1(\eta)$ for each $\eta \in \mathcal{L}_1(\boldsymbol{\omega})^*$. Then by Theorems 5.3 and 4.4 of [P4] and by Theorem 6.4.6, $f = f_\eta$ for each $\eta \in \mathcal{L}_1(\boldsymbol{\omega})^*$. Now, for $x^* \in X^*$, $\eta = x^* \circ \boldsymbol{\omega}$ belongs to $\mathcal{L}_1(\boldsymbol{\omega})^*$ by Theorem 6.5.2 as $(x^* \circ \boldsymbol{\omega})$ is σ-additive and $\mathcal{R}$-regular and $\int_T |f| dv(x^* \circ \boldsymbol{\omega}, \mathcal{R}) \le ||u_{x^*}|| \boldsymbol{\omega}_1^\bullet(f, T)$, where $u_{x^*}(f) = \int_T f d(x^* \circ \boldsymbol{\omega})$ for $f \in \mathcal{L}_1(\boldsymbol{\omega})$ (see Lemma 6.5.1). Hence $f \in \mathcal{L}_1(x^* \circ \boldsymbol{\omega})$ for each $x^* \in X^*$. As $c_0 \not\subset X$, by the last part of Theorem 3.1.8, by Definition 3.1.9 and by Notation 3.3.7 of Chapter 3, $f \in \mathcal{L}_1(\boldsymbol{\omega})$. Hence $L_1(\boldsymbol{\omega})$ is a weakly sequentially complete Banach space. $\square$

We close this chapter by considering the following example.

Example 6.5.10. Let $(x_i)_{i \in I}$ be a family of nonzero elements in a Banach space X. Let I_F be the δ-ring of all finite subsets of I. For $A \in I_F$, let $\mathbf{m}(A) = \sum_{i \in A} x_i \in X$. Clearly, $\mathbf{m}$ is σ-additive on I_F. With respect to the discrete topology on I, clearly $\mathbf{m}$ is $\delta(\mathcal{C})$-regular, where $\delta(\mathcal{C}) = I_F$. Hence a function $f : I \to \mathbb{K}$ is $x^* \circ \mathbf{m}$-integrable in I for $x^* \in X^*$ if and only if $|\sum_{i \in J} f(i)(x^* \circ \mathbf{m})(i)| = |\sum_{i \in J} f(i) x^*(x_i)| < \infty$ for each subset J of I. Then by the Orlicz-Pettis theorem if and only if $(f(i)x_i)_{i \in I}$ is summable in X. Then by Theorem 6.3.4, f is $\mathbf{m}$-integrable in I if and only if $(f(i)x_i)_{i \in I}$ is summable in X. (Compare with Example 3.12 of [T].)

Chapter 7

Complements to the Thomas Theory of Vectorial Radon Integration

7.1 Integration of complex functions with respect to a Radon operator

Thomas developed in §1 of [T] a theory of vectorial Radon integration of real functions with respect to a Radon operator u on $\mathcal{K}(T, \mathbb{R})$ (see Notation 5.3.1 and Definition 5.3.3 of Chapter 5). In this section we indicate briefly how the results in §1 of [T] can be extended to complex functions in $\mathcal{K}(T)$.

In this section we extend Definition 1.1 of Thomas [T] to Radon operators on $\mathcal{K}(T)$ with values in a normed space X over $\mathbb{C}$. Using the definitions and results given in the earlier part of this section, we study the vectorial Radon integration when the Radon operator assumes values in an lcHs over $\mathbb{C}$.

Definition 7.1.1. Let $u : \mathcal{K}(T) \to X$ be a Radon operator in the sense of Definition 5.3.3, where X is a normed space over $\mathbb{C}$. We define

$$u^{\bullet}(f) = \sup_{|\varphi| \leq f, \varphi \in \mathcal{K}(T)} |u(\varphi)|$$

for $f \in \mathcal{I}^{+}$, where $\mathcal{I}^{+}$ is the set of all nonnegative lower semicontinuous functions on T. When $f : T \to [0, \infty]$ has compact support we define

$$u^{\bullet}(f) = \inf_{f \leq g, g \in \mathcal{I}^{+}} u^{\bullet}(g)$$

and when $f : T \to [0, \infty]$ is arbitrary, we define

$$u^\bullet(f) = \sup_{h \le f} u^\bullet(h)$$

where $h : T \to [0, \infty]$ has compact support. This definition is similar to that in §1 in Chapter V of [B].

$u^\bullet(f)$ is called the semivariation of f with respect to u. For $A \subset T$, we define $u^\bullet(A) = u^\bullet(\chi_A)$. If $u^\bullet(A) = 0$, we say that A is u-null and use the expression u-almost everywhere (briefly, u-a.e.) correspondingly. A function $f : T \to [0, \infty]$ is said to be u-null if $u^\bullet(f) = 0$.

It is easy to verify that the definition is consistent.

Thomas [T] uses the terminology of Radon measure u instead of our terminology of Radon operator u.

Unless otherwise stated, X will denote a normed space over $\mathbb{C}$ and $u : \mathcal{K}(T) \to X$ will denote a Radon operator.

Proposition 7.1.2. *For $f \in \mathcal{I}^+$,*

$$u^\bullet(f) = \sup_{0 \le \varphi \le f, \varphi \in \mathcal{K}(T)} u^\bullet(\varphi).$$

Proof. By Definition 7.1.1 we have

$$
\begin{aligned}
u^\bullet(f) &= \sup_{|\varphi| \le f, \varphi \in \mathcal{K}(T)} |u(\varphi)| \\
&= \sup_{\Psi \in \mathcal{K}(T), |\Psi| \le f} \sup_{|\varphi| \le |\Psi|, \varphi \in \mathcal{K}(T)} |u(\varphi)| \\
&= \sup_{\Psi \in \mathcal{K}(T), |\Psi| \le f} u^\bullet(|\Psi|) = \sup_{0 \le \varphi \le f, \varphi \in \mathcal{K}(T)} u^\bullet(\varphi).
\end{aligned}
$$

Hence the proposition holds. □

We recall the following definition from [B].

Definition 7.1.3. Each element $u \in \mathcal{K}(T)^*$ is called a complex Radon measure and is sometimes identified with the complex measure μ_u induced by u in the sense of Definition 4.3 of [P3]. $|u|$ is the positive linear functional in $\mathcal{K}(T)^*$ given by (12) on p. 55 of [B] and $|u|^*(f)$ for $f \in \mathcal{I}^+$ is given by Definition 1 on p. 107 of Chapter IV of [B].

Proposition 7.1.4. *If $u \in \mathcal{K}(T)^*$ and $f \in \mathcal{I}^+$, then*

$$u^\bullet(f) = \sup_{0 \le \Psi \le f, \Psi \in \mathcal{K}(T)} |u|(\Psi) = |u|^\bullet(f) = |u|^*(f).$$

Consequently,

$$u^\bullet(f) = |u|^\bullet(f)$$

for $f : T \to [0, \infty]$.

Proof. For $f \in \mathcal{I}^+$, by Definition 7.1.1 we have

$$u^\bullet(f) = \sup_{|\varphi|\leq f, \varphi\in\mathcal{K}(T)} |u(\varphi)|$$

$$= \sup_{0\leq\Psi\leq f, \Psi\in\mathcal{K}(T)} \sup_{|\varphi|\leq\Psi, \varphi\in\mathcal{K}(T)} |u(\varphi)|$$

$$= \sup_{0\leq\Psi\leq f, \Psi\in\mathcal{K}(T)} |u|(\Psi) \qquad (7.1.4.1)$$

by (12) on p. 55 of Chapter III of [B].

Therefore, by (7.1.4.1),

$$|u|^\bullet(f) = \sup_{|\varphi|\leq f, \varphi\in\mathcal{K}(T)} ||u|(\varphi)|$$

$$= \sup_{0\leq\Psi\leq f, \Psi\in\mathcal{K}(T)} |u|(\Psi) = u^\bullet(f).$$

Moreover, by the definition on p. 107 of Chapter IV of [B]

$$|u|^\bullet(f) = \sup_{0\leq\Psi\leq f, \Psi\in\mathcal{K}(T)} |u|(\Psi) = |u|^*(f).$$

Now the last part is evident from Definition 7.1.1. □

Definition 7.1.5. $\mathcal{F}^0(u) = \{f : T \to \mathbb{K}, u^\bullet(|f|) < \infty\}$ and we define $u^\bullet(f) = u^\bullet(|f|)$ if $f : T \to \mathbb{K}$.

Clearly, $u^\bullet$ is a seminorm on $\mathcal{F}^0(u)$ and hence $\mathcal{F}^0(u)$ is a seminormed space with respect to $u^\bullet(\cdot)$.

The complex versions of Proposition 1.3 and of Lemmas 1.4 and 1.5 of [T] hold and consequently, we have the following definition.

Definition 7.1.6. The space $\mathcal{L}_1(u)$ of u-integrable functions is the closure of $\mathcal{K}(T)$ in the space $\mathcal{F}^0(u)$. Thus, a complex function f belongs to $\mathcal{L}_1(u)$, if given $\epsilon > 0$, there exists $\varphi \in \mathcal{K}(T)$ such that $u^\bullet(|\varphi - f|) < \epsilon$.

Remark 7.1.7. A complex function f on T belongs to $\mathcal{L}_1(u)$ if and only if, given $\epsilon > 0$, there exists $g \in \mathcal{L}_1(u)$ such that $u^\bullet(|f - g|) < \epsilon$. Also note that if the complex function $f = g$ u-a.e. in T and if $g \in \mathcal{L}_1(u)$, then $f \in \mathcal{L}_1(u)$.

Convention 7.1.8. Let $f : T \to [0, \infty]$. Then f is said to be u-integrable if there exists a complex function $g \in \mathcal{L}_1(u)$ such that $f = g$ u-a.e. in T.

The complex analogues of Proposition 1.7, Theorem 1.8 and Remark following it in [T] hold.

Definition 7.1.9. As $\mathcal{K}(T)$ is dense in $\mathcal{L}_1(u)$ with respect to $u^\bullet$, the continuous linear extension of u to $\mathcal{L}_1(u)$ with values in $\widetilde{X}$, the completion of X, is denoted by $\int du$. Thus, if $f \in \mathcal{L}_1(u)$, then $\int f du \in \tilde{X}$.

Then the complex analogues of **1.10**, of Theorem 1.11 and of **1.12** of [T] hold.

The following result is the complex analogue of Proposition on p. 70 of [T].

Proposition 7.1.10. *Let X be a normed space and $u : \mathcal{K}(T) \to X$ be a Radon operator. Let H be a norm determining set in X^* so that*

$$|x| = \sup_{x^* \in H} |\langle x, x^* \rangle| \text{ for } x \in X.$$

Then

$$u^{\bullet}(|f|) = \sup_{x^* \in H} |u_{x^*}|^{\bullet}(|f|) \tag{7.1.10.1}$$

for $f \in \mathcal{L}_1(u)$ or for $f \in \mathcal{I}^+$, where $u_{x^*} = x^* \circ u$.

Proof. By Lemma 5.2.13, $H \subset \{x^* \in X^* : |x^*| \leq 1\}$. In view of the complex version of Lemma 1.4 of [T], the proof of **1.13** of [T] as given in [T] holds for complex functions too and hence the proposition holds. ☐

Lemma 7.1.11. *For $\mu \in \mathcal{K}(T)^*$,*

$$\mu^{\bullet}(|f|) = |\mu|(|f|)) = \sup_{|\varphi| \leq 1, \varphi \in \mathcal{K}(T)} \left| \int_T \varphi f d\mu \right|$$

for $f \in \mathcal{L}_1(\mu)$ where $|\mu|$ is given by (12) on p. 55 of Chapter III of [B].

Proof. Let $\nu_f(\varphi) = \int_T \varphi f d\mu$ for $f \in \mathcal{L}_1(\mu)$ and $\varphi \in \mathcal{K}(T)$. Then

$$|\nu_f(\varphi)| \leq ||\varphi||_T \int_T |f| d|\mu|$$

so that $\nu_f \in \mathcal{K}(T)^*_b$. Hence by Theorem 3.3 of [P4] we have

$$|\nu_f|(T) = v(\mu_{\nu_f}|_{\mathcal{B}(T)}, \mathcal{B}(T))(T) = ||\nu_f|| = \sup_{|\varphi| \leq 1, \varphi \in \mathcal{K}(T)} |\nu_f(\varphi)|$$

$$= \sup_{|\varphi| \leq 1, \varphi \in \mathcal{K}(T)} \left| \int_T \varphi f d\mu \right|. \tag{7.1.11.1}$$

As noted in the beginning of the proof of Theorem 6.4.6,

$$|\nu_f|(T) = \int_T |f| d|\mu|.$$

Then by (7.1.11.1) and by Proposition 7.1.4 we have

$$|\nu_f|(T) = \int_T |f| d|\mu| = \sup_{|\varphi| \leq 1, \varphi \in \mathcal{K}(T)} \left| \int_T \varphi f d\mu \right| = |\mu|(|f|) = \mu^{\bullet}(|f|)$$

for $f \in \mathcal{L}_1(\mu)$. ☐

Theorem 7.1.12. *For $f \in \mathcal{L}_1(u)$,*

$$u^{\bullet}(|f|) = \sup_{|\varphi| \leq 1, \varphi \in \mathcal{K}(T)} \left| \int_T \varphi f \, du \right|.$$

Proof. Let H be a norm determining set in X^*. For $x^* \in H$, by Lemma 7.1.11 we have

$$u^{\bullet}_{x^*}(|f|) = \sup_{\varphi \in \mathcal{K}(T), |\varphi| \leq 1} \left| \int_T \varphi f \, du_{x^*} \right|$$

for $f \in \mathcal{L}_1(u)$. Therefore, by Propositions 7.1.10 and 7.1.4 we have

$$u^{\bullet}(|f|) = \sup_{x^* \in H} u^{\bullet}_{x^*}(|f|) = \sup_{x^* \in H} \sup_{\varphi \in \mathcal{K}(T), |\varphi| \leq 1} \left| \int_T \varphi f \, du_{x^*} \right|$$

$$= \sup_{\varphi \in \mathcal{K}(T), |\varphi| \leq 1} \sup_{x^* \in H} \left| \int_T \varphi f \, du_{x^*} \right|$$

$$= \sup_{\varphi \in \mathcal{K}(T), |\varphi| \leq 1} \left| \int_T \varphi f \, du \right|. \qquad \square$$

Definition 7.1.13. Let $u : \mathcal{K}(T) \to X$ be a Radon operator where X is a normed space. If Y is a topological space and $f : T \to Y$, then f is said to be u-measurable if, for every compact $K \subset T$ and $\epsilon > 0$, there exists a compact $K_1 \subset K$ such that $u^{\bullet}(K \backslash K_1) < \epsilon$ and $f|_{K_1}$ is continuous.

Replacing $\|\mathbf{m}\|$ by $u^{\bullet}$ and arguing as in the proof of Theorem 6.2.4, we obtain the following theorem.

Theorem 7.1.14. *Let u, X, f and Y be as in Definition 7.1.13. Then f is u-measurable if and only if, given $K \in \mathcal{C}$, there exist a u-null set $N \subset K$ and a countable disjoint family $(K_i)_1^{\infty} \subset \mathcal{C}$ such that $K \backslash N = \bigcup_1^{\infty} K_i$ and $f|_{K_i}$ is continuous for each $i \in \mathbb{N}$.*

The set of u-measurable complex functions is evidently stable under the usual algebraic operations and under composition with a continuous function. Every continuous function on T is clearly u-measurable.

The proof given in Appendix III of [T] holds good for complex normed spaces too and hence Lemma 1.19 of [T] holds for complex spaces too. The proofs of Propositions 1.20 and 1.21, of Remark on p. 74, of Theorem 1.22 and of Lemma 1.23 of [T] hold for complex functions too. Lemmas 1.24, 1.25 and 1.25 bis of [T] hold for the absolute value of complex functions too.

Proposition 7.1.15. *Let X be a normed space and let H be a norm determining set in X^* so that $|x| = \sup_{x^* \in H} |\langle x, x^* \rangle|$ for $x \in X$. Then for every u-measurable bounded positive function f,*

$$u^{\bullet}(f) = \sup_{x^* \in H} |u_{x^*}|^{\bullet}(f).$$

Proof. By the complex version of Lemma 1.23 of [T], $f\chi_K \in \mathcal{L}_1(u)$ for $K \in \mathcal{C}$. Then by Proposition 7.1.10 we have

$$u^\bullet(f) = \sup_{K\in\mathcal{C}} u^\bullet(f\chi_K) = \sup_{K\in\mathcal{C}} \sup_{x^*\in H} |u_{x^*}|^\bullet(f\chi_K)$$
$$= \sup_{x^*\in H} \sup_{K\in\mathcal{C}} |u_{x^*}|^\bullet(f\chi_K) = \sup_{x^*\in H} |u_{x^*}|^\bullet(f). \qquad \square$$

Corollary 7.1.16. *Under the hypothesis of Proposition 7.1.15, for a u-measurable set A,*

$$u^\bullet(A) = \sup_{x^*\in H} |u_{x^*}|^\bullet(A).$$

Corollary 7.1.17. *If f is locally u-integrable (i.e., if φf is u-integrable for each $\varphi \in \mathcal{K}(T)$), then*

$$u^\bullet(f) = \sup_{x^*\in H} |u_{x^*}|^\bullet(f)$$

where H is a norm determining set in X^*.

Proof. By Urysohn's lemma, for each $K \in \mathcal{C}$, there exists $\varphi_K \in \mathcal{K}(T)$ such that $\chi_K \le \varphi_K \le 1$. Then

$$u^\bullet(f) = \sup_{K\in\mathcal{C}} u^\bullet(|f|\chi_K) \le \sup_{K\in\mathcal{C}} u^\bullet(|f|\varphi_K) \le \sup_{0\le\varphi\le1,\varphi\in\mathcal{K}(T)} u^\bullet(|f|\varphi)$$
$$\le u^\bullet(|f|) = u^\bullet(f) \qquad (7.1.17.1)$$

and hence by Proposition 7.1.15, by Definition 7.1.5 and by (7.1.17.1) we have

$$u^\bullet(f) = \sup_{0\le\varphi\le1,\varphi\in\mathcal{K}(T)} u^\bullet(f\varphi) = \sup_{x^*\in H} \sup_{0\le\varphi\le1,\varphi\in\mathcal{K}(T)} |u_{x^*}|^\bullet(\varphi f)$$
$$= \sup_{x^*\in H} |u_{x^*}|^\bullet(f).$$

Thus the corollary holds. $\qquad \square$

Definition 7.1.18. Let u be a Radon operator with values in an lcHs X over $\mathbb{C}$. Let q be a continuous seminorm on X. We denote $q(x)$ by $|x|_q$. The semivariation of u with respect to q for $f \in \mathcal{I}^+$ is defined by

$$u_q^\bullet(f) = \sup_{|\varphi|\le f,\varphi\in\mathcal{K}(T)} |u(\varphi)|_q$$

and one completes the definition for $f : T \to [0,\infty]$ as in Definition 7.1.1 given in the case of a normed space.

For $q \in \Gamma$, let $X_q = X/q^{-1}(0)$ and let $\widetilde{X_q}$ be the Banach space completion of X_q with respect to $|\cdot|_q$. Let $\Pi_q : X \to X_q \subset \widetilde{X_q}$ be the canonical quotient map. (See the beginning of Section 1.2 of Chapter 1.) Let $u_q = \Pi_q \circ u$. Then $|u_q(\varphi)| = |u(\varphi)|_q$ so that $u_q^\bullet$ is the semivariation of u_q.

Definition 7.1.19. For the Radon operator u on $\mathcal{K}(T)$ with values in the lcHs X and for $q \in \Gamma$, the family of continuous seminorms on X, we define

$$\mathcal{F}^0(u) = \{f : \mathcal{K}(T) \to \mathbb{K}, \ u_q^{\bullet}(|f|) < \infty \text{ for each } q \in \Gamma\}.$$

Then by the complex version of Lemma 1.5 of [T], $\mathcal{K}(T) \subset \mathcal{F}^0(u)$ and this permits the following definition.

Definition 7.1.20. Let $\mathcal{F}^0(u)$ be provided with the seminorms $u_q^{\bullet}(\cdot)$ for $q \in \Gamma$. The space $\mathcal{L}_1(u)$ of u-integrable functions is the closure of $\mathcal{K}(T)$ in the space $\mathcal{F}^0(u)$. Hence a function $f : \mathcal{K}(T) \to \mathbb{K}$ is u-integrable if, for each $q \in \Gamma$ and $\epsilon > 0$, there exists $\varphi_q \in \mathcal{K}(T)$ such that $u_q^{\bullet}(|\varphi_q - f|) < \epsilon$.

Thus $\mathcal{L}_1(u)$ is the intersection of the spaces $\mathcal{L}_1(u_q)$ provided with the smallest topology permitting the injections $\mathcal{L}_1(u) \subset \mathcal{L}_1(u_q)$ continuous for each $q \in \Gamma$.

Definition 7.1.21. Let u be a Radon operator on $\mathcal{K}(T)$ with values in an lcHs X over $\mathbb{C}$. Let Y be a topological space. We say that a function $f : T \to Y$ is u-measurable if f is u_q-measurable for each $q \in \Gamma$ and is u-null if it is u_q-null for each $q \in \Gamma$. (See Definitions 7.1.1 and 7.1.13.)

With these definitions, the complex versions of Propositions 1.7, 1.20 and 1.21 and Theorem 1.22 of [T] hold without any modifications for complex lcHs-valued Radon operators u. Thus a function f is u-integrable if and only if it is u-measurable and is dominated in modulus by a u-integrable function. The Hausdorff space $L_1(u)$ associated with $\mathcal{L}_1(u)$ consists of classes of functions in which two functions equal u-a.e in T are identified.

Definition 7.1.22. For a function $f \in \mathcal{L}_1(u)$, we denote by $u(f)$ or by $\int f du$ the value in f of the continuous linear extension of u to $\mathcal{L}_1(u)$. Thus this is an element in the completion $\widetilde{X}$ of X. Then the mapping $f \to \int f du$ is a continuous linear mapping of $\mathcal{L}_1(u)$ in $\widetilde{X}$.

Hereafter, by lcHs we mean a complex lcHs, i.e., an lcHs over $\mathbb{C}$. Then Propositions 1.28 and 1.30 and results **1.31**, **1.32**, **1.33** and **1.34** of [T] hold for complex lcHs-valued u on $\mathcal{K}(T)$.

If X is a projective limit of Banach spaces X_i, then $\int f du$ is identified with the element $(\int f du_i)_i$ (i.e., with the projective limit of $(\int f du_i)_i$). Consequently, if $u : \mathcal{K}(T) \to X$, X an lcHs over $\mathbb{C}$, then for $f \in \mathcal{L}_1(u)$, $\int f du = \varprojlim \int f du_q$.

Lemma 7.1.23. *Suppose X is a quasicomplete lcHs. If $A \subset X$ is bounded and if x_0 belongs to the closure of A in $\widetilde{X}$, then $x_0 \in X$.*

Proof. Let τ be the lcHs topology of X. By hypothesis there exists a net $(x_\alpha) \subset A$ such that $x_\alpha \to x_0$ in $\tilde{\tau}$, the topology of the completion $\widetilde{X}$. Thus (x_α) is Cauchy in

$\tilde{\tau}$. As A is τ-bounded and as X is quasicomplete, the τ-closure of A is τ-complete. Since $\tilde{\tau}|_A = \tau$, it follows that (x_α) is τ-Cauchy. Hence there exists x_1 in the τ-closure of A (so that $x_1 \in X$) such that $x_\alpha \to x_1$ in τ and hence in $\tilde{\tau}$. Since $\tilde{\tau}$ is Hausdorff, $x_0 = x_1 \in X$.   □

Theorem 7.1.24. *Let u be a Radon operator on $\mathcal{K}(T)$ with values in a quasicomplete lcHs X. Then for each $f \in \mathcal{L}_1(u)$, $\int f du$ belongs to X. In other words, if X is an lcHs, then $\int f du$ belongs to the quasicompletion of X for each $f \in \mathcal{L}_1(u)$.*

Proof. By Lemma 7.1.23 above, it suffices to show that $\int f du$ belongs to the closure in $\widetilde{X}$ of a τ-bounded set $A \subset X$ whenever $f \in \mathcal{L}_1(u)$.

For the sake of completeness, we give the proof of this result and we follow the proof of Theorem 1.35 of [T].

Case 1. Suppose f is bounded with compact support. Let ω be a relatively compact open set in T such that f is null in $T\backslash\omega$ and let $|f| \le 1$.

If $\left|\langle \int \varphi du, x^* \rangle\right| \le 1$ for $\varphi \in \mathcal{K}(T)$ with $|\varphi| \le \chi_\omega$, then

$$1 \ge \sup_{|\varphi| \le \chi_\omega, \varphi \in \mathcal{K}(T)} \left|\left\langle \int \varphi du, x^* \right\rangle\right| = \sup_{|\varphi| \le \chi_\omega, \varphi \in \mathcal{K}(T)} |u_{x^*}(\varphi)|$$
$$= u_{x^*}^\bullet(\chi_\omega) = |u_{x^*}|(\omega)$$

by Definition 7.1.1 and by Proposition 7.1.4. As $|f| \le \chi_\omega$ and f is u-integrable by hypothesis we have

$$\left|\left\langle \int f du, x^* \right\rangle\right| = \left|\int f du_{x^*}\right| \le \int |f| d|u_{x^*}| \le |u_{x^*}|(\omega) \le 1$$

and hence $\int f du \in A^{00}$ where $A = \{\int \varphi du : \varphi \in \mathcal{K}(T), |\varphi| \le \chi_\omega\} \subset X \subset \widetilde{X}$.

Since A is absolutely convex, by the bipolar theorem 8.1.5 of [E] A^{00} is the $\sigma(\widetilde{X}, X^*)$-closure of A, and hence by Theorem 3.12 of [Ru2] is also the $\tilde{\tau}$-closure of A. As A is weakly bounded, it is τ-bounded by Theorem 3.18 of [Ru2]. Hence by Lemma 7.1.23, $\int f du \in X$.

Case 2. Let f be null outside a compact set.

First let us consider the case $f \ge 0$. Let $f_n = \min(f, n)$. Then by Case 1, $\int f_n du \in X$ for all n and by the lcHs analogue of Lemma 1.25 bis of [T] and by the complex analogue of **1.10** of [T], $\int f du = \lim_n \int f_n du$. Thus $\int f du$ belongs to the closure of $(\int f_n du)_{n=1}^\infty$ in $\widetilde{X}$ and as $(\int f_n)_{n=1}^\infty$ is convergent, it is bounded in X. Hence by Lemma 7.1.23, $\int f du \in X$. If f is complex-valued and vanishes outside a compact set, then by the complex analogue of Proposition 1.7 of [T], $\int f du$ belongs to X in this case too.

Case 3. f is an arbitrary element in $\mathcal{L}_1(u)$.

By the above cases, $\int f\varphi du \in X$ for each $\varphi \in \mathcal{K}(T)$. The set

$$B = \left\{ \int f\varphi du \right\}_{|\varphi| \le 1, \varphi \in \mathcal{K}(T)}$$

is weakly bounded since

$$\sup_{|\varphi| \le 1, \varphi \in \mathcal{K}(T)} \left| \left\langle \int f\varphi du, x^* \right\rangle \right| \le \int |f| d|u_{x^*}| < \infty$$

for each $x^* \in X^*$. Then by Theorem 3.18 of [Ru2], B is τ-bounded.

If $|\langle \int f\varphi du, x^* \rangle| \le 1$ for $\varphi \in \mathcal{K}(T)$ with $|\varphi| \le 1$, then by Theorem 7.1.12, $u^{\bullet}_{x^*}(|f|) \le 1$ and hence $|\langle \int f du, x^* \rangle| \le 1$. Therefore, $\int f du \in B^{00}$. Then arguing as in Case 1 and appealing to Lemma 7.1.23, we conclude that $\int f du \in X$. This completes the proof of the theorem. □

The proof of Proposition on p. 84 of [T] holds here for metrizable lcHs and hence we have:

Theorem 7.1.25. *If X is a metrizable lcHs and if $u : \mathcal{K}(T) \to X$ is a Radon operator, then the space $\mathcal{L}_1(u)$ is pseudo metrizable and complete.*

7.2 Integration with respect to a weakly compact bounded Radon operator

The aim of this section is to improve the results in Section 2 of Thomas [T]. Remark 2 on p. 161 of [G] and Theorem 6 of [G] when T is compact, play a key role in [T] to develop the theory of vectorial Radon integration with respect to a weakly compact bounded Radon operator or a prolongable Radon operator. Grothendieck comments in the said remark that his techniques developed in earlier sections of [G] are textually valid to prove Theorem 6 of [G] for $C_0(T)$, where T is a locally compact Hausdorff space. But, as shown in [P10], his techniques can be used to prove the said remark if and only if T is further σ-compact. However, by different methods, we established in [P9] and [P11] the validity of Theorem 6 of [G] for $C_0(T)$ when T is an arbitrary locally compact Hausdorff space, thereby restoring the validity of the Thomas theory in [T]. The proposition given in Complements of Section 2 of [T] improves Theorem 2.7 and Theorem 2.7 bis of [T]. But, we obtain here results which further improve the said proposition of [T]. See Theorem 7.2.19 and Corollary 7.2.20.

Definition 7.2.1. Let X be a quasicomplete lcHs and let $u : \mathcal{K}(T) \to X$ be a Radon operator (see Definition 5.3.3 of Chapter 5). Then u is said to be bounded if $u : (C_c(T), ||\cdot||_T) \to X$ is continuous.

Notation 7.2.2. Whenever X is a quasicomplete lcHs, Γ denotes the family of continuous seminorms on X.

Proposition 7.2.3. *Let X be a quasicomplete lcHs. The Radon operator $u : \mathcal{K}(T) \to X$ is bounded if and only if $u_q^{\bullet}(T) < \infty$ for each $q \in \Gamma$.*

Proof. Let u be bounded. Then, for each $q \in \Gamma$, by Definition 7.2.1 there exists a constant M_q such that $|u_q(\varphi)| = |u(\varphi)|_q = q(u(\varphi)) \le M_q \|\varphi\|_T$ for each $\varphi \in \mathcal{K}(T)$. Then

$$u_q^{\bullet}(T) = u_q^{\bullet}(\chi_T) = \sup_{|\varphi| \le 1, \varphi \in \mathcal{K}(T)} q(u(\varphi)) \le M_q < \infty$$

and hence $u_q^{\bullet}(T) < \infty$ for each $q \in \Gamma$.

Conversely, if $u_q^{\bullet}(T) = M_q < \infty$ for each $q \in \Gamma$, then for $\varphi \in \mathcal{K}(T)$, we have $|\varphi| \le \|\varphi\|_T \chi_T$ and hence

$$q(u(\varphi)) = |u_q(\varphi)|_q \le \|\varphi\|_T u_q^{\bullet}(\chi_T) = M_q \|\varphi\|_T$$

for $q \in \Gamma$. Hence u is bounded. $\qquad\qquad\qquad\qquad\qquad\qquad\qquad\qquad\qquad\square$

Convention 7.2.4. Let X be a quasicomplete lcHs and let $u : \mathcal{K}(T) \to X$ be a bounded Radon operator. Then u has a continuous linear extension to the whole of $C_0(T)$ with values in X and hence we shall always assume that $u : (C_0(T), \|\cdot\|_T) \to X$ is continuous whenever $u : \mathcal{K}(T) \to X$ is a bounded Radon operator.

Proposition 7.2.5. *Let X be a quasicomplete lcHs. If $u : \mathcal{K}(T) \to X$ is a bounded Radon operator, then $C_0(T) \subset \mathcal{L}_1(u)$. Moreover, for $f \in C_0(T)$, $uf = \int f du$.*

Proof. Let $f \in C_0(T)$ and let $q \in \Gamma$. Then there exists $(\varphi_n)_1^{\infty} \subset \mathcal{K}(T)$ such that $\|\varphi_n - f\|_T \to 0$. By hypothesis and by Proposition 7.2.3, $u_q^{\bullet}(T) < \infty$ and hence

$$u_q^{\bullet}(|\varphi_n - f|) \le u_q^{\bullet}(\|f - \varphi_n\|_T \chi_T) \le \|f - \varphi_n\|_T u_q^{\bullet}(T) \to 0$$

as $n \to \infty$. Hence $f \in \mathcal{L}_1(u_q)$. As q is arbitrary in Γ, it follows that $f \in \mathcal{L}_1(u)$. Thus $C_0(T) \subset \mathcal{L}_1(u)$.

Moreover, $uf = \lim_n u\varphi_n$ and $\int f du = \lim_n u\varphi_n$. Hence $uf = \int f du$. $\qquad\square$

For a bounded Radon operator u, it is possible that $\mathcal{L}_1(u) = C_0(T)$ as shown below.

Example 7.2.6. Let $u : \mathcal{K}(T) \to C_0(T)$ be the identity operator. Then $u : (C_c(T), \|\cdot\|_T) \to (C_0(T), \|\cdot\|_T)$ is continuous and has a unique continuous extension to $C_0(T)$. Then $\mathcal{L}_1(u) = C_0(T)$.

In fact, if $f \in \mathcal{I}^+$, then

$$u^{\bullet}(f) = \sup_{|\varphi| \le f, \varphi \in \mathcal{K}(T)} |u(\varphi)| = \sup_{|\varphi| \le f, \varphi \in \mathcal{K}(T)} \|\varphi\|_T = \|f\|_T \qquad (7.2.6.1)$$

since $f = \sup_{0 \le \varphi \le f, \varphi \in \mathcal{K}(T)} \varphi$.

If $f : T \to [0, \infty]$ has compact support, then by (7.2.6.1) and by Definition 7.1.1 we have

$$\|f\|_T \le \inf_{f \le g \in \mathcal{I}^+} \|g\|_T = u^{\bullet}(f).$$

Let K be the support of f. By Urysohn's lemma there exists $\Psi \in C_c(T)$ with $0 \le \Psi \le 1$ and $\Psi|_K = 1$. Then $0 \le f \le ||f||_T \Psi \in \mathcal{I}^+$ and

$$u^\bullet(||f||_T \Psi) = ||f||_T u^\bullet(\Psi) = ||f||_T \, ||\Psi||_T = ||f||_T$$

by (7.2.6.1) as $\Psi \in \mathcal{I}^+$. Hence $u^\bullet(f) = ||f||_T$.

When $f : T \to [0, \infty]$ is arbitrary, then

$$u^\bullet(f) = \sup_{K \in \mathcal{C}} u^\bullet(f \chi_K) = \sup_{K \in \mathcal{C}} ||f \chi_K||_T = ||f||_T. \qquad (7.2.6.2)$$

If $f \in \mathcal{L}_1(u)$, then given $\epsilon > 0$, there exists $\varphi \in \mathcal{K}(T)$ such that $u^\bullet(|f - \varphi|) < \epsilon$. Then by (7.2.6.2), $||f - \varphi||_T < \epsilon$ and hence $f \in C_0(T)$. Therefore, $\mathcal{L}_1(u) = C_0(T)$.

Definition 7.2.7. Let X be a quasicomplete lcHs and let $u : \mathcal{K}(T) \to X$ be a bounded Radon operator. (See Convention 7.2.4.) Then u is called a weakly compact bounded Radon operator if $\{u\varphi : \varphi \in C_0(T), ||\varphi||_T \le 1\}$ is relatively weakly compact in X.

Remark 7.2.8. Bounded Radon operators and weakly compact bounded Radon operators are respectively called bounded Radon measures and weakly compact bounded Radon measures in [T].

Lemma 7.2.9. *Let X be a Banach space and let $u : \mathcal{K}(T) \to X$ be a continuous linear map. Then for each open set ω in T,*

$$u^\bullet(\omega) = \sup_{|x^*| \le 1} |x^* u|(\omega) \qquad (7.2.9.1)$$

where $|x^* u|(\omega) = \mu_{|x^* u|}(\omega)$ and $\mu_{|x^* u|}$ is the (complex) Radon measure induced by $|x^* u|$ in the sense of Definition 4.3 of [P3].

Proof. By Definition 7.1.1 and by Proposition 7.1.4 we have

$$u^\bullet(\omega) = \sup_{|\varphi| \le \chi_\omega, \varphi \in \mathcal{K}(T)} |u(\varphi)|$$

$$= \sup_{|\varphi| \le \chi_\omega, \varphi \in \mathcal{K}(T)} \sup_{|x^*| \le 1} |x^* u(\varphi)|$$

$$= \sup_{|x^*| \le 1} \sup_{|\varphi| \le \chi_\omega, \varphi \in \mathcal{K}(T)} |x^* u(\varphi)|$$

$$= \sup_{|x^*| \le 1} (x^* u)^\bullet(\omega)$$

$$= \sup_{|x^*| \le 1} |x^* u|^*(\omega).$$

Since ω is $|x^* u|^*$-measurable, we have

$$u^\bullet(\omega) = \sup_{|x^*| \le 1} |x^* u|(\omega)$$

and hence (7.2.9.1) holds. $\qquad \square$

Using the above lemma and [P9], an improved version of Theorem 2.2 of [T] for complex functions is given below.

Theorem 7.2.10. *Let X be a Banach space and let $u : \mathcal{K}(T) \to X$ be a bounded Radon operator. Then the following statements are equivalent:*

(i) *Every bounded Borel (complex) function belongs to $\mathcal{L}_1(u)$.*

(ii) *Every bounded σ-Borel (complex) function belongs to $\mathcal{L}_1(u)$.*

(iii) *Every bounded (complex) Baire function belongs to $\mathcal{L}_1(u)$.*

(iv) *For every open set ω in T, the weak integral $\int_\omega$ belongs to X; i.e., there exists a vector x_ω in X such that*

$$\int_\omega d(x^*u) = x^*(x_\omega)$$

for each $x^* \in X^*$ and we say that the weak integral $\int_\omega du = x_\omega$.

(v) *For every σ-Borel open set ω in T, the weak integral $\int_\omega$ belongs to X.*

(vi) *For every open Baire set ω in T, the weak integral $\int_\omega du$ belongs to X.*

(vii) *u is a weakly compact bounded Radon operator (so that by Convention 7.2.4, $u : C_0(T) \to X$ is weakly compact).*

Proof. Clearly, (i)$\Rightarrow$(ii)$\Rightarrow$(iii).

(iii)$\Rightarrow$(vi) By (iii), for each open Baire set ω in T, there exists $x_\omega \in X$ such that $\int_\omega du = x_\omega$ and hence

$$x^*(x_\omega) = \int_\omega d(x^*u)$$

for $x^* \in X^*$. Hence (vi) holds.

(vi)$\Rightarrow$(vii) By (vi), for each open Baire set ω in T there exists $x_\omega \in X$ such that

$$\int_\omega d(x^*u) = x^*(x_\omega) \tag{7.2.10.1}$$

for $x^* \in X^*$. As $x^*u \in C_0(T)^* = M(T)$, the complex Radon measure μ_{x^*u} induced by x^*u in the sense of Definition 4.3 of [P3] is a $\mathcal{B}(T)$-regular complex measure on $\mathcal{B}(T)$. Let K be a compact G_δ in T. Then by Theorem 55.A of [H] there exists $(\varphi_n)_1^\infty \subset C_0(T)$ such that $0 \le \varphi_n \searrow \chi_K$. Then by LDCT and by 5.2.10 we have

$$\mu_{x^*u}(K) = \lim_n \int_T \varphi_n d\mu_{x^*u} = \lim_n (x^*u)(\varphi_n) = \lim_n \int_T \varphi_n d(x^* \circ \mathbf{m}) = (x^* \circ \mathbf{m})(K)$$

where $\mathbf{m}$ is the representing measure of u. Then by the Baire regularity of $\mu_{x^*u}|_{\mathcal{B}_0(T)}$ and of $(x^* \circ \mathbf{m})|_{\mathcal{B}_0(T)}$ we have $\mu_{x^*u}|_{\mathcal{B}_0(T)} = (x^* \circ \mathbf{m})|_{\mathcal{B}_0(T)}$ and consequently, by Theorem 2.4 of [P4] and by the Borel regularity of μ_{x^*u} and of $x^* \circ \mathbf{m}$ on $\mathcal{B}(T)$, we conclude that

$$\mu_{x^*u} = x^* \circ \mathbf{m} \text{ on } \mathcal{B}(T). \tag{7.2.10.2}$$

Then by (7.2.10.1) and (7.2.10.2) we have

$$x^*(x_\omega) = \int_\omega d(x^*u) = \mu_{x^*u}(\omega) = (x^* \circ \mathbf{m})(\omega) = (x^* \circ u^{**})(\chi_\omega)$$

for $x^* \in X^*$. Since $u^{**}(\chi_\omega) \in X^{**}$, we conclude that $\mathbf{m}(\omega) = u^{**}(\chi_\omega) = x_\omega \in X$. Consequently, by Theorem 3(vii) of [P9], u is a weakly compact operator on $C_0(T)$ and hence (vii) holds.

Clearly, (iv)$\Rightarrow$(v)$\Rightarrow$(vi). (vii)$\Rightarrow$(iv) (resp. (vii)$\Rightarrow$(v), (vii)$\Rightarrow$(vi)) By (vii) and by Theorem 2(ii) of [P9], $u^{**}(\chi_A) \in X$ for each $A \in \mathcal{B}(T)$ and hence $u^{**}(\chi_\omega) \in X$ for each open set (resp. σ-Borel open set, open Baire set) ω in T. Let $u^{**}(\chi_\omega) = x_\omega \in X$. Then by arguing as in the proof of (vi)$\Rightarrow$(vii) we note that (7.2.10.2) holds and hence by (7.2.10.2) we have

$$x^*(x_\omega) = x^*u^{**}(\chi_\omega) = (x^* \circ \mathbf{m})(\omega) = \mu_{x^*u}(\omega) = \int_\omega d(x^*u)$$

and hence (iv) (resp. (v), (vi)) holds.

(vii)$\Rightarrow$(i) Since $u : C_0(T) \to X$ is weakly compact, u^* is also weakly compact and hence $\{\mu_{u^*x^*} : |x^*| \le 1\} = \{\mu_{x^*u} : |x^*| \le 1\}$ is relatively weakly compact in $M(T)$. Then by Theorem 1 of [P8], given a Borel set A in T and $\epsilon > 0$, there exist a compact set K and an open set U in T such that $K \subset A \subset U$ and $\sup_{|x^*|\le 1} |\mu_{x^*u}|(U\backslash K) < \epsilon$. Then by Lemma 7.2.9 we have $u^\bullet(U\backslash K) < \epsilon$. Now choose $\varphi \in \mathcal{K}(T)$ such that $\chi_K \le \varphi \le \chi_U$ so that $u^\bullet(\chi_U - \varphi) \le u^\bullet(\chi_U - \chi_K) = u^\bullet(U\backslash K) < \epsilon$. Then

$$u^\bullet(|\chi_A - \varphi|) \le u^\bullet(\chi_U - \chi_K) = u^\bullet(U\backslash K) < \epsilon.$$

Therefore, by Definition 7.1.6, $\chi_A \in \mathcal{L}_1(u)$.

Consequently, every Borel simple function $s \in \mathcal{L}_1(u)$. If f is a bounded Borel (complex) function, then there exists a sequence (s_n) of Borel simple functions such that $||s_n - f||_T \to 0$. Then

$$u^\bullet(|f - s_n|) \le ||f - s_n||_T u^\bullet(T) \to 0$$

as $n \to \infty$, since $u^\bullet(T)$ is finite by Proposition 7.2.3. Hence $f \in \mathcal{L}_1(u)$ and thus (i) holds.

Hence the statements (i)–(vii) are equivalent. $\square$

The following theorem is an improved complex version of Proposition 2.5 of Thomas [T].

Theorem 7.2.11. *Let X be a Banach space and let $u : \mathcal{K}(T) \to X$ be a bounded Radon operator. Then the following statements are equivalent:*

(i) *u is weakly compact (see Convention 7.2.4).*

(ii) *Given $\epsilon > 0$, for each open set ω in T, there exists a compact $K \subset \omega$ such that $u^\bullet(\omega \backslash K) < \epsilon$.*

(iii) *Given $\epsilon > 0$, for each $A \in \mathcal{B}(T)$, there exist a compact K and an open set ω in T such that $K \subset A \subset \omega$ and $u^\bullet(\omega \backslash K) < \epsilon$.*

(iv) *Given $\epsilon > 0$, for each compact K in T there exists an open set U in T such that $K \subset U$ with $u^\bullet(U \backslash K) < \epsilon$ and there exists a compact C in T such that $u^\bullet(T \backslash C) < \epsilon$.*

(v) *Given $\epsilon > 0$, for each σ-Borel open set $\omega \subset T$ there exists a compact $K \subset \omega$ such that $u^\bullet(\omega \backslash K) < \epsilon$ and there exists a compact C in T such that $u^\bullet(T \backslash C) < \epsilon$.*

(vi) *Given $\epsilon > 0$, for each $A \in \mathcal{B}_c(T)$ there exist a compact K and a σ-Borel open set ω in T such that $K \subset A \subset \omega$ and $u^\bullet(\omega \backslash K) < \epsilon$.*

(vii) *Given $\epsilon > 0$, for each compact K in T there exists a σ-Borel open set U in T such that $K \subset U$ with $u^\bullet(U \backslash K) < \epsilon$ and there exists a compact G_δ set C such that $u^\bullet(T \backslash C) < \epsilon$.*

(viii) *Given $\epsilon > 0$, for each open Baire set ω in T there exists a compact G_δ $K \subset \omega$ such that $u^\bullet(\omega \backslash K) < \epsilon$ and there exists a compact C such that $u^\bullet(T \backslash C) < \epsilon$.*

(ix) *Given $\epsilon > 0$, for each Baire set $A \subset T$ there exist a compact $G_\delta K$ and an open Baire set ω in T such that $K \subset A \subset \omega$ and $u^\bullet(\omega \backslash K) < \epsilon$.*

(x) *Given $\epsilon > 0$, for each compact $G_\delta K$ in T there exists an open Baire set U in T such that $K \subset U$ with $u^\bullet(U \backslash K) < \epsilon$ and there exists a compact $G_\delta C$ such that $u^\bullet(T \backslash C) < \epsilon$.*

Proof. Since u is a bounded operator on $C_0(T)$ by Convention 7.2.4, the set $F = \{\mu_{x^* u} : |x^*| \leq 1\} = \{\mu_{u^* x^*} : |x^*| \leq 1\}$ is bounded in $M(T)$. Let $|F| = \{\mu_{|x^* u|} : |x^*| \leq 1\}$. Then by Theorem 1 of [P8], F is relatively weakly compact in $M(T)$ if and only if $|F|$ is so (resp. if and only if u is weakly compact (as u is weakly compact if and only if u^* is weakly compact)). Moreover, by Theorem 4.11 of [P3] and by Theorem 3.3 of [P4], $\mu_{|x^* u|}(A) = v(\mu_{x^* u}, \mathcal{B}(T))(A) = |\mu_{x^* u}|(A)$ for $A \in \mathcal{B}(T)$. Then by Lemma 7.2.9 and by (vi) (resp. (vi)$'$, (vi)$''$) of Proposition 1 of [P9], (ix) (resp. (iii), (vi)) holds if and only if u is weakly compact. Hence (i)$\Leftrightarrow$(iii)$\Leftrightarrow$(vi) $\Leftrightarrow$(ix).

(iii)$\Rightarrow$(ii) obviously.

(ii)$\Rightarrow$(i) Let ω be an open set in T and let $\epsilon > 0$. Then by hypothesis, there exists a compact K such that $K \subset \omega$ and $u^\bullet(\omega \backslash K) < \epsilon$. By Lemma 7.2.9, this means $\sup_{|x^*| \leq 1} \mu_{|x^* u|}(\omega \backslash K) < \epsilon$. Then by Theorem 4.11 of [P3] and by Theorem 3.3 of [P4], $\sup_{|x^*| \leq 1} v(\mu_{x^* u}, \mathcal{B}(T))(\omega \backslash K) < \epsilon$ and hence by Theorem 2 of [P8], u is weakly compact. Thus (i) holds.

(i)$\Leftrightarrow$(iv) By Theorem 4.11 of [P3], $|F| = \{|\mu_{x^* u}| : |x^*| \leq 1\}$ and hence by Theorem 4.22.1 of [E], $|F|$ is relatively weakly compact in $M(T)$ if and only if , given $\epsilon > 0$, for each compact K in T there exists an open set U in T such that $K \subset U$ and $\sup_{|x^*| \leq 1} |\mu_{x^*u}|(U \backslash K) < \epsilon$ and there exists a compact C in T such that $\sup_{|x^*| \leq 1} |\mu_{x^*u}|(T \backslash C) < \epsilon$. Consequently, by Lemma 7.2.9, $|F|$ is relatively weakly compact in $M(T)$ if and only if (iv) holds and hence if and only if u is weakly compact.

Similarly, using Theorem 4.22.1 of [E], Theorem 4.11 of [P3], Theorem 50.D of [H] and Lemma 7.2.9 one can show that (i)$\Leftrightarrow$(vii) and (i)$\Leftrightarrow$(x).

By Proposition 1(iii) of [P9], by Theorem 4.11 of [P3] and by Lemma 7.2.9, (i)$\Leftrightarrow$(v) and (i)$\Leftrightarrow$(viii).

Hence the statements (i)–(x) are equivalent.   □

Definition and Notation 7.2.12. Let X be a Banach space and let $u : \mathcal{K}(T) \to X$ be a Radon operator. Let H be a subset of X^* separating the points of X. A function $f : T \to \mathbb{K}$ is said to be u-integrable with respect to the topology $\sigma(X, H)$ if f is x^*u-integrable for each $x^* \in H$. Then the integral of f with respect to $\sigma(X, H)$ is an element in the completion of $(X, \sigma(X, H))$ which is identified with $\langle H \rangle^{\text{alg}}$, the set of all linear functionals on the linear span $\langle H \rangle$ of H and the integral is denoted by $\int f d\tilde{u}$. This is identified with the function $x^* \to \int f d(x^*u)$ for $x^* \in H$. Thus

$$\left\langle x^*, \int f d\tilde{u} \right\rangle = \int f du_{x^*} \qquad \text{for } x^* \in H.$$

The Orlicz property of a set H in X^* (see Definition 5.2.9 of Chapter 5) plays a key role in the sequel.

The following result improves the complex version of Theorem 2.7 of Thomas [T].

Theorem 7.2.13. *Let X be a Banach space and let H be a norm determining subset of X^*. Suppose H possesses the Orlicz property. Let $u : \mathcal{K}(T) \to X$ be a bounded Radon operator. Then $u : C_0(T) \to (X, \sigma(X, H))$ is continuous (here we use Convention 7.2.4) and u is weakly compact on $C_0(T)$ if and only if $\int_\omega d\tilde{u} \in X$ for each open Baire set ω in T, where $\tilde{u}$ is the Radon operator obtained from u on providing X with the topology $\sigma(X, H)$.*

Proof. Arguing as in the proof of Theorem 2.7 of [T] and using Theorem 1 of [P8] instead of Appendix I: C_2 of [T], we observe that the condition is sufficient.

Conversely, if u is weakly compact with its representing measure $\mathbf{m}$ in the sense of 5.2.10 then by Theorem 2(ii) of [P9], $u^{**}(\chi_\omega) = \mathbf{m}(\omega) = x_\omega$ (say) $\in X$ for each open Baire set ω in T and hence

$$x^*(x_\omega) = \int_\omega d(x^*u) = x^* \left(\int_\omega d\tilde{u} \right)$$

for $x^* \in X^*$ and hence for $x^* \in H$. Then $\langle x^*, \int_\omega d\tilde{u} \rangle = \int_\omega du_{x^*} = \int_\omega d(x^*u) = x^*(x_\omega)$ for $x^* \in H$. Hence $\sup_{x^* \in H} |x^*(\int_\omega d\tilde{u} - x_\omega)| = 0$. As H is norm determining, we have $|| \int_\omega d\tilde{u} - x_\omega || = 0$ and hence $\int_\omega d\tilde{u} = x_\omega \in X$. Therefore the condition is also necessary. □

To improve the complex version of Theorem 2.12 of [T] we need the following lemma.

Lemma 7.2.14. *Let X be a Banach space and let $u : C_0(T) \to X$ be a continuous linear mapping with the representing measure $\mathbf{m}$ (in the sense of 5.2.10). Then*

$$\int_A d(x^*u) = (x^* \circ \mathbf{m})(A) \qquad (7.2.14.1)$$

for $x^* \in X^*$ and for $A \in \mathcal{B}(T)$. If $\chi_A \in \mathcal{L}_1(u)$, then

$$\int_A du = \mathbf{m}(A) \qquad (7.2.14.2)$$

and consequently (7.2.14.2) holds for $A \in \mathcal{B}(T)$ if u is weakly compact.

Proof. By the proof of (vi)$\Rightarrow$(vii) of Theorem 7.2.10 (without using (7.2.10.1)), $\mu_{x^*u} = x^* \circ \mathbf{m}$ on $\mathcal{B}(T)$ for $x^* \in X^*$. Hence

$$\int_A d(x^*u) = \int_A d\mu_{x^*u} = \mu_{x^*u}(A) = (x^* \circ \mathbf{m})(A)$$

for $A \in \mathcal{B}(T)$ and for $x^* \in X^*$.

If $\chi_A \in \mathcal{L}_1(u)$, there exists $x_A \in X$ such that $\int_A du = x_A \in X$. Consequently, by (7.2.14.1) we have

$$x^*(x_A) = \int_A d(x^*u) = (x^* \circ \mathbf{m})(A)$$

for $x^* \in X^*$. As $\mathbf{m}$ has range in X^{**}, we conclude that $\mathbf{m}(A) = x_A = \int_A du$ for $A \in \mathcal{L}_1(u)$. If u is weakly compact, $\mathcal{B}(T) \subset \mathcal{L}_1(u)$ by Theorem 7.2.10(i) and hence the last part holds. □

Using the above lemma we obtain the following improvement of the complex version of Theorem 2.12 of [T].

Theorem 7.2.15. *Let X be a Banach space and let $u_n : \mathcal{K}(T) \to X$ be a weakly compact bounded Radon operator for $n \in \mathbb{N}$. If for every open Baire set ω in T, the sequence $(\int_\omega du_n)_1^\infty$ is convergent in X, then there exists a weakly compact bounded Radon operator u on $\mathcal{K}(T)$ with values in X such that*

$$\lim_n \int f du_n = \int f du$$

for each bounded (complex) Borel function f on T.

Proof. Let $\mathbf{m}_n$ be the representing measure of u_n in the sense of 5.2.10. Then by hypothesis, by Theorem 7.2.10(iii) and by Lemma 7.2.14,

$$\lim_n \mathbf{m}_n(\omega) \text{ exists in } X \qquad (7.2.15.1)$$

for each open Baire set ω in T. Moreover, by Theorems 6(xix) and 2 of [P9], $\mathbf{m}_n$ is Borel regular and σ-additive in the topology τ of X for each $n \in \mathbb{N}$. By Lemma 5.2.19, φ is $\mathbf{m}_n$-integrable in T and by 5.2.10 we have

$$x^* u_n(\varphi) = \int_T \varphi d(x^* \circ \mathbf{m}_n) = x^* \left(\int_T \varphi d\mathbf{m}_n \right)$$

for $x^* \in X^*$ and for $\varphi \in C_0(T)$. Then by the Hahn-Banach theorem

$$u_n(\varphi) = \int_T \varphi d\mathbf{m}_n \text{ for } \varphi \in C_0(T).$$

Consequently, by (7.2.15.1) and by Lemma 5.2.20 there exists an X-valued continuous linear mapping u on $C_0(T)$ such that

$$\lim_n u_n(\varphi) = u(\varphi)$$

for $\varphi \in C_0(T)$. Moreover, by (7.2.15.1) and by Lemma 5.2.18, $\lim_n \mathbf{m}_n(U) \in X$ for each open set U in T. Consequently, by the complex version of Proposition 2.11 of [T], $\lim_n \int f du_n = \int f du \in X$ for each bounded complex Borel function f on T. Then particularly,

$$\lim_n \int \chi_A du_n = \int \chi_A du \in X \qquad (7.2.15.2)$$

for each $A \in \mathcal{B}(T)$. Then by Theorem 7.2.13, u is weakly compact.

This completes the proof of the theorem. $\qquad\qquad\qquad\qquad\square$

Remark 7.2.16. The proofs of Propositions 2.13, 2.14, 2.17 and 2.20, of Corollary 2.18 and of Lemma 2.21 of [T] hold for complex spaces too.

We need the following lemma to generalize Theorem 7.2.10 to quasicomplete lcHs.

Lemma 7.2.17. *Let u be a Radon operator on $\mathcal{K}(T)$ with values in an lcHs X. For $q \in \Gamma$, let $u_q = \Pi_q \circ u$ where Γ and Π_q are as in the beginning of Section 1.2 of Chapter 1. Then for each open set ω in T,*

$$u_q^\bullet(\omega) = \sup_{x^* \in U_q^0} |x^* u|^\bullet(\omega) = \sup_{x^* \in U_q^0} |x^* u|(\omega)$$

where U_q^0 is as in Notation 1.2.11 and $u_q^\bullet$ is as in Definition 7.1.18.

Proof. Let $q \in \Gamma$. Then $u_q : \mathcal{K}(T) \to \widetilde{X_q}$ is a continuous linear map and hence by Lemma 7.2.9 and by Propositions 1.2.15 and 7.1.4 we have

$$u_q^\bullet(\omega) = \sup_{x^* \in U_q^0} |\Psi_{x^*} u_q|^\bullet(\omega) = \sup_{x^* \in U_q^0} |\Psi_{x^*} \circ (\Pi_q \circ u)|^\bullet(\omega)$$

$$= \sup_{x^* \in U_q^0} |x^* u|^\bullet(\omega) = \sup_{x^* \in U_q^0} |x^* u|^*(\omega) = \sup_{x^* \in U_q^0} |x^* u|(\omega)$$

since $\Psi_{x^*}(\Pi_q \circ u)(x) = \Psi_{x^*}(ux + q^{-1}(0)) = x^* ux$ for $x \in X$ and for $x^* \in U_q^0$ and since the open set ω is $|x^* u|^*$-measurable. $\square$

The following theorem generalizes Theorem 7.2.10 to quasicomplete lcHs.

Theorem 7.2.18. *Let X be a quasicomplete lcHs and let $u : \mathcal{K}(T) \to X$ be a bounded Radon operator. Then the following statements are equivalent:*

(i) *Every bounded (complex) Borel function belongs to $\mathcal{L}_1(u)$.*

(ii) *Every bounded (complex) σ-Borel function belongs to $\mathcal{L}_1(u)$.*

(iii) *Every bounded (complex) Baire function belongs to $\mathcal{L}_1(u)$.*

(iv) *For every open set ω in T the weak integral $\int_\omega du$ belongs to X; i.e., there exists a vector x_ω in X such that*

$$\int_\omega d(x^* u) = x^*(x_\omega)$$

for each $x^* \in X^*$. Then we say that the weak integral $\int_\omega du = x_\omega$.

(v) *For every σ-Borel open set ω in T, the weak integral $\int_\omega du$ belongs to X.*

(vi) *For every open Baire set ω in T, the weak integral $\int_\omega$ belongs to X.*

(vii) *u is weakly compact (see Convention 7.2.4).*

Proof. Since the results mentioned in the proof of Theorem 7.2.10 hold for lcHs-valued continuous linear transformations on $C_0(T)$ and since the theorems in [P9] used in the proof of Theorem 7.2.10 are valid not only for Banach spaces but also for quasicomplete lcHs, the proof of the latter theorem excepting that of (vii)$\Rightarrow$(i) continues to be valid when X is a quasicomplete lcHs.

Now we shall show that (vii)$\Rightarrow$(i). Let $q \in \Gamma$ and let $U_q = \{x \in X : q(x) \leq 1\}$. Then U_q^0, the polar of U_q is equicontinuous and hence by Corollary 9.3.2 of [E] or by Proposition 4 of [P9], $u^*(U_q^0)$ is relatively weakly compact in $M(T)$. Then by Theorem 1 of [P8], given $A \in \mathcal{B}(T)$ and $\epsilon > 0$, there exist a compact set K and an open set U in T such that $K \subset A \subset U$ and $\sup_{x^* \in U_q^0} |x^* u|(U \backslash K) < \epsilon$. Consequently, by Lemma 7.2.17, $u_q^\bullet(U \backslash K) < \epsilon$. Then arguing as in the last part of the proof of (vii)$\Rightarrow$(i) of Theorem 7.2.10 we have a $\varphi_q \in \mathcal{K}(T)$ such that $\chi_K \leq \varphi_q \leq \chi_U$ so that

$$u_q^\bullet(|\chi_A - \varphi_q|) \leq u_q^\bullet(\chi_U - \chi_A) \leq u_q^\bullet(\chi_U - \chi_K) = u_q^\bullet(U \backslash K) < \epsilon.$$

Since q is arbitrary in Γ, by Definition 7.1.20, $\chi_A \in \mathcal{L}_1(u)$.

Then every Borel simple function s belongs to $\mathcal{L}_1(u)$. If f is a bounded Borel (complex) function, then there exists a sequence (s_n) of Borel simple functions such that $||s_n - f||_T \to 0$. Then, for each $q \in \Gamma$, we have

$$u_q^{\bullet}(|f - s_n|) \leq ||f - s_n||_T u_q^{\bullet}(T) \to 0$$

as $n \to \infty$, since $u_q^{\bullet}(T)$ is finite by Proposition 7.2.3 and by the hypothesis that $u_q : (C_c(T), || \cdot ||_T) \to \widetilde{X}_q$ is continuous.

This completes the proof of the theorem. $\square$

The following theorem gives an improvement of the complex version of Theorem 2.7 bis of [T].

Theorem 7.2.19. *Let X be an lcHs and let H be a subset of X^* such that the topology τ of X is the same as the topology of uniform convergence in the equicontinuous subsets of H. Suppose H has the Orlicz property. Let u be a bounded Radon operator on $\mathcal{K}(T)$ with values in X. Then u is weakly compact if and only if, for every open Baire set ω in T, the ultra weak integral $\int_{\omega} d\tilde{u}$ (relative to the topology $\sigma(X, H)$) belongs to $\widetilde{X}$, the lcHs completion of X.*

Proof. By hypothesis, τ is generated by the seminorms $\{q_E : E \in H_{\mathcal{E}}\}$ where $H_{\mathcal{E}} = \{E \subset H : E \text{ is equicontinuous}\}$ and $q_E(x) = \sup_{x^* \in E} |x^*(x)|$. As observed in the proof of Theorem 5.2.16, $\sigma(X, H)$ is Hausdorff.

Suppose, for each open Baire set ω in T, there exists a vector $x_{\omega} \in \widetilde{X}$ such that

$$\int_{\omega} d(x^*u) = \mu_{x^*u}(\omega) = x^*(x_{\omega}) \qquad (7.2.19.1)$$

for each $x^* \in H$ and hence for each $x^* \in \langle H \rangle$, the linear span of H. Then, given a disjoint sequence $(U_n)_1^{\infty}$ of open Baire sets in T, for each subsequence P of $\mathbb{N}$, by (7.2.19.1) we have

$$\sum_{n \in P} x^*(x_{U_n}) = \sum_{n \in P} \mu_{x^*u}(U_n) = \mu_{x^*u}\left(\bigcup_{n \in P} U_n\right) \in \mathbb{K}$$

and hence $\sum_1^{\infty} x^*(x_{U_n})$ is subseries convergent for each $x^* \in H$.

Since $(X, \sigma(X, \langle H \rangle))^* = \langle H \rangle$ by Theorem 5.3.9 of [DS1], $\sum_1^{\infty} x_{U_n}$ is subseries convergent in $\sigma(X, H)$. As H has the Orlicz property by hypothesis, $\sum_1^{\infty} x_{U_n}$ is unconditionally convergent in τ. Hence

$$\lim_n q_E(x_{U_n}) = 0. \qquad (7.2.19.2)$$

Then by (7.2.19.1) and by (7.2.19.2) we have

$$\lim_n q_E(x_{U_n}) = \lim_n \sup_{x^* \in E} |x^*(x_{U_n})| = \lim_n \sup_{x^* \in E} |\mu_{x^*u}(U_n)| = 0. \qquad (7.2.19.3)$$

Since E is equicontinuous and since $u : C_0(T) \to X$ is continuous, by Lemma 2 of [P9] and by 5.2.10, $u^*E = \{u^*x^* : x^* \in E\} = \{x^*u : x^* \in E\} = \{\mu_{x^*u} : x^* \in E\}$ is bounded in $M(T)$. Then by (7.2.19.3) and by Theorem 1 of [P8]

$$\{\mu_{x^*u} : x^* \in E\} \text{ is relatively weakly compact in } M(T). \tag{7.2.19.4}$$

For $E \in H_{\mathcal{E}}$, $\Pi_{q_E} : \widetilde{X} \to \widetilde{X_{q_E}} \subset \widetilde{(X_{q_E})}$. If Ψ_{x^*} is as in Proposition 1.2.13(i) for $x^* \in E$, then $\{\Psi_{x^*} : x^* \in E\}$ is a norm determining subset of the closed unit ball of $(\widetilde{X_{q_E}})^*$ for $\widetilde{X_{q_E}}$ by Proposition 1.2.13(iii) for $x^* \in E$. Then by Proposition 1.2.13(i) and by 5.2.10 we have

$$(\Psi_{x^*} \circ \Pi_{q_E} \circ u)(\varphi) = (x^*u)(\varphi) \tag{7.2.19.5}$$

for $\varphi \in C_0(T)$ and hence $\Psi_{x^*} \circ \Pi_{q_E} \circ u \in \mathcal{K}(T)_b^* = (C_0(T), \|\cdot\|_T)^* = M(T)$. Then by (7.2.19.4) and (7.2.19.5), $\{\mu_{(\Psi_{x^*} \circ \Pi_{q_E} \circ u)} : x^* \in E\}$ is relatively weakly compact in $M(T)$. Then by Corollary 5.2.15, $\Pi_{q_E} \circ u$ is weakly compact for $E \in H_{\mathcal{E}}$. Consequently, by the complex analogue of Lemma 2.21 of [T], u is weakly compact.

Conversely, if u is weakly compact, then by Theorem 7.2.18(vi) the weak integral $\int_\omega f\,du$ belongs to X for each open Baire set ω in T and hence there exists a vector $x_\omega \in X$ and hence in $\widetilde{X}$ such that

$$x^*\left(\int_\omega du\right) = \int_\omega d(x^*u) = x^*(x_\omega)$$

for each $x^* \in X^*$ and hence for each $x^* \in H$. Thus $\int_\omega d\tilde{u}$ (relative to the topology $\sigma(X, H)$) belongs to $\widetilde{X}$.

This completes the proof of the theorem. $\square$

Corollary 7.2.20. *Under the hypothesis of Theorem 7.2.19 for X, H and the topology τ, a bounded Radon operator $u : \mathcal{K}(T) \to X$ is weakly compact (see Convention 7.2.4) if for each open set ω in T which is a countable union of closed sets, the ultra weak integral $\int_\omega d\tilde{u}$ (relative to the topology $\sigma(X, H)$) belongs to $\widetilde{X}$, the lcHs completion of X.*

Proof. By Lemma 5.2.3 and by Theorem 7.2.19, the corollary holds. $\square$

Remark 7.2.21. Corollary 7.2.20 is obtained directly in the proposition on p. 98 of [T]. But Theorem 7.2.19 is much stronger than the said proposition of [T].

7.3 Integration with respect to a prolongable Radon operator

Following Thomas [T] we study the integration of complex functions with respect to a prolongable Radon operator u on $\mathcal{K}(T)$ (Thomas calls them prolongable Radon measures) and improve most of the principal results such as the complex versions of Theorems 3.3, 3.4, 3.11, 3.13 and 3.20 of [T].

Definition 7.3.1. Let u be a Radon operator on $\mathcal{K}(T)$ with values in an lcHs. Then we say that u is prolongable if every bounded (complex) Borel function with compact support is u-integrable.

Notation 7.3.2. Let ω be an open set in T and let u be a Radon operator on $\mathcal{K}(T)$. We define $\hat{f}$ on T by $\hat{f}(t) = f(t)$ for $t \in \omega$ and 0 for $t \in T\backslash\omega$. For $\varphi \in \mathcal{K}(\omega)$, $\hat{\varphi} \in \mathcal{K}(T)$ and we identify $\mathcal{K}(\omega)$ with the set of functions in $\mathcal{K}(T)$ whose support is contained in ω. The Radon operator u_ω is defined as the restriction of u to $\mathcal{K}(\omega)$, i.e., $u_\omega(\varphi) = u(\hat{\varphi})$. u_ω is called the Radon operator induced by u on $\mathcal{K}(\omega)$.

Lemma 7.3.3. *Let u be a Radon operator on $\mathcal{K}(T)$ with values in a normed space, ω an open subset of T and u_ω the Radon operator induced by u on $\mathcal{K}(\omega)$. Then:*

(i) *For $f \in \mathcal{I}^+(\omega)$, $\hat{f} \in \mathcal{I}^+(T)$ and $(u_\omega)^\bullet(f) = u^\bullet(\hat{f})$.*

(ii) *For $f \geq 0$ with compact support in ω, $(u_\omega)^\bullet(f) = u^\bullet(\hat{f})$.*

(iii) *If f is a (complex) function with compact support in ω belonging to $\mathcal{L}_1(u_\omega)$, then $\hat{f}$ belongs to $\mathcal{L}_1(u)$ and $\int f du_\omega = \int \hat{f} du$ and the last conclusion also extends to Radon operators with values in an lcHs.*

For the proof of the above lemma we refer to the proof of Lemma 3.2 of Thomas [T] given in Appendix III of [T] which holds for complex functions too.

Remark 7.3.4. If $u^\bullet(\omega) < \infty$ for ω in Lemma 7.3.3, then u_ω is a bounded Radon operator on $\mathcal{K}(\omega)$ in the sense that $u_\omega : (C_c(\omega), ||\cdot||_\omega)$ is continuous. Hence particularly if ω is relatively compact in T, then u_ω is a bounded Radon operator on $\mathcal{K}(\omega)$.

Since $u^\bullet(\omega) = \sup_{|\varphi|\leq 1, \varphi\in\mathcal{K}(\omega)} |u_\omega(\varphi)|$ and since u_ω is continuous on $(C_c(\omega), ||\cdot||_\omega)$, the above remark holds.

The following theorem improves Theorem 3.3 of Thomas [T].

Theorem 7.3.5. *Let u be a Radon operator on $\mathcal{K}(T)$ with values in a quasicomplete lcHs X. Then the following statements are equivalent:*

(i) *u is prolongable in the sense of Definition 7.3.1.*

(ii) *Every bounded σ-Borel (complex) function with compact support belongs to $\mathcal{L}_1(u)$.*

(iii) *Every bounded complex Baire function with compact support belongs to $\mathcal{L}_1(u)$.*

(iv) *For each relatively compact open set ω in T, the weak integral $\int_\omega du$ belongs to X; i.e., there exists $x_\omega \in X$ such that $\int_\omega d(x^*u) = x^*(x_\omega)$ for $x^* \in X^*$.*

(v) *For each relatively compact open Baire set ω in T, the weak integral $\int_\omega du$ belongs to X.*

(vi) *If ω is a relatively compact open set in T, then $u|_{\mathcal{K}(\omega)}$ is a weakly compact bounded Radon operator.*

(vii) *For each relatively compact open Baire set ω in T, $u|_{\mathcal{K}(\omega)}$ is a weakly compact bounded Radon operator.*

(viii) *For each compact K in T, the weak integral $\int_K du$ belongs to X.*

(ix) *For each compact G_δ K in T, the weak integral $\int_K du$ belongs to X.*

(x) *A set $A \subset \mathcal{K}(T)$ is said to be bounded in $\mathcal{K}(T)$ if there exists $K \in \mathcal{C}$ such that supp $\varphi \subset K$ for each $\varphi \in A$ and $\sup_{\varphi \in A} \|\varphi\|_T < \infty$. For each relatively compact open set ω in T, u transforms bounded subsets of $\mathcal{K}(\omega)$ into relatively weakly compact subsets of X.*

(xi) *For each compact K, $\lim_{\omega \searrow K} u^\bullet(\omega \backslash K) = 0$ where ω is open in T.*

(xii) *For every compact G_δ K, $\lim_{\omega \searrow K} u^\bullet(\omega \backslash K) = 0$ where ω is open in T.*

Proof. Let $\mathcal{E}$ be the family of all equicontinuous subsets of X^*.

(i)$\Leftrightarrow$(ii) As shown in the proof of (1)$\Rightarrow$(18) of Theorem 5.3.12, a Borel function with compact support is σ-Borel and a σ-Borel function is obviously Borel. Hence (i)$\Leftrightarrow$(ii).

(ii)$\Rightarrow$(iii) Obvious.

(iii)$\Rightarrow$(v) If ω is a relatively compact open Baire set in T, then χ_ω is a bounded Baire function with compact support and hence by (iii), $\chi_\omega \in \mathcal{L}_1(u)$. Then by Theorem 7.1.24 there exists a vector $x_\omega \in X$ such that $\int_\omega du = x_\omega$ and hence $\int_\omega d(x^*u) = x^*(x_\omega)$ for $x^* \in X^*$. Therefore, (v) holds.

(v)$\Rightarrow$(vi) Let ω be a relatively compact open set in T. Let $(\omega_n)_1^\infty$ be a disjoint sequence of open Baire sets in ω. Then $(\omega_n)_1^\infty \subset \mathcal{B}_0(T)$ as shown in the proof of (5)$\Rightarrow$(1) in the proof of Theorem 5.3.12. Let $P \subset \mathbb{N}$ and let $\omega_P = \bigcup_{n \in P} \omega_n$. Then by (v) and by Theorem 7.1.24 there exist $x_P \in X$ and $(x_{\omega_n})_{n \in P} \subset X$ such that

$$x^*(x_{\omega_n}) = \int_{\omega_n} d(x^*u) \qquad (7.3.5.1)$$

and

$$x^*(x_P) = \int_{\omega_P} d(x^*u) = \sum_{n \in P} \int_{\omega_n} d(x^*u) = \sum_{n \in P} x^*(x_{\omega_n})$$

for $x^* \in X^*$. Thus $\sum_1^\infty x^*(x_{\omega_n})$ is subseries convergent for each $x^* \in X^*$ and hence by the Orlicz-Pettis theorem, $\sum_1^\infty x_{\omega_n}$ is unconditionally convergent in X.

Therefore, by (7.3.5.1) we have

$$
\begin{aligned}
0 = \lim_n q_E(x_{\omega_n}) &= \lim_n \sup_{x^* \in E} |x^*(x_{\omega_n})| = \lim_n \sup_{x^* \in E} |(x^*u)(\omega_n)| \\
&= \lim_n \sup_{x^* \in E} |(u^*x^*)(\omega_n)| = \lim_n \sup_{\mu \in u^*E} |\mu(\omega_n)|.
\end{aligned}
$$

Therefore the bounded set u^*E (see Lemma 2 of [P9]) is relatively weakly compact in $M(T)$ by Theorem 1 of [P8]. Since E is arbitrary in $\mathcal{E}$, by Proposition 4 of [P9] $u|_{\mathcal{K}(\omega)}$ is a weakly compact bounded Radon operator (see Convention 7.2.4 with respect to $\mathcal{K}(\omega)$). Hence (vi) holds.

(vi)$\Rightarrow$(vii) Obvious.

(vii)$\Rightarrow$(i)

Claim. If A is a relatively compact Borel set, then $\chi_A \in \mathcal{L}_1(u)$ and consequently, each Borel simple function with compact support belongs to $\mathcal{L}_1(u)$.

In fact, $\bar{A} = K \in \mathcal{C}$ and hence by Theorem 50.D of [H], there exists a relatively compact open Baire set ω_0 such that $K \subset \omega_0$. By (vii), $u_{\omega_0} = u|_{C_0(\omega_0)}$ is weakly compact. Let $E \in \mathcal{E}$. Then $(u_{\omega_0})_{q_E} = \Pi_{q_E} \circ u_{\omega_0} : C_0(\omega_0) \to \widetilde{X_{q_E}}$ is weakly compact and hence, given $\epsilon > 0$, by Theorem 7.2.11 there exist a compact C and an open set ω in ω_0 (hence open in T) such that $C \subset A \subset \omega$ and $(u_{\omega_0})^\bullet q_E(\omega \backslash C) < \epsilon$. By Urysohn's lemma there exists $\varphi \in C_c(\omega)$ such that $0 \le \varphi \le 1$ and $\varphi|_C = 1$. Then $(u_{\omega_0})^\bullet_{q_E}(|\chi_A - \varphi|) \le (u_{\omega_0})^\bullet_{q_E}(\omega \backslash C) < \epsilon$. Since E is arbitrary in $\mathcal{E}$, this shows that $\chi_A \in \mathcal{L}_1(u_{\omega_0})$. Since $\hat{\chi}_A = \chi_A$, by Lemma 7.3.3(iii) $\chi_A \in \mathcal{L}_1(u)$ and consequently, each Borel simple function with compact support belongs to $\mathcal{L}_1(u)$. Hence the claim holds.

Let f be a bounded (complex) Borel function with compact support. Then there exists a sequence (s_n) of Borel simple functions such that $|s_n| \nearrow |f|$ and $s_n \to f$ uniformly in T. Then supp $s_n \subset \text{supp} f = K$ (say) for all n. Let $\omega \in \mathcal{V}$ such that $K \subset \omega$. Then by the above claim, $(s_n)_1^\infty \subset \mathcal{L}_1(u)$. Then, given $\epsilon > 0$, choose n_0 such that

$$
\|s_n - f\|_T (u_\omega)^\bullet_{q_E}(\chi_\omega) < \epsilon
$$

for $n \ge n_0$. Then

$$
(u_\omega)^\bullet_{q_E}(|s_n - f|) \le \|s_n - f\|_T (u_\omega)^\bullet_{q_E}(\chi_\omega) < \epsilon
$$

for $n \ge n_0$. Thus $f \in \mathcal{L}_1(u_\omega)$ and consequently, $\hat{f} \in \mathcal{L}_1(u)$ by Lemma 7.3.3(iii). Since $\hat{f}(t) = f(t)$ for $t \in \omega$ and $\hat{f}(t) = 0$ for $t \in T \backslash \omega$ and $K \subset \omega$, $\hat{f} = f$ and hence $f \in \mathcal{L}_1(u)$. Thus (i) holds.

(i)$\Rightarrow$(iv) Let $\omega \in \mathcal{V}$. Then by Definition 7.3.1, $\chi_\omega \in \mathcal{L}_1(u)$ and hence $\int_\omega du = x_\omega$ belongs to X. Then

$$
\int_\omega d(x^*u) = x^* \left(\int_\omega du \right) = x^*(x_\omega)
$$

and hence (iv) holds.

(iv)⇒(v) Obvious.

As shown above, (iii)⇒(v)⇒(vi)⇒(vii)⇒(i), (i)⇔(iv)⇔(v) and (i)⇒(iv)⇒(v)
⇒(vi)⇒(vii)⇒(i) so that (i)⇔(iv)⇔(v).

(i)⇒(viii) Obvious.

(viii)⇒(v) Let ω be a relatively compact open Baire set in T. Then $\bar{\omega}$ and
the boundary of ω are compact. Hence there exist vectors $x_{\bar{\omega}}$ and y in X such that
$x^*(x_{\bar{\omega}}) = \int_{\bar{\omega}} d(x^*u)$ and $x^*(y) = \int_A d(x^*u)$ for $x^* \in X^*$, where A is the boundary
of ω. Then $x^*(x_{\bar{\omega}} - y) = \int_{\omega} d(x^*u)$ for $x^* \in X^*$. Hence (v) holds.

(viii)⇒(ix) Obvious.

(ix)⇒(v) Let V be a relatively compact open Baire set in T. Then arguing as
in the proof of (7)⇒(5) of Theorem 5.3.12, $V = K\backslash(K\backslash V)$ with $K \in \mathcal{C}_0$ and $K\backslash V \in$
$\mathcal{C}_0$. Then by (ix) there exist x_K and $x_{K\backslash V}$ in X such that $\int_K d(x^*u) = x^*(x_K)$
and $\int_{K\backslash V} d(x^*u) = x^*(x_{K\backslash V})$ for $x^* \in X^*$. Then $x^*(x_K - x_{K\backslash V}) = \int_V d(x^*u)$ for
$x^* \in X^*$. Hence (v) holds.

(vi)⇒(x) Let $\omega \in \mathcal{V}$ and let $A \subset \mathcal{K}(\omega)$ be bounded. Then there exists $K \in \mathcal{C}$
such that supp $\varphi \subset K$ for $\varphi \in A$ and $\sup_{\varphi \in A} ||\varphi||_T = M < \infty$. Then by Theorem
50.D of [H] there exists a relatively compact open set ω in T such that $K \subset \omega$.
Then $A \subset C_c(\omega)$. Then by (vi), $u(A)$ is relatively weakly compact in X.

(x)⇒(vi) Let $\omega \in \mathcal{V}$ and let $A = \{\varphi \in C_0(\omega) : ||\varphi||_\omega \leq 1\}$. Then A is
bounded in $\mathcal{K}(T)$ and hence by (x), $u(A)$ is relatively weakly compact in X. Hence
$u_\omega = u|_{\mathcal{K}(\omega)}$ is a weakly compact bounded Radon operator. Hence (vi) holds.

(xi)⇒(viii) Suppose (xi) holds. Then given $\epsilon > 0$, there exists an open set
$\omega \supset K$ such that $u^\bullet(\omega\backslash K) < \epsilon$. Then by Urysohn's lemma, there exists $\varphi \in C_c(T)$
such that $\chi_K \leq \varphi \leq \chi_\omega$ so that $u^\bullet(\varphi - \chi_K) < \epsilon$. Hence $\chi_K \in \mathcal{L}_1(u)$ so that
$\int_K du \in X$. Then the weak integral $\int_K d\tilde{u} \in X$ and hence (viii) holds.

(xi)⇒(xii) Obvious.

(xii)⇒(ix) Let K be a compact G_δ in T. Then given $\epsilon > 0$, by (xii) there
exists an open set ω in T such that $K \subset \omega$ and $u^\bullet(\omega\backslash K) < \epsilon$. By Urysohn's lemma
there exists $\varphi \in c_c(T)$ such that $\chi_K \leq \varphi \leq \chi_\omega$ so that $u^\bullet(\varphi - \chi_K) < \epsilon$. Hence
$\chi_K \in \mathcal{L}_1(u)$ so that $\int_K du \in X$. Particularly, the weak integral $\int_K du \in X$ and
hence (ix) holds.

Thus (i)–(xii) are equivalent and this completes the proof of the theorem.   □

The following theorem is analogous to Theorems 7.2.13 and improves Theorem 3.4 of Thomas [T].

Theorem 7.3.6.

(a) *Let u be a Radon operator on $\mathcal{K}(T)$ with values in a Banach space (X, τ)
and let $H \subset X^*$ be a norm determining set for X with the Orlicz property
(respectively,*

(b) *let u be a Radon operator on $\mathcal{K}(T)$ with values in a quasicomplete lcHs (X, τ) and let $H \subset X^*$ have the Orlicz property and let the topology τ of X be identical with the topology of uniform convergence in the equicontinuous subsets of H). Let $\tilde{u}$ be the operator obtained from u on providing X with the topology $\sigma(X, H)$ and let $\widetilde{X}$ be the lcHs completion of (X, τ). Then the following are equivalent:*

(i) *u is a prolongable Radon operator.*

(ii) *For each $\omega \in \mathcal{V}$ the ultra weak integral $\int_\omega d\tilde{u}$ belongs to $\widetilde{X}$; i.e., there exists $x_\omega \in \widetilde{X}$ such that*

$$x^*(x_\omega) = \int_\omega d(x^*u)$$

for $x^* \in H$.

(iii) *Similar to (ii) with $\omega \in \mathcal{B}_0(T) \bigcap \mathcal{V}$.*

(iv) *For each $K \in \mathcal{C}$, the ultra weak integral $\int_K d\tilde{u}$ belongs to $\widetilde{X}$ (see (ii)).*

(v) *Similar to (iv) with $K \in \mathcal{C}_0$.*

Proof. Since (b) subsumes (a), we shall prove (b).

Let $\mathcal{E}_H$ be the family of equicontinuous subsets of H. Then (i)$\Rightarrow$(ii) by Theorem 7.3.5(iv).

Clearly, (ii)$\Rightarrow$(iii).

Let (iii) hold. Let $E \in \mathcal{E}_H$. For each $\omega \in \mathcal{B}_0(T) \bigcap \mathcal{V}$, there exists $x_\omega \in \widetilde{X}$ such that

$$x^*(x_\omega) = \int_\omega d(x^*u) \tag{7.3.6.1}$$

for $x^* \in H$. Arguing as in the proof of Theorem 7.2.19, given a disjoint sequence $(\omega_n)_1^\infty$ of open Baire sets with $\bigcup_1^\infty \omega_n \subset \omega$ and using (7.3.6.1) in place of (7.2.19.1), we have

$$\lim_n q_E(x_{\omega_n}) = 0. \tag{7.3.6.2}$$

Since E is equicontinuous and $u : C_0(\omega) \to X$ is continuous, by Lemma 2 of [P9] u^*E is bounded in $M(T)$. Then arguing as in the proof of Theorem 7.2.19, we conclude that $\Pi_{q_E} \circ u : \mathcal{K}(\omega) \to \widetilde{X_{q_E}}$ is a weakly compact bounded Radon operator for $E \in \mathcal{E}_H$. Then by the complex version of Lemma 2.21 of [T], u is prolongable. Hence (iii)$\Rightarrow$(i).

Thus (i)$\Leftrightarrow$(ii)$\Leftrightarrow$(iii).

Let u be prolongable and let $K \in \mathcal{C}$. Then by Theorem 50.D of [H] there exists $U \in \mathcal{V}$ such that $K \subset U$. Then by (ii), the ultra weak integrals $\int_U d\tilde{u} = x_U$

and $\int_{U\backslash K} d\tilde{u} = x_{U\backslash K}$ belong to $\widetilde{X}$. Then

$$\int_K d(x^*u) = \int_U d(x^*u) - \int_{U\backslash K} d(x^*u) = x^*(x_U - x_{U\backslash K})$$

for $x^* \in H$. Thus the ultra weak integral $\int_K d\tilde{u}$ belongs to $\widetilde{X}$ and hence (iv) holds.

(iv)⇒(v) Obvious.

Let (v) hold. Let ω be a relatively compact open Baire set. Then $\bar{\omega} \in \mathcal{C}$ and hence by Theorem 50.D of [H] there exists $K \in C_0$ such that $\bar{\omega} \subset K$. Then $\omega = K\backslash(K\backslash\omega)$ and $K\backslash\omega \in \mathcal{C}_0$ by Theorem 51.D of [H]. Then by hypothesis, there exist vectors x_K and $x_{K\backslash\omega}$ in $\widetilde{X}$ such that $\int_K d(x^*u) = x^*(x_K)$ and $\int_{K\backslash\omega} d(x^*u) = x^*(x_{K\backslash\omega})$ for $x^* \in H$. Consequently, $x^*(x_K - x_{K\backslash\omega}) = \int_\omega d(x^*u)$ for $x^* \in H$. Therefore, (v)⇒(iii).

Let (ii) hold. Then, particularly, for each open Baire set $\omega \in \mathcal{V}$, $u|_{\mathcal{K}(\omega)}$ is a weakly compact bounded Radon operator by Theorem 7.2.19. Then by Theorem 7.3.5(vii), u is prolongable.

This completes the proof of the theorem.   □

Remark 7.3.7. The complex versions of Theorem 3.5, Corollary 3.6, Proposition 3.7 and Lemma 3.10 of [T] hold in virtue of Remark 7.2.16 above.

Proposition 7.3.8. *Let $u : \mathcal{K}(T) \to X$ be a bounded Radon operator where X is a quasicomplete lcHs. Then u is a weakly compact bounded Radon operator if and only if u is prolongable and the function 1 belongs to $\mathcal{L}_1(u)$.*

Proof. If u is a weakly compact bounded Radon operator, then by Theorem 7.2.18 every bounded complex Borel function belongs to $\mathcal{L}_1(u)$ and hence the function 1 belongs to $\mathcal{L}_1(u)$ and moreover, every bounded complex Borel function with compact support belongs to $\mathcal{L}_1(u)$. Hence u is prolongable and the function 1 belongs to $\mathcal{L}_1(u)$. Conversely, if u is prolongable and the function 1 is u-integrable, then by the complex version of Corollary 3.6 of [T] every bounded Borel function is u-measurable and hence every bounded complex Borel function is u-integrable by the complex version of Theorem 1.22 of [T]. Consequently, by Theorem 7.2.18, u is a weakly compact bounded Radon operator.   □

Theorem 7.3.9. *Let X be a Banach space (resp. a quasicomplete lcHs) and let $u : \mathcal{K}(T) \to X$ be a prolongable Radon operator. Then a scalar function f on T is u-integrable if and only if f is weakly u-integrable and for every open Baire set ω in T, the weak integral $\int_\omega f du$ belongs to X (resp. the weak integral $\int_\omega f d\tilde{u} \in X$ where $\tilde{u}$ is the Radon operator obtained from u on providing X with the topology $\sigma(X, X^*)$).*

Proof. First we prove the theorem when X is a Banach space. By the complex versions of Theorem 1.22 and Corollary 3.6 of [T], for a bounded Borel function g, gf is u-integrable whenever f is u-integrable and hence the weak integral of gf

belongs to X. Hence the condition is necessary. Let $\tilde{u}$ be the operator obtained from u on providing X with the topology $\sigma(X, X^*)$ and let the hypothesis hold for each open Baire set in T. Then $\int_\omega f d\tilde{u} \in X$ for each open Baire set ω in T.

Let $\mathcal{F}$ be the complex vector space generated by the characteristic functions of open Baire sets in T, provided with the supremum norm. Then arguing as in the proof of Theorem 3.11 of [T] we have

$$\sup_{|g|\le1, g\in\mathcal{F}} \left| \int gf d\tilde{u} \right| < \infty.$$

Consequently, the mapping $\Phi : \mathcal{F} \to X$ given by

$$\Phi(g) = \int gf d\tilde{u}$$

is continuous. Then by Claim 4 in the proof of Theorem 6.3.3, $C_0(T) \subset \bar{\mathcal{F}}$, where $\bar{\mathcal{F}}$ is the closure of $\mathcal{F}$ in the Banach space of all bounded complex functions on T.

If Ψ is the continuous extension of Φ to $C_0(T)$, then

$$\langle \Psi(\varphi), x^* \rangle = \int \varphi f du_{x^*}$$

for $x^* \in X^*$, as there exists $(g_n)_1^\infty \subset \mathcal{F}$ such that $\|g_n - \varphi\|_T \to 0$ and $\Psi(g_n) = \Phi(g_n) = \int g_n f d\tilde{u}$. Thus

$$\Psi(\varphi) = \int \varphi f d\tilde{u},$$

$\int \varphi f d\tilde{u}$ belongs to X and Ψ is continuous on $C_0(T)$. Thus Ψ is a bounded Radon operator on $\mathcal{K}(T)$. Let ω be an open Baire set in T. Then by § 14 of [Din1] there exists $(\varphi_n)_1^\infty \subset C_0(T)$ such that $\varphi_n \nearrow \chi_\omega$. Then by LDCT we have

$$\int_\omega d(x^*\Psi) = \lim_n \int_T \varphi_n f d(x^*u) = \int_\omega f d(x^*u)$$

for $x^* \in X^*$. Thus $\int_\omega d\tilde{\Psi} = \int_\omega f d\tilde{u}$ which belongs to X by hypothesis. Then by the equivalence of (vi) and (vii) of Theorem 7.2.10, Ψ is weakly compact. Since $d\Psi_{x^*} = f du_{x^*}$ and since by hypothesis f is u_{x^*}-measurable for $x^* \in X^*$, the function f is Ψ_{x^*}-measurable for $x^* \in X^*$. Then by the complex version of Theorem 1.28 of [T], f is Ψ-measurable and consequently, by Theorem 7.2.11(vii), given $\epsilon > 0$, there exists a compact K such that $\Psi^\bullet(T\backslash K) < \frac{\epsilon}{2}$. As f is Ψ-measurable, there exists a compact $K_0 \subset K$ such that $f|_{K_0}$ is continuous and $\Psi^\bullet(K\backslash K_0) < \frac{\epsilon}{2}$. Then $\Psi^\bullet(T\backslash K_0) < \epsilon$. By Proposition 7.1.15, by Lemma 7.2.9 and by the fact that $d\Psi_{x^*} = f du_{x^*}$ we have

$$\epsilon > \Psi^\bullet(T\backslash K_0) = \sup_{|x^*|\le1} |\Psi_{x^*}|^\bullet(T\backslash K_0) = \sup_{|x^*|\le1} |\Psi_{x^*}|(T\backslash K_0)$$

$$= \sup_{|x^*|\le1} \int_{T\backslash K_0} |f| d|u_{x^*}| = \sup_{|x^*|\le1} \int_T |f - \chi_{K_0}f| d|u_{x^*}|$$

since $T\backslash K_0$ is open in T.

Since $\chi_{K_0}f$ is continuous in K_0, $\chi_{K_0}f$ is bounded (as K_0 is compact) and as u is prolongable, by Definition 7.3.1 $\chi_{K_0}f$ is u-integrable. Then by the complex version of Lemma 3.10 of [T], $f \in \mathcal{L}_1(u)$.

Now let X be a quasicomplete lcHs. For $q \in \Gamma$, let $\Pi_q : X \to X_q \subset \widetilde{X_q}$. Then Π_q is linear and continuous. If $y^* \in X_q^*$, then $y^* \circ \Pi_q \in X^*$. As f is weakly u-integrable, $f \in \mathcal{L}_1(x^*u)$ for each $x^* \in X^*$. Moreover, by hypothesis for each open Baire set $\omega \in T$ there exists a vector x_ω belonging to X such that

$$\int_\omega f d\tilde{u} = x_\omega$$

and hence

$$(y^* \circ \Pi_q)(x_\omega) = \int_\omega f d(y^* \circ \Pi_q)u$$

for $y^* \in X_q^*$. Hence $\Pi_q(\int_\omega f d\tilde{u}) = \int_\omega f d(\widetilde{\Pi_q \circ u}) = \Pi_q(x_\omega) \in \widetilde{X_q}$ for $q \in \Gamma$. Hence by the Banach space case, $f \in \mathcal{L}_1(\Pi_q \circ u)$ for each $q \in \Gamma$. Consequently, by the complex version of Proposition 1.28 of [T], f is u-integrable and therefore, by Theorem 7.1.24, $\int f du \in X$.

By the complex versions of Theorem 1.22 extended to lcHs (see p. 77 of [T]) and of Corollary 3.6 of [T] and by **1.12** of [T] the conditions are also necessary.

This completes the proof of the theorem.					$\square$

Lemma 7.3.10. *Let $(\theta_n)_1^\infty \subset \mathcal{K}(T)^*$ be such that for each $\varphi \in \mathcal{K}(T)$,*

$$\sum_1^\infty |\theta_n(\varphi)| < \infty.$$

Let $u(\varphi) = (\theta_n(\varphi))_1^\infty$ for $\varphi \in \mathcal{K}(T)$. Then u is a prolongable Radon operator with values in $\ell_1(\mathbb{N})$. Let f be a complex function which is θ_n-integrable for each $n \in \mathbb{N}$ such that

$$\sum_1^\infty |\int_\omega f d\mu_n| < \infty$$

for each open Baire set ω in T, where μ_n is the complex Radon measure induced by θ_n in the sense of Definition 4.3 of [P3]. Then $f \in \mathcal{L}_1(u)$. If $s(\varphi) = \sum_1^\infty \int \varphi d\mu_n$ for $\varphi \in \mathcal{K}(T)$, then $s \in \mathcal{K}(T)^*$, f is s-integrable and

$$\int_\omega f ds = \sum_1^\infty \int_\omega f d\mu_n$$

for each open Baire set ω in T.

Proof. The proof of Lemma 3.14 of [T] holds here for the complex case too. The only change is that we have to use Corollary 5.2.5 in place of Appendix I, T4 of [T]. The details are left to the reader.					$\square$

The following theorem improves the complex version of Theorem 3.13 of [T].

Theorem 7.3.11. *Let u be a prolongable Radon operator on $\mathcal{K}(T)$ with values in a Banach space X and let H be a norm determining set in X^*. Suppose H has the Orlicz property. Let $\tilde{u}$ be the operator obtained from u on providing X with the topology $\sigma(X, H)$. Then a complex function f on T is u-integrable if and only if f is $\tilde{u}$-integrable (i.e., if and only if f is u_{x^*}-integrable for each $x^* \in H$) and for each open Baire set ω in T, the integral $\int_\omega f d\tilde{u}$ belongs to X.*

Proof. The proof of Theorem 3.13 of [T] holds here verbatim excepting that we have to apply Corollary 5.2.5 instead of Appendix I T4 of [T]. The details are left to the reader. □

The following theorem improves the complex version of Theorem 3.20 of [T].

Theorem 7.3.12. *Let u be a Radon operator on $\mathcal{K}(T)$ with values in a complete lcHs X. Let H be a subset of X^* with the Orlicz property and let the topology τ of X be the same as the topology of uniform convergence in the equicontinuous subsets of H. Let $\tilde{u}$ be the operator obtained from u on providing X with the topology $\sigma(X, H)$. Then a complex function f on T is u-integrable if and only if f is u_{x^*}-integrable for each $x^* \in H$ and for each open Baire set ω in T, $\int_\omega d\tilde{u} \in X$ (i.e., there exists $x_\omega \in X$ such that $x^*(x_\omega) = \int_\omega f d(x^* u)$ for $x^* \in H$).*

Proof. The proof of Theorem 3.20 of [T] holds here verbatim excepting that we have to use Theorem 7.2.19 in place of Theorem 2.7 bis of [T]. The details are left to the reader. □

7.4 Baire versions of Proposition 4.8 and Theorem 4.9 of [T]

Using the Baire version of the Dieudonné-Grothendieck theorem we give a complex Baire version of Proposition 4.8 and Theorem 4.9 of [T] including that of the remark on p. 128 of [T]. For this we start with the following two lemmas.

Lemma 7.4.1. *Let u be a prolongable Radon operator on $\mathcal{K}(T)$ with values in a Banach space X and let $f \in \mathcal{L}_1(u)$. Then the operator $\Psi : C_0(T) \to X$ given by*

$$\Psi(\varphi) = \int \varphi f du \ for \varphi \in C_0(T)$$

is weakly compact.

Proof. By the complex version of Theorem 1.22 of [T], $\varphi f \in \mathcal{L}_1(u)$ for $\varphi \in C_0(T)$ and hence Ψ is well defined. Clearly, Ψ is linear and

$$|\Psi(\varphi) \leq ||\varphi||_T u^\bullet(f)$$

for $\varphi \in C_0(T)$. Hence Ψ is continuous.

Since $f \in \mathcal{L}_1(u)$, by the complex versions of Theorem 1.22 and Corollary 3.6 of [T], $\chi_\omega f \in \mathcal{L}_1(u)$ for each open Baire set ω in T and thus

$$x_\omega = \int_\omega f du \in X. \tag{7.4.1.1}$$

Then

$$x^*(x_\omega) = \int_\omega f d(x^*u) \tag{7.4.1.2}$$

for each open Baire set ω in T.

Let $\mathcal{F}$ be the vector space spanned by the characteristic functions of open Baire sets in T and let it be provided with the supremum norm. Then by (7.4.1.1), for each $g \in \mathcal{F}$ there exists $x_g \in X$ such that

$$x_g = \int g f du \tag{7.4.1.3}$$

so that

$$x^*(x_g) = \int g f d(x^*u) \tag{7.4.1.4}$$

for $x^* \in X^*$.

Let $\Phi(g) = x_g$ for $g \in \mathcal{F}$. Then $\Phi : \mathcal{F} \to X$ is linear and

$$|\Phi(g)| = \left| \int g f du \right| \le \|g\|_T \|u^\bullet(f).$$

Hence Φ is continuous. Therefore, Φ has a unique continuous linear extension $\hat{\Phi}$ on the closure $\bar{\mathcal{F}}$ in the Banach space of all bounded complex functions on T with the supremum norm. Then $C_0(T) \subset \bar{\mathcal{F}}$ by Claim 4 in the proof of Theorem 6.3.3 and hence let $\Phi_0 = \hat{\Phi}|_{C_0(T)}$. Thus $\Phi_0 : C_0(T) \to X$ is continuous and linear and hence by Theorem 1 of [P9] its representing measure η is given by $\Phi_0^{**}|_{\mathcal{B}(T)}$. Moreover, by the same theorem, $(x^* \circ \eta) \in M(T)$ for each $x^* \in X^*$ and

$$x^*\Phi_0(\varphi) = \int \varphi d(x^* \circ \eta) \tag{7.4.1.5}$$

for $\varphi \in C_0(T)$ and for $x^* \in X^*$.

Let $\varphi \in C_0(T)$. Then by Claim 4 in the proof of Theorem 6.3.3 there exists $(g_n)_1^\infty \subset \mathcal{F}$ such that $g_n \to \varphi$ uniformly in T so that $\Phi_0(\varphi) = \lim_n \Phi(g_n)$. Then by (7.4.1.4) we have

$$x^*\Phi_0(\varphi) = \lim_n x^*\Phi(g_n) = \lim_n x^*(x_{g_n}) = \lim_n \int f g_n d(x^*u) \tag{7.4.1.6}$$

for $x^* \in X^*$. Since $f \in \mathcal{L}_1(u)$, $f \in \mathcal{L}_1(x^*u)$ for $x^* \in X^*$ and hence $\int_T |f| d|x^*u| < \infty$ for each $x^* \in X^*$. Moreover, by the complex version of Theorem 1.22 of [T],

$f\varphi \in \mathcal{L}_1(u)$ and hence $f\varphi \in \mathcal{L}_1(x^*u)$ for $x^* \in X^*$. Since $f \in \mathcal{L}_1(u)$, $f \in \mathcal{L}_1(x^*u)$ for $x^* \in X^*$ and hence

$$\int |f| d|x^*u| < \infty. \tag{7.4.1.7}$$

Consequently, by (7.4.1.7) we have

$$\left| \int f\varphi d(x^*u) - \int f g_n d(x^*u) \right| \leq |\varphi - g_n||_T \int |f| d|x^*u| \to 0$$

as $n \to \infty$ and therefore

$$\int f\varphi d(x^*u) = \lim_n \int f g_n d(x^*u) \tag{7.4.1.8}$$

for $x^* \in X^*$. Then by (7.4.1.6) and (7.4.1.8) we have

$$x^* \Phi_0(\varphi) = \int f\varphi d(x^*u) \text{ for } x^* \in X^* \text{and for } \varphi \in C_0(T). \tag{7.4.1.9}$$

Let ω be an open Baire set in T. Then by § 14 of [Din 1] there exists $(\varphi_n)_1^\infty \subset C_0(T)$ such that $\varphi_n \nearrow \chi_\omega$. Consequently, by LDCT, by (7.4.1.2), by (7.4.1.5) and by (7.4.1.9) we have

$$(x^* \circ \boldsymbol{\eta})(\omega) = \int \chi_\omega d(x^* \circ \boldsymbol{\eta}) = \lim_n \int \varphi_n d(x^* \circ \boldsymbol{\eta})$$

$$= \lim_n x^* \Phi_0(\varphi_n) = \lim_n \int f\varphi_n d(x^*u)$$

$$= \int \chi_\omega f d(x^*u) = x^*(x_\omega).$$

Thus

$$(x^* \circ \boldsymbol{\eta})(\omega) = x^*(x_\omega) \text{ for } x^* \in X^*$$

and for each open Baire set ω in T. Thus $\boldsymbol{\eta}(\omega) = x_\omega$ and hence by Theorem 3(vii) of [P9], Φ_0 is weakly compact.

On the other hand, by (7.4.1.9) and by the definition of Ψ (see the statement of Lemma 7.4.1)) we have

$$x^* \Psi(\varphi) = \int f\varphi d(x^*u) = x^* \Phi_0(\varphi)$$

for $\varphi \in C_0(T)$ and for $x^* \in X^*$. Then by the Hahn-Banach theorem, $\Psi = \Phi_0$ and hence Ψ is weakly compact. □

Lemma 7.4.2. *Let* u, f, X *and* Ψ *be as in Lemma 7.4.1 and let* $M_\Psi = \{A \subset T : \chi_A \in \mathcal{L}_1(\Psi)\}$. *Then:*

(i) M_Ψ *is a* σ*-algebra in* T.

(ii) $\mathcal{B}(T) \subset M_\Psi$.

(iii) *If $\boldsymbol{\mu_\Psi}(A) = \int_A d\Psi$ for $A \in M_\Psi$, then $\boldsymbol{\mu_\Psi}(\omega) = \int_\omega f du$ for each open Baire set ω in T.*

(iv) *If $\boldsymbol{\lambda}(A) = \int_A f du$ for $A \in \mathcal{B}(T)$, then $\boldsymbol{\lambda}$ is σ-additive on $\mathcal{B}(T)$ and for each $x^* \in X^*$, $x^*\boldsymbol{\lambda}(\cdot)$ is Borel regular.*

(v) *$\boldsymbol{\mu_\Psi}(A) = \boldsymbol{\lambda}(A)$ for $A \in \mathcal{B}(T)$ and consequently, $\boldsymbol{\lambda}$ is Borel regular.*

(vi) *For a bounded complex Borel function g on T,*

$$\int g d\Psi = \int g f du. \tag{7.4.2.1}$$

Proof. Since Ψ is weakly compact by Lemma 7.4.1, (i) and (ii) hold by Theorem 7.5.4 (see the next section).

(iii) For $\varphi \in C_0(T)$, by Lemma 7.4.1 $\Psi(\varphi) = \int f\varphi du$ and hence

$$x^*\Psi(\varphi) = \int \varphi f d(x^*u) \tag{7.4.2.2}$$

for $x^* \in X^*$. Then by § 14 of [Din 1], there exists $(\varphi_n)_1^\infty \subset C_0(T)$ such that $\varphi_n \nearrow \chi_\omega$. Then by Theorem 4.7 (LDCT) of [T] and by Proposition 7.2.5 we have

$$\Psi(\omega) = \int \chi_\omega d\Psi = \lim_n \int \varphi_n d\Psi \tag{7.4.2.3}$$

and hence by (7.4.2.2) and (7.4.2.3) and by LDCT for complex measures we have

$$x^*\Psi(\omega) = \lim_n \int \varphi_n d(x^*\Psi) = \lim_n (x^*\Psi)(\varphi_n) = \lim_n \int \varphi_n f d(x^*u) = \int \chi_\omega f d(x^*u)$$

for $x^* \in X^*$. Therefore

$$(x^*\Psi)(\omega) = \int_\omega f d(x^*u) \text{ for } x^* \in X^*.$$

Since $f \in \mathcal{L}_1(u)$, by the complex analogue of Theorem 1.22 of [T], $\chi_\omega f \in \mathcal{L}_1(u)$ and hence we have

$$\int_\omega f d(x^*u) = x^* \left(\int_\omega f du \right).$$

Therefore,

$$(x^*\Psi)(\omega) = x^* \left(\int_\omega f du \right) \text{ for } x^* \in X^*.$$

Then by the Hahn-Banach theorem and by the definition of $\boldsymbol{\mu_\Psi}$ we have

$$\boldsymbol{\mu_\Psi}(\omega) = \int_\omega f du$$

for open Baire sets ω in T. Hence (iii) holds.

(iv) As u is prolongable, by the complex version of Corollary 3.6 of [T] χ_A is u-measurable for each $A \in \mathcal{B}(T)$. As $|\chi_A f| \leq |f|$ for $A \in \mathcal{B}(T)$ and as $f \in \mathcal{L}_1(u)$, by the complex version of Theorem 1.22 of [T], $\chi_A f$ is u-integrable for each $A \in \mathcal{B}(T)$. Let $(A_n)_1^\infty$ be a disjoint sequence of Borel sets in T with $A = \bigcup_1^\infty A_n$. Then $\sum_{k=1}^n \chi_{A_k} \nearrow \chi_A$ and hence $(\sum_1^n \chi_{A_k})|f| \leq \chi_A |f| \in \mathcal{L}_1(u)$ and $(\sum_1^n \chi_{A_k})f \to \chi_A f$ in T. Hence by the complex version of Theorem 4.7 of [T],

$$\sum_1^\infty \int \chi_{A_k} f \, du = \sum_{k=1}^\infty \int_{A_k} f \, du = \int_A f \, du$$

and hence $\boldsymbol{\lambda}(\cdot)$ is σ-additive on $\mathcal{B}(T)$. Moreover,

$$x^* \boldsymbol{\lambda}(\cdot) = \int_{(\cdot)} f \, d(x^* u)$$

is Borel regular on $\mathcal{B}(T)$ by Theorem 6.4.6 (see the beginning of the proof of Theorem 6.4.6) and hence (iv) holds.

(v) By (iii), $\boldsymbol{\lambda}(\omega) = \boldsymbol{\mu}_\Psi(\omega)$ for open Baire sets ω in T and hence by the Baire regularity of $\boldsymbol{\lambda}|_{\mathcal{B}_0(T)}$ and $\boldsymbol{\mu}_\Psi|_{\mathcal{B}_0(T)}$ we conclude that $\boldsymbol{\lambda}(A) = \boldsymbol{\mu}_\Psi(A)$ for $A \in \mathcal{B}_0(T)$. For $x^* \in X^*$, $x^* \boldsymbol{\lambda}(\cdot) = \int_{(\cdot)} f \, d(x^* u)$ is Borel regular and σ-additive on $\mathcal{B}(T)$ by Theorem 6.4.6. Since $\boldsymbol{\mu}_\Psi$ is Borel regular and σ-additive on $\mathcal{B}(T)$ by Theorem 7.5.4 (see Section 7.5), $x^* \boldsymbol{\mu}_\Psi$ is Borel regular and σ-additive on $\mathcal{B}(T)$, and hence by the uniqueness part of Proposition 1 of [DP1],

$$x^* \boldsymbol{\lambda}(A) = x^* \boldsymbol{\mu}_\Psi(A)$$

for $A \in \mathcal{B}(T)$ and for $x^* \in X^*$. Hence by the Hahn-Banach theorem, $\boldsymbol{\lambda}(A) = \boldsymbol{\mu}_\Psi(A)$ for $A \in \mathcal{B}(T)$. Consequently, $\boldsymbol{\lambda}$ is Borel regular on $\mathcal{B}(T)$.

(vi) By (iv) and (v),

$$\int s \, d\boldsymbol{\mu}_\Psi = \int s f \, du \qquad (7.4.2.4)$$

for a Borel simple function s. Given a bounded complex Borel function g, there exists $(s_n)_1^\infty$ of Borel simple functions such that $s_n \to g$ uniformly in T. Then by the complex version of **1.10** of [T] we have

$$\left| \int s_n f \, du - \int g f \, du \right| \leq ||s_n - g||_T u^\bullet(f) \to 0 \qquad (7.4.2.5)$$

as $n \to \infty$. Since Ψ is a bounded Radon operator, by Proposition 7.2.3, $\Psi^\bullet(T) < \infty$. Since $| \int g \, d\Psi - \int s_n d\Psi | = | \int (g - s_n) d\Psi | \leq ||g - s_n||_T \Psi^\bullet(T) \to 0$, by (iv) and (v) we have $\int g \, d\Psi = \lim_n \int s_n d\Psi = \lim_n \int s_n d\boldsymbol{\mu}_\Psi = \lim_n \int f s_n du = \int f g \, du$ by (7.4.2.4) and (7.4.2.5). Hence (vi) holds. $\qquad \square$

Theorem 7.4.3. *Let u be a prolongable Radon operator on $\mathcal{K}(T)$ with values in a Banach space (resp. a quasicomplete lcHs) X. Let $(f_n)_1^\infty$ be a sequence of u-integrable complex functions converging u-a.e. to a function f in T. If the sequence $\int_\omega f_n du$ converges in τ (the topology of X) (respectively, converges weakly) in X for all open Baire sets ω in T, then the function f is u-integrable and $\int_\omega f_n du$ converges in τ (resp. weakly) to $\int_\omega f du \in X$ for each open Baire set ω in T. Moreover, for each bounded complex Borel function g on T,*

$$\int f_n g du \to \int f g du \text{ in } \tau \text{ in } X$$

(resp.

$$\int f_n g du \to \int f g du \text{ weakly in } X.)$$

Proof. Let ω be an open Baire set in T. By hypothesis, there exists a vector $x_\omega \in X$ such that

$$\lim_n \int_\omega f_n du = x_\omega \text{ in } \tau \tag{7.4.3.1}$$

(resp.

$$\lim_n \int_\omega f_n du = x_\omega \text{ weakly.} \tag{7.4.3.1'})$$

In both the cases, by **1.34** of [T],

$$\lim_n \int_\omega f_n du_{x^*} = x^*(x_\omega) \tag{7.4.3.2}$$

for $x^* \in X^*$. On the other hand, by hypothesis and by Theorem 6.4.6 we have

$$\lim_n \int_\omega f_n du_{x^*} = \int_\omega f du_{x^*} \tag{7.4.3.3}$$

for $x^* \in X^*$, since $u_{x^*} = x^*u \in \mathcal{K}(T)^*$. Then by (7.4.3.2) and (7.4.3.3) we have

$$x^*(x_\omega) = \int_\omega f du_{x^*} \tag{7.4.3.4}$$

for each open Baire set ω in T and for each $x^* \in X^*$. Consequently, by the hypothesis that $f_n \to f$ u-a.e. in T so that $f_n \to f$ u_{x^*}-a.e. in T for $x^* \in X^*$ and by Theorem 6.4.6 we have

$$f \in \mathcal{L}_1(x^*u) \tag{7.4.3.5}$$

for $x^* \in X^*$. Then by (7.4.3.4) and (7.4.3.5) and by Theorem 7.3.9, f is u-integrable in both cases of X.

By the complex analogue of Theorem 1.22 of [T] $f\chi_\omega$ is u-integrable for each open Baire set ω in T and hence by (7.4.3.4) we have

$$x^*\left(\int_\omega f du\right) = \int_\omega f d(x^*u) = \int_\omega f du_{x^*} = x^*(x_\omega)$$

for each $x^* \in X^*$. Since $\int_\omega f du \in X$, by the Hahn-Banach theorem we have

$$\int_\omega f du = x_\omega$$

and hence by (7.4.3.1) (resp. by (7.4.3.1'))

$$\int_\omega f_n du \to \int_\omega f du \text{ in } \tau \qquad (7.4.3.6)$$

(resp.

$$\int_\omega f_n du \to \int_\omega f du \text{ weakly.})$$

Let $x^* \in X^*$, X a quasicomplete lcHs and g be a bounded u-measurable complex function. Then g is an $x^* u$-measurable function. Clearly, $\theta = x^* u \in \mathcal{K}(T)^*$ and hence by Theorem 6.4.6 we have

$$\lim_n \int g f_n d(x^* u) = \int g f d(x^* u). \qquad (7.4.3.7)$$

On the other hand, by the complex version of Theorem 1.22 of [T], $g f_n$ and $g f$ belong to $\mathcal{L}_1(u)$ and hence by (7.4.3.7) we have

$$\lim_n \left(x^* \int g f_n du \right) = \lim_n \int g f_n d(x^* u) = \int g f d(x^* u) = x^* \left(\int g f du \right)$$

for each $x^* \in X^*$. Hence

$$\int g f_n du \to \int g f du \text{ weakly.}$$

To prove the result for the convergence in τ, let $\Psi_n : C_0(T) \to X$ be given by $\Psi_n(\varphi) = \int \varphi f_n du$ for $\varphi \in C_0(T)$ and let $\Psi : C_0(T) \to X$ be given by $\Psi(\varphi) = \int \varphi f du$ for $\varphi \in C_0(T)$. Then by Lemma 7.4.1, Ψ_n and Ψ are weakly compact. By hypothesis, by Lemma 7.4.2(iii) and by (7.4.3.6) we have

$$\int_\omega d\Psi_n = \int_\omega f_n du \to \int_\omega f du = \int_\omega d\Psi \text{ in } \tau. \qquad (7.4.3.8)$$

Case 1. X is a Banach space.

By (7.4.3.8), by Theorem 7.2.15 and by Lemma 7.4.2,

$$\lim_n \int g d\Psi_n = \lim_n \int g f_n du \to \int g d\Psi = \int f g du$$

in τ for each bounded complex Borel function g on T.

Case 2. X is a quasicomplete lcHs.

For each $q \in \Gamma$, by Lemma 7.4.1 and by the continuity of Π_q, $\Pi_q \circ \Psi$ and $\Pi_q \circ \Psi_n$ are weakly compact. By hypothesis and by the first part of the theorem, $\int_\omega f_n du \to \int_\omega f du$ in τ for each open Baire set in T. Then by Lemma 7.4.2,

$$\Pi_q \left(\int_\omega d\Psi_n \right) = \Pi_q \left(\int_\omega f_n du \right) = \int_\omega f_n d(\Pi_q \circ u) \to \int_\omega f d(\Pi_q \circ u) \text{ in } \widetilde{X}_q.$$

Hence by the case of Banach spaces, we have

$$\lim_n \int g d(\Pi_q \circ \Psi_n) = \lim_n \int g f_n d(\Pi_q \circ u) \to \int g f d(\Pi_q \circ u) = \int g d(\Pi_q \circ \Psi).$$

Hence

$$q \left(\int g f_n du - \int g f du \right) = \left| \int g d(\Pi_q \circ \Psi_n) - \int g d(\Pi_q \circ \Psi) \right|_q \to 0$$

for each $q \in \Gamma$ and hence

$$\int g f_n du \to \int g f du \text{ in } \tau.$$

This completes the proof of the theorem. $\qquad\qquad\qquad\qquad\qquad\qquad\square$

Theorem 7.4.4. *Let X be a Banach space with topology τ. Let u be a prolongable Radon operator on $\mathcal{K}(T)$ with values in X. Let $(f_n)_1^\infty$ be a sequence of u-integrable complex functions and suppose the sequence $(\int_\omega f_n du)_1^\infty$ converges in τ (resp. converges weakly) in X for each open Baire set ω in T. Then there exists a function $f \in \mathcal{L}_1(u)$, u-essentially unique, such that $\int_\omega f_n du \to \int_\omega f du$ in τ (resp. weakly) for each open Baire set ω in T. Moreover, when $(\int_\omega f_n du)$ converges in τ for each open Baire set ω in T, then for each bounded complex Borel function g,*

$$\int f_n g du \to \int f g du \text{ in } \tau$$

as well as

$$\int f_n g du \to \int f g du \text{ weakly}$$

in X.

Proof. Let $x^* \in X^*$. By hypothesis, in both the cases of convergence, $\int_\omega f_n d(x^* u)$ converges in $\mathbb{K}$ for each open Baire set ω in T. Hence by the Baire version of the Dieudonné-Grothendieck theorem (i.e., Theorem 5.2.6),

$$\text{the sequence } (f_n) \text{ converges weakly in } \mathcal{L}_1(u_{x^*}) \qquad\qquad (7.4.4.1)$$

for each $x^* \in X^*$.

Then by the complex version of Lemma 1 on p. 126 of [T] one can suppose that the f_n are null in the complement of $\bigcup_1^\infty K_p$, where $(K_p)_1^\infty \subset \mathcal{C}$. By the complex version of Lemma 2 on pp. 126–127 of [T], with the sequence $(K_p)_1^\infty$ we can associate a sequence $(x_i^*)_1^\infty \subset X^*$ with the property mentioned in the lemma. Then by the complex version of Lemma 3 on p. 127 of [T], there exists a sequence of barycenters g_n of the f_n given by

$$g_n = \sum_{i=n}^{N(n)} \alpha_i^{(n)} f_i, \quad \alpha_i^{(n)} \geq 0, \text{ and } \sum_{i=n}^{N(n)} \alpha_i^{(n)} = 1 \qquad (7.4.4.2)$$

such that (g_n) converges in mean in $\mathcal{L}_1(x_i^* u)$ and converges $(x_i^* u)$-a.e. in T for each $i \in \mathbb{N}$. Thus, for each $i \in \mathbb{N}$, there exists $N_i \subset T$ such that $|x_i^* u|^\bullet(N_i) = 0$ and $(g_n(t))_1^\infty$ is convergent in $T\backslash N_i$. (Compare with the proof of Theorem 6.4.12.) Thus, if $N = \bigcap_1^\infty N_i$, then $(g_n(t))_1^\infty$ is convergent in $T\backslash N = \bigcup_1^\infty(T\backslash N_i)$ and $|x_i^* u|^\bullet(N) = 0$ for $i \in \mathbb{N}$. Therefore, by the complex version of Lemma 2 on pp. 126–127 of [T], g_n converges u-a.e. in T. Let f be the u-a.e. limit of the sequence (g_n). As (f_n) converges weakly in $\mathcal{L}_1(u_{x^*})$ for each $x^* \in X^*$ by (7.4.4.1) and as (g_n) is given by (7.4.4.2), it follows that (g_n) also converges weakly in $\mathcal{L}_1(u_{x^*})$ for each $x^* \in X^*$. Then by Theorem 6.4.6 (taking $\theta = x^* u \in \mathcal{K}(T)^*$), $f \in \mathcal{L}_1(u_{x^*})$ for each $x^* \in X^*$ and

$$\lim_n \int gf_n du_{x^*} = \int gf du_{x^*} \qquad (7.4.4.3)$$

for $x^* \in X^*$ and for every bounded complex Borel function g on T. Thus f is weakly u-integrable and by (7.4.4.3) we have

$$\lim_n \int_\omega f_n d(x^* u) = \int_\omega f d(x^* u) \qquad (7.4.4.4)$$

for each open Baire set ω in T and for $x^* \in X^*$. But by hypothesis, in both cases of convergence, there exists $x_\omega \in X$ such that

$$\lim_n \int_\omega f_n d(x^* u) = x^*(x_\omega) \qquad (7.4.4.5)$$

for each $x^* \in X^*$. Thus by (7.4.4.4) and (7.4.4.5) we have

$$\int_\omega f d(x^* u) = x^*(x_\omega) \qquad (7.4.4.6)$$

for $x^* \in X^*$. Then by Theorem 7.3.9, $f \in \mathcal{L}_1(u)$ and by (7.4.4.6) we have

$$x^*\left(\int_\omega f du\right) = \int_\omega f d(x^* u) = x^*(x_\omega) \qquad (7.4.4.7)$$

for $x^* \in X^*$. Then by (7.4.4.5) and (7.4.4.7), $\int_\omega f_n du \to \int_\omega f du$ weakly.

If $\int_\omega f_n du$ converges to x_ω in τ, then $\int_\omega f du = x_\omega$ by (7.4.4.7) and by the Hahn-Banach theorem and hence

$$\int_\omega f_n du \rightarrow \int_\omega f du \text{ in } \tau.$$

Suppose there exists $h \in \mathcal{L}_1(u)$ such that

$$\int_\omega f_n du \rightarrow \int_\omega h du$$

in τ (resp. weakly). Then

$$\int_\omega f_n du_{x^*} \rightarrow \int_\omega h du_{x^*}$$

and hence by Theorem 6.4.6, $f = h \ u_{x^*}$-a.e. in T for each $x^* \in X^*$. Then by the complex version of Proposition 3.7 of [T], $f = h \ u$-a.e. in T and hence f is u-essentially unique.

Let $\Psi_n : C_0(T) \rightarrow X$ be given by

$$\Psi_n(\varphi) = \int \varphi f_n du, \ \ \varphi \in C_0(T).$$

Then by Lemma 7.4.1, Ψ_n is weakly compact for each n and by hypothesis and by Lemma 7.4.2(iii), $\lim_n \int_\omega d\Psi_n \in X$ in τ for each open Baire set ω in T. Then by Theorem 7.2.15 and by Lemma 7.4.2 there exists a bounded weakly compact operator Ψ on $\mathcal{K}(T)$ with values in X such that

$$\lim_n \int g d\Psi_n = \lim_n \int g f_n du = \int g d\Psi \tag{7.4.4.8}$$

in τ for each bounded complex Borel function g on T. Then by the fact that $f \in \mathcal{L}_1(u)$ and g is bounded, $gf \in \mathcal{L}_1(u)$. Moreover, by (7.4.4.3) and (7.4.4.8) and by Theorem 6.4.12 we have

$$\lim_n \int g f_n \, du = \int g f du$$

and consequently, $\lim_n x^* \int g f_n du = \lim_n \int g f_n du_{x^*} = \int g f du_{x^*} = x^* \left(\int g f du \right) = x^* \left(\int g d\Psi \right)$ for $x^* \in X^*$. Consequently, by the Hahn-Banach theorem

$$\int g d\Psi = \int g f du$$

and hence by (7.4.4.8) we have

$$\lim_n \int g f_n du = \int g f du \text{ in } \tau$$

for each bounded complex Borel function g. Since gf_n and gf are u-integrable, (7.4.4.3) implies

$$\lim_n x^* \left(\int gf_n du \right) = \lim_n \int gf du_{x^*} = \int fg du_{x^*} = x^* \left(\int gf du \right)$$

and hence

$$\int f_n g du \to \int fg du \text{ weakly}.$$

This completes the proof of the theorem. $\qquad\qquad\qquad\qquad\qquad\qquad\square$

7.5 Weakly compact and prolongable Radon vector measures

If u is a bounded Radon operator with values in a quasicomplete lcHs, we define $M_u = \{A \subset T : \chi_A \in \mathcal{L}_1(u)\}$ and $\boldsymbol{\mu}_u(A) = \int_A du$ for $A \in M_u$. When u is a weakly compact bounded Radon operator, we show that M_u is a σ-algebra containing $\mathcal{B}(T)$ and $\boldsymbol{\mu}_u = \widetilde{\mathbf{m}_u}$, where $\mathbf{m}_u$ is the representing measure of u in the sense of 5.2.10, and $\widetilde{\mathbf{m}_u}$ is the generalized Lebesgue completion of $\mathbf{m}_u = u^{**}|_{\mathcal{B}(T)}$ with respect to $\mathcal{B}(T)$. In Theorem 7.5.7 (resp. Theorem 7.5.8) we give several characterizations of a weakly compact bounded Radon (resp. a prolongable Radon) operator u. Theorems 7.5.9 and 7.5.11 study the regularity properties of $\boldsymbol{\mu}_u$ when u is a weakly compact bounded Radon operator or a prolongable Radon operator, respectively. Then we study the outer measure $\boldsymbol{\mu}_u^*$ of $\boldsymbol{\mu}_u$ in the sense of [Si] when u is a weakly compact bounded Radon operator and give the relation between M_u and $\boldsymbol{\mu}_u^*$-measurable sets in Theorem 7.5.20, where we also show that $M_u = M_{\boldsymbol{\mu}_u^*}$ and $\boldsymbol{\mu}_u^*(E) = \boldsymbol{\mu}_u(E)$ for $E \in M_u$. We introduce the concepts of Lebesgue-Radon completion and localized Lebesgue-Radon completion and in terms of them we generalize Theorems 4.4 and 4.6 of [P4].

Definition 7.5.1. Let X be a quasicomplete lcHs and $u : \mathcal{K}(T) \to X$ be a Radon operator. Let $M_u = \{A \subset T : \chi_A \in \mathcal{L}_1(u)\}$ and let $\boldsymbol{\mu}_u(A) = \int_A du$ for $A \in M_u$. Then $\boldsymbol{\mu}_u$ is called the Radon vector measure induced by u and M_u is called the domain of $\boldsymbol{\mu}_u$. $\boldsymbol{\mu}_u$ is called a weakly compact (resp. prolongable) Radon vector measure if u is a weakly compact bounded (resp. a prolongable) Radon operator on $\mathcal{K}(T)$.

Theorem 7.5.2. Let X, u, M_u and $\boldsymbol{\mu}_u$ be as in Definition 7.5.1. Then M_u is a ring of u-measurable sets and $\boldsymbol{\mu}_u$ is σ-additive on M_u.

Proof. Since $0 \in \mathcal{L}_1(u)$, $\emptyset \in M_u$. Let $q \in \Gamma$. For $A_1, A_2 \in M_u$ and $K \in \mathcal{C}$, by the complex version of Proposition 1.21 of [T] and by Theorem 7.1.14 there exist disjoint sequences $(K_{i,n})_{n=1}^{\infty} \subset \mathcal{C}$, $i = 1, 2$ and sets N_1 and N_2 with $u_q^\bullet(N_1 \cup N_2) = 0$

such that

$$K = \bigcup_{n=1}^{\infty} K_{i,n} \cup N_i$$

with $K_{i,n} \subset K \cap A_i$ or $K_{i,n} \subset K \backslash A_i$, $i = 1, 2$.

Let $J_i = \{n : K_{i,n} \subset K \cap A_i\}$ and $I_i = \{n : K_{i,n} \subset K \backslash A_i\}$, $i = 1, 2$. Then

$$K \cap (A_1 \backslash A_2) = \left(\bigcup_{n \in J_1} K_{1,n} \cup F_1 \right) \cap \left(\bigcup_{n \in I_2} K_{2,n} \cup F_2 \right)$$

and

$$K \backslash (A_1 \backslash A_2) = \left(\bigcup_{n \in I_1} K_{1,n} \cup F_3 \right) \cup \left(\bigcup_{n \in J_2} K_{2,n} \cup F_4 \right)$$

with $u_q^{\bullet}(F_i) = 0$, $i = 1, 2, 3, 4$. Consequently, $A_1 \backslash A_2$ is u_q-measurable. As $\chi_{A_1 \backslash A_2} \leq \chi_{A_1} \in \mathcal{L}_1(u)$, by the complex version of Theorem 1.22 of [T], $A_1 \backslash A_2 \in M_u$. Since $\chi_{A_1 \cup A_2} = \chi_{A_2} + \chi_{A_1 \backslash A_2}$, $A_1 \cup A_2 \in M_u$. Hence M_u is a ring of u-measurable sets.

For $x^* \in X^*$, $x^* u \in \mathcal{K}(T)^*$ and hence by the complex version of Proposition 1.30 of [T], $\mathcal{L}_1(u) \subset \mathcal{L}_1(x^* u)$. Then $\mu_{x^* u}$ is the complex Radon measure induced by $x^* u$ in the sense of Definition 4.3 of [P3]. Thus

$$\mu_{x^* u}(A) = (x^* u)(\chi_A) = x^* \boldsymbol{\mu}_u(A) \qquad (7.5.2.1)$$

for $A \in M_u$. Let $(A_i)_1^{\infty}$ be a disjoint sequence in M_u with $A = \bigcup_1^{\infty} A_i \in M_u$. Then by (7.5.2.1) we have

$$x^* \boldsymbol{\mu}_u(A) = \mu_{x^* u}(A) = \sum_1^{\infty} \mu_{x^* u}(A_i) = \sum_1^{\infty} x^* \boldsymbol{\mu}_u(A_i)$$

for each $x^* \in X^*$, since $\mu_{x^* u}$ is σ-additive on $M_{x^* u}$ and since $M_u \subset M_{x^* u}$. Now by the Orlicz-Pettis theorem which holds for lcHs by McArthur [McA] we conclude that

$$\boldsymbol{\mu}_u(A) = \sum_1^{\infty} \boldsymbol{\mu}_u(A_i)$$

and hence $\boldsymbol{\mu}_u$ is σ-additive on M_u. $\square$

Remark 7.5.3. It is possible that $M_u = \{\emptyset\}$. For example, let u be the identity operator on $\mathcal{K}([0, 1))$. Then by Example 7.2.6, $C_0([0, 1)) = \mathcal{L}_1(u)$ and clearly, 0 is the only idempotent function in $C_0([0, 1))$.

Theorem 7.5.4. *Let X be a quasicomplete lcHs and let $u : \mathcal{K}(T) \to X$ be a weakly compact bounded Radon operator. Then the following statements hold:*

(i) *M_u is a σ-algebra in T.*

(ii) *$\mathcal{B}(T) \subset M_u$.*

(iii) *If $\mathbf{m}_u$ is the representing measure of the continuous extension of u on $C_0(T)$ (see 5.2.10 of Chapter 5), then $\boldsymbol{\mu}_u|_{\mathcal{B}(T)} = \mathbf{m}_u$ and hence $\boldsymbol{\mu}_u|_{\mathcal{B}(T)}$ is $\mathcal{B}(T)$-regular.*

(iv) *For each $A \in M_u$,*

$$\boldsymbol{\mu}_u(A) = \lim_{K \in \mathcal{C}} \boldsymbol{\mu}_u(A \cap K)$$

where $\mathcal{C}$ is directed by the relation $K_1 \leq K_2$ if $K_1 \subset K_2$, $K_1, K_2 \in \mathcal{C}$.

Proof. Let $q \in \Gamma$. Then $U_q^0 = \{x^* \in X^* : |x^*(x)| \leq 1 \text{ for } x \in U_q\}$ is equicontinuous in X^*. As u is a weakly compact operator on $C_0(T)$ (see Convention 7.2.4 and Definition 7.2.7), by Corollary 9.3.2 of Edwards [E], by 5.2.10 and by Lemma 2 of [P9], $u^*(U_q^0) = \{u^*x^* : x^* \in U_q^0\} = \{x^* \circ u : x^* \in U_q^0\}$ is relatively weakly compact in $M(T)$. Let $\epsilon > 0$. Then by Theorem 4.22.1 of [E] there exists a compact set K_q in T such that

$$\sup_{x^* \in U_q^0} |x^* \circ u|(T \backslash K_q) < \epsilon \tag{7.5.4.1}$$

and by Proposition 1 of [P9], given an open set U in T there exists a compact $C_q \subset U$ such that

$$\sup_{x^* \in U_q^0} |x^* \circ u|(U \backslash C_q) < \epsilon. \tag{7.5.4.2}$$

(i) By Theorem 7.5.2, M_u is a ring of u-measurable sets. By (7.5.4.1) and by Lemma 7.2.17, $u_q^\bullet(T \backslash K_q) < \epsilon$. By Urysohn's lemma there exists $\varphi_q \in K(T)$ such that $\chi_{K_q} \leq \varphi_q \leq \chi_T$ so that $u_q^\bullet(1 - \varphi_q) \leq u_q^\bullet(T \backslash K_q) < \epsilon$. As q is arbitrary in Γ, this shows that $1 \in \mathcal{L}_1(u)$ and hence $T \in M_u$.

Let $(A_n)_1^\infty$ be a disjoint sequence in M_u with $A = \bigcup_1^\infty A_n$. Then $\sum_{k=1}^n \chi_{A_k} \nearrow \chi_A \leq \chi_T \in \mathcal{L}_1(u)$. Since $\sum_1^n \chi_{A_k} = \chi_{\bigcup_1^n A_k} \in \mathcal{L}_1(u)$, by the complex version of Theorem 4.7 of [T], $\chi_A \in \mathcal{L}_1(u)$ and hence $A \in M_u$. Therefore, (i) holds.

(ii) Let U be an open set in T. Then by Lemma 7.2.17 and by (7.5.4.2) there exists a compact C_q in T such that $C_q \subset U$ and $u_q^\bullet(U \backslash C_q) < \epsilon$. By Urysohn's lemma there exists $\varphi_q \in \mathcal{K}(T)$ such that $\chi_{C_q} \leq \varphi_q \leq \chi_U$ and hence $u_q^\bullet(\chi_U - \varphi_q) \leq u_q^\bullet(U \backslash C_q) < \epsilon$. Therefore, $U \in M_u$. As M_u is a σ-algebra in T by (i), we conclude that $\mathcal{B}(T) \subset M_u$.

(iii) By 5.2.10, $x^*u = u^*x^* = x^* \circ \mathbf{m}_u$ for $x^* \in X^*$. Given an open set U in T and an $\epsilon > 0$, as in the beginning of the proof choose a compact $K_q \subset U$ for which (7.5.4.2) holds with K_q in place of C_q. Then by Urysohn's lemma there exists $\varphi_q \in \mathcal{K}(T)$ such that $\chi_{K_q} \leq \varphi_q \leq \chi_U$ so that by (7.5.4.2) and by Lemma 7.2.17 we have

$$u_q^\bullet(\chi_U - \varphi_q) \leq u_q^\bullet(U \backslash K_q) < \epsilon. \tag{7.5.4.3}$$

Then

$$\left| \mathbf{m}_u(U) - \int_T \varphi_q d\mathbf{m}_u \right|_q = \sup_{x^* \in U_q^0} \left| \left(x^* \left(\mathbf{m}_u(U) - \int_T \varphi_q d\mathbf{m}_u \right) \right) \right|$$

$$= \sup_{x^* \in U_q^0} \left| \int_T \chi_U d(x^* u) - \int_T \varphi_q d(x^* \mathbf{m}_u) \right|$$

$$= \sup_{x^* \in U_q^0} \left| \int_T \chi_U d(x^* u) - \int_T \varphi_q d(x^* u) \right|$$

$$\leq \sup_{x^* \in U_q^0} \int_T |\chi_U - \varphi_q| dv(x^* u)$$

$$\leq \sup_{x^* \in U_q^0} |x^* u|(U \backslash K_q)$$

$$= u_q^\bullet(U \backslash K_q) < \epsilon \qquad (7.5.4.4)$$

by Lemma 7.2.17 and (7.5.4.3). Consequently, by 5.2.10 we have

$$|\mathbf{m}_u(U) - u(\varphi_q)|_q = \sup_{x^* \in U_q^0} \left| x^* \left(\mathbf{m}_u(U) - \int_T \varphi_q d\mathbf{m}_u \right) \right|$$

$$= \left| \mathbf{m}_u(U) - \int_T \varphi_q d\mathbf{m}_u \right|_q < \epsilon \qquad (7.5.4.5)$$

since φ_q is $\mathbf{m}_u$-integrable in T. On the other hand,

$$|\boldsymbol{\mu}_u(U) - u(\varphi_q)|_q = \left| \int \chi_U du - u(\varphi_q) \right|_q$$

$$= \sup_{x^* \in U_q^0} \left| \int_T \chi_U d(x^* u) - x^* u(\varphi_q) \right|$$

$$\leq \sup_{x^* \in U_q^0} \left| \int_T |\chi_U - \varphi_q| dv(x^* u) \right.$$

$$\leq u_q^\bullet(U \backslash K_q) < \epsilon \qquad (7.5.4.6)$$

by (7.5.4.4).

Therefore, by (7.5.4.5) and (7.5.4.6) we have

$$|\mathbf{m}_u(U) - \boldsymbol{\mu}_u(U)|_q < 2\epsilon.$$

Since ϵ is arbitrary,

$$|\mathbf{m}_u(U) - \boldsymbol{\mu}_u(U)|_q = 0.$$

Now, as q is arbitrary in Γ and as X is Hausdorff, $\mathbf{m}_u(U) = \boldsymbol{\mu}_u(U)$.

If U_1, U_2 are open sets in T with $U_1 \subset U_2$, then

$$\mathbf{m}_u(U_2 \backslash U_1) = \mathbf{m}_u(U_2) - \mathbf{m}_u(U_1) = \boldsymbol{\mu}_u(U_2) - \boldsymbol{\mu}_u(U_1) = \boldsymbol{\mu}_u(U_2 \backslash U_1)$$

and consequently, $\boldsymbol{\mu}_u(E) = \mathbf{m}_u(E)$ for E in the ring generated by $\mathcal{U}$, the family of open sets in T.

Let $\mathcal{M} = \{A \in \mathcal{B}(T) : \boldsymbol{\mu}_u(A) = \mathbf{m}_u(A)\}$. If $(E_n)_1^\infty$ is a monotone sequence in $\mathcal{M}$, by the σ-additivity of $\boldsymbol{\mu}_u$ on $\mathcal{B}(T)$ by Theorem 7.5.2 and by (ii) above and by the σ-additivity of $\mathbf{m}_u$ on $\mathcal{B}(T)$ by Theorem 2 of [P9] as u is weakly compact, we have $\boldsymbol{\mu}_u(\lim_n E_n) = \mathbf{m}_u(\lim_n E_n)$ and hence $\mathcal{M}$ is a monotone class. Then by Theorem 6.B of [H], $\boldsymbol{\mu}_u = \mathbf{m}_u$ on $\mathcal{B}(T)$. Consequently, by Theorem 6 of [P9], $\boldsymbol{\mu}_u|_{\mathcal{B}(T)}$ is Borel regular. Hence (iii) holds.

(iv) By (ii), $\mathcal{C} \subset M_u$ and hence $A \cap K \in M_u$ for $A \in M_u$ and $K \in \mathcal{C}$. Given $q \in \Gamma$, by the lcHs complex version of Lemma 1.24 of [T] we have

$$\lim_{K \in \mathcal{C}} u_q^\bullet(\chi_{A \setminus K}) = \lim_{K \in \mathcal{C}} u_q^\bullet(A \setminus (A \cap K)) = 0.$$

As q is arbitrary in Γ, this shows that $\chi_{A \cap K} \to \chi_A$ in the topology of $\mathcal{L}_1(u)$ and consequently,

$$\int \chi_A du = \lim_{K \in \mathcal{C}} \int \chi_{A \cap K} du,$$

$$i.e., \boldsymbol{\mu}_u(A) = \lim_{K \in \mathcal{C}} \boldsymbol{\mu}_u(A \cap K).$$

This completes the proof of the theorem. $\qquad\square$

Theorem 7.5.5. *Let X be a quasicomplete lcHs and let $u : C_0(T) \to X$ be a continuous linear mapping. As in 5.2.10, let $\mathbf{m}_u = u^{**}|_{\mathcal{B}(T)}$, the representing measure of u and let $\widetilde{\mathcal{B}(T)}$ be the generalized Lebesgue completion of $\mathcal{B}(T)$ with respect to $\mathbf{m}_u$ in the sense of Definition 1.2.3. Let $\widetilde{\mathbf{m}}_u(A \cup N) = \mathbf{m}_u(A)$ where $A \in \mathcal{B}(T)$ and $N \subset M \in \mathcal{B}(T)$ with $||\mathbf{m}_u||_q(M) = 0$ for all $q \in \Gamma$. Then $\widetilde{\mathbf{m}}_u$ is well defined and σ-additive on $\widetilde{\mathcal{B}(T)}$, $\widetilde{\mathcal{B}(T)}$ is a σ-algebra containing $\mathcal{B}(T)$ and $\widetilde{\mathbf{m}}_u|_{\mathcal{B}(T)} = \mathbf{m}_u$.*

Proof. If $A_1 \cup N_1 = A_2 \cup N_2$, where $A_j \in \mathcal{B}(T)$, $N_j \subset M_j \in \mathcal{B}(T)$ with $||\mathbf{m}_u||_q(M_j) = 0$ for $q \in \Gamma$ and for $j = 1, 2$, let $M = M_1 \cup M_2$. Then obviously, $A_1 \cup M = A_2 \cup M$ and $\mathbf{m}_u(A_1) = \mathbf{m}_u(A_1 \cup M) - \mathbf{m}_u(M \setminus A_1) = \mathbf{m}_u(A_1 \cup M)$ since $|\mathbf{m}_u(M \setminus A_1)| \leq ||\mathbf{m}_u||(M) = 0$. Similarly, $\mathbf{m}_u(A_2) = \mathbf{m}_u(A_2 \cup M) = \mathbf{m}_u(A_1 \cup M) = \mathbf{m}_u(A_1)$. Hence $\widetilde{\mathbf{m}}_u$ is well defined.

Let $(E_n)_1^\infty$ be a sequence in $\widetilde{\mathcal{B}(T)}$. Let $E = \bigcup_1^\infty E_n$. Then there exist F_n, N_n and M_n such that $E_n = F_n \cup N_n$, $F_n \in \mathcal{B}(T)$, $N_n \subset M_n \in \mathcal{B}(T)$ and $||\mathbf{m}_u||_q(M_n) = 0$ for all $q \in \Gamma$ and for $n \in \mathbb{N}$. Let $F = \bigcup_1^\infty F_n$ and $M = \bigcup_1^\infty M_n$. Then F and M belong to $\mathcal{B}(T)$, $\bigcup_1^\infty N_n \subset \bigcup_1^\infty M_n$ and $||\mathbf{m}_u||_q(M) = 0$ for all $q \in \Gamma$. Then $E = F \cup N$, $N \subset M$ and hence $E \in \widetilde{\mathcal{B}(T)}$. If $(E_n)_1^\infty$ are mutually disjoint, then $\widetilde{\mathbf{m}}_u(E) = \mathbf{m}_u(\bigcup_1^\infty F_n) = \sum_1^\infty \mathbf{m}_u(F_n) = \sum \widetilde{\mathbf{m}}_u(E_n)$ and hence $\widetilde{\mathbf{m}}_u$ is σ-additive on $\widetilde{\mathcal{B}(T)}$. Clearly, $\widetilde{\mathbf{m}}_u|_{\mathcal{B}(T)} = \mathbf{m}_u$. $\qquad\square$

Theorem 7.5.6. *Let X be a quasicomplete lcHs, let $u : \mathcal{K}(T) \to X$ be a weakly compact bounded Radon operator and let $q \in \Gamma$. By Convention 7.2.4, $u : C_0(T) \to X$ is continuous and let $\mathbf{m}_u$ be the representing measure of u with $\widetilde{\mathbf{m}_u}$ its generalized Lebesgue completion with respect to $\mathcal{B}(T)$. By Theorem 6 of $[P9]$ $\mathbf{m}_u$ is Borel regular on $\mathcal{B}(T)$. Then:*

(i) *For $M \in \mathcal{B}(T)$, $u_q^\bullet(M) = ||\mathbf{m}_u||_q(M)$. (Recall $u_q = \Pi_q \circ u$ from Lemma 7.2.17.)*

(ii) *For $A \in \widetilde{\mathcal{B}(T)}$, the generalized Lebesgue completion of $\mathcal{B}(T)$ with respect to $\mathbf{m}_u$, let $A = B \cup N$, $B \in \mathcal{B}(T)$ and $N \subset M \in \mathcal{B}(T)$ with $||\mathbf{m}_u||_q(M) = 0$ for all $q \in \Gamma$. Then $u_q^\bullet(N) = ||\widetilde{\mathbf{m}}_u||_q(N) = 0$ for all $q \in \Gamma$.*

(iii) *A set A in T is u-integrable if and only if it is $\widetilde{\mathbf{m}_u}$-integrable in T and hence $\widetilde{\mathcal{B}(T)} = M_u$. Consequently, for $A \in M_u$, $\boldsymbol{\mu}_u(A) = \widetilde{\mathbf{m}_u}(A)$. (See Definition 7.5.5.)*

(iv) *A function $f : T \to \mathbb{K}$ is u-measurable if and only if it is $\mathbf{m}_u$-measurable.*

Proof. (i) Since $\mathcal{B}(T) \subset M_u$ by Theorem 7.5.4(ii), the set M is u-measurable and u-integrable. Then by the complex version of Theorem 1.11(b) of $[T]$, M is $\Pi_q \circ u$- integrable and hence $\chi_M \in \mathcal{L}_1(u_q)$ for $q \in \Gamma$. Let $\Psi_{x^*}(x + q^{-1}(0)) = x^*(x)$ for $x \in X$ and $x^* \in U_q^0$. Then Ψ_{x^*} is well defined, linear and continuous and $\{\Psi_{x^*} : x^* \in U_q^0\}$ is a norm determining subset of the closed unit ball of $(X_q)^*$ by Proposition 1.2.15 of Chapter 1. Consequently, by the complex version of **1.13** of $[T]$, by 5.2.10, by Lemma 7.1.11 and by Proposition 1.2.15 we have

$$u_q^\bullet(M) = u_q^\bullet(\chi_M) = \sup_{x^* \in U_q^0} |x^* u|^\bullet(\chi_M)$$

$$= \sup_{x^* \in U_q^0} |u^* x^*|^*(\chi_M)$$

$$= \sup_{x^* \in U_q^0} |u^* x^*|(\chi_M)$$

$$= \sup_{x^* \in U_q^0} |(x^* \circ \mathbf{m}_u)|(M)$$

$$= ||\mathbf{m}_u||_q(M).$$

Hence (i) holds.

(ii) By (i), $u_q^\bullet(M) = 0$ implies $||\mathbf{m}_u||_q(M) = 0$. As $N \subset M$, $u_q^\bullet(N) \le u_q^\bullet(M) = 0$ and $||\widetilde{\mathbf{m}_u}||_q(N) \le ||\mathbf{m}_u||_q(M) = 0$. Thus $u_q^\bullet(N) = ||\widetilde{\mathbf{m}_u}||_q(N) = 0$ for all $q \in \Gamma$.

(iii) Suppose $A \subset T$ is u-integrable. Then it is u-measurable and hence by the analogue of Theorem 6.2.9, given a compact K in T and $q \in \Gamma$, there exists a disjoint sequence $(K_n^{(q)})_1^\infty$ of compact sets and a set $N^{(q)}$ contained in K such that $K \backslash N^{(q)} = \bigcup_1^\infty K_n^{(q)}$, $u_q^\bullet(N^{(q)}) = 0$ and $K_n^{(q)} \subset K \cap A$ or $K_n^{(q)} \subset K \backslash A$ for each n. Then $N^{(q)} = K \backslash (K \backslash N^{(q)}) = K \backslash \bigcup_1^\infty K_n^{(q)} \in \mathcal{B}(T)$. Consequently, by (i) we have

$||\mathbf{m}_u||_q(N^{(q)}) = 0$. Then by Theorem 6.2.9 and by the arbitrariness of $q \in \Gamma$ we conclude that A is Lusin $\mathbf{m}_u$-measurable. Consequently, by Theorem 6.2.5, A is $\mathbf{m}_u$-measurable and hence $A \in \widetilde{\mathcal{B}(T)}$. (See Definition 1.2.6 of Chapter 1.) Thus $M_u \subset \widetilde{\mathcal{B}(T)}$.

Conversely, let χ_A be $\widetilde{\mathbf{m}_u}$-integrable in T. Then A is $\mathbf{m}_u$-measurable. Therefore, by Theorem 6.2.9, given a compact K and $q \in \Gamma$, there exist a disjoint sequence $(K_n^{(q)})_1^\infty$ of compact sets and a set $N^{(q)} \subset K$ such that $K \backslash N^{(q)} = \bigcup_1^\infty K_n^{(q)}$ with $K_i^{(q)} \subset A$ or $K_i^{(q)} \subset (K \backslash A)$ and with $||\mathbf{m}_u||_q(N^{(q)}) = 0$. Then $N^{(q)} \in \mathcal{B}(T)$ and by (i), $u_q^\bullet(N^{(q)}) = 0$. Since q is arbitrary in Γ, by the analogue of Theorem 6.2.9, A is u-measurable. Since $1 \in \mathcal{L}_1(u)$ by Theorem 7.5.4(i), by the domination principle χ_A is u-integrable and hence $A \in M_u$. Thus $\mathcal{B}(T) \subset M_u$. Therefore, $M_u = \widetilde{\mathcal{B}(T)}$.

For $A \in M_u = \widetilde{\mathcal{B}(T)}$, $\boldsymbol{\mu}_u(A) = \int \chi_A du$ and hence by 5.2.10 we have

$$x^* \boldsymbol{\mu}_u(A) = \int_A d(x^* u) = \int_A d(u^* x^*) = \int_A d(x^* \circ \mathbf{m}_u) = x^* \widetilde{\mathbf{m}_u}(A)$$

for $x^* \in X^*$. As $\widetilde{\mathbf{m}_u}(A)$ and $\boldsymbol{\mu}_u(A)$ belong to X, by the Hahn-Banach theorem $\boldsymbol{\mu}_u(A) = \widetilde{\mathbf{m}_u}(A)$ for $A \in M_u$. Hence (iii) holds.

(iv) If f is u-measurable, given a compact K in T and $q \in \Gamma$, then by Theorem 7.1.14 there exist a disjoint sequence $(K_i^{(q)})_1^\infty \subset \mathcal{C}$ and a set $N^{(q)}$ with $u_q^\bullet(N^{(q)}) = 0$ such that $K \backslash N^{(q)} = \bigcup_1^\infty K_i^{(q)}$ and with $f|_{K_i^{(q)}}$ continuous for each i. Since $N^{(q)} \in \mathcal{B}(T)$, by (i), $||\mathbf{m}_u||_q(N^{(q)}) = u_q^\bullet(N^{(q)}) = 0$ and hence by Theorem 6.2.4, f is Lusin $\mathbf{m}_u$-measurable. Then by Theorem 6.2.5, f is $\mathbf{m}_u$-measurable. Conversely, if f is $\mathbf{m}_u$-measurable, then by Theorem 6.2.5, f is Lusin $\mathbf{m}_u$-measurable and hence reversing the above argument and using (i), one can easily show that f is u-measurable.

This completes the proof of the theorem. $\square$

The following theorem generalizes Theorem 3.3 of [P4] to weakly compact bounded Radon operators on $\mathcal{K}(T)$ and improves the first part of Theorem 9.13 of [P13] by replacing open sets by open Baire sets. Moreover, Theorem 9.13 of [P13] was announced earlier in [P6].

Theorem 7.5.7. *Let X be a quasicomplete lcHs and let $u : \mathcal{K}(T) \to X$ be a bounded Radon operator. Let M_u and $\widetilde{\mathbf{m}_u}$ be given as in Definitions 7.5.1 and 7.5.5, respectively. Then the following statements are equivalent:*

(i) *u is a weakly compact bounded Radon operator.*

(ii) *M_u is a σ-algebra in T and $\mathcal{C}_0 \subset M_u$.*

(iii) *$\mathcal{B}(T) \subset M_u$.*

(iv) *$\mathcal{B}_c(T) \subset M_u$.*

   (v) $\mathcal{B}_0(T) \subset M_u$.

  (vi) *Every bounded u-measurable complex function f belongs to $\mathcal{L}_1(\mathbf{m}_u)$.*

 (vii) *Every bounded complex Borel function f belongs to $\mathcal{L}_1(\mathbf{m}_u)$.*

(viii) *Every bounded complex σ-Borel function f belongs to $\mathcal{L}_1(\mathbf{m}_u)$.*

  (ix) *Every bounded complex Baire function f belongs to $\mathcal{L}_1(\mathbf{m}_u)$.*

   (x) *Every bounded u-measurable complex function f belongs to $\mathcal{L}_1(u)$.*

  (xi) *Every bounded complex Borel function f belongs to $\mathcal{L}_1(u)$.*

 (xii) *Every bounded complex σ-Borel function f belongs to $\mathcal{L}_1(u)$.*

(xiii) *Every bounded complex Baire function f belongs to $\mathcal{L}_1(u)$.*

 (xiv) *For every open set U in T there exists a vector $x_U \in X$ such that the weak integral $\int_U du = x_U$ in the sense that*

$$x^*(x_U) = \int_U d(x^*u)$$

for each $x^* \in X^*$, where x^*u is treated as a complex Radon measure in T.

  (xv) *Similar to (xiv) except that the open set U is σ-Borel.*

 (xvi) *Similar to (xiv) except that the open set U is an open Baire set.*

(xvii) *Every bounded u-measurable complex function f belongs to $\mathcal{L}_1(\boldsymbol{\mu}_u)$.*

(xviii) *Every bounded complex Borel function f belongs to $\mathcal{L}_1(\boldsymbol{\mu}_u)$.*

 (xix) *Every bounded complex σ-Borel function f belongs to $\mathcal{L}_1(\boldsymbol{\mu}_u)$.*

  (xx) *Every bounded complex Baire function f belongs to $\mathcal{L}_1(\boldsymbol{\mu}_u)$.*

Proof. By Theorem 7.5.4, (i)$\Rightarrow$(ii) and (i)$\Rightarrow$(iii)$\Rightarrow$(iv)$\Rightarrow$(v). Obviously, (ii)$\Rightarrow$(v).

(v)$\Rightarrow$(i) Since $\mathcal{B}_0(T) \subset M_u$, every open Baire set U is u-integrable and hence there exists a vector $x_U \in X$ such that $\int \chi_U du = x_U$. Consequently, the weak integral $\int_U du$ belongs to X and therefore, by (vi)$\Rightarrow$(vii) of Theorem 7.2.10, (i) holds.

Thus (i)$\Leftrightarrow$(ii)$\Leftrightarrow$(iii)$\Leftrightarrow$(iv)$\Leftrightarrow$(v).

(i)$\Rightarrow$(vi) By the hypothesis (i) and by Theorem 7.5.6, $A \in M_u$ if and only if A is $\widetilde{\mathbf{m}_u}$-integrable in T. If f is a bounded u-measurable complex function, then by Theorem 7.5.4(i), χ_T is u-integrable and hence by the complex lcHs version of Theorem 1.22 of [T], $f \in \mathcal{L}_1(u)$. Thus (i)$\Rightarrow$(x). Moreover, by (x) and by Theorem 7.6.13 (see Section 7.6) $f \in \mathcal{L}_1(\mathbf{m}_u)$ and

$$\int f du = \int f \, d\mathbf{m}_u. \qquad (7.5.7.1)$$

Hence (vi) holds. Thus (i)$\Rightarrow$(x)$\Rightarrow$(vi).

Obviously, by the complex version of Corollary 2.18 of [T], (x)⇒(xi)⇒(xii)⇒ (xiii). Clearly, (xi)⇒(xiv) (resp. (xii)⇒(xv), (xiii)⇒(xvi)) and consequently, (xiv (resp. (xv), (xvi)) implies by Theorem 7.2.18 that u is a weakly compact bounded Radon operator. Clearly, by Theorem 7.5.6.(iv), (vi)⇒(vii) and obviously, (vii)⇒ (viii)⇒(ix). (ix) implies that $\int_U d\mathbf{m}_u = x_U \in X$ for each open Baire set U in T. Then by (7.5.7.1), $\int_U du = \int_U d\mathbf{m}_u = x_u$ for each open Baire set U in T and hence by Theorem (7.2.18), (ix)⇒(i).

Thus (i)⇔(vi)⇔(vii)⇔(viii)⇔(ix).

By (7.5.7.1), $\int_U d_u = \int_U d\mathbf{m}_u = x_U \in X$ for each open Baire set U in T and hence by Theorem 7.2.18, (xiii)⇒(i).

Hence (i)⇔(x)⇔(xi)⇔(xii)⇔(xiii).

(iii)⇒(xiv)⇒(xv) (obvious) ⇒(xvi) (obvious) and (xvi)⇒(i) by Theorem 7.2.18.

Hence (i)⇔(iii)⇔(xiv)⇔(xv)⇔(xvi).

Since $\widetilde{\mathbf{m}}_u = \boldsymbol{\mu}_u$ on M_u by Theorem 7.5.6 and since for each $\widetilde{\mathcal{B}(T)}$-simple function $\tilde{s}$ there exists a $\mathcal{B}(T)$-simple function s such that $s = \tilde{s}$ m-a.e. in T, (xvii)⇔(vi); (xviii)⇔(vii); (xix)⇔(viii) and (xx)⇔(ix).

Hence all the statements are equivalent.

This completes the proof of the theorem.   □

The following theorem has been given without proof in Theorem 9.14 of [P13]. It was also announced earlier in [P6].

Theorem 7.5.8. *Let X be a quasicomplete lcHs and let $u : \mathcal{K}(T) \to X$ be a Radon operator. Then the following statements are equivalent:*

(i) *u is prolongable.*

(ii) *$\delta(\mathcal{C}) \subset M_u$.*

(iii) *$\delta(\mathcal{C}_0) \subset M_u$.*

(iv) *M_u is a δ-ring containing all relatively compact open sets in T.*

(v) *M_u is a δ-ring containing $\mathcal{C}$.*

(vi) *M_u is a δ-ring containing $\mathcal{C}_0$.*

(vii) *Every bounded complex Borel function with compact support belongs to $\mathcal{L}_1(u)$.*

(viii) *For every compact K in T, there exists $x_K \in X$ such that $\mu_{x^*u}(K) = x^*(x_K)$ for $x^* \in X^*$, where μ_{x^*u} is the complex Radon measure induced by x^*u in the sense of Definition 4.3 of [P3].*

(ix) *Similar to (viii) for each relatively compact open set U instead of K.*

(x) *Similar to (viii) with K compact G_δ.*

Proof. (i)⇒(ii) Suppose u is prolongable. Let $V \in \mathcal{V}$. Then by (i), $u : \mathcal{K}(V) \to X$ is a weakly compact bounded Radon operator and hence by Theorem 7.5.4, $\mathcal{B}(V) \subset$

M_{uv} where u_V is $u|_{\mathcal{K}(V)}$. Since V is arbitrary in $\mathcal{V}$ and since $\delta(\mathcal{C}) = \bigcup_{V \in \mathcal{V}} \mathcal{B}(V)$, by Lemma 7.3.3(iii), (ii)holds.

(ii)$\Rightarrow$(iii) Obvious.

(i)$\Rightarrow$(iv) Let (A_n) be a decreasing sequence in M_u with $A_n \searrow A$. Since $\chi_{A_n} \to \chi_A$ and $\chi_A \leq \chi_{A_n}$ for all n, by the complex lcHs analogue of Theorem 4.7 of [T] $\chi_A \in \mathcal{L}_1(u)$ and hence $A \in M_u$. Since M_u is a ring of sets by Theorem 7.5.2, M_u is a δ-ring of sets. Since (i)$\Rightarrow$(ii), $\mathcal{V} \subset M_u$ by the complex analogue of Theorem 1.22 of [T] and hence (i)$\Rightarrow$(iv).

(iv)$\Rightarrow$(v) If $C \in \mathcal{C}$, then by Theorem 50.D of [H] there exists $V \in \mathcal{V}$ such that $C \subset V$. Then $C = V \backslash (V \backslash C) \in M_u$ by (iv). Hence (iv)$\Rightarrow$(v).

(v)$\Rightarrow$(vi) Obvious.

(vi)$\Rightarrow$(iii) Obvious.

(i)$\Rightarrow$(vii) Let f be a bounded complex Borel function with supp $f = K \in \mathcal{C}$. Let $U \in \mathcal{V}$ with $K \subset U$. By (i), $u|_{\mathcal{K}(U)}$ is a weakly compact bounded Radon operator and hence by Theorem 7.5.7(xi), f is $u|_{\mathcal{K}(U)}$-integrable. Therefore, given $\epsilon > 0$ and $q \in \Gamma$, there exists $\varphi \in \mathcal{K}(U)$ such that $(u|_{\mathcal{K}(U)})_q^{\bullet}(|f - \varphi|) < \epsilon$. Now $\hat{\varphi} = \varphi$ and $\hat{f} = f$ in Lemma 7.3.3 and hence by the said lemma we have

$$u_q^{\bullet}(|f - \varphi|) = u_q^{\bullet}(|\hat{f} - \hat{\varphi}|) = (u|_{\mathcal{K}(U)})_q^{\bullet}(|f - \varphi|) < \epsilon.$$

Hence f is u-integrable and thus (vii) holds.

(vii)$\Rightarrow$(viii) Let $K \in \mathcal{C}$. Then by (vii), $\chi_K \in \mathcal{L}_1(u)$ and hence $\int_K du = x_K$(say) $\in X$. Then $x^*(x_K) = x^*(\int_K du) = \int_K d(x^*u) = \mu_{x^*u}(K)$.

(vii)$\Rightarrow$(ix) Let $U \in \mathcal{V}$. Then by (vii), $\chi_U \in \mathcal{L}_1(u)$ and hence $\int_U du = x_U$(say) $\in X$. Then $x^*(x_U) = x^*(\int_U du) = \int_U d(x^*u) = \mu_{x^*u}(U)$.

(vii)$\Rightarrow$(x) The proof is similar to that of (vii)$\Rightarrow$(viii).

(viii)$\Rightarrow$(i) as (iv)$\Leftrightarrow$(i) of Theorem 5.3.13 and as $(x^* \circ \mathbf{m}_u) = x^*u$.

(ix)$\Rightarrow$(i) as (ii)$\Leftrightarrow$(i) of Theorem 5.3.13 and as $x^* \circ \mathbf{m}_u = x^*u$.

(x)$\Rightarrow$(i) as (ix)$\Leftrightarrow$(i) of Theorem 5.3.13 and as $x^* \circ \mathbf{m}_u = x^*u$.

(iii) $\Rightarrow$(i) By (iii), every $K \in \mathcal{C}_0$ belongs to M_u and hence is u-integrable. Hence $\int_K du = x_K$ (say) $\in X$. Then $x^*(x_K) = \int_K d(x^*u) = \mu_{x^*u}(K)$. But by (5.3.4.3), $\mu_{x^*u} = x^* \circ \mathbf{m}_u$ and hence $\mu_{x^*u}(K) = x^* \circ \mathbf{m}_u(K) = x^*(x_K)$ for $x^* \in X^*$ and hence by Theorem 5.3.13(v), (i) holds.

Hence the statements (i)–(x) are equivalent and this completes the proof of the theorem. $\square$

Theorem 7.5.9. *Let X be a quasicomplete lcHs and suppose $u : \mathcal{K}(T) \to X$ is a weakly compact bounded Radon operator. Then:*

(i) $\boldsymbol{\mu}_u$ is M_u-regular. That is, given $A \in M_u$ and a neighborhood W of 0 in X, there exist a compact $C \subset A$ and an open set $U \supset A$ such that $\boldsymbol{\mu}_u(F) \in W$ for all $F \in M_u$ with $F \subset U \backslash C$.

(ii) For $A \in M_u$,

$$\boldsymbol{\mu}_u(A) = \lim_{K \in \mathcal{C}(A)} \boldsymbol{\mu}_u(K) = \lim_{U \in \mathcal{U}(A)} \boldsymbol{\mu}_u(U) = \lim_{K \in \mathcal{C}(A)} \mathbf{m}_u(K) = \lim_{U \in \mathcal{U}(A)} \mathbf{m}_u(U)$$

where $\mathcal{C}(A) = \{K \in \mathcal{C} : K \subset A\}$ is directed by the relation $K_1 \leq K_2$ if $K_1 \subset K_2$ and $\mathcal{U}(A) = \{U \in \mathcal{U} : A \subset U\}$ is directed by the relation $U_1 \leq U_2$ if $U_2 \subset U_1$.

Proof. Let W be a neighborhood of 0 in X and let W_0 be a balanced neighborhood of 0 in X such that $W_0 \subset W$. Then there exist an $\epsilon > 0$ and a finite family $(q_i)_1^n$ in Γ such that

$$\bigcap_{i=1}^n \{x : q_i(x) < \epsilon\} \subset W_0. \tag{7.5.9.1}$$

Let $A \in M_u$. Then by Theorem 7.5.6(iii) A is of the form $A = B \cup N$ with $N \subset M \in \mathcal{B}(T)$ and $u_q^\bullet(M) = 0$. Since $\mathbf{m}_u$ is Borel regular by Theorem 6 of [P9], there exist an open set U_i in T and a compact K_i in T such that $K_i \subset B \subset B \cup M \subset U_i$ with $\|\mathbf{m}_u\|_{q_i}(F) < \epsilon$ for all $F \in \mathcal{B}(T)$ with $F \subset U_i \backslash K_i$ for $i = 1, 2, \ldots, n$. Consequently, for all $F \subset U_i \backslash K_i$ with $F \in M_u$ we have $|\mathbf{m}_u(F)|_{q_i} < \epsilon$ for $i = 1, 2, \ldots, n$ since $M_u = \widetilde{\mathcal{B}(T)}$ by Theorem 7.5.6(iii). Hence $\mathbf{m}_u(F) \in W_0$ for such $F \in M_u$. Then by Theorem 7.5.6(iii) we conclude that $\boldsymbol{\mu}_u$ is M_u-regular. Hence (i) holds.

(ii) Given a neighborhood W of 0 in X, by (i) there exist $K_0 \in \mathcal{C}$ and $U_0 \in \mathcal{U}$ such that $K_0 \subset A \subset U_0$ and $\boldsymbol{\mu}_u(F) \in W$ for all $F \in M_u$ with $F \subset U_0 \backslash K_0$. Let $K \in \mathcal{C}$ with $K_0 \subset K \subset A$ and $U \in \mathcal{U}$ with $A \subset U \subset U_0$. Such K and U exist by the regularity of $\boldsymbol{\mu}_u$ in M_u. Then

$$\boldsymbol{\mu}_u(A) - \boldsymbol{\mu}_u(K) = \boldsymbol{\mu}_u(A \backslash K) \in W \tag{7.5.9.2}$$

as $A \backslash K \subset A \backslash K_0 \subset U_0 \backslash K_0$ and

$$\boldsymbol{\mu}_u(U) - \boldsymbol{\mu}_u(A) = \boldsymbol{\mu}_u(U \backslash A) \in W \tag{7.5.9.3}$$

as $U \backslash A \subset U_0 \backslash A \subset U_0 \backslash K_0$.

Since $\boldsymbol{\mu}_u = \mathbf{m}_u$ by Theorem 7.5.6(iii), (ii) holds by (7.5.9.2) and (7.5.9.3). $\square$

For studying the regularity properties of a prolongable Radon operator we need the following lemma.

Lemma 7.5.10. *Let X be a quasicomplete lcHs and let $u : \mathcal{K}(T) \to X$ be a prolongable Radon operator. Let U be a relatively compact open set in T. Let $v = u|_{\mathcal{K}(U)}$. Then for each compact set $K \subset U$ and $E \in M_u$, $E \cap K \in M_v$. Moreover, given $q \in \Gamma$ and $\epsilon > 0$, there exist open sets O and V in T such that $E \cap K \subset O \subset V \subset \bar{V} \subset U$, $\{F \subset O \backslash (E \cap K) : F \in M_u\} = \{F \subset O \backslash (E \cap K) : F \in M_v\}$, $\boldsymbol{\mu}_u(F) = \boldsymbol{\mu}_v(F)$ for such F and $u_q^\bullet(F) = v_q^\bullet(F) \le v_q^\bullet(O \backslash (E \cap K)) = u_q^\bullet(O \backslash (E \cap K)) < \epsilon.$*

Proof. Since $K \in \mathcal{C}$ and $K \subset U$, by Theorem 50.D of [H] there exists an open set V such that $K \subset V \subset \bar{V} \subset U$. Since $E \in M_u$ and since $\delta(\mathcal{C}) \subset M_u$ by Theorem 7.5.8(ii), $E \cap K \in M_u$.

Claim. For each $\varphi \in \mathcal{K}(T)$, $\varphi\chi_K \in \mathcal{L}_1(v)$.   (7.5.10.1)

In fact, as v is weakly compact on $C_0(U)$, given $q \in \Gamma$ and $\epsilon > 0$, by Theorem 6 of [P9] there exists an open set G in T such that $K \subset G \subset V$ with

$$\sup_{x^* \in U_q^0} |x^* \circ v|(G \backslash K) < \frac{\epsilon}{\|\varphi\|_T}.$$

Then by Lemma 7.2.17 we have $v_q^\bullet(G \backslash K) < \frac{\epsilon}{\|\varphi\|_T}$. By Urysohn's lemma there exists $g \in \mathcal{K}(U)$ such that $\chi_K \le g \le \chi_G$. Then $\varphi g \in \mathcal{K}(U)$ and

$$v_q^\bullet(|\varphi\chi_K - \varphi g|) \le \|\varphi\|_T v_q^\bullet(|\chi_K - g|) \le \|\varphi\|_T v_q^\bullet(G \backslash K) < \epsilon.$$

This shows that $\varphi\chi_K \in \mathcal{L}_1(v)$.

As $E \in M_u$, there exists $\Psi \in \mathcal{K}(T)$ such that $u_q^\bullet(|\chi_E - \Psi|) < \epsilon$. By the above claim $\Psi\chi_K \in \mathcal{L}_1(v)$ and $v_q^\bullet(|\chi_{E \cap K} - \Psi\chi_K|) = v_q^\bullet(\chi_K(|\chi_E - \Psi|)) = u_q^\bullet(\chi_K|\chi_E - \Psi|) \le u_q^\bullet(|\chi_E - \Psi|) < \epsilon$ by Lemma 7.3.3 since $\chi_K|\chi_E - \Psi|$ has compact support contained in U. Thus $E \cap K \in M_v$.

Given $q \in \Gamma$ and $\epsilon > 0$, by Theorem 7.5.9 there exists an open set O in T with $O \in M_v$ such that $E \cap K \subset O \subset U$ with $v_q^\bullet(O \backslash (E \cap K)) < \epsilon$. Replacing $O \cap V$ by O we have $E \cap K \subset O \subset V \subset \bar{V} \subset U$ and $v_q^\bullet(O \backslash (E \cap K)) < \epsilon$ and $O \in M_v$ since $\mathcal{B}(U) \subset M_v$ by Theorem 7.5.7. Then each $F \in M_v$ with $F \subset O \backslash (E \cap K)$ has compact support contained in U and hence by Lemma 7.3.3 $F \in M_u$, $\boldsymbol{\mu}_u(F) = \boldsymbol{\mu}_v(F)$ and $u_q^\bullet(F) = v_q^\bullet(F) \le v_q^\bullet(O \backslash (E \cap K)) < \epsilon.$

Conversely, let us suppose $F \subset O \backslash (E \cap K)$ and $F \in M_u$. Then given $q \in \Gamma$ and $\epsilon > 0$, let $\varphi \in K(T)$ such that $u_q^\bullet(|\varphi - \chi_F|) < \epsilon$. Clearly, as $F \subset O \subset V \subset \bar{V} \subset U$, we have

$$|\varphi\chi_{\bar{V}} - \chi_F| = |\varphi\chi_{\bar{V}} - \chi_{\bar{V}}\chi_F| \le |\varphi - \chi_F|,$$

$\varphi\chi_{\bar{V}} \in \mathcal{L}_1(v)$ by the above claim with K replaced by $\bar{V}$ and by Lemma 7.3.3 we have

$$v_q^\bullet(|\chi_F - \varphi\chi_{\bar{V}}|) \le u_q^\bullet(|\chi_F - \varphi|) < \epsilon.$$

Thus $\chi_F \in \mathcal{L}_1(v)$ and hence $F \in M_v$. $\square$

Theorem 7.5.11. *Let X be a quasicomplete lcHs and let $u : \mathcal{K}(T) \to X$ be a prolongable Radon operator. Then:*

(i) *$\boldsymbol{\mu}_u$ is restrictedly M_u-outer regular in the sense that given $E \in M_u$, $K \in \mathcal{C}$ and a neighborhood W of 0 in X, there exists a relatively compact open set O in T such that $E \cap K \subset O$ with $\boldsymbol{\mu}_u(F) \in W$ for all $F \subset O \backslash (E \cap K)$ with $F \in M_u$.*

(ii) *$\boldsymbol{\mu}_u$ is $\delta(\mathcal{C})$-regular.*

(iii)   (a) *For $E \in \delta(\mathcal{C})$,*

$$\boldsymbol{\mu}_u(E) = \lim_{K \subset E, K \in \mathcal{C}} \boldsymbol{\mu}_u(K) = \lim_{K \in \mathcal{C}} \lim_{E \cap K \subset O, O \in \mathcal{U} \cap \delta(\mathcal{C})} \boldsymbol{\mu}_u(O);$$

   and

   (b) *for $E \in M_u$,*

$$\boldsymbol{\mu}_u(E) = \lim_{K \in \mathcal{C}} \lim_{E \cap K \subset O, O \in \mathcal{U} \cap \delta(\mathcal{C})} \boldsymbol{\mu}_u(O)$$

   where $\mathcal{C}$ is directed by the relation $K_1 \leq K_2$ if $K_1 \subset K_2$ and $\mathcal{U} \cap \delta(\mathcal{C})$ is directed by the relation $O_1 \leq O_2$ if $O_2 \subset O_1$.

Proof. Let W_0 be a balanced neighborhood of 0 in X such that $W_0 \subset W$. Let $(q_i)_1^n \subset \Gamma$ and $\epsilon > 0$ such that $\bigcap_{i=1}^n \{x : q_i(x) < \epsilon\} \subset W_0$.

(i) Let $E \in M_u$ and $K \in \mathcal{C}$. Let U be a relative compact open set in T such that $K \subset U$. Then by Lemma 7.5.10 there exist open sets O and V in T such that $E \cap K \subset O \subset V \subset \bar{V} \subset U$, $\{F \subset O \backslash (E \cap K) : F \in M_u\} = \{F \subset O \backslash (E \cap K) : F \in M_v\}$, $\boldsymbol{\mu}_u(F) = \boldsymbol{\mu}_v(F)$ for such F and $u_{q_i}^\bullet(O \backslash (E \cap K)) = v_{q_i}^\bullet(O \backslash (E \cap K)) < \epsilon$ for $i = 1, 2, \ldots, n$ where $v = u|_{\mathcal{K}(U)}$. This shows that

$$|\boldsymbol{\mu}_u(F)|_{q_i} \leq u_{q_i}^\bullet(O \backslash (E \cap K)) < \epsilon$$

for $i = 1, 2, \ldots, n$ and for $F \in M_u$ with $F \subset O \backslash (E \cap K)$. Hence $\boldsymbol{\mu}_u(F) \in W_0 \subset W$ for such F and hence (i) holds.

(ii) Let $E \in \delta(\mathcal{C})$. Then there exists a relatively compact open set U_0 in T such that $E \subset U_0$. By hypothesis, $v = u|_{\mathcal{K}(U_0)}$ is a weakly compact bounded Radon operator and hence by Theorem 7.5.9(i) there exist a compact set C and an open set U_1 in T such that $C \subset E \subset U_1 \subset U_0$ and such that for all $F \subset U_1 \backslash C$ with $F \in M_v$ we have $\boldsymbol{\mu}_v(F) \in W$. Clearly, $\widehat{\chi_F} = \chi_F$ for such F where $\widehat{\chi_F}$ is given as in Notation 7.3.2 and consequently, by Lemma 7.3.3.(iii) we have $F \in M_u$ and

$$\boldsymbol{\mu}_v(F) = \int \chi_F dv = \int \widehat{\chi_F} du = \int \chi_F du = \boldsymbol{\mu}_u(F). \tag{7.5.11.1}$$

Then for $F \subset U_1 \backslash C$ with $F \in \delta(\mathcal{C})$ we have $F \in M_v \cap M_u$ and $\boldsymbol{\mu}_u(F) = \boldsymbol{\mu}_v(F) \in W$ by (7.5.11.1). Hence $\boldsymbol{\mu}_u|_{\delta(\mathcal{C})}$ is $\delta(\mathcal{C})$-regular.

(iii) Let $E \in \delta(\mathcal{C})$ (resp. $E \in M_u$). Then E is u-integrable as $\delta(\mathcal{C})$ is contained in M_u by Theorem 7.5.8 (resp. by Definition 7.5.1) and hence by the complex lcHs version of Lemma 1.24 of [T] there exists $K_0 \in \mathcal{C}$ such that

$$u_{q_i}^{\bullet}(E \backslash (E \cap K)) < \epsilon$$

for $K_0 \subset K \in \mathcal{C}$ and for $i = 1, 2, \ldots, n$. Then

$$|\boldsymbol{\mu}_u(E) - \boldsymbol{\mu}_u(E \cap K)|_{q_i} = |\boldsymbol{\mu}_u(E \backslash (E \cap K)|_{q_i} \leq u_{q_i}^{\bullet}(E \backslash E \cap K) < \epsilon$$

for $i = 1, 2, \ldots, n$ and hence

$$\boldsymbol{\mu}_u(E) = \lim_{K \in \mathcal{C}} \boldsymbol{\mu}_u(E \cap K). \tag{7.5.11.2}$$

By (i), $\boldsymbol{\mu}_u(F) \in W$ for all $F \subset O \backslash (E \cap K)$ with $F \in M_u$ and hence by (i)

$$\boldsymbol{\mu}_u(E \cap K) = \lim_{E \cap K \subset O, O \in \mathcal{U} \cap \delta(\mathcal{C})} \boldsymbol{\mu}_u(O). \tag{7.5.11.3}$$

Then by (7.5.11.2) and (7.5.11.3) we have

$$\boldsymbol{\mu}_u(E) = \lim_{K \in \mathcal{C}} \boldsymbol{\mu}_u(E \cap K) = \lim_{K \in \mathcal{C}} \lim_{K \cap E \subset O \in \mathcal{U} \cap \delta(\mathcal{C})} \boldsymbol{\mu}_u(O)$$

and hence (iii) (b) holds for $E \in M_u$. As $\boldsymbol{\mu}_u|_{\delta(\mathcal{C})}$ is $\delta(\mathcal{C})$-inner regular by (ii),

$$\boldsymbol{\mu}_u(E) = \lim_{K \in \mathcal{C}, K \subset E} \boldsymbol{\mu}_u(K)$$

for $E \in \delta(\mathcal{C})$ and hence (iii)(a) also holds.

This completes the proof of the theorem.   □

Definition 7.5.12. Let X be a quasicomplete lcHs and let $u : \mathcal{K}(T) \to X$ be a weakly compact bounded Radon operator with $\boldsymbol{\mu}_u$ as in Definition 7.5.1. For each open set U in T, by Theorem 7.5.9(ii),

$$\boldsymbol{\mu}_u(U) = \lim_{K \in \mathcal{C}, K \subset U} \boldsymbol{\mu}_u(K) \tag{7.5.12.1}$$

since $U \in M_u$ by Theorem 7.5.4. Let $A \subset T$ and let

$$\boldsymbol{\mu}_u^*(A) = \lim_{A \subset U \in \mathcal{U}} \boldsymbol{\mu}_u(U)$$

whenever the limit exists, where $\mathcal{U}$ is directed by the relation $U_1 \leq U_2$ if $U_2 \subset U_1$.

Theorem 7.5.13. *Let X, u and $\boldsymbol{\mu}_u^*$ be as in Definition 7.5.12. Then $\boldsymbol{\mu}_u^*(A)$ exists in X for each $A \subset T$.*

Proof. $\boldsymbol{\mu}_u$ is σ-additive on M_u by Theorem 7.5.2 and M_u is a σ-algebra in T by Theorem 7.5.4(i). Therefore, the range of $\boldsymbol{\mu}_u$ is relatively weakly compact in X by the Theorem on Extension of [K3] (or by Corollary 2 of [P7]) and hence is bounded in the lcHs topology τ of X. Since $\mathcal{B}(T) \subset M_u$ by Theorem 7.5.4(ii) and since $\boldsymbol{\mu}_u$ is σ-additive on M_u, for each increasing sequence $(K_n)_1^\infty \subset \mathcal{C}$, $\lim_n \boldsymbol{\mu}_u(K_n) \in X$. Moreover, by Theorem 7.5.9(i), $\boldsymbol{\mu}_u$ is M_u-regular and hence, given $K \in \mathcal{C}$ and a neighborhood W of 0 in X, there exists $U \in \mathcal{U}$ such that $K \subset U$ and for each compact C with $K \subset C \subset U$, $\boldsymbol{\mu}_u(C) - \boldsymbol{\mu}_u(K) \in W$. Thus conditions 6.1 of Sion [Si] are satisfied by $T, \mathcal{C}, \mathcal{U}$ and $\boldsymbol{\mu}_u$, excepting that X is a quasicomplete lcHs so that every bounded closed set in X is complete. Since $\boldsymbol{\mu}_u$ is σ-additive on M_u and since $\mathcal{U} \subset M_u$, for every monotone sequence $(U_n)_1^\infty$ of open sets in T, $\lim_n \boldsymbol{\mu}_u(U_n) \in X$. Consequently, by Lemma 2.5 of Sion [Si], $\{\boldsymbol{\mu}_u(U)\}_{A \subset U \in \mathcal{U}}$ is a Cauchy net in X. Since X is quasicomplete and since the range of $\boldsymbol{\mu}_u$ is bounded, $\lim_{A \subset U \in \mathcal{U}} \boldsymbol{\mu}_u(U)$ exists in X and hence $\boldsymbol{\mu}_u^*(A) \in X$ for each $A \subset T$. $\square$

Theorem 7.5.14. *Let X be a quasicomplete lcHs and let $u : \mathcal{K}(T) \to X$ be a weakly compact bounded Radon operator. Then a subset A of T is u-integrable if and only if, for each $q \in \Gamma$ and $\epsilon > 0$, there exist a compact C and an open set U in T such that $C \subset A \subset U$ and $u_q^\bullet(U \backslash C) < \epsilon$.*

Proof. Suppose A is u-integrable.

Claim 1. $x^* \boldsymbol{\mu}_u(E) = \mu_{x^*u}(E)$ for $E \in M_u$ and for $x^* \in X^*$. $\qquad$ (7.5.14.1)

In fact, as $E \in M_u$, $E \in M_{x^*u}$. If $f \in L_1(u)$, then $f \in \mathcal{L}_1(x^*u)$ and hence $\mu_{x^*u}(f) = \int f d(x^*u) = x^*(\int f du)$ and hence $\mu_{x^*u}(E) = x^* \boldsymbol{\mu}_u(E)$. Therefore, Claim 1 holds.

By Theorem 7.5.9(i), given $q \in \Gamma$ and $\epsilon > 0$, there exist $C \in \mathcal{C}$ and $U \in \mathcal{U}$ such that $C \subset A \subset U$ and such that $q(\boldsymbol{\mu}_u(F)) < \frac{\epsilon}{4}$ for all $F \in M_u$ with $F \subset U \backslash C$. Then by Lemma 7.2.17, by Lemma 1.2.15(i), by Proposition 1.1.6 and by Claim 1 we have

$$u_q^\bullet(U \backslash C) = \sup_{x^* \in U_q^0} |x^* \circ u|(U \backslash C) \leq 4 \sup_{x^* \in U_q^0, F \in \mathcal{B}(T), F \subset U \backslash C} |\mu_{x^*u}(F)|$$
$$= 4 \sup_{x^* \in U_q^0, F \subset U \backslash C, F \in \mathcal{B}(T)} |x^* \boldsymbol{\mu}_u(F)| = 4 \sup_{F \subset U \backslash C, F \in \mathcal{B}(T)} q(\boldsymbol{\mu}_u(F))$$
$$< 4 \frac{\epsilon}{4} = \epsilon$$

since $\mathcal{B}(T) \subset M_u$ by Theorem 7.5.4(ii).

Conversely, let us suppose that the conditions are satisfied for each $q \in \Gamma$ and for each $\epsilon > 0$. For $q \in \Gamma$ and $\epsilon > 0$, let $C \subset A \subset U$, $C \in \mathcal{C}$, $U \in \mathcal{U}$ with $u_q^\bullet(U \backslash C) < \epsilon$. Then by Urysohn's lemma, there exists $\varphi \in \mathcal{K}(T)$ such that $\chi_C \leq \varphi \leq \chi_U$. Then $|\chi_A - \varphi| \leq \chi_U - \chi_C$ and hence we have

$$u_q^\bullet(\chi_A - \varphi) \leq u_q^\bullet(\chi_{U \backslash C}) = u_q^\bullet(U \backslash C) < \epsilon.$$

Therefore, $\chi_A \in \mathcal{L}_1(u)$ and hence $A \in M_u$. $\square$

Definition 7.5.15. Let X, u and $\boldsymbol{\mu}_u^*$ be as in Definition 7.5.12. Let $M_{\boldsymbol{\mu}_u^*} = \{E \subset T : \boldsymbol{\mu}_u^*(A) = \boldsymbol{\mu}_u^*(A \cap E) + \boldsymbol{\mu}_u^*(A \backslash E) \text{ for all } A \subset T\}$. Members of $M_{\boldsymbol{\mu}_u^*}$ are called Carathéodory-Sion $\boldsymbol{\mu}_u^*$-measurable sets.

Theorem 7.5.16. *Let X, u and $\boldsymbol{\mu}_u^*$ be as in Definition 7.5.12. Then $M_{\boldsymbol{\mu}_u^*} \subset M_u$ and $\boldsymbol{\mu}_u^*(A) = \boldsymbol{\mu}_u(A)$ for $A \in M_{\boldsymbol{\mu}_u^*}$. Consequently, each $A \in M_{\boldsymbol{\mu}_u^*}$ is u-measurable and u-integrable.*

Proof. Since the range of $\boldsymbol{\mu}_u$ is bounded and since X is quasicomplete, Theorem 6.3 of Sion [Si] holds here and hence $\mathcal{B}(T) \subset M_{\boldsymbol{\mu}_u^*}$. Let $A \in M_{\boldsymbol{\mu}_u^*}$, $q \in \Gamma$ and $\epsilon > 0$. Then by (3) of Theorem 6.3 of Sion [Si] there exist $K \in \mathcal{C}$ and $U \in \mathcal{U}$ such that $K \subset A \subset U$ and $q(\boldsymbol{\mu}_u^*(F)) < \frac{\epsilon}{4}$ for all $F \in M_{\boldsymbol{\mu}_u^*}$ with $F \subset U \backslash K$. As $\mathcal{B}(T) \subset M_u$ by Theorem 7.5.4(ii), by Definition 7.5.12 and by Theorem 7.5.9(ii) we have $\boldsymbol{\mu}_u^*(A) = \lim_{A \subset U \in \mathcal{U}} \boldsymbol{\mu}_u(U) = \boldsymbol{\mu}_u(A)$ for $A \in \mathcal{B}(T)$ and hence $q(\boldsymbol{\mu}_u(F)) < \frac{\epsilon}{4}$ for all $F \in \mathcal{B}(T)$ with $F \subset U \backslash K$. Since $U \backslash K$ is open, by Lemma 7.2.17 and by Claim 1 in the proof of Theorem 7.5.14 and by Propositions 1.1.6 and 1.2.15(i) we have

$$u_q^\bullet(U \backslash K) = \sup_{x^* \in U_q^0} |x^* \circ u|(U \backslash K) \leq 4 \sup_{x^* \in U_q^0, F \subset U \backslash K, F \in \mathcal{B}(T)} |\mu_{x^* u}(F)|$$

$$= 4 \sup_{x^* \in U_q^0, F \subset U \backslash K, F \in \mathcal{B}(T)} |(x^* \boldsymbol{\mu}_u)(F)| = 4 \sup_{F \subset U \backslash K, F \in \mathcal{B}(T)} q(\boldsymbol{\mu}_u(F))$$

$$< 4\frac{\epsilon}{4} = \epsilon$$

as $\mathcal{B}(T) \subset M_u$ by Theorem 7.5.4(ii).

Consequently, by Theorem 7.5.14, $A \in M_u$ and hence $M_{\boldsymbol{\mu}_u^*} \subset M_u$. Then by Theorem 7.5.9(ii) it follows that $\boldsymbol{\mu}_u^*(A) = \boldsymbol{\mu}_u(A)$ for $A \in M_{\boldsymbol{\mu}_u^*}$. The last part is evident from Definition 7.5.1.

This completes the proof of the theorem.   □

To prove that $M_{\boldsymbol{\mu}_u^*} = M_u$, we proceed as below and prove the following lemmas.

Let θ be a positive linear functional on $\mathcal{K}(T)$. For $E \subset T$, let

$$\mu_\theta^*(E) = \inf_{\chi_E \leq g \in \mathcal{I}^+} \sup\{\theta(\Psi) : \Psi \leq g, \Psi \in C_c^+(T)\}.$$

Then by Rudin [Ru1], μ_θ^* is an outer measure on $\mathcal{P}(T)$, the family of all subsets of T. Let $M_{\mu_\theta^*} = \{E \subset T : E \text{ is } \mu_\theta^* \text{-measurable} - \text{that is, } \mu_\theta^*(A) = \mu_\theta^*(A \cap E) + \mu_\theta^*(A \backslash E) \text{ for all } A \subset T\}$. Then $M_{\mu_\theta^*}$ is a σ-algebra in T and contains $\mathcal{B}(T)$. (See Theorem 2.2 of [P3]).

Lemma 7.5.17. *Let $\theta \in \mathcal{K}(T)^*$. For $A \subset T$ with $\mu_{|\theta|}^*(A) < \infty$, let*

$$\mu_\theta^*(A) = \{(\mu_{\theta_1^+}^* - \mu_{\theta_1^-}^*) + i(\mu_{\theta_2^+}^* - \mu_{\theta_2^-}^*)\}(A)$$

where $\theta = Re\theta + iIm\theta$, $Re\theta = \theta_1^+ - \theta_1^-$, and $Im\theta = \theta_2^+ - \theta_2^-$. Let $E \in M_\theta = \{A \in M_{\mu_{\theta_1^+}^*} \cap M_{\mu_{\theta_1^-}^*} \cap M_{\mu_{\theta_2^+}^*} \cap M_{\mu_{\theta_2^-}^*}$ with $\mu_{|\theta|}(A) < \infty\}$. Then

$$\mu_\theta^*(A) = \mu_\theta^*(A \cap E) + \mu_\theta^*(A\backslash E). \tag{7.5.17.1}$$

Proof. Let $E \in M_\theta$. Then for $A \subset T$, we have

$$\mu_{\theta_i^+}^*(A) = \mu_{\theta_i^+}^*(A \cap E) + \mu_{\theta_i^+}^*(A\backslash E) \tag{7.5.17.2}$$

and

$$\mu_{\theta_i^-}^*(A) = \mu_{\theta_i^-}^*(A \cap E) + \mu_{\theta_i^-}^*(A\backslash E) \tag{7.5.17.3}$$

for $i = 1,2$. If $\mu_{|\theta|}^*(A) < \infty$, then $\mu_{\theta_i^+}^*(A) < \infty$ and $\mu_{\theta_i^-}^*(A) < \infty$ for $i = 1,2$ and hence by (7.5.17.2) and (7.5.17.3) and by the definition of $\mu_\theta^*(A)$, (7.5.17.1) holds. □

Lemma 7.5.18. *If $\theta \in \mathcal{K}(T)^*$ and is bounded, (i.e., $\theta \in \mathcal{K}(T)_b^*$ of [P4]), then for $A \subset T$,*

$$\mu_\theta^*(A) = \lim_{A \subset U \in \mathcal{U}} \mu_\theta^*(U)$$

where for $U_1, U_2 \in \mathcal{U}$, $U_1 \geq U_2$ if $U_1 \subset U_2$.

Proof. By Theorem 2.2 of [P3],

$$\mu_{\theta_i^+}^*(A) = \inf\{\mu_{\theta_i^+}^*(U) : A \subset U \in \mathcal{U}\}$$
$$= \lim_{A \subset U \in \mathcal{U}} \mu_{\theta_i^+}^*(U)$$

for $i = 1,2$ and for $A \subset T$. A similar expression holds for $\mu_{\theta_i^-}^*(A)$ for $i = 1,2$. Then by the definition of $\mu_\theta^*(A)$ as given in Lemma 7.5.17, the lemma holds. □

Lemma 7.5.19. *Let θ, M_θ, θ_1 and θ_2 be as in Lemma 7.5.17 and let $\mu_1 = \mu_{\theta_1^+}$, $\mu_2 = \mu_{\theta_1^-}$, $\mu_3 = \mu_{\theta_2^+}$ and $\mu_4 = \mu_{\theta_2^-}$. For $A \subset T$ and $E \in M_\theta$,*

$$\mu_j^*(A \cap E) = \inf_{A \cap E \subset U \in \mathcal{U}} \mu_j^*(U) = \lim_{A \cap E \subset U \in \mathcal{U}} \mu_j^*(U)$$
$$= \inf_{A \subset U \in \mathcal{U}} \mu_j^*(U \cap E) = \lim_{A \subset U \in \mathcal{U}} \mu_j^*(U \cap E) \tag{7.5.19.1}$$

and

$$\mu_j^*(A\backslash E) = \inf_{A\backslash E \subset U \in \mathcal{U}} \mu_j^*(U) = \lim_{A\backslash E \subset U \in \mathcal{U}} \mu_j^*(U)$$
$$= \inf_{A \subset U \in \mathcal{U}} \mu_j^*(U\backslash E) = \lim_{A \subset U \in \mathcal{U}} \mu_j^*(U\backslash E) \tag{7.5.19.2}$$

for $j = 1,2,3,4$. Consequently,

$$\mu_\theta^*(A \cap E) + \mu_\theta^*(A\backslash E) = \lim_{A \subset U \in \mathcal{U}} \mu_\theta^*(U \cap E) + \lim_{A \subset U \in \mathcal{U}} \mu_\theta^*(U\backslash E). \tag{7.5.19.3}$$

Proof.

$$\mu_j^*(A \cap E) = \inf_{A \cap E \subset U \in \mathcal{U}} \mu_j^*(U) \geq \inf_{A \subset U \in \mathcal{U}} \mu_j^*(U \cap E) \geq \mu_j^*(A \cap E)$$

and since $\{\mu_j^*(U) : A \cap E \subset U \in \mathcal{U}\} \subset \{\mu_j^*(U) : A \subset U \in \mathcal{U}\}$, $\inf\{\mu_j^*(U) : A \cap E \subset U \in \mathcal{U}\} \geq \inf\{\mu_j^*(U) : A \subset U \in \mathcal{U}\} \geq \inf\{\mu_j^*(U \cap E) : A \subset U \in \mathcal{U}\}$. Hence

$$\mu_j^*(A \cap E) = \inf_{A \cap E \subset U \in \mathcal{U}} \mu_j^*(U) = \lim_{A \cap E \subset U \in \mathcal{U}} \mu_j^*(U)$$

$$= \inf_{A \subset U \in \mathcal{U}} \mu_j^*(U \cap E) = \lim_{A \subset U \in \mathcal{U}} \mu_j^*(U \cap E)$$

for $j = 1, 2, 3, 4$ and hence (7.5.19.1) holds. Similarly, (7.5.19.2) holds. Then (7.5.19.3) holds by (7.5.19.1) and (7.5.19.2). □

Theorem 7.5.20. *Let X be a quasicomplete lcHs and let $u : \mathcal{K}(T) \to X$ be a weakly compact bounded Radon operator. Then $M_{\boldsymbol{\mu}_u^*} = M_u$ and $\boldsymbol{\mu}_u^*(A) = \widetilde{\mathbf{m}_u}(A) = \boldsymbol{\mu}_u(A)$ for $A \in M_u$.*

Proof. In the light of Theorem 7.5.16, to prove the first part it suffices to show that $M_u \subset M_{\boldsymbol{\mu}_u^*}$. Let $E \in M_u$. Then $x^* u \in K(T)^*$ and is bounded. Moreover,

$$E \in M_{x^* u} \text{ for each } x^* \in X^* \tag{7.5.20.1}$$

since E is u-integrable and hence is $x^* u$-integrable for each $x^* \in X^*$. Then by Definition 7.5.12 we have

$$\boldsymbol{\mu}_u^*(A) = \lim_{A \subset U \in \mathcal{U}} \boldsymbol{\mu}_u(U) \text{ for } A \subset T. \tag{7.5.20.2}$$

Then by (7.5.20.2) and by (7.5.14.1) we have

$$x^* \boldsymbol{\mu}_u^*(A) = \lim_{A \subset U \in \mathcal{U}} x^* \boldsymbol{\mu}_u(U)$$

$$= \lim_{A \subset U \in \mathcal{U}} \mu_{x^* u}(U)$$

$$= \lim_{A \subset U \in \mathcal{U}} \mu_{x^* u}(U \cap E) + \lim_{A \subset U \in \mathcal{U}} \mu_{x^* u}(U \backslash E) \tag{7.5.20.3}$$

by (7.5.20.1) since E belongs to $M_{x^* u}$. Then by (7.5.20.3) and by (7.5.19.3) we have

$$x^* \boldsymbol{\mu}_u^*(A) = \mu_{x^* u}^*(A \cap E) + \mu_{x^* u}^*(A \backslash E). \tag{7.5.20.4}$$

On the other hand, by (7.5.19.1) and (7.5.14.1) we have

$$\mu_{x^* u}^*(A \cap E) = \lim_{A \subset U \in \mathcal{U}} \mu_{x^* u}(U \cap E) = \lim_{A \subset U \in \mathcal{U}} x^* \boldsymbol{\mu}_u(U \cap E) \tag{7.5.20.5}$$

since $U \in \mathcal{B}(T) \subset M_u$ by Theorem 7.5.4 and since $E \in M_u$ by hypothesis.

Similarly,

$$\mu_{x^* u}^*(A \backslash E) = \lim_{A \subset U \in \mathcal{U}} \mu_{x^* u}(U \backslash E) = \lim_{A \subset U \in \mathcal{U}} x^* \boldsymbol{\mu}_u(U \backslash E) \tag{7.5.20.6}$$

and hence by (7.5.20.3), (7.5.20.4), (7.5.20.5), (7.5.20.6), by Definition 7.5.12 and by Lemma 7.5.19 we have

$$
\begin{aligned}
x^* \boldsymbol{\mu}_u^*(A) &= \lim_{A \subset U \in \mathcal{U}} \mu_{x^* u}(U \cap E) + \lim_{A \subset U \in \mathcal{U}} \mu_{x^* u}(U \backslash E) \\
&= \lim_{A \cap E \subset U \in \mathcal{U}} \mu_{x^* u}(U) + \lim_{A \backslash E \subset U \in \mathcal{U}} \mu_{x^* u}(U) \\
&= x^* \lim_{A \cap E \subset U \in \mathcal{U}} \boldsymbol{\mu}_u(U) + x^* \lim_{A \backslash E \subset U \in \mathcal{U}} \boldsymbol{\mu}_u(U) \\
&= x^* \boldsymbol{\mu}_u^*(A \cap E) + x^* \boldsymbol{\mu}_u^*(A \backslash E)
\end{aligned}
$$

for $x^* \in X^*$. Therefore, by the Hahn-Banach theorem we have

$$
\boldsymbol{\mu}_u^*(A) = \boldsymbol{\mu}_u^*(A \cap E) + \boldsymbol{\mu}_u^*(A \backslash E)
$$

for $A \subset T$ and hence $E \in M_{\boldsymbol{\mu}_u^*}$. Then by Theorem 7.5.16, $M_u = M_{\boldsymbol{\mu}_u^*}$. Moreover, $\boldsymbol{\mu}_u(E) = \widetilde{\mathbf{m}_u}(E) = \boldsymbol{\mu}_u^*(E)$ for $E \in M_u$ by Theorems 7.5.6 and 7.5.16.

This completes the proof of the theorem. $\qquad \square$

Theorem 7.5.21. *Let X be a quasicomplete lcHs. Let $u : \mathcal{K}(T) \to X$ be a weakly compact bounded Radon operator and let $q \in \Gamma$. Given $A \in M_u$, there exist a σ-compact set F and a G_δ set G such that $F \subset A \subset G$ and $u_q^\bullet(G \backslash F) = 0$. Conversely, given $A \subset T$, suppose for each $q \in \Gamma$ there exist a σ-compact F and a G_δ G such that $F \subset A \subset G$ and $u_q^\bullet(G \backslash F) = 0$. Then $A \in M_u$.*

Proof. Let $A \in M_u$. Let $q \in \Gamma$ be given. For $\epsilon = \frac{1}{n}$, by Theorem 7.5.14 there exist a compact K_n and an open set U_n in T such that $K_n \subset A \subset U_n$ with $u_q^\bullet(U_n \backslash K_n) < \frac{1}{n}$. Then $F = \bigcup_1^\infty K_n$ is σ-compact, $G = \bigcap_1^\infty U_n$ is G_δ and $F \subset A \subset G$. Clearly, $u_q^\bullet(G \backslash F) = 0$, since $u_q^\bullet(G \backslash F) \leq u_q^\bullet(U_n \backslash K_n) < \frac{1}{n}$ for all $n \in \mathbb{N}$.

Conversely, let $A \subset T$ be such that for each $q \in \Gamma$ there exist a σ-compact F and a G_δ G such that $F \subset A \subset G$ with $u_q^\bullet(G \backslash F) = 0$. Without loss of generality we shall assume that $F = \bigcup C_n$, $(C_n)_1^\infty \subset \mathcal{C}$, $C_n \nearrow F$, $G = \bigcap_1^\infty U_n$, $(U_n)_1^\infty \subset \mathcal{U}$ and $U_n \searrow G$. Since $U_n \backslash C_n \searrow G \backslash F$, $(U_n \backslash C_n)_1^\infty \subset M_u$ and $u_q^\bullet(\cdot) = \|\mathbf{m}_u\|_q(\cdot)$ on $\mathcal{B}(T)$ by Theorem 7.5.6(i) and since $\|\mathbf{m}_u\|_q$ is continuous as $\mathbf{m}_u$ is σ-additive, we have

$$
0 = u_q^\bullet(G \backslash F) = \|\mathbf{m}_u\|_q(G \backslash F) = \lim_n \|\mathbf{m}_u\|_q(U_n \backslash C_n).
$$

Thus, given $\epsilon > 0$, there exists n_0 such that $\|\mathbf{m}_u\|_q(U_n \backslash C_n) < \epsilon$ for $n \geq n_0$. Let $U = U_{n_0}$ and $C = C_{n_0}$. Then $u_q^\bullet(U \backslash C) = \|\mathbf{m}_u\|_q(U \backslash C) < \epsilon$ and hence by Theorem 7.5.14, $A \in M_u$.

This completes the proof of the theorem. $\qquad \square$

Definition 7.5.22. Let X be a quasicomplete lcHs. Let $\mathcal{D}$ be a δ-ring containing $\mathcal{C}$ and let $\boldsymbol{\mu} : \mathcal{D} \to X$ be σ-additive. If $\boldsymbol{\mu}$ is the restriction of an X-valued weakly compact Radon vector measure $\boldsymbol{\mu}_u$ (see Definition 7.5.1), then the Lebesgue-Radon completion $\widetilde{\mathcal{D}}$ of $\mathcal{D}$ with respect to $\boldsymbol{\mu}_u$ is defined as the family $\{E \subset T : \text{given } q \in$

Γ there exist a σ-compact F and a G_δ G such that $F \subset E \subset G$ with $u_q^\bullet(G\backslash F) = 0\}$ and the Lebesgue-Radon completion $\widetilde{\boldsymbol{\mu}_u}$ of $\boldsymbol{\mu}_u$ with respect to $\mathcal{D}$ is said to exist on $\widetilde{\mathcal{D}}$ if

$$\widetilde{\boldsymbol{\mu}_u}(E) = \lim_{K\in\mathcal{C}, K\subset E} \boldsymbol{\mu}(K)$$

exists in X for each $E \in \widetilde{\mathcal{D}}$.

The following theorem generalizes the bounded case of Theorem 4.4 of [P4].

Theorem 7.5.23. *Let X be a quasicomplete lcHs. Let $\boldsymbol{\mu} : \delta(\mathcal{C}) \to X$ be σ-additive. Then $\boldsymbol{\mu}$ is the restriction of an X-valued weakly compact Radon vector measure $\boldsymbol{\mu}_u$ if and only if $\boldsymbol{\mu}$ is $\delta(\mathcal{C})$-regular and its range is relatively weakly compact. In that case, u is unique and is called the weakly compact bounded Radon operator determined by $\boldsymbol{\mu}$. Moreover, $M_u = \widetilde{\delta(\mathcal{C})}$, the Lebesgue-Radon completion of $\delta(\mathcal{C})$ with respect to $\boldsymbol{\mu}_u$. The Lebesgue-Radon completion $\widetilde{\boldsymbol{\mu}_u}$ of $\boldsymbol{\mu}_u$ with respect to $\delta(\mathcal{C})$ exists on $\widetilde{\delta(\mathcal{C})}$ and coincides with $\boldsymbol{\mu}_u$.*

Proof. Suppose $u : \mathcal{K}(T) \to X$ is a weakly compact bounded Radon operator and suppose $\boldsymbol{\mu} = \boldsymbol{\mu}_u|_{\delta(\mathcal{C})}$. Then $\boldsymbol{\mu}_u$ is M_u-regular by Theorem 7.5.9(i) and $\delta(\mathcal{C}) \subset M_u$ by Theorem 7.5.4(ii) since $\delta(\mathcal{C}) \subset \mathcal{B}(T)$. Let $E \in \delta(\mathcal{C})$. As $\boldsymbol{\mu}_u$ is M_u-regular, given a neighborhood W of 0 in X, there exist a compact $K \subset E$ and an open set U in T such that $U \supset E$ and such that $\boldsymbol{\mu}_u(F) \in W$ for all $F \in M_u$ with $F \subset U\backslash K$. Since u is also prolongable, $\boldsymbol{\mu}_u|_{\delta(\mathcal{C})}$ is $\delta(\mathcal{C})$-regular by Theorem 7.5.11(iii). Therefore, $\boldsymbol{\mu}$ is $\delta(\mathcal{C})$-regular. Since M_u is a σ-algebra in T by Theorem 7.5.4(i) and since $\boldsymbol{\mu}_u$ is σ-additive on M_u by Theorem 7.5.2, the range of $\boldsymbol{\mu}_u$ and hence that of $\boldsymbol{\mu}$ is relatively weakly compact by the Theorem on Extension of [K3] or by Corollary 2 of [P7].

If $\boldsymbol{\mu}$ is also equal to $\boldsymbol{\mu}_v|_{\delta(\mathcal{C})}$ for another weakly compact bounded Radon operator v on T, then by the uniqueness part of Theorem 4.4(i) of [P4] and by Claim 1 in the proof of Theorem 7.5.14, $\mu_{x^*u} = \mu_{x^*v}$ on $\delta(\mathcal{C})$ and hence $x^*u = x^*v$ on $\mathcal{K}(T)$ for each $x^* \in X^*$. Then by the Hahn-Banach theorem, $u = v$. Therefore, u is unique.

Conversely, let $\boldsymbol{\mu}$ be σ-additive and $\delta(\mathcal{C})$-regular on $\delta(\mathcal{C})$ with its range relatively weakly compact. Then by the Theorem on Extension of [K3] or by Corollary 2 of [P7], $\boldsymbol{\mu}$ has a unique σ-additive extension $\boldsymbol{\mu}_c$ on $\mathcal{B}_c(T)$ with values in X. If $\boldsymbol{\mu}_0 = \boldsymbol{\mu}_c|_{\mathcal{B}_0(T)}$, then by Theorem 1 of [DP1] , $\boldsymbol{\mu}_0$ has a unique X-valued Borel (resp. σ-Borel) regular σ-additive extension $\hat{\boldsymbol{\mu}}$ (resp. $\hat{\boldsymbol{\mu}}_c$) on $\mathcal{B}(T)$ (resp. on $\mathcal{B}_c(T)$) and $\hat{\boldsymbol{\mu}}|_{\mathcal{B}_c(T)} = \hat{\boldsymbol{\mu}}_c$. Since $\hat{\boldsymbol{\mu}}_c$ and $\boldsymbol{\mu}_c$ are σ-additive and extend $\boldsymbol{\mu}$ to $\mathcal{B}_c(T)$, by the uniqueness part of Proposition 1 of [DP1] $x^*\boldsymbol{\mu}_c = x^*\hat{\boldsymbol{\mu}}_c$ for $x^* \in X^*$ and consequently, by the Hahn-Banach theorem $\hat{\boldsymbol{\mu}}_c = \boldsymbol{\mu}_c$ on $\mathcal{B}_c(T)$. Therefore, $\boldsymbol{\mu}_c$ is $\mathcal{B}_c(T)$-regular. Thus $\boldsymbol{\mu}$ has a unique $\mathcal{B}_c(T)$-regular σ-additive extension $\boldsymbol{\mu}_c$ on $\mathcal{B}_c(T)$ and $\boldsymbol{\mu}_c = \hat{\boldsymbol{\mu}}_c = \hat{\boldsymbol{\mu}}|_{\mathcal{B}_c(T)}$.

Let

$$uf = \int_T f d\boldsymbol{\mu}_c, \quad f \in C_0(T). \tag{7.5.23.1}$$

Clearly, $u : C_0(T) \to X$ is linear and continuous. Then by Theorem 1 of [P7], u is weakly compact and let $\mathbf{m}_u$ be the representing measure of u in the sense of 5.2.10 of Chapter 5. Then $\mathbf{m}_u$ has range in X by Theorem 2 of [P9] and by Theorem 1 of [P9] and by (7.5.23.1) we have

$$x^* u f = \int_T f d(x^* \mathbf{m}_u) = \int_T f d(x^* \boldsymbol{\mu}_c)$$

for $f \in C_0(T)$ and for $x^* \in X^*$. Moreover, $\mathbf{m}_u|_{\mathcal{B}(T)}$ is Borel regular. Then by Theorem 6.1.12, $\mathbf{m}_u|_{\mathcal{B}_c(T)}$ is $\mathcal{B}_c(T)$-regular. Consequently, by the uniqueness part of the Riesz representation theorem (σ-Borel version) we conclude that $x^* \boldsymbol{\mu}_c = x^* \mathbf{m}_u$ on $\mathcal{B}_c(T)$ for each $x^* \in X^*$ and hence by the Hahn-Banach theorem, $\boldsymbol{\mu}_c = \mathbf{m}_u|_{\mathcal{B}_c(T)}$. Since $\boldsymbol{\mu}_c|_{\delta(\mathcal{C})} = \boldsymbol{\mu}$, $\boldsymbol{\mu}$ is the restriction of $\mathbf{m}_u$ to $\delta(\mathcal{C})$. Then by Theorem 7.5.6(iii), $\boldsymbol{\mu}$ is the restriction of $\boldsymbol{\mu}_u$ to $\delta(\mathcal{C})$.

$M_u = \widetilde{\delta(\mathcal{C})}$, the Lebesgue-Radon completion of $\delta(\mathcal{C})$ with respect to $\boldsymbol{\mu}_u$ by Definition 7.5.22 and by Theorem 7.5.21. Then the Lebesgue-Radon completion $\widetilde{\boldsymbol{\mu}_u}$ of $\boldsymbol{\mu}_u$ exists on $\widetilde{\delta(\mathcal{C})}$ by Theorem 7.5.9(ii). Moreover, by the same theorem, $\widetilde{\boldsymbol{\mu}_u} = \boldsymbol{\mu}_u$.

This completes the proof of the theorem. □

The following theorem generalizes the bounded case of Theorem 4.6 of [P4].

Theorem 7.5.24. *Let X be a quasicomplete lcHs. Let $\boldsymbol{\mu}_0 : \delta(\mathcal{C}_0) \to X$ be σ-additive with range relatively weakly compact. Then $\boldsymbol{\mu}_0$ admits a unique X-valued $\delta(\mathcal{C})$-regular σ-additive extension $\boldsymbol{\mu} : \delta(\mathcal{C}) \to X$. Moreover, the following assertions hold:*

(i) *$\boldsymbol{\mu}_0$ is the restriction of a weakly compact Radon vector measure $\boldsymbol{\mu}_u$ and such u is unique. We say that $\boldsymbol{\mu}_0$ determines the weakly compact bounded Radon operator u.*

(ii) *If $\boldsymbol{\mu}$ is as above, then $\boldsymbol{\mu}_0$ and $\boldsymbol{\mu}$ determine the same weakly compact bounded Radon operator u (see Theorem 7.5.23).*

(iii) *If u is as in (i) and (ii), then*

$$M_u = \text{the Lebesgue-Radon completion } \widetilde{\delta(\mathcal{C})} \text{ with respect to } \boldsymbol{\mu}_u$$

and

$$\boldsymbol{\mu}_u(E) = \widetilde{\boldsymbol{\mu}_u}(E) \text{ for each } E \in M_u$$

where $\widetilde{\boldsymbol{\mu}_u}$ is the Lebesgue-Radon completion of $\boldsymbol{\mu}_u$ with respect to $\delta(\mathcal{C})$.

Proof. By Theorem on Extension of [K3] or by Corollary 2 of [P7], by Proposition 1 of [DP1] and by the Hahn-Banach theorem, $\boldsymbol{\mu}_0$ has a unique σ-additive X-valued Baire extension $\boldsymbol{\nu} : \mathcal{B}_0(T) \to X$. Let

$$u f = \int_T f d\boldsymbol{\nu}, \qquad f \in C_0(T).$$

Clearly, $u : C_0(T) \to X$ is linear and continuous. Then by Theorem 1 of [P7], u is weakly compact and hence the Borel restriction of its representing measure (see 5.2.10 of Chapter 5) $\mathbf{m}_u$ is the restriction of $\boldsymbol{\mu}_u$ to $\mathcal{B}(T)$ by Theorem 7.5.6(iii). Moreover, by 5.2.10 of Chapter 5,

$$x^* u f = \int_T f d(x^* \boldsymbol{\nu}) = \int_T f d(x^* \mathbf{m}_u) \qquad \text{for } f \in C_0(T) \text{ and for } x^* \in X^*.$$

Then by the uniqueness part of the Riesz representation theorem (Baire version) we have $x^* \boldsymbol{\nu} = (x^* \mathbf{m}_u)|_{\mathcal{B}_0(T)}$ for each $x^* \in X^*$. Consequently, by the Hahn-Banach theorem we have $\boldsymbol{\nu} = \mathbf{m}_u|_{\mathcal{B}_0(T)}$. Since $\mathbf{m}_u = \boldsymbol{\mu}_u|_{\mathcal{B}(T)}$ by Theorem 7.5.6(iii), we have $\boldsymbol{\nu} = \boldsymbol{\mu}_u|_{\mathcal{B}_0(T)}$. Let $\boldsymbol{\mu} = \boldsymbol{\mu}_u|_{\delta(\mathcal{C})}$. Then by Theorem 7.5.11(ii), $\boldsymbol{\mu} = \boldsymbol{\mu}_u|_{\delta(\mathcal{C})} = \mathbf{m}_u|_{\delta(\mathcal{C})}$ is σ-additive and $\delta(\mathcal{C})$-regular and extends $\boldsymbol{\mu}_0$. If $\boldsymbol{\mu}_1 : \delta(\mathcal{C}) \to X$ is σ-additive, $\delta(\mathcal{C})$-regular and extends $\boldsymbol{\mu}_0$, then by the uniqueness part of Theorem 4.1(i) of [P4], $x^* \boldsymbol{\mu}_1 = x^* \boldsymbol{\mu}$ for each $x^* \in X^*$ and hence by the Hahn-Banach theorem, $\boldsymbol{\mu}_1 = \boldsymbol{\mu}$. Hence $\boldsymbol{\mu}_0$ admits a unique $\delta(\mathcal{C})$-regular σ-additive extension $\boldsymbol{\mu} : \delta(\mathcal{C}) \to X$.

(i) If there exists another weakly compact bounded Radon operator $v : \mathcal{K}(T) \to X$ such that $\boldsymbol{\mu}_v|_{\delta(\mathcal{C}_0)} = \boldsymbol{\mu}_0$, then by the uniqueness part of Theorem 1 of [DP1] and by the uniqueness of σ-additive extension of $\boldsymbol{\mu}_0$ to $\mathcal{B}_0(T)$, $\boldsymbol{\mu}_u = \boldsymbol{\mu}_v$, and hence this implies that $x^* \boldsymbol{\mu}_u = x^* \boldsymbol{\mu}_v$ for $x^* \in X^*$. Consequently, by (7.5.14.1) and by Theorem 7.5.4 we have $\mu_{x^* u} = \mu_{x^* v}$ on $\mathcal{B}(T)$ for each $x^* \in X^*$. Thus $x^* u(\varphi) = x^* v(\varphi)$ for $\varphi \in \mathcal{K}(T)$ and for $x^* \in X^*$. Consequently, by the Hahn-Banach theorem, $u = v$. Hence u is unique.

(ii) Let $\boldsymbol{\omega}$ be the X-valued σ-additive extension of $\boldsymbol{\mu}$ to $\mathcal{B}_c(T)$. This exists by hypothesis and by Corollary 2 of [P7]. Then $\boldsymbol{\omega}$ also extends $\boldsymbol{\mu}_0$ to $\mathcal{B}_c(T)$. Then $uf = \int_T f d\boldsymbol{\nu} = \int_T f d\boldsymbol{\omega}$ for $f \in C_0(T)$. Then $\boldsymbol{\mu}$ and $\boldsymbol{\mu}_0$ determine the same weakly compact bounded Radon operator u. Hence (ii) holds.

(iii) By Theorem 7.5.23, $M_u = \widetilde{\delta(\mathcal{C})}$, the Lebesgue-Radon completion of $\delta(\mathcal{C})$ with respect to $\boldsymbol{\mu}_u$. Hence (iii) holds.

By Theorem 7.5.9(ii) and by Theorem 7.5.6(iii), for $E \in M_u$ we have

$$\boldsymbol{\mu}_u(E) = \widetilde{\mathbf{m}_u}(E) = \lim_{K \in \mathcal{C}, K \subset E} \boldsymbol{\mu}_u(K) = \widetilde{\boldsymbol{\mu}_u}(E).$$

This completes the proof of the theorem.   □

Theorem 7.5.25. Let u_1 and u_2 be prolongable Radon operators on $\mathcal{K}(T)$. If $\boldsymbol{\mu}_{u_1}|_{\delta(\mathcal{C}_0)} = \boldsymbol{\mu}_{u_2}|_{\delta(\mathcal{C}_0)}$, then $u_1 = u_2$ so that $M_{u_1} = M_{u_2}$.

Proof. Let U be a relatively compact open Baire set in T. Then by Theorem 7.5.8(iii), $\mathcal{B}_0(U) \subset \delta(\mathcal{C}_0) \subset M_{u_i}$, $i = 1, 2$. Then $\boldsymbol{\mu}_{u_1}|_{\mathcal{B}_0(U)} = \boldsymbol{\mu}_{u_2}|_{\mathcal{B}_0(U)} = \mathbf{m}_U$ (say). Then $\mathbf{m}_U : \mathcal{B}_0(U) \to X$ is σ-additive and the linear transformation $\boldsymbol{\omega}_U : C_0(U) \to X$ given by

$$\boldsymbol{\omega}_U f = \int_U f d\mathbf{m}_U, \qquad f \in C_0(U)$$

is continuous and linear and hence weakly compact by Theorem 1 of [P7]. Then for $f \in C_0(U)$,

$$\boldsymbol{\omega}_U f = \int_U f d\mathbf{m}_U = \int_U f d\boldsymbol{\mu}_{u_i}|_{\mathcal{B}_0(U)} = \int_T f d\boldsymbol{\mu}_{u_i}|_{\mathcal{B}_0(U)} = u_i f \text{ for } i = 1, 2.$$

Thus

$$u_1 f = u_2 f, \quad f \in C_c(U).$$

Since each $f \in \mathcal{K}(T)$ belongs to $C_c(U)$ for some relatively compact open Baire set U in T by Theorem 50.D of [H], we conclude that $u_1 = u_2$ on $\mathcal{K}(T)$.

This completes the proof of the theorem. $\qquad\square$

Remark 7.5.26. One can also use Theorem 4.6(i) of [P4] and (7.5.4.1) to prove the above result since $\mu_{x^* u_1} = \mu_{x^* u_2}$ on $\delta(\mathcal{C}_0)$ for each $x^* \in X^*$.

The following theorem generalizes Theorem 4.4 of [P4].

Theorem 7.5.27. *Let X be a quasicomplete lcHs. Let $\boldsymbol{\mu} : \delta(\mathcal{C}) \to X$ be σ-additive. Then $\boldsymbol{\mu}$ is the restriction of an X-valued prolongable Radon vector measure $\boldsymbol{\mu}_u$ if and only if $\boldsymbol{\mu}$ is $\delta(\mathcal{C})$-regular. In that case, u is unique and u is called the prolongable Radon operator determined by $\boldsymbol{\mu}$. Moreover, the Radon vector measure $\boldsymbol{\mu}_u$ on M_u is given by*

$$\boldsymbol{\mu}_u(E) = \lim_{K \in \mathcal{C}} \lim_{E \cap K \subset O, O \in \mathcal{U} \cap \delta(\mathcal{C})} \boldsymbol{\mu}_u(O) \qquad (7.5.27.1)$$

for $E \in M_u$, where $\mathcal{C}$ is directed by the relation $K_1 \leq K_2$ if $K_1 \subset K_2$ and $\mathcal{U} \cap \delta(\mathcal{C})$ is directed by the relation $O_1 \leq O_2$ if $O_2 \subset O_1$. The localized Lebesgue-Radon completion $\ell(\widetilde{\delta(\mathcal{C})})$ of $\delta(\mathcal{C})$ with respect to $\boldsymbol{\mu}$ is defined by $\ell(\widetilde{\delta(\mathcal{C})}) = \{E \subset T : E \cap K \in M_u \text{for each } K \in \mathcal{C} \text{and given} q \in \Gamma, \lim_{K \in \mathcal{C}} v_q^\bullet(E \backslash (E \cap K)) = 0\}$. The localized Lebesgue-Radon completion $\hat{\boldsymbol{\mu}}$ of $\boldsymbol{\mu}$ with respect to $\delta(\mathcal{C})$ is said to exist on $\ell(\widetilde{\delta(\mathcal{C})})$ if $\hat{\boldsymbol{\mu}}(E) = \lim_{K \in \mathcal{C}} \lim_{E \cap K \subset O, O \in \mathcal{U} \cap \delta(\mathcal{C})} \boldsymbol{\mu}_u(O)$ exists in X for each $E \in \ell(\widetilde{\delta(\mathcal{C})})$. Then $\ell(\widetilde{\delta(\mathcal{C})}) = M_u$ and $\hat{\boldsymbol{\mu}}(E)$ exists in X for each $E \in \ell(\widetilde{\delta(\mathcal{C})})$ and $\hat{\boldsymbol{\mu}}(E) = \boldsymbol{\mu}_u(E)$ for $E \in M_u$.

Proof. If u is a prolongable Radon operator on $\mathcal{K}(T)$, then by Theorems 7.5.2, 7.5.8 and 7.5.11(ii), $\delta(\mathcal{C}) \subset M_u$ and $\boldsymbol{\mu}|_{\delta(\mathcal{C})}$ is σ-additive and $\delta(\mathcal{C})$-regular. Conversely, let $\boldsymbol{\mu} : \delta(\mathcal{C}) \to X$ be σ-additive and $\delta(\mathcal{C})$-regular. Let U be a relatively compact open set in T and let $\mathbf{m}_U = \boldsymbol{\mu}|_{\mathcal{B}(U)}$. Let $V_U : C_0(U) \to X$ be given by

$$V_U f = \int_U f d\mathbf{m}_U, \quad f \in C_0(U).$$

Then by Theorem 1 of [P7], V_U is weakly compact. On the other hand, if $uf = \int_T f d\boldsymbol{\mu}$ for $f \in \mathcal{K}(T)$, then $u|_{\mathcal{K}(U)} = V_U|_{\mathcal{K}(U)}$ is continuous and hence the unique continuous extension of $u|_{\mathcal{K}(U)}$ to $C_0(U)$ coincides with V_U which is weakly compact. Hence u is prolongable.

Let $\boldsymbol{\mu} : \delta(\mathcal{C}) \to X$ be σ-additive and $\delta(\mathcal{C})$-regular and let $U \in \mathcal{V}$ with $\mathbf{m}_U = \boldsymbol{\mu}|_{\mathcal{B}(U)}$. Since $\mathbf{m}_U$ is $\mathcal{B}(U)$-regular by the hypothesis, by Lemma 5.2.19 $\mathbf{m}_U$ is the representing measure of the weakly compact operator V_U given above and since $u|_{C_0(U)} = V_U$, it follows by Theorem 7.5.4 that $\boldsymbol{\mu}_u|_{\mathcal{B}(U)} = \mathbf{m}_U = \boldsymbol{\mu}|_{\mathcal{B}(U)}$. Since U is an arbitrary relatively compact open set in T, it follows that $\boldsymbol{\mu}_u|_{\delta(\mathcal{C})} = \boldsymbol{\mu}$. In fact, given $E \in \delta(\mathcal{C})$, let U be a relatively compact open set such that $E \subset U$. Then $E \in \mathcal{B}(U)$ and hence $\boldsymbol{\mu}_u(E) = \boldsymbol{\mu}(E)$.

The uniqueness of u follows from Theorem 7.5.25 and thus u is uniquely determined by $\boldsymbol{\mu}$. Moreover, by Theorem 7.5.11(iii)(b),

$$\boldsymbol{\mu}_u(E) = \lim_{K \in \mathcal{C}} \lim_{E \cap K \subset O, O \in \mathcal{U} \cap \delta(\mathcal{C})} \boldsymbol{\mu}_u(O)$$

for $E \in M_u$. Hence $\boldsymbol{\mu}_u$ is given by (7.5.27.1).
Let

$$\mathcal{R} = \{E \subset T : \text{for each } K \in \mathcal{C}, E \cap K \in M_u$$
$$\text{and for } q \in \Gamma \lim_{K \in \mathcal{C}} u_q^\bullet(E \backslash (E \cap K)) = 0\}.$$

Let $E \in M_u$. Let $K \in \mathcal{C}$ and $q \in \Gamma$. Then by Theorem 7.5.8, $E \cap K \in M_u$. Since $\chi_E \in \mathcal{L}_1(u)$, by the complex lcHs version of Lemma 1.24 of [T],

$$\lim_{K \in \mathcal{C}} u_q^\bullet(\chi_{E \backslash K}) = \lim_{K \in \mathcal{C}} u_q^\bullet(E \backslash (E \cap K)) = 0$$

for each $q \in \Gamma$. Hence $M_u \subset \mathcal{R}$.

To prove the reverse inclusion, let $E \in \mathcal{R}$. Then $E \cap K \in M_u$ for each $K \in \mathcal{C}$. Moreover, given $q \in \Gamma$, by hypothesis $\lim_{K \in \mathcal{C}} u_q^\bullet(E \backslash (E \cap K)) = 0$. Since $\mathcal{L}_1(u)$ is closed in $\mathcal{F}^0(u)$ by Definition 7.1.20, it follows that $E \in M_u$ and hence $M_u = \mathcal{R}$. Thus $M_u = \ell(\widetilde{\delta(\mathcal{C})})$, the localized Lebesgue-Radon completion of $\delta(\mathcal{C})$. Finally, $\widehat{\boldsymbol{\mu}} = \boldsymbol{\mu}_u$ is immediate by the definition of $\widehat{\boldsymbol{\mu}}$ and from (7.5.27.1).

This completes the proof of the theorem. □

The following theorem generalizes Theorem 4.6 of [P4].

Theorem 7.5.28. *Let X be a quasicomplete lcHs. Let $\boldsymbol{\mu}_0 : \delta(\mathcal{C}_0) \to X$ be σ-additive. Then $\boldsymbol{\mu}_0$ is the restriction of a unique X-valued prolongable Radon vector measure $\boldsymbol{\mu}_u$ and u is called the prolongable Radon operator determined by $\boldsymbol{\mu}_0$. Then $\boldsymbol{\mu}_0$ admits a unique $\delta(\mathcal{C})$-regular σ-additive extension $\boldsymbol{\mu} : \delta(\mathcal{C}) \to X$ and $\boldsymbol{\mu}$ and $\boldsymbol{\mu}_0$ determine the same prolongable Radon operator u.*

Proof. By Theorem of Dinculeanu and Kluvánek [DK], $\boldsymbol{\mu}_0$ has a unique σ-additive $\delta(\mathcal{C})$-regular extension $\boldsymbol{\mu} : \delta(\mathcal{C}) \to X$.

Since each $f \in \mathcal{K}(T)$ is $\boldsymbol{\mu}_0$-integrable in T,

$$uf = \int_T f d\boldsymbol{\mu}_0, \qquad f \in \mathcal{K}(T)$$

is well defined, linear and has values in X. Moreover, for a relatively compact open set U in T, by Theorem 50.D of [H] there exists a relatively compact open Baire set U_0 such that $U \subset U_0$. Then for $f \in C_c(U)$ and for $q \in \Gamma$,

$$q(uf) = q\left(\int_T f d\boldsymbol{\mu}_0\right) \leq ||f||_U ||\boldsymbol{\mu}_0||_q(U_0)$$

and hence u is a Radon operator. Moreover, the operator

$$v : C_0(U) \to X$$

given by

$$vf = \int_U f d\boldsymbol{\mu}_0, \quad f \in C_0(U)$$

is continuous and is the restriction of

$$u_U : C_0(U_0) \to X$$

given by

$$u_U f = \int_U f d\boldsymbol{\mu}_0 = \int_U f d\boldsymbol{\mu}, \quad f \in C_0(U_0)$$

which is weakly compact by Lemma 5.2.19. Therefore, u is prolongable. Since $\boldsymbol{\mu}$ is uniquely determined by $\boldsymbol{\mu}_0$ and since $\boldsymbol{\mu}$ determines u, $\boldsymbol{\mu}_0$ and $\boldsymbol{\mu}$ determine u uniquely.

This completes the proof of the theorem. $\square$

7.6 Relation between $\mathcal{L}_p(u)$ and $\mathcal{L}_p(\mathbf{m}_u)$, u a weakly compact bounded Radon operator or a prolongable Radon operator

Let X be a Banach space (resp. a quasicomplete lcHs) and let $u : \mathcal{K}(T) \to X$ be a weakly compact bounded Radon operator. Then by Convention 7.2.4, $u : C_0(T) \to X$ is continuous and weakly compact. Let $\mathbf{m}_u : \mathcal{B}(T) \to X$ be the representing measure of u in the sense of 5.2.10 of Chapter 5. Then $\mathbf{m}_u$ is σ-additive on $\mathcal{B}(T)$, Borel regular and

$$u(\varphi) = \int_T \varphi d\mathbf{m}_u, \quad \varphi \in C_0(T)$$

where the integral is a (BDS)-integral. See Definition 3 and Theorems 2 and 6 of [P9]. Hereafter, $\mathbf{m}_u$ will denote the representing measure of u. In the first part of this section, we show that $f \in \mathcal{L}_1(u)$ if and only if $f \in \mathcal{L}_1(\mathbf{m}_u)$ and in that case, $\int f du = \int_T f d\mathbf{m}_u$. For such u, we also show that $\mathcal{L}_p(u)$ is the same as $\mathcal{L}_p(\mathbf{m}_u)$ for $1 \leq p < \infty$. (See Theorems 7.6.7, 7.6.9, 7.6.13 and 7.6.15.)

In the second part we consider a prolongable Radon operator v on $\mathcal{K}(T)$ with values in a quasicomplete lcHs X. If ω is a relatively compact open set in T and $1 \leq p < \infty$, then we show that $\mathcal{L}_p(v|_{\mathcal{K}(\omega)}) = \mathcal{L}_p(\mathbf{m}_v|_{\mathcal{B}(\omega)})$. (See Theorem 7.6.19.) Thus questions (Q5) and (Q6) mentioned in the Preface are also answered in the affirmative.

Let X be a Banach space and $u : \mathcal{K}(T) \to X$ be a weakly compact bounded Radon operator. Let $g : T \to \mathbb{K}$ be $\mathbf{m}_u$-measurable. Then by Theorem 3.1.3 of Chapter 3 we have

$$(\mathbf{m}_u)_p^{\bullet}(g, T) = \sup_{|x^*| \leq 1} \left(\int_T |g|^p dv(x^* \mathbf{m}_u) \right)^{\frac{1}{p}} \tag{$*$}$$

for $1 \leq p < \infty$.

Definition 7.6.1. Let X be a Banach space and $u : \mathcal{K}(T) \to X$ be a weakly compact bounded Radon operator. Let $g : T \to \mathbb{K}$ be u-measurable. For $1 \leq p < \infty$, let

$$u_p^{\bullet}(g) = \sup_{|x^*| \leq 1} \left(\int_T |g|^p d|x^* u| \right)^{\frac{1}{p}} \tag{7.6.1.1}$$

where $|x^* u|$ is given by (12) on p. 55 of [B].

The following theorem gives the relation between $(\mathbf{m}_u)_p^{\bullet}(g, T)$ and $u_p^{\bullet}(g)$ for $1 \leq p < \infty$.

Theorem 7.6.2. *Let X be a Banach space and $u : \mathcal{K}(T) \to X$ be a weakly compact bounded Radon operator. Let $\mathbf{m}_u$ be the representing measure of u as in 5.2.10 of Chapter 5. Then a function $g : T \to \mathbb{K}$ is u-measurable if and only if g is $\mathbf{m}_u$-measurable. Moreover, for a u-measurable scalar function g,*

$$u_p^{\bullet}(g) = (\mathbf{m}_u)_p^{\bullet}(g, T) \tag{7.6.2.1}$$

for $1 \leq p < \infty$. Also we have

$$u_p^{\bullet}(f + g) \leq u_p^{\bullet}(f) + u_p^{\bullet}(g), \tag{7.6.2.2}$$

$$u_p^{\bullet}(\alpha f) = |\alpha| u_p^{\bullet}(f), \ \alpha \in \mathbb{K}, \tag{7.6.2.3}$$

and

$$u_1^{\bullet}(fg) \leq u_{p_1}^{\bullet}(f) \cdot u_{p_2}^{\bullet}(g) \tag{7.6.2.4}$$

for u-measurable scalar functions f and g on T whenever $1 < p_1, p_2 < \infty$ with $\frac{1}{p_1} + \frac{1}{p_2} = 1$.

Proof. By Theorem 7.5.6(iv), g is $\mathbf{m}_u$-measurable if and only if it is u-measurable.

By (7.6.1.1) and by 5.2.10 of Chapter 5 we have

$$u_p^\bullet(g) = \sup_{|x^*| \le 1} \left(\int_T |g|^p d|x^* u| \right)^{\frac{1}{p}}$$

$$= \sup_{|x^*| \le 1} \left(\int_T |g|^p d|u^* x^*| \right)^{\frac{1}{p}}$$

$$= \sup_{|x^*| \le 1} \left(\int_T |g|^p d|x^* \circ \mathbf{m}_u| \right)^{\frac{1}{p}}$$

$$= \sup_{|x^*| \le 1} \left(\int_T |g|^p dv(x^* \circ \mathbf{m}_u) \right)^{\frac{1}{p}}$$

by Notation 4.4, Theorem 4.7(vi) and Theorem 4.11 of [P3] and by Theorem 3.3 of [P4] where $v(x^* \circ \mathbf{m}_u) = v(x^* \circ \mathbf{m}_u, \mathcal{B}(T))$ on $\mathcal{B}(T)$. Hence by $(*)$ we have

$$u_p^\bullet(g) = (\mathbf{m}_u)_p^\bullet(g, T).$$

Now by (7.6.2.1) and by Theorem 3.1.13 we have

$$u_p^\bullet(f + g) = (\mathbf{m}_u)_p^\bullet(f + g, T)$$
$$\le (\mathbf{m}_u)_p^\bullet(f, T) + (\mathbf{m}_u)_p^\bullet(g, T)$$
$$= u_p^\bullet(f) + u_p^\bullet(g)$$

and

$$u_p^\bullet(\alpha f) = |\alpha| u_p^\bullet(|f|)$$

for $1 \le p < \infty$ and for $\alpha \in \mathbb{K}$ whenever $f, g : T \to \mathbb{K}$ are u-measurable. Moreover, if $1 < p_1, p_2 < \infty$ with $\frac{1}{p_1} + \frac{1}{p_2} = 1$, then by Theorem 3.1.13(iii) and by (7.6.2.1) we have

$$u_1^\bullet(fg) = (\mathbf{m}_u)_1^\bullet(fg, T)$$
$$\le (\mathbf{m}_u)_{p_1}^\bullet(f, T) \cdot (\mathbf{m}_u)_{p_2}^\bullet(g, T)$$
$$= u_{p_1}^\bullet(f) \cdot u_{p_2}^\bullet(g).$$

This completes the proof of the theorem. □

Definition 7.6.3. Let X be a Banach space and let $1 \le p < \infty$. Let $u : \mathcal{K}(T) \to X$ be a weakly compact bounded Radon operator. Let $\mathcal{F}_p^0(u) = \{f : T \to \mathbb{K}, f u\text{-}$ measurable and, $u^\bullet(|f|^p) < \infty\}$. Let $I_p(u) = \{f : T \to \mathbb{K}, f u\text{-measurable and } |f|^p \in \mathcal{L}_1(u)\}$. Let

$$I(u) = \{f : T \to \mathbb{K}, f u\text{-measurable and } u\text{-integrable}\}.$$

Theorem 7.6.4. *Under the hypothesis of Definition 7.6.3, $I_1(u) = I(u)$.*

Proof. If $f \in I(u)$, then $f \in \mathcal{L}_1(u)$. Thus, given $\epsilon > 0$, there exists $\varphi \in \mathcal{K}(T)$ such that $u^{\bullet}(|f - \varphi|) < \epsilon$. Since $u^{\bullet}(||f| - |\varphi|) \leq u^{\bullet}(|f - \varphi|) < \epsilon$, $|f| \in \mathcal{L}_1(u)$ and hence $f \in I_1(u)$. Conversely, if $f \in I_1(u)$, then f is u-measurable and $|f|$ is u-integrable. Then by the complex analogue of Theorem 1.22 of [T], f is u-integrable. Hence $I_1(u) = I(u)$. □

Definition 7.6.5. Let X be a Banach space, $u : \mathcal{K}(T) \to X$ be a weakly compact bounded Radon operator and $1 \leq p < \infty$. Let $\mathcal{L}_p(u) = \{f \in I_p(u) : u^{\bullet}(|f|^p) < \infty\}$.

Theorem 7.6.6. *Let X, u and p be as in Definition 7.6.5. Then $\mathcal{L}_p(u) = I_p(u) \subset \mathcal{F}_p^0(u)$.*

Proof. If $f \in I_p(u)$, then f is u-measurable and $|f|^p \in \mathcal{L}_1(u)$. Then by the complex analogues of Definition 1.6 and Lemma 1.5 of [T], $u^{\bullet}(|f|^p) < \infty$. □

Theorem 7.6.7. *Let X be a Banach space and $u : \mathcal{K}(T) \to X$ be a weakly compact bounded Radon operator. Then a function $f : T \to \mathbb{K}$ is u-integrable if and only if f is $\mathbf{m}_u$-integrable in T and in that case*

$$\int f du = \int_T f d\mathbf{m}_u.$$

Moreover, for $f \in \mathcal{L}_1(u)$, $u_1^{\bullet}(f) = (\mathbf{m}_u)_1^{\bullet}(f, T)$.

Proof. Let $\varphi \in C_0(T)$. Then by the proof of Lemma 5.2.19, φ is $\mathbf{m}_u$-integrable in T. Then by 5.2.10 of Chapter 5 we have

$$x^* u(\varphi) = \int_T \varphi d(x^* u) = \int_T \varphi d(u^* x^*) = \int_T \varphi d(x^* \circ \mathbf{m}_u) = x^* \left(\int_T \varphi d\mathbf{m}_u \right)$$

for $x^* \in X^*$. Hence by the Hahn-Banach theorem

$$u(\varphi) = \int \varphi du = \int_T \varphi d\mathbf{m}_u \tag{7.6.7.1}$$

for $\varphi \in C_0(T)$.

Let f be u-integrable. Then there exists $(\varphi_n)_1^{\infty} \subset C_c(T)$ such that $u^{\bullet}(|f - \varphi_n|) \to 0$ as $n \to \infty$ and hence by the complex analogue of **1.10** of [T] we have

$$\left| \int f du - u(\varphi_n) \right| = \left| \int (f - \varphi_n) du \right| \leq u^{\bullet}(|f - \varphi_n|) \to 0$$

as $n \to \infty$ and hence

$$\int f du = \lim_n u(\varphi_n) = \lim_n \int \varphi_n du. \tag{7.6.7.2}$$

Then by Proposition 7.1.10, by Lemma 7.1.11 and by 5.2.10 we have

$$
\begin{aligned}
u^\bullet(|f-\varphi_n|) &= \sup_{|x^*|\le 1} |u_{x^*}|^\bullet(|f-\varphi_n|)\\
&= \sup_{|x^*|\le 1} |u_{x^*}|(|f-\varphi_n|)\,\text{(by 7.1.11)}\\
&= \sup_{|x^*|\le 1} |x^*u|(|f-\varphi_n|)\\
&= \sup_{|x^*|\le 1} |u^*x^*|(|f-\varphi_n|)\\
&= \sup_{|x^*|\le 1} \int |f-\varphi_n|d|x^*\circ\mathbf{m}_u|\\
&= \sup_{|x^*|\le 1} \int |f-\varphi_n|dv(x^*\circ\mathbf{m}_u) \qquad (**)
\end{aligned}
$$

by Notation 4.4 and by Theorems 4.7(vi) and 4.11 of [P3] and by Theorem 3.3 of [P4] where $v(x^*\circ\mathbf{m}_u)=v(x^*\circ\mathbf{m}_u,\mathcal{B}(T))$.

Therefore, by $(**)$ we have

$$
u^\bullet(|f-\varphi_n|) = \sup_{|x^*|\le 1}\int_T |f-\varphi_n|dv(x^*\circ\mathbf{m}_u) = (\mathbf{m}_u)_1^\bullet(f-\varphi_n,T). \qquad (7.6.7.3)
$$

As $u^\bullet(|f-\varphi_n|)\to 0$, by (7.6.7.3) we have $(\mathbf{m}_u)_1^\bullet(f-\varphi_n,T)\to 0$. Consequently, by Theorem 6.1.10, $f\in\mathcal{L}_1(\mathbf{m}_u)$. Moreover, by (7.6.7.1), (7.6.7.3), (3.1.3.1) and (7.6.7.2) we have

$$
\int_T f d\mathbf{m}_u = \lim_n \int_T \varphi_n d\mathbf{m}_u = \lim_n \int \varphi_n du = \int f du.
$$

Thus f is $\mathbf{m}_u$-integrable in T if f is u-integrable and

$$
\int f du = \int_T f d\mathbf{m}_u. \qquad (7.6.7.4)
$$

Conversely, let f be $\mathbf{m}_u$-integrable in T. Then by Theorem 6.1.10 there exists $(\varphi_n)_1^\infty\subset C_c(T)$ such that $(\mathbf{m}_u)_1^\bullet(f-\varphi_n,T)\to 0$ so that by (3.1.3.1) we have

$$
\int_T f d\mathbf{m}_u = \lim_n \int_T \varphi_n d\mathbf{m}_u.
$$

But by (7.6.7.1) we have

$$
\int_T \varphi_n d\mathbf{m}_u = \int \varphi_n du
$$

and hence

$$
\int f d\mathbf{m}_u = \lim_n \int \varphi_n du. \qquad (7.6.7.5)
$$

As $(\mathbf{m}_u)_1^\bullet(f-\varphi_n,T) = u^\bullet(|f-\varphi_n|)$ by (7.6.2.1), $u^\bullet(|f-\varphi_n|) \to 0$ and hence $f \in \mathcal{L}_1(u)$ by the complex version of Definition 1.6 of [T]. Then by the complex analogue of **1.10** of [T] we have

$$\left| \int f\,du - \int \varphi_n\,du \right| \leq u^\bullet(|f-\varphi_n|)) \to 0$$

as $n \to \infty$ and hence by (7.6.7.1) and (7.6.7.5) we have

$$\int f\,du = \lim_n \int \varphi_n\,du = \lim_n \int_T \varphi_n\,d\mathbf{m}_u = \int_T f\,d\mathbf{m}_u.$$

Thus $\mathcal{L}_1(u) = \mathcal{L}_1(\mathbf{m}_u)$ and for $f \in \mathcal{L}_1(u)$,

$$\int f\,du = \int_T f\,d\mathbf{m}_u$$

whenever u is a weakly compact bounded Radon operator on $\mathcal{K}(T)$ with values in a Banach space X. Moreover, for $f \in \mathcal{L}_1(u)$, $u_1^\bullet(f) = (\mathbf{m}_u)_1^\bullet(f,T)$ by (7.6.2.1).

This completes the proof of the theorem. □

Theorem 7.6.8. *Let X, u and p be as in Definition 7.6.5. Then $\mathcal{L}_p(u)$ is a semi-normed space.*

Proof. Let $f, g \in \mathcal{L}_p(u)$ and α be a scalar. Then $|f|^p, |g|^p \in \mathcal{L}_1(u)$. Since $|f+g|^p \leq 2^p \max(|f|^p,|g|^p) \leq 2^p(|f|^p+|g|^p)$, since $|f+g|^p$ is u-measurable and since $|f|^p+|g|^p \in \mathcal{L}_1(u)$, by the complex version of Theorem 1.22 of [T], $|f+g|^p \in \mathcal{L}_1(u)$ and hence $f+g \in \mathcal{L}_p(u)$. Clearly, $|\alpha f|^p \in \mathcal{L}_1(u)$ for $\alpha \in \mathbb{K}$ and hence $\mathcal{L}_p(u)$ is a vector space over $\mathbb{K}$. Moreover, by (7.6.2.2) and (7.6.2.3) and by Theorem 7.6.6, $\mathcal{L}_p(u)$ is a seminormed space. □

Theorem 7.6.9. *Let X, u and p be as in Definition 7.6.5. Then $\mathcal{L}_p(u) = \mathcal{L}_p(\mathbf{m}_u)$ and hence is complete for $1 \leq p < \infty$.*

Proof. By Definition 7.6.5 and Theorem 7.6.6, $\mathcal{L}_p(u) = I_p(u)$ for $1 \leq p < \infty$. Moreover, by Theorem 7.6.2, for $f \in \mathcal{L}_p(u)$,

$$u_p^\bullet(f) = (\mathbf{m}_u)_p^\bullet(f,T). \tag{7.6.9.1}$$

Then by Theorems 7.6.6 and 7.6.7, $f \in \mathcal{L}_p(u)$ if and only if $f \in \mathcal{L}_p(\mathbf{m}_u)$. Consequently, by (7.6.9.1) and by Theorem 3.2.8 of Chapter 3, $\mathcal{L}_p(u)$ is complete.

This completes the proof of the theorem. □

Definition 7.6.10. Let X be a quasicomplete lcHs and $u : K(T) \to X$ be a weakly compact bounded Radon operator. A u-measurable function $f : T \to \mathbb{K}$ is said to be u-integrable in T if it is $u_q = \Pi_q \circ u$-integrable in T with values in $\widetilde{X_q}$

(considering $u_q : \mathcal{K}(T) \to X_q \subset \widetilde{X_q}$) for each $q \in \Gamma$ (see Definition 7.1.18). In that case, using Notation 1.2.17 of Chapter 1, we define

$$\int f du = \varprojlim \int f du_q.$$

Definition 7.6.11. Let X be a quasicomplete lcHs and $u : \mathcal{K}(T) \to X$ be a weakly compact bounded Radon operator. Let $1 \le p < \infty$. For $q \in \Gamma$ and $g : T \to \mathbb{K}$ u-measurable, let

$$(u_q)^{\bullet}_p(g) = \sup_{x^* \in U^0_q} \left(\int_T |g|^p dv(x^* u) \right)^{\frac{1}{p}}$$

where $U^0_q = \{ x^* \in X^* : |x^*(x)| \le 1 \text{ for } x \in U_q \}$.

Theorem 7.6.12. *Under the hypothesis of Definition 7.6.11,*

$$(u_q)^{\bullet}_p(g) = ((\mathbf{m}_u)_q)^{\bullet}_p(g, T)$$

where $\mathbf{m}_u$ is the representing measure of u in the sense of 5.2.10 of Chapter 5.

Proof. By Proposition 1.2.15(ii)(b) and by the definition of Ψ_{x^*} as given in Proposition 1.2.15(ii)(a) of Chapter 1, $\{\Psi_{x^*} : x^* \in U^0_q\}$ is a norm determining subset of the closed unit ball of $(X_q)^*$ and for $x^* \in U^0_q$, $x^*(\Pi_q \circ u) = \Psi_{x^*} u_q = x^* u_q = x^* u$ by (ii)(a) of the said proposition. Then by 5.2.10 of Chapter 5 and by Lemma 3.1.2(ii) of Chapter 3 and by (7.6.2.1) we have

$$(\Psi_{x^*} u_q)^{\bullet}_p(g) = \left(\int_T |g|^p dv(\Psi_{x^*} u_q) \right)^{\frac{1}{p}} = \left(\int_T |g|^p dv(x^* u) \right)^{\frac{1}{p}}$$

$$= \left(\int_T |g|^p dv(x^* \circ \mathbf{m}_u) \right)^{\frac{1}{p}}$$

for $x^* \in U^0_q$ and hence

$$(u_q)^{\bullet}_p(g) = \sup_{x^* \in U^0_q} (\Psi_{x^*} u_q)^{\bullet}_p(g) = \sup_{x^* \in U^0_q} \left(\int_T |g|^p dv(x^* \circ \mathbf{m}_u) \right)^{\frac{1}{p}} = ((\mathbf{m}_u)_q)^{\bullet}_p(g, T)$$

by Theorem 4.3.2 of Chapter 4. $\square$

Theorem 7.6.13. *Let X be a quasicomplete lcHs and $u : \mathcal{K}(T) \to X$ be a weakly compact bounded Radon operator. Then a function $f : T \to \mathbb{K}$ is u-integrable if and only if it is $\mathbf{m}_u$-integrable in T and in that case,*

$$\int f du = \int_T f d\mathbf{m}_u \in X.$$

Proof. By Theorem 7.6.12, $(u_q)_1^\bullet(g) = ((\mathbf{m}_u)_q)_1^\bullet(g,T)$ for $q \in \Gamma$. Let f be u-integrable. Then f is u_q-integrable for each $q \in \Gamma$ and hence by Theorem 7.6.7, f is $(\mathbf{m}_u)_q$-integrable in T and therefore, f is $\mathbf{m}_u$-integrable in T. Similarly, if f is $\mathbf{m}_u$-integrable in T, then f is u-integrable. If f is u-integrable, then given $\epsilon > 0$ and $q \in \Gamma$, there exists $\varphi \in C_c(T)$ such that $u_q^\bullet(|\varphi - f|) < \epsilon$. Then by Theorem 7.6.12, $((\mathbf{m}_u)_q)_1^\bullet(f - \varphi, T) < \epsilon$ and hence f is $(\mathbf{m}_u)_q$-integrable in T. Thus there exists $x_q \in \widetilde{X_q}$ such that $\int f du_q = \int_T f d(\mathbf{m}_u)_q = x_q$ for each $q \in \Gamma$. Thus

$$\int f du = \lim_{\leftarrow} \int f du_q = \lim_{\leftarrow} x_q = \lim_{\leftarrow} \int f d(\mathbf{m}_u)_q = \int f d\mathbf{m}_u \in X$$

by Definition 4.2.1 and Theorem 4.2.3 of Chapter 4. Similarly, it can be shown that if f is $\mathbf{m}_u$-integrable in T, then f is u-integrable and $\int f d\mathbf{m}_u = \int f du \in X$. Hence the theorem holds. □

Definition 7.6.14. Let X be a quasicomplete lcHs and $u : \mathcal{K}(T) \to X$ be a weakly compact bounded Radon operator. Let $1 \leq p < \infty$. Let $\mathcal{F}_p^{0(u)} = \{f : T \to \mathbb{K}, fu$-measurable and $(u_q)_p^\bullet(f) < \infty$ for each $q \in \Gamma\}$. Then we define $\mathcal{L}_p(u) = \{f \in \mathcal{F}_p^{0(u)}$ and $|f|^p u$-integrable (with values in $X)\}$.

Theorem 7.6.15. *Under the hypothesis of Definition 7.6.14, $\mathcal{L}_p(u) = \mathcal{L}_p(\mathbf{m}_u)$ for $1 \leq p < \infty$ and for $f \in \mathcal{L}_p(\mathbf{m}_u)$, $\int |f|^p du = \int_T |f|^p d\mathbf{m}_u$ and conversely.*

Proof. Let $1 \leq p < \infty$ and let $f \in \mathcal{L}_p(u)$. Then $(u_q)_p^\bullet(f) < \infty$ and $|f|^p$ is u_q-integrable for each $q \in \Gamma$. Then by Theorem 7.6.9, $|f|^p \in \mathcal{L}_1(\mathbf{m}_{u_q})$ and consequently, by Theorem 7.6.7, $|f|^p \in \mathcal{L}_1(u_q)$ and

$$\int |f|^p du_q = \int_T |f|^p d(\mathbf{m}_u)_q$$

for $q \in \Gamma$. Therefore, $|f|^p$ is $\mathbf{m}_u$-integrable in T and

$$\int |f|^p du = \lim_{\leftarrow} \int |f|^p du_q = \lim_{\leftarrow} \int_T |f|^p d(\mathbf{m}_{u_q}) = \int_T |f|^p d\mathbf{m}_u.$$

Thus $f \in \mathcal{L}_p(\mathbf{m}_u)$ and $\int |f|^p du = \int_T |f|^p d\mathbf{m}_u$. By reversing the argument, one can prove the converse. □

Using Theorem 7.6.12 and adapting the proof of Theorem 4.5.3(i) of Chapter 4 one can prove the following theorem. The details are left to the reader.

Theorem 7.6.16. *Let X be a quasicomplete lcHs, $u : \mathcal{K}(T) \to X$ be a weakly compact bounded Radon operator and $1 \leq p < \infty$. Let $f_n^{(q)}$, $n \in \mathbb{N}$, be u_q-measurable scalar functions on T for $q \in \Gamma$. Let $K^{(q)}$ be a finite constant such that $|f_n^{(q)}| \leq K^{(q)}$ u_q-a.e. in T for each n. If $f_n^{(q)} \to f$ u_q-a.e. in T where f is a scalar function on*

T, then f, $f_n^{(q)}$, $n \in \mathbb{N}$ belong to $\mathcal{L}_p(u_q)$ and $\lim_n(u_q)_p^\bullet(f_n^{(q)} - f) = 0$ for $q \in \Gamma$. Consequently, $f \in \mathcal{L}_p(u)$. When $p = 1$, f is u-integrable and

$$\lim_n \left| q \left(\int f du \right) - \left| \int f_n^{(q)} du_q \right|_q \right| = 0.$$

Remark 7.6.17. By using the complex version of Lemma 2.21 of [T] and Theorem 7.1.24, one can give an alternative proof of the above theorem.

Now we consider the generalization of the above results when u is a prolongable Radon operator on $\mathcal{K}(T)$.

Lemma 7.6.18. *Let X be a Banach space and $v : \mathcal{K}(T) \to X$ be a prolongable Radon operator. Let $f \in \mathcal{L}_1(v)$ with $\operatorname{supp} f = K \in \mathcal{C}$. If ω is a relatively compact open set in T with $K \subset \omega$, then the following hold:*

(i) $\varphi \chi_\omega \in C_c(\omega)$ *for each* $\varphi \in C_c(T)$.
(ii) *There exists* $(\varphi_n)_1^\infty \subset C_c(\omega)$ *such that* $v^\bullet(|f - \varphi_n|) \to 0$ *as* $n \to \infty$.
(iii) $\int f dv = \lim_n v(\varphi_n \chi_\omega) = \lim_n \int \varphi_n \chi_\omega d\mathbf{m}_v = \lim_n \int_\omega \varphi_n d\mathbf{m}_v$.
(iv) $\lim_n (\mathbf{m}_v|_{\mathcal{B}(\omega)})^\bullet(f - \varphi_n \chi_\omega, \omega) = 0$.
(v) f *is* $\mathbf{m}_v$-*integrable in T and* $\int f dv = \int_T f d\mathbf{m}_v$.

Proof. (i) Let $t \in \omega$ and let $(t_\alpha)_{\alpha \in I} \subset \omega$ be a net converging to t. Then $\varphi(t_\alpha) \to \varphi(t)$ and hence $(\varphi \chi_\omega)(t_\alpha) = \varphi(t_\alpha) \to \varphi(t) = (\varphi \chi_\omega)(t)$. Hence (i) holds.

(ii) Since $f \in \mathcal{L}_1(v)$, by Definition 7.1.6 there exists a sequence $(\varphi_n) \subset \mathcal{K}(T)$ such that $v^\bullet(|f - \varphi_n|) \to 0$ as $n \to \infty$. Then by (i), $\varphi_n \chi_\omega \in C_c(\omega)$ and $v^\bullet(|f - \varphi_n|\chi_\omega) \le v^\bullet(|f - \varphi_n|) \to 0$ as $n \to \infty$. Hence (ii) holds.

(iii) By the complex analogue of **1.10** of [T] and by (ii) we have

$$\left| \int f dv - v(\varphi_n) \right| = \left| \int (f - \varphi_n) dv \right| \le v^\bullet(|f - \varphi_n|) \to 0$$

as $n \to \infty$ and hence we have

$$\int f dv = \lim_n v(\varphi_n) = \lim_n \int \varphi_n dv. \qquad (7.6.18.1)$$

Moreover, $f = f\chi_\omega$ and hence

$$|f - \varphi_n \chi_\omega| = |f\chi_\omega - \varphi_n \chi_\omega| = |f - \varphi_n|\chi_\omega \le |f - \varphi_n|.$$

Therefore,

$$v^\bullet(|f - \varphi_n \chi_\omega|) = v^\bullet(|f\chi_\omega - \varphi_n \chi_\omega|) = v^\bullet(|f - \varphi_n|\chi_\omega) \le v^\bullet(|f - \varphi_n|)$$

for $n \in \mathbb{N}$. Therefore,

$$v^\bullet(|f - \varphi_n \chi_\omega|) \le v^\bullet(|f - \varphi_n|) \to 0 \qquad (7.6.18.2)$$

as $n \to \infty$. Then by the complex analogue of **1.10** of [T] and by (7.6.18.2) we have

$$\left| \int f dv - v(\varphi_n \chi_\omega) \right| = \left| \int (f - \varphi_n \chi_\omega) dv \right| \leq v^\bullet(|f - \varphi_n \chi_\omega|) \to 0$$

as $n \to \infty$. Hence

$$\int f dv = \lim_n v(\varphi_n \chi_\omega). \tag{7.6.18.3}$$

As $\varphi_n \chi_\omega \in \mathcal{K}(\omega)$ by (i), by (7.6.18.3) we have

$$\int f dv = \lim_n \int \varphi_n \chi_\omega dv. \tag{7.6.18.4}$$

As $v|_{\mathcal{K}(\omega)}$ is a weakly compact bounded Radon operator by Definition 5.3.8, by Theorem 7.6.7 and by (7.6.18.4) we conclude that

$$\int f dv = \lim_n \int \varphi_n \chi_\omega dv = \lim_n \int_\omega \varphi_n d\mathbf{m}_v. \tag{7.6.18.5}$$

Hence (iii) holds.

(iv) By Proposition 7.1.10, by Lemma 7.1.11 and by 5.2.10 of Chapter 5 we have

$$v^\bullet(|f - \varphi_n \chi_\omega|) = \sup_{|x^*| \leq 1} |v_{x^*}|^\bullet(|f - \varphi_n \chi_\omega|) = \sup_{|x^*| \leq 1} |v_{x^*}|(|f - \varphi_n \chi_\omega|)$$

$$= \sup_{|x^*| \leq 1} |v^* x^*|(|f - \varphi_n \chi_\omega|)$$

$$= \sup_{|x^*| \leq 1} \int |f - \varphi_n \chi_\omega| dv(x^* \circ \mathbf{m}_v|_{\mathcal{B}(\omega)})$$

$$= (\mathbf{m}_v|_{\mathcal{B}(\omega)})_1^\bullet(f - \varphi_n \chi_\omega, \omega). \tag{7.6.18.6}$$

Hence by (7.6.18.2) and (7.6.18.6), (iv) holds.

(v) By (iv) and by Theorem 6.1.10 applied to ω in place of T we conclude that $f \in \mathcal{L}_1(\mathbf{m}_v|_{\mathcal{B}(\omega)})$. As $(\varphi_n \chi_\omega)_1^\infty \subset \mathcal{K}(\omega)$ and as $v|_{\mathcal{K}(\omega)}$ is a weakly compact bounded Radon operator by the hypothesis on v, by Theorem 7.6.7 we conclude that

$$v(\varphi_n \chi_\omega) = \int_\omega \varphi_n d(\mathbf{m}_v|_{\mathcal{B}(\omega)}) = \int_\omega \varphi_n d\mathbf{m}_v \tag{7.6.18.7}$$

for $n \in \mathbb{N}$, since $\varphi_n(t) = 0$ for $t \in T \backslash \omega$. Then by (3.1.3.1), (7.6.18.5), (7.6.18.6) and (7.6.18.7) we have

$$\int_\omega f d\mathbf{m}_v = \lim_n \int \varphi_n \chi_\omega d\mathbf{m}_v = \lim_n v(\varphi_n \chi_\omega) = \lim_n \int \varphi_n \chi_\omega dv = \int f dv$$

and hence $\int f dv = \int_\omega f d\mathbf{m}_v$. Thus (v) holds. $\qquad \square$

Theorem 7.6.19. *Let X be a quasicomplete lcHs and $v : \mathcal{K}(T) \to X$ be a prolongable Radon operator. If ω is a relatively compact open set in T and $1 \leq p < \infty$, then $\mathcal{L}_p(v|_{\mathcal{K}(\omega)}) = \mathcal{L}_p(\mathbf{m}_v|_{\mathcal{B}(\omega)})$ so that for $f \in \mathcal{L}_p(\mathbf{m}_v)$ with $\mathrm{supp}f = K \in \mathcal{C}$ and with $K \subset \omega$,*

$$\int_\omega |f|^p d\mathbf{m}_v = \int |f\chi_\omega|^p dv.$$

For $p = 1$, and for $f \in \mathcal{L}_1(\mathbf{m}_v)$, $\int_\omega f d\mathbf{m}_v = \int f\chi_\omega dv$.

Proof. Let $1 \leq p < \infty$ and let $f \in \mathcal{L}_p(\mathbf{m}_v)$. Then for each $q \in \Gamma$, $f \in \mathcal{L}_p(\mathbf{m}_{v_q})$ and hence $|f|^p \in \mathcal{L}_1(\mathbf{m}_{v_q})$. Consequently, $|f|^p \chi_\omega \in \mathcal{L}_1(\mathbf{m}_{v_q})$. Therefore, by Theorem 7.6.16, $|f|^p \chi_\omega$ is v_q-integrable and consequently, by Lemma 7.6.18 we have

$$\int |f|^p \chi_\omega dv_q = \int_\omega |f|^p d(\mathbf{m}_{v_q}) = \int_\omega |f|^p d(\mathbf{m}_v)_q$$

for each $q \in \Gamma$. Therefore, $|f|^p \chi_\omega$ is v-integrable and

$$\int |f|^p \chi_\omega dv = \lim_{\leftarrow} \int |f|^p \chi_\omega dv_q = \lim_{\leftarrow} \int_\omega |f|^p d(\mathbf{m}_v)_q = \int_\omega |f|^p d\mathbf{m}_v.$$

The proof of the second result is similar and is left to the reader. □

The following theorem is analogous to Theorem 7.6.16 for prolongable Radon operators.

Theorem 7.6.20. *Let X be a quasicomplete lcHs, $v : \mathcal{K}(T) \to X$ be a prolongable Radon operator and $1 \leq p < \infty$. Let $f_n^{(q)}, n \in \mathbb{N}$, be v_q-measurable scalar functions on T for $q \in \Gamma$. Let $g^{(q)} \in \mathcal{L}_p(v_q)$ such that $|f_n^{(q)}| \leq g^{(q)}$ v_q-a.e. in T for each n. If $f_n^{(q)} \to f$ v_q-a.e. in T where f is a scalar function on T, then $f, f_n^{(q)}, n \in \mathbb{N}$, belong to $\mathcal{L}_p(v_q)$ and $\lim_n (v_q)^\bullet (f_n^{(q)} - f) = 0$ for $q \in \Gamma$. Consequently, $f \in \mathcal{L}_p(v)$. When $p = 1$, f is v-integrable and*

$$\lim_n \left| q \left(\int f dv \right) - \left| \int f_n^{(q)} dv_q \right|_q \right| = 0.$$

Remark 7.6.21. In Examples 1.36, 1.37 and 1.38 of [T] with the scalar field $\mathbb{C}$ instead of $\mathbb{R}$, $\int f d\mu$ is described for a μ-integrable function f. Example 2.8 of [T] gives a characterization for μ to be a weakly compact bounded Radon operator. Example 2.9 of [T] says that any Radon operator with values in $l^1(I)$ is always weakly compact. Example 3.9 of [T] deals with prolongable or weakly compact Radon operators on $l_p(I)$ for $1 \leq p < \infty$. Example 3.12 of [T] deals with the discrete Radon operators. Example 3.16 of [T] deals with the characterization of μ-integrability of f where $\mu = (\mu_i)_{i \in I}$ is a measure with values in $l_p(I)$, $1 \leq p < \infty$. Example 3.17 of [T] deals with Radon operators in a real Hilbert space and Example 3.18 of [T] deals with Radon operators in $C(S)$, S a compact Hausdorff space.

Bibliography

[B]   N. Bourbaki, Integration, Chapitres I–IV, Chapitre V, Hermann, Paris, 1965.

[Be]   S.K. Berberian, Measure and Integration, Chelsea, New York, 1965.

[Bo]   F. Bombal, Medidas Vectoriales y Espacios de Funciones Continuas, Lecture Notes, Univ. Complutense, Facultad de Mat., Madrid (Spanish), 1984.

[Br]   J.K. Brooks, *On the existence of a control measure for strongly bounded vector measures*, Bull. Amer. Math. Soc., **17** (1971), 999–1001.

[BD1]   J.K. Brooks, and N. Dinculeanu, *Strong additivity, absolute continuity, and compactness in spaces of measures*, J. Math. Anal. Appl. **45**(1974), 156–175.

[BD2]   J.K. Brooks, and N. Dinculeanu, *Lebesgue-type spaces for vector integration, linear operators, weak completeness and weak compactness*, J. Math. Anal. Appl. **54**(1976), 348–389.

[BDS]   R.G. Bartle, N. Dunford, and J.T. Schwartz, *Weak compactness and vector measures*, Canad. J. Math. **7**(1955), 289–305.

[C1]   G.P. Curbera, *Operators into L^1 of a vector measure and applications to Banach lattices*, Math. Ann. **292**(1992), 317–330.

[C2]   G.P. Curbera, *When L^1 of a vector measure is an AL-space*, Pacific J. Math., **162**(1994), 287–303.

[C3]   G.P. Curbera, *Banach space properties of L^1 of a vector measure*, Proc. Amer. Math. Soc., **33**(1995), 3797–3806.

[C4]   G.P. Curbera, *Volterra convolution operators with values in rearrangement invariant spaces*, J. London Math. Soc., **60**(1999), 258–268.

[CR]   G.P. Curbera and W.J. Ricker, *Banach lattices with the Fatou property and optimal domains of kernel operators*, Indag. Math. (N.S.), (in press).

[D]   M.M. Day, Normed Linear spaces, Ergebenisse der Mathematik und ihrer Grenzgebeite, Berlin, 1962, 2nd edition, printing corrected.

[Del1]   O. Delgado, *Cómo obtener subespacios de un espacio de Banach de funciones?*, preprint.

[Del2]   O. Delgado, *Banach function subspaces of L_1 of a vector measure and related Orlicz spaces*, Indag. Math. (N.S.) **15**(2004), 485–495.

[Del3] O. Delgado, *L_1-spaces of vector measures defined on δ-rings*, Arch. Math. **84**(2005), 432–443.

[Del4] O. Delgado, *Optimal domains for kernel operators on $[0, \infty]x[0, \infty]$*, Studia Math. **174**, (2006), 131–145.

[Die]   J. Dieudonné, *Sur la convergence des suites de mesures de Radon*, Anais. Acad. Bras. Ciencias **23**(1951), 21–38.

[Din1] N. Dinculeanu, Vector Measures, Pergamon Press, Berlin, 1967.

[Din2] N. Dinculeanu, Vector Integration and Stochastic Integration in Banach spaces, John Wiley, New York, 1999.

[DK]   N. Dinculeanu, and I. Kluvánek, *On vector measures*, Proc. London Math. Soc. **17**(1967), 505–512.

[DL]   N. Dinculeanu, and P.W. Lewis, *Regularity of Baire measures*, Proc. Amer. Math. Soc. **26**(1970), 92–94.

[Do1]   I. Dobrakov, *On integration in Banach spaces, I*, Czechoslovak Math. J. **20**(1970), 511–536.

[Do2]   I. Dobrakov, *On integration in Banach spaces, II*, Czechoslovak Math. J. **20**(1970), 680–695.

[Do3]   I. Dobrakov, *On integration in Banach spaces, IV*, Czechoslovak Math. J. **30**(1980), 259–279.

[Do4]   I. Dobrakov, *On integration in Banach spaces, V*, Czechoslovak Math. J., **30**(1980), 610–628.

[DP1]   I. Dobrakov, and T.V. Panchapagesan, *A simple proof of the Borel extension theorem and weak compactness of operators*, Czechoslovak Math. J. **52**(2002), 691–703.

[DP2]   I. Dobrakov, and T.V. Panchapagesan, *A generalized Pettis measurability criterion and integration of vector functions*, Studia Math., **164**(2004), 205–229.

[DS1]   N. Dunford, and J.T. Schwartz, Linear Operators, Part I, General Theory, Interscience, New York, 1957.

[DS2]   N. Dunford, and J.T. Schwartz, Linear Operators, Part III, Spectral Operators, Wiley-Interscience, New York, 1971.

[DU]   J. Diestel, and J.J. Uhl, Vector Measures, Amer. Math. Soc., Providence, R.I., 1977.

[E]   R.E. Edwards, Functional Analysis, Theory and Applications, Holt Rinehart and Winston, New York, 1965.

[FMNSS1] A. Fernández, F. Mayoral, F. Naranjo, C. Saez, and E.A. Sánchez-Pérez, *Vector measure Mayrey-Rosenthal-type factorizations and ℓ-sums of L^1-spaces*, J. Functional Anal. **220**(2005), 460–485.

[FMNSS2] A. Fernández, F. Mayoral, F. Naranjo, C. Saez, and E.A. Sánchez-Pérez, *Spaces of p-integrable functions with respect to a vector measures*, to appear in Positivity.

[FN1] A. Fernández, and F. Naranjo, *Operator and the space of integrable func-tions with respect to a Fréchet space-valued measures*, J. Austral Math. Soc. (Series A), **65**(1998), 176–193.

[FN2] A. Fernández and F. Naranjo, *AL- and AM-spaces of integrable scalar func-tions with respect to a Fréchet-valued measure*, Questiones Math. **23**(2000), 247–258.

[FNR] A. Fernández, F. Naranjo, and W.J. Ricker, *Completeness of L^1-spaces for measures with values in vector spaces*, J. Math. Anal. Appl., **223**(1998), 76–87.

[G]   A. Grothendieck, *Sur les applications linéaires faiblement compactes d'es-paces du type $C(K)$*, Canad. J. Math., **5**(1953), 129–173.

[H]   P.R. Halmos, Measure Theory, Van Nostrand, New York, 1950.

[Ho]   J. Horváth, Topological Vector Spaces and Distributions, Addison-Wesley, London, 1966.

[HR]   E. Hewitt, and K.A. Ross, Abstract Harmonic Analysis, Vol. 1, Springer-Verlag, New York, 1963.

[HS]   E. Hewitt, and K. Stromberg, Real and Abstract Analysis, Springer-Verlag, New York, 1965.

[JO]   B. Jefferies, and S. Okada, *Bilinear integration in tensor products*, Rocky Mount. J. Math., **28**(1998), 517–545.

[Ke]   J.L. Kelley, General Topology, D. Van Nostrand, New York, 1955.

[KN]   J.L. Kelley and I. Namioka, Linear Topological Spaces, Van Nostrand, New Jersey, 1959.

[K1]   I. Kluvánek, *Characterizations of Fourier-Stieltjes transforms of vector and operator-valued measures*, Czechoslovak Math. J. **17**(1967), 261–277.

[K2]   I. Kluvánek, *Fourier transforms of vector-valued functions and measures*, Studia Math., **37**(1970), 1–12.

[K3]   I. Kluvánek, *The extension and closure of vector measures* in "Vector and Operator Valued Measures and Applications", Academic Press, New York, (1973), 175–190.

[K4]   I. Kluvánek, *Applications of vector measures in integration, topology and geometry in linear spaces*, Proc. Chapel Hill /(N.C.(1979), Contemp. Math. 2(1980), 102–134.

[KK]   I. Kluvánek, and G. Knowles, Vector Measures and Control Systems, North Holland, American Elsevier, 1975.

[KÖ]   G. Köthe, Topological Vector Spaces I, Springer-Verlag, New York, 1969.

[L1]   D.R. Lewis, *Integration with respect to vector measures*, Pacific J. Math., **33**(1970), 157–165.

[L2]   D.R. Lewis, *On integrability and summability in vector spaces*, Illinois J. Math., **16**(1972), 294–307.

[McA] C.W. McArthur, *On a theorem of Orlicz and Pettis*, Pacific J. Math., **22**(1967), 297–302.

[McS] E.J. McShane, Integration, Princeton Univ. Press, Princeton, 1944.

[MB] E.J. McShane, and T.A. Botts, Real Analysis, Van Nostrand, Princeton, New Jersey, 1959.

[MN1] P.R. Masani and H. Niemi, *The integration theory of Banach space-valued measures and the Tonelli-Fubini theorems. I. Scalar-valued measures on δ-rings*, Adv. Math. **73**(1989), 204–241.

[MN2] P.R. Masani, and H. Niemi, *The integration theory of Banach space-valued measures and Tonelli-Fubini theorems, II. Pettis integration*, Advances Math., **75**(1989), 121–167.

[MP] Félix Martínez-Giménez and E.A. Sánchez- Pérez, *Vector measure range duality and factorizations of (D,p)-summing operators from Banach function spaces*, Bull. Braz. Math. Soc. New Series, **35**(2004), 51–69.

[Nai] M.A. Naimark, Normed Rings, Noordhoff, Groningen, 1959.

[Nar] F. Naranjo, *Espacios de funciones integrables respecto de una medida vectorial con valores en un espacio de Fréchet*, Ph.D. Thesis, Universidad de Sevilla, 1997.

[O] S. Okada, *The dual space of $\mathcal{L}_1(\mu)$ for a vector measure μ*, J. Math. Anal. Appl., **177**(1993), 583–599.

[OR1] S. Okada, and W.J. Ricker, *Non-weak compactness of the integration map for vector measures*, J. Austral. Math. Soc. (Series A), **51**(1993), 287–303.

[OR2] S. Okada, and W.J. Ricker, *Criteria for weak compactness of vector-valued integration maps*, Comment. Math. Univ. Carolinae, **35**(1994), 485–495.

[OR3] S. Okada, and W.J. Ricker, *Compactness properties of vector-valued integration maps in locally convex spaces*, Colloq. Math., **67** (1994), 1–14.

[OR4] S. Okada, and W.J. Ricker, *Vector measures and integration in non-complete spaces*, Arch. Math. (Basel), **63**(1994), 344–353.

[OR5] S. Okada, and W.J. Ricker, *Compact integration operator for Fréchet-valued measures,* Indag. Math. (N.S.), **13**(2002), 209–227.

[OR6] S. Okada, and W.J. Ricker, *Fréchet-space-valued measures and the AL-property*, Revista de la Real Academia Sciences Ser. A Mathemáticas RAC-SAM, **97**(2003), 305–314.

[PSV] S. Oltra, E.A. Sánchez-Pérez, and O. Valero, *Spaces $L_2(\lambda)$ of a positive vector measure λ and generalized Fourier coefficients*, Rocky Mount. J. Math., **35**(2005), 211–225.

[P1] T.V. Panchapagesan, Medida e Integracin, Parte I, Teoría de la Medida, Tomos 1 y 2, Univ. de los Andes, Facultad de Ciencias, Mérida, Venezuela (Spanish), 1991.

[P2]   T.V. Panchapagesan, Integral de Radon en espacios localmente compactos y de Hausdorff, IV Escuela Venezolana de Mat., Universidad de los Andes, Facultad de Ciencias, Mérida, venezuela (Spanish), 1991.

[P3]   T.V. Panchapagesan, *On complex Radon measures I*, Czechoslovak Math. J., **42**(1992), 599–612.

[P4]   T.V. Panchapagesan, *On complex Radon measures II*, Czechoslovak Math. J., **43**(1993), 65–82.

[P5]   T.V. Panchapagesan, *On the distinguishing features of the Dobrakov integral*, Divulgaciones Mat., **3**(1995), 79–114.

[P6]   T.V. Panchapagesan, *On Radon vector measures*, Real Analysis Exchange, **21**(1995/96), 75–76.

[P7]   T.V. Panchapagesan, *Applications of a theorem of Grothendieck to vector measures*, J. Math. Anal. Appl., **214**(1997), 89–101.

[P8]   T.V. Panchapagesan, *Baire and σ-Borel characterizations of weakly compact sets in $M(T)$*, Trans. Amer. Math. Soc., **350**(1998), 4839–4847.

[P9]   T.V. Panchapagesan, *Characterizations of weakly compact operators on $C_0(T)$*, Trans. Amer. Math. Soc., **350**(1998), 4849–4867.

[P10]   T.V. Panchapagesan, *On the limitations of the Grothendieck techniques*, Real Acad. Cienc. Exact. Fis. Natur. Madrid, **94**(2000), 437–440.

[P11]   T.V. Panchapagesan, *A simple proof of the Grothendieck theorem on the Dieudonné property of $C_0(T)$*, Proc. Amer. Math. Soc., **129** (2001), 823–831.

[P12]   T.V. Panchapagesan, *Weak compactness of unconditionally convergent operators on $C_0(T)$*, Math. Slovaca, **52**(2002), 57–66.

[P13]   T.V. Panchapagesan, *Positive and complex Radon measures in locally compact Hausdorff spaces*, Handbook of Measure theory, chapter 26, Elsevier, Amsterdam, (2002), 1056–1090.

[P14]   T.V. Panchapagesan, *A Borel extension approach to weakly compact operators on $C_0(T)$*, Czechoslovak Math. J., **52**(2002), 97–115.

[Ri1]   W.J. Ricker, *Criteria for closedness of vector measures*, Proc. Amer. Math. Soc., **91**(1984), 75–80.

[Ri2]   W.J. Ricker, *Completeness of the L^1-space of a vector measure*, Proc. Edinburgh Math. Soc., **33**(1990), 71–78.

[Ri3]   W.J. Ricker, *Separability of the L^1-space of a vector measure*, Glasgow Math. J., **34**(1992), 1–9.

[Ri4]   W.J. Ricker, *Rybakov's theorem in Fréchet spaces and completeness of L^1-spaces*, J. Austral. Math. Soc. (Series A), **64**(1998), 247–252.

[Ri5]   W.J. Ricker, Operator Algebras Generated by Commuting Projections: A Vector Measure Approach, Lecture Notes, 1711, Springer-Verlag, New York, 1999.

[Ri6]   W.J. Ricker, *Weak compactness of the integration map associated with a spectral measure*, Indag. Math. (N.S.), **135**(1994) 353–364.

[Ru1]   W. Rudin, Real and Complex Analysis, McGraw-Hill, New York, 1966.

[Ru2]   W. Rudin, Functional Analysis, McGraw-Hill, New York, 1973.

[S]   E. Saab, *On the Radon-Nikodym property in a class of locally convex spaces*, Pacific J. Math., **75**(1978), 281–291.

[Scha] H.H. Schaefer with M.P. Wolf, Topological Vector Spaces, Springer-Verlag, Second Edition, New York, 1999.

[SK]   S. Singh Khurana, *Weak compactness of certain sets of measures*, Proc. Amer. Math. Soc., **131**(2003), 3251–3255.

[Shu1] A. Shuchat, *Vector measures and scalar operators in locally convex spaces*, Michigan Math. J., **24**(1977), 303–310.

[Shu2] A. Shuchat, *Spectral measures and homomorphisms*, Rev. Roum. Math. Pures et Appl., Tome XXIII(1978), 939–945.

[Si]   M. Sion, *Outer measures with values in topological groups*, Proc. London Math. Soc., **19**(1969), 89–106.

[SP1]   E.A. Sánchez-Pérez, *Compactness arguments for spaces of p-integrable functions with respect to a vector measure and factorization of operators through Lebesgue-Bochner spaces*, Illinois J. Math., **45**(2001), 907–923.

[SP2]   E.A. Sánchez-Pérez, Spaces of integrable functions with respect to vector measures of convex range and factorization of operators from L_p-spaces, Pacific J. Math., **207**(2002), 489–495.

[SP3]   E.A. Sánchez-Pérez, *Vector measure orthonormal functions and best approximation for the 4-norm*, Arch. Math. **80**(2003), 177–190.

[SP4]   E.A. Sánchez-Pérez, *Vector measure duality and tensor product representations of L_p-spaces of vector measures*, Proc. Amer. Math. Soc., **132**(2004), 3319–3326.

[St]   G.F. Stefansson, L_1 *of a vector measure*, Le Mathematiche **48**(1993), 219–234.

[T]   E. Thomas, *L'intégration par rapport a une mesure de Radon vectorielle*, Ann. Inst. Fourier (Grenoble), **20**(1970), 55–191.

[Tu]   Ju.B. Tumarkin, *On locally convex spaces with basis*, Dokl. Akad. Nauk. SSSR, **11**(1970), 1672–1675.

[W]   H. Weber, *Fortsetzung von Massen mit Werten in uniformen Halbgruppen*, Arch. Math. XXVII(1976), 412–423.

[Y]   K. Yosida, Functional Analysis, Springer-Verlag, Fourth Edition, New York, 1974.

Acknowledgment

The author primarily wishes to express his indeptedness to Professor M. Raja-gopalan, an internationally reputed topologist, without whose help he could not have moved to Venezuela, to work in an environment conducive to his extensive research work on vector measures resulting in the publication of the present monograph.

The author wishes to express his gratitude to the referee for pointing out some recent developments in this field, enabling him to include them in the Bibliography. Also to Professors A. Fernández and E.A. Sánchez Pérez for providing him with photocopies of some of their papers. The help received from his colleague Professor Osmin Monsalve in preparation of the earlier versions of the manuscript in LaTeX is gratefully acknowledged. Also the author acknowledges the untiring help received from his younger daughter Ms. Srividhya Panchapagesan in preparing the LaTeX version of the monograph, especially the part containing the index and list of symbols.

The work is supported partially by the C.D.C.H.T. project No.1079-B-2003 of the Universidad de los Andes, Mérida, Venezuela.

List of Symbols

Index

Monografie Matematyczne

[1]   S. Banach, *Théorie des opérations linéaires*, 1932
[2]   S. Saks, *Théorie de l'integrale*, 1933
[3]   C. Kuratowski, *Topologie* I, 1933
[4]   W. Sierpiński, *Hypothèse de continu*, 1934
[5]   A. Zygmund, *Trigonometrical Series*, 1935
[6]   S. Kaczmarz, H. Steinhaus, *Theorie der Orthogonalreihen*, 1935
[7]   S. Saks, *Theory of the integral*, 1937
[8]   S. Banach, *Mechanika*, T. I, 1947
[9]   S. Banach, *Mechanika*, T. II, 1947
[10] S. Saks, A. Zygmund, *Funkcje analityczne*, 1948
[11] W. Sierpiński, *Zasady algebry wyższej*, 1946
[12] K. Borsuk, *Geometria analityczna w n wymiarach*, 1950
[13] W. Sierpiński, *Działania nieskończone*, 1948
[14] W. Sierpiński, *Rachunek różniczkowy poprzedzony badaniem funkcji elementarnych*, 1947
[15] K. Kuratowski, *Wykłady rachunku różniczkowego i całkowego*, T. I, 1948
[16] E. Otto, *Geometria wykreślna*, 1950
[17] S. Banach, *Wstęp do teorii funkcji rzeczywistych*, 1951
[18] A. Mostowski, *Logika matematyczna*, 1948
[19] W. Sierpiński, *Teoria liczb*, 1950
[20] C. Kuratowski, *Topologie* I, 1948
[21] C. Kuratowski, *Topologie* II, 1950
[22] W. Rubinowicz, *Wektory i tensory*, 1950
[23] W. Sierpiński, *Algèbre des ensembles*, 1951
[24] S. Banach, *Mechanics*, 1951
[25] W. Nikliborc, *Równania różniczkowe*, Cz. I, 1951
[26] M. Stark, *Geometria analityczna*, 1951
[27] K. Kuratowski, A. Mostowski, *Teoria mnogości*, 1952
[28] S. Saks, A. Zygmund, *Analytic functions*, 1952
[29] F. Leja, *Funkcje analityczne i harmoniczne*, Cz. I, 1952
[30] J. Mikusiński, *Rachunek operatorów*, 1953
* [31] W. Ślebodziński, *Formes extérieures et leurs applications*, 1954
[32] S. Mazurkiewicz, *Podstawy rachunku prawdopodobieństwa*, 1956
[33] A. Walfisz, *Gitterpunkte in mehrdimensionalen Kugeln*, 1957
[34] W. Sierpiński, *Cardinal and ordinal numbers*, 1965
[35] R. Sikorski, *Funkcje rzeczywiste*, 1958
[36] K. Maurin, *Metody przestrzeni Hilberta*, 1959
[37] R. Sikorski, *Funkcje rzeczywiste*, T. II, 1959
[38] W. Sierpiński, *Teoria liczb* II, 1959

* [39]   J. Aczél, S. Gołąb, *Funktionalgleichungen der Theorie der geometrischen Objekte*, 1960

[40]   W. Ślebodziński, *Formes extérieures et leurs applications*, II, 1963

[41]   H. Rasiowa, R. Sikorski, *The mathematics of metamathematics*, 1963

[42]   W. Sierpiński, *Elementary theory of numbers*, 1964

* [43]   J. Szarski, *Differential inequalities*, 1965

[44]   K. Borsuk, *Theory of retracts*, 1967

[45]   K. Maurin, *Methods of Hilbert spaces*, 1967

[46]   M. Kuczma, *Functional equations in a single variable*, 1967

[47]   D. Przeworska-Rolewicz, S. Rolewicz, *Equations in linear spaces*, 1968

[48]   K. Maurin, *General eigenfunction expansions and unitary representations of topological groups*, 1968

[49]   A. Alexiewicz, *Analiza funkcjonalna*, 1969

* [50]   K. Borsuk, *Multidimensional analytic geometry*, 1969

* [51]   R. Sikorski, *Advanced calculus. Functions of several variables*, 1969

[52]   W. Ślebodziński, *Exterior forms and their applications*, 1971

[53]   M. Krzyżański, *Partial differential equations of second order*, vol. I, 1971

[54]   M. Krzyżański, *Partial differential equations of second order*, vol. II, 1971

[55]   Z. Semadeni, *Banach spaces of continuous functions*, 1971

[56]   S. Rolewicz, *Metric linear spaces*, 1972

[57]   W. Narkiewicz, *Elementary and analytic theory of algebraic numbers*, 1974

[58]   Cz. Bessaga, A. Pełczyński, *Selected topics in infinite dimensional topology*, 1975

* [59]   K. Borsuk, *Theory of shape*, 1975

[60]   R. Engelking, *General topology*, 1977

[61]   J. Dugundji, A. Granas, *Fixed point theory*, 1982

* [62]   W. Narkiewicz, *Classical problems in number theory*, 1986

The volumes marked with * are available at the exchange department of the library of the Institute of Mathematics, Polish Academy of Sciences.

Monografie Matematyczne, New Series (MMNS)

Edited by
Przemysław Wojtaszczyk, IMPAN and Warsaw University, Poland

Starting in the 1930s with volumes written by such distinguished mathematicians as Banach, Saks, Kuratowski, and Sierpinski, the original series grew to comprise 62 excellent monographs up to the 1980s. In cooperation with the Institute of Mathematics of the Polish Academy of Sciences (IMPAN), Birkhäuser now resumes this tradition to publish high quality research monographs in all areas of pure and applied mathematics.

■ **Vol. 68: Grigoryan, S.A.**, Kazan State University, Russia / **Tonev, T.V.**, University of Montana, USA

Shift-invariant Uniform Algebras on Groups

2006. 294 pages. Hardcover.
ISBN 978-3-7643-7606-2

The central subject of the book – the theory of shift-invariant algebras – is an outgrowth of the established theory of generalized analytic functions. Associated subalgebras of almost periodic functions of real variables and of bounded analytic functions on the unit disc are carried along within the general framework. In particular, it is shown that the algebra of almost periodic functions with spectrum in a semigroup of the reals does not have a half-plane-corona if and only if all non-negative semicharacters of the semigroup are monotone decreasing, or equivalently, if and only if the strong hull of the semigroup coincides with the positive half of its group envelope. Under the same conditions the corresponding subalgebra of bounded analytic functions on the disc has neither a half-plane-corona nor a disc-corona. There are given characterizations of semigroups such that classical theorems of complex analysis hold on the associated shift-invariant algebras. Bourgain algebras, orthogonal measures, and primary ideals of big disc algebras are described. The notion of a harmonic function is extended on compact abelian groups, and corresponding Fatou-type theorems are proven. Important classes of inductive limits of standard uniform algebras, including Blasche algebras, are introduced and studied. In particular, it is shown that algebras of hyper-analytic functions, associated with families of inner functions, do not have a big-disc-corona.

■ **Vol. 67: Zoladek, H.**, University of Warsaw, Poland

The Monodromy Group

2006. 592 pages. Hardcover.
ISBN 978-3-7643-7535-5

■ **Vol. 66: Müller, P.F.X.**, Johannes-Kepler-University Linz, Austria

Isomorphisms between H^1 Spaces

2005. 472 pages. Hardcover.
ISBN 978-3-7643-2431-5

■ **Vol. 65: Badescu, L.**, Università degli Studi di Genova, Italy

Projective Geometry and Formal Geometry

2004. 228 pages. Hardcover.
ISBN 978-3-7643-7123-4

■ **Vol. 64: Walczak, P.**, University of Łódź, Poland

Dynamics of Foliations, Groups and Pseudogroups

2004. 240 pages. Hardcover.
ISBN 978-3-7643-7091-6

■ **Vol. 63: Schürmann, J.**, Westfälische Wilhelms-Universität Münster, Germany

Topology of Singular Spaces and Constructible Sheaves

2003. 464 pages. Hardcover.
ISBN 978-3-7643-2189-5